7th EDITION

AGING

RELATED TITLES FROM SAGE

OTHER SAGE TITLES OF INTEREST

7th EDITION

AGING

CONCEPTS AND CONTROVERSIES

HARRY R. MOODY
Director of Academic Affairs, AARP

JENNIFER R. SASSER
Marylhurst University

Los Angeles | London | New Delhi
Singapore | Washington DC

Los Angeles | London | New Delhi
Singapore | Washington DC

FOR INFORMATION:

SAGE Publications, Inc.
2455 Teller Road
Thousand Oaks, California 91320
E-mail: order@sagepub.com

SAGE Publications Ltd.
1 Oliver's Yard
55 City Road
London EC1Y 1SP
United Kingdom

SAGE Publications India Pvt. Ltd.
B 1/I 1 Mohan Cooperative Industrial Area
Mathura Road, New Delhi 110 044
India

SAGE Publications Asia-Pacific Pte. Ltd.
33 Pekin Street #02-01
Far East Square
Singapore 048763

Acquisitions Editor: David Repetto
Associate Editor: Maggie Stanley
Assistant Editor: Terri Accomazzo
Editorial Assistant: Lydia Balian
Production Editor: Laureen Gleason
Copy Editor: Melinda Masson
Typesetter: C&M Digitals (P) Ltd.
Proofreader: Kate Peterson
Indexer: Diggs Publication Services
Cover Designer: Gail Buschman
Marketing Manager: Erica DeLuca
Permissions Editor: Karen Ehrmann

Printed in the United States of America

Library of Congress Cataloging-in-Publication Data

Moody, Harry R.

Aging: concepts and controversies/Harry R. Moody, Jennifer R. Sasser.—7th ed.

p. cm.
Includes bibliographical references and index.

ISBN 978-1-4522-0309-6 (pbk.: acid-free paper)

1. Gerontology—United States. 2. Aging—United States. I. Sasser, Jennifer R. II. Title.

HQ1064.U5M665 2012
305.260973—dc23   2011039048

This book is printed on acid-free paper.

12 13 14 15 10 9 8 7 6 5 4 3 2

Brief Contents

Detailed Contents

Basic Concepts II. Aging, Health Care, and Society   129

Controversy 4. Should We Ration Health Care for Older People?   163

PREFACE

This seventh edition of *Aging: Concepts and Controversies* appears at a distinctive historical moment. The oldest members of the baby boom generation are now collecting Social Security, and in the coming years, the process of population aging will dramatically accelerate. Given current demographic trends, it is likely that tens of thousands of Americans born after the year 2000 will live to see the dawn of another century, the 22nd. Students who are reading this book will spend the greatest part of their lives experiencing dramatic changes already evident in telecommunications, biotechnology, and genetics. Within recent years, global upheavals in finance have dramatically underscored the choices and risks that we all face over the life course. This ever-accelerating change will produce even more debate and controversy about how we are to live in an aging society and in the 21st century.

This volume responds to public debates and new social conditions with the same unique approach that inspired earlier versions of the book. This approach is to present key ideas and content from gerontology as an opportunity for critical thinking. Memorizing facts is not enough. The approach of this text is to encourage each student to grasp basic ideas and to reflect more deeply about issues raised by the study of aging.

As we move further into the 21st century with a population growing older, we all have a stake in developing a better understanding of the subject. This book consciously focuses on issues of interest to all of us as citizens and as educated human beings, not just as potential gerontologists or professional service providers. The book takes a similarly broad view toward what aging is all about. From the opening chapter, students are encouraged to see aging not as a fixed period of life, but as a process beginning at birth and extending over the entire life course. This open-ended quality of human aging is a theme woven throughout the book: from biological experiments on extending the life span to difficult choices about allocation of health care resources.

The multiple possibilities for how we might age both as individuals and as a society create complex choices that are important for all of us. New thinking is needed if we are to grasp the issues at stake. That is why the pedagogical design of this book focuses on controversies and questions, rather than on assimilating facts or coming up with a single "correct" view about aging or older people. The supplemental readings are selected to accentuate contrast and conflict and to stimulate faculty and students to think more deeply about what is at stake in the debates presented here. In contrast to most textbooks, this volume directs the student's attention toward original sources and encourages teachers to provide the tools to respond to claims made in those original texts. The goal is nothing more, or less, than liberal education for gerontology.

The point is not to find the single "right answer" raised in the debates in this book. Rather, as students become engaged in the debates, they will appreciate the need for having the factual background necessary to make responsible judgments and interpretations. That is the purpose of the three major essays, the Basic Concepts sections, around which the book's controversies are organized. The data and conceptual frameworks offered in these essays will help students make sense of the controversies, understand their origin, engage in critical thinking, and, finally, develop their own views. The introductions preceding each controversy and the questions that follow serve to reinforce the essential link between factual knowledge and interpretation at the heart of the book. This book, then, can best be seen as a textbook constructed to provide drama and compelling interest for the reader. It is structured to encourage a style of teaching and learning that goes beyond conveying facts and methods.

Other, more specific features of the book reinforce this pedagogical approach. The **Focus on Practice** sections demonstrate the relevance of the controversies for human services work in our society. The **Focus on the Future** sections make us ever mindful of the accelerating pace of change in our society and its implications. The **Global Perspective** and **Urban Legends of Aging** sections provide additional opportunities for expansive and critical thinking. The appendix offers guidance for researching and writing term papers on aging, and the online resources provided as part of the book's ancillary package open up access to tools for tapping the World Wide Web. Whether students reading this book go on to specialized professional work or whether they never take another course in gerontology, our aim is directed at issues of compelling human importance, now and in the future. By returning again and again to those questions of perennial human interest, we express our hope that both teachers and students will find new excitement in questions that properly concern us all, whatever our age.

WHAT IS NEW TO THIS EDITION?

This new edition builds on the unique approach adopted in earlier editions. There is a close link between concepts and controversies in each of the three broad domains of human aging: the life course, socioeconomic trends, and health care. This link has proved to be so teachable in earlier editions that this organization has been reinforced. We have also completely updated and augmented the figures and graphics in the book, using an effective illustration wherever appropriate. Information cited has been made as up-to-date as possible to reflect the most recent data and perspectives available. In addition, each chapter of controversies contains a feature section highlighting comparable issues in different countries around the world. These feature sections acknowledge the way in which aging is increasingly a global phenomenon with lessons of international significance. Also, we've added a controversy that focuses on the new aging marketplace, that is, the emergence of services and products geared to older clients. Readers of this book can also make use of a Web-based online appendix to get even more up-to-date information and explore topics in depth. To access this appendix, simply go to the website for the book at the publisher's address: www.sagepub.com. In addition, since the last edition, teachers can now receive at no cost a monthly electronic newsletter, *Teaching Gerontology,* which provides current guidance and resources on how to approach the concepts and controversies featured in this book. A subscription to this e-newsletter is available by contacting hrmoody@aarp.org.

ANCILLARIES

For the Instructor

The password-protected Instructor Site at www.sagepub.com/moody7e gives instructors access to a full complement of resources to support and enhance their courses. The following assets are available on the Instructor Site:

- A **test bank** with multiple-choice, true/false, short-answer, and essay questions. The test bank is provided on the site in Word format as well as in Respondus.
- **PowerPoint** slides for each chapter, for use in lecture and review. Slides are integrated with the book's distinctive features and incorporate key tables, figures, and photos.
- **Video resources** that enhance the information in each chapter.
- **SAGE journal articles** for each chapter that provide extra content on important topics from SAGE's sociology journals.
- **Web resources** that provide links to websites to encourage additional learning on specific topics.
- **Highlights from the *Teaching Gerontology* newsletter** that provide additional information on teaching the topics.

For the Student

To maximize students' understanding and promote critical thinking and active learning, we have provided the following chapter-specific student resources on the open-access portion of www.sagepub.com/moody7e:

- **Flash Cards** that reiterate key chapter terms and concepts.
- **Quizzes**, including multiple-choice and true/false questions.
- **Video resources** that enhance the information in each chapter.
- **SAGE journal articles** for each chapter that provide extra content on important topics from SAGE's sociology journals.
- **Web resources** that provide links to websites to encourage additional learning on specific topics.

ACKNOWLEDGMENTS

In preparing this seventh edition, I have been helped enormously by the many professors who have used earlier editions and have thoughtfully offered ideas on how to improve the book. In my role as Director of Academic Affairs at AARP, it has been a privilege to listen to faculty from around the country, and I am indebted to them although they are not named here. I also acknowledge past colleagues at the Brookdale Center for Healthy Aging and Longevity at Hunter College who have helped me over many years to refine the ideas found in these pages. I am especially grateful to Rose Dobrof for her guidance and inspiration. Let me express my gratitude to AARP and to John Rother, its Director of Policy and Strategy, who has shared with me his thoughtful reflections on the future of aging in America. Professor Jennifer Sasser, who is now coauthor, has once again been indispensable. Above all, I thank my wife, Elizabeth, patient reader and thoughtful editor, and my children, Carolyn and Roger, who have made all the difference in my life.

—H.R.M.

I am most grateful to Rick Moody for extending to me the opportunity to work with him as coauthor on this and future editions of this book, but most especially for his mentoring and friendship, to my students and colleagues at Marylhurst University, and especially to my elder friends at Mary's Woods, as well as to the Intentional Aging Collective, the Marylhurst Gerontology Association, and, last but certainly not least, the Gero-Babes: unending thanks for ongoing inspiration and solidarity. To my daughter Isobel, my mommy Susie, and Erica and Simeon—where would I be, without your love?

—J.R.S.

SAGE and the authors gratefully acknowledge the contributions of the following reviewers:

Clifford Garoupa
Fresno City College

Mario Garrett
California College San Diego

Robert Hard
Albertus Magnus College

Pamela Nadash
University of Massachusetts Boston

Wendy Pank
Bismarck State College

Daniel J. Van Dussen
Youngstown State University

PROLOGUE

It is no secret that the number of people over age 65 in the United States is growing rapidly, a phenomenon recognized as the "graying of America" (Himes, 2001). The numbers are staggering. There has been a 30-fold increase in older people in the United States since 1870: from 1 million up to 35 million in 2000—a number now larger than the entire population of Canada. During recent decades, the 65+ group has been increasing twice as fast as the rest of the population.

As a result, the U.S. population looks different than it did earlier this century. Life expectancy at birth was 47 in 1900, but is now close to 78. A hundred years ago, only 4% of the population was over the age of 65; today, that figure has jumped to 13%. The pace of growth continued in the first decade of the 21st century, and in 2011 the huge baby boom generation—those born between 1946 and 1964—moved into the ranks of senior citizens. By 2030, the proportion of the population over age 65 will reach 20%. This rate of growth in the older population is unprecedented in human history. Within a few decades, one in five Americans will be eligible for Social Security and Medicare, contrasted with one in eight today.

We usually think of aging as strictly an individual matter. But we can also describe an entire population as aging or growing older, although to speak that way is metaphorical. In literal terms, only organisms, not populations, grow older. Still, the average age of the population is increasing, and the proportion of the population made up of people over age 65 is rising. This change in the demographic structure of the population is referred to as **population aging** (R. Clark et al., 2004).

Population aging results more from a drop in the numbers of children than from people living longer. In 1900, America was a relatively "young" population: The percentage of children and teenagers in the population was 40%. By 1990, the proportion of youth had dropped to 24%. By contrast, those over age 65 increased from 4% in 1900 to 13% in 2000, with larger increases still to come. During the next several decades, overall population growth in the United States will be concentrated among middle-aged and older Americans.

The United States is not the only country undergoing population aging (Bosworth & Burtless, 1998). For example, average life expectancy at birth in Japan is currently 82 years, the highest in the world. In Germany, Italy, and Japan, the population is aging because of low birthrates. Think of the state of Florida today as a model for population aging: a population in which nearly one of five people is already over the age of 65. We can ask: How long will it take different nations to reach the condition of "Florida-ization"? The answer is that Italy already looked like Florida by the year 2003, Japan by 2005, and Germany by

Exhibit P.1 Life Expectancy at Birth and at 65 Years of Age by Race and Sex: United States, 1970–2004

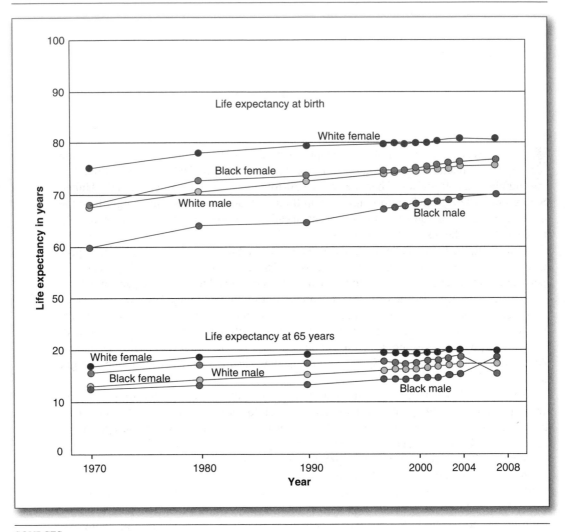

SOURCES:

Centers for Disease Control and Prevention.

Health, United States, 2010, U.S. Dept. of Health and Human Services, Feb. 2011, DHHS Pub. no. 2011-1232.

NOTE: Life expectancies prior to 1997 are from decennial life tables based on census data and deaths for a 3-year period around the census. Beginning in 1997, the annual life tables are complete life tables based on methodology similar to that used for decennial life tables.

2006. France and Great Britain will resemble Florida in 2016, whereas the United States in general will not reach "Florida-ization" until 2023.

Population aging also shows up as an increase in the **median age** for the entire population, that is, the age at which half the population is older and half the population is younger.

The median age of the U.S. population in 1820 was only 17 years; it had risen by 1900 to 23 and by 2000 to 35. It is estimated that the median age of the American population by 2030 will be 42 years. This shift is a measure of the dramatic impact of population aging.

It is clear, then, that populations "age" for reasons different than individuals do, and the reasons have to do with demographic trends. In the first place, population aging occurs because birthrates go down. With a smaller proportion of children in the population, the average age of the population goes up. Population aging can also come about because of improvements in life expectancy—people living longer on average. Finally, the process of population aging can be influenced for a time because of birth **cohorts**. A cohort is a group of people born over a particular time who thereby experience common **life events** during the same historical period. For example, the cohort born during the Great Depression of the 1930s was relatively small and thus has had minimal impact on the average age of the population. By contrast, the baby boomers born after World War II are a large cohort. Because of this cohort's size, the middle-aged baby boomers are dramatically hastening the aging of the U.S. population.

In summary, then, trends in birthrates, death rates, and the flow of cohorts all contribute to population aging. What makes matters complicated is that all three trends can be happening simultaneously, as they have been in the United States in recent decades. Casual observers sometimes suggest that the U.S. population is aging mainly because people are living longer. But that observation is not quite accurate because it fails to take into account multiple trends defined by demographic factors of fertility, mortality, and flow of cohorts.

A demographic description tells us what the population looks like, but it does not explain the reason that population trends happen in the first place. We need to ask: Why has this process of population aging occurred? The rising proportion of older people in the population can be explained by **demographic transition theory**, which points to a connection between population change and the economic process of industrialization. In preindustrial societies, there is a generally stable population because both birthrates and death rates remain high. With industrialization, death rates tend to fall, whereas birthrates remain high for a period, and so the total population grows. But at a certain point, at least in advanced industrial societies, birthrates begin to fall back in line with death rates. Eventually, when the rate of fertility is exactly balanced by the rate of mortality, we have a condition of stability known as zero population growth (Chu, 1997). The population is neither growing nor shrinking.

The Western industrial revolution of the 19th century brought improved agricultural production, improved standards of living, and also an increase in population size. Over time, there came a shift in the age structure of the population, known to demographers as the demographic transition. This was a shift away from a population with high fertility and high mortality to one of low fertility and low mortality. That population pattern is what we see today in the United States, Europe, and Japan. The result in all industrialized societies has been population aging: a change in the age distribution of the population.

Most developing countries in the Third World—in Africa, Asia, and Latin America—still have fertility rates and death rates much higher than those of advanced industrialized countries. For the United States in 1800, as for most Third World countries today, that population distribution can be represented as a population pyramid: many births (high fertility) and relatively few people surviving to old age (high mortality). For countries that are approaching zero population growth, that pyramid is replaced by a cylinder: Each cohort becomes approximately the same in size.

Exhibit P.2   Demographic Transition

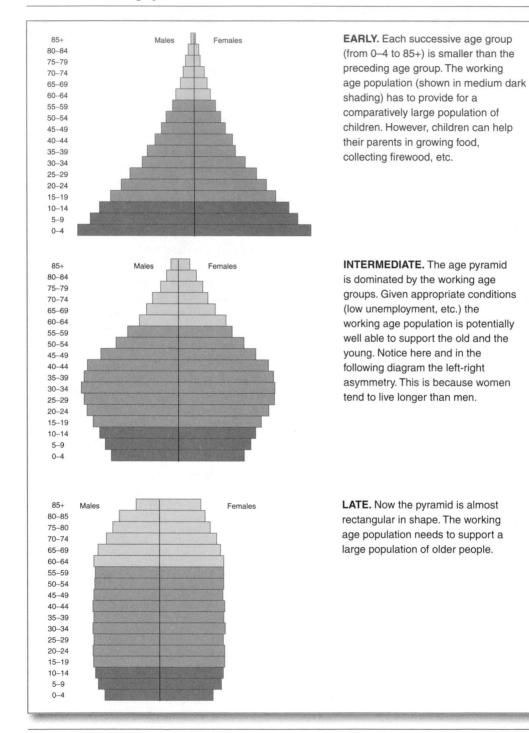

EARLY. Each successive age group (from 0–4 to 85+) is smaller than the preceding age group. The working age population (shown in medium dark shading) has to provide for a comparatively large population of children. However, children can help their parents in growing food, collecting firewood, etc.

INTERMEDIATE. The age pyramid is dominated by the working age groups. Given appropriate conditions (low unemployment, etc.) the working age population is potentially well able to support the old and the young. Notice here and in the following diagram the left-right asymmetry. This is because women tend to live longer than men.

LATE. Now the pyramid is almost rectangular in shape. The working age population needs to support a large population of older people.

SOURCE: J. F. Barker, June 2004.

Exhibit P.3   Birthrates and Death Rates

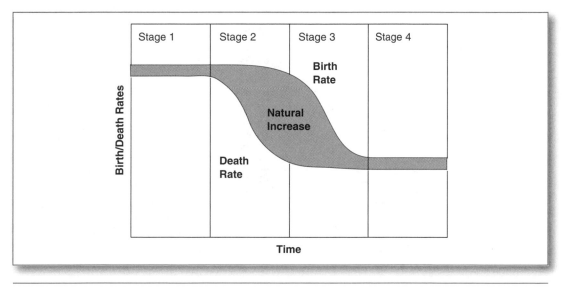

SOURCE: Figure 12, "The Classic Stages of Demographic Transition," from McFalls, 2003.

Exhibit P.4   Distribution of the Projected Older Population by Age for the United States: 2010 to 2050

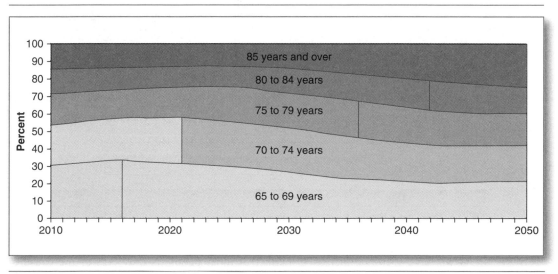

SOURCE: U.S. Census Bureau, 2008.

NOTE: Line indicates the year that each age group is the largest proportion of the older population.

Exhibit P.5   The Dramatic Aging of America, 1900–2030

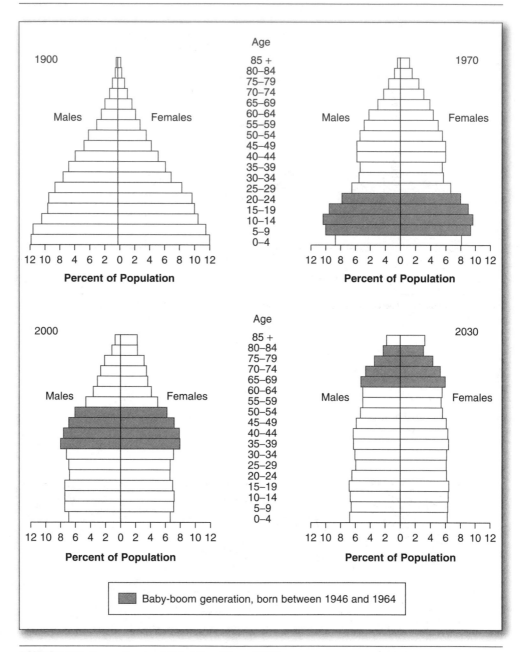

SOURCE: U.S. Census Bureau. Adapted from Himes (2001).

As we have seen, the increased number of older people is only part of the cause of population aging. It is important to remember that overall population aging has actually been brought about much more by declines in fertility than by reductions in mortality. The

Exhibit P.6   Actual and Projected Increases in the Population of Adults Aged 65 and
Over: 2000 to 2050 (in millions)

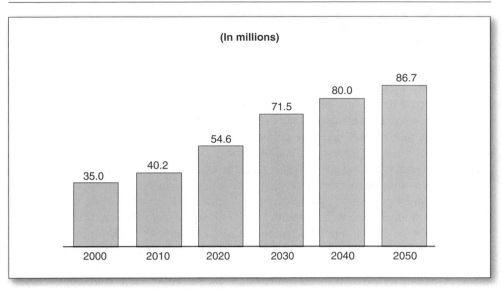

(In millions)

2000	2010	2020	2030	2040	2050
35.0	40.2	54.6	71.5	80.0	86.7

SOURCE: Year 2000, U.S. Census Bureau, 2001; Years 2010 to 2050, U.S. Census Bureau, 2004.

NOTE: The reference population for these data is the resident population.

trend toward declining fertility in America can be traced back to the early 19th century, and
the process of population aging has causes that date back even longer. To complete the
demographic picture, we need to point to other factors that influence population size and
composition, such as improvements in the chance of survival of people at different ages, or
the impact of immigration into the United States, largely by younger people. But one conclusion is inescapable. Today's increased proportion of people over age 65 springs from
causes that are deeply rooted in American society. Population aging is a long-range trend
that will characterize our society into the 21st century. It is a force we all will cope with for
the rest of our lives.

But how is American society coping with population aging? How are the major institutions of society—government, the economy, the family—responding to the aging of a large
number of individuals? The answer, in simplified terms, is rooted in a basic difference
between individual and population aging. As human beings, we are all familiar with the life
course process of individual aging. It is therefore not surprising that, as a society, we have
devised many policies and practices to take into account changes that predictably occur in
the later years, such as planning for retirement, medical interventions for chronic illnesses,
and familiar government programs such as Social Security and Medicare.

Whether it involves changes in biological functioning or changes in work roles, individual aging is tangible and undeniable, a pattern we observe well enough in our parents
and family members, not to mention in ourselves. But population aging is more subtle and

less easily observed. We have many institutional policies and programs to deal with individual aging, but our society is just beginning to wrestle with the controversies generated by the population aging trends now emerging, with the prospect of even more dramatic debate and change in the decades ahead. These demographic changes are significant and are stimulating tremendous ferment in our society's fundamental institutions. For that reason, this book is organized around controversies along with the facts and basic concepts that stand behind them.

Our society's response to population aging can best be summed up in the aphorism that generals prepare for the next war by fighting the old one over again. That is to say, in our individual and social planning, we tend to look back to past experience to guide our thinking about the future. Thus, when the railroad was first introduced, it was dubbed "the iron horse." But it wasn't a horse at all, and the changes that rail transport brought to society were revolutionary, beyond anything that could be expected by looking to the past.

The same holds true for population aging. We cannot anticipate the changes that will be brought about by population aging by looking backward. Population aging is historically unprecedented among the world's societies. Moreover, we should not confuse population aging with the process of individual aging. An aging society, after all, is not like an individual with a fixed life span. Why is it that people are so often fearful when they begin to think about the United States' future as an aging society? Part of the reason is surely that many of us are locked into images of decline that are based on prejudice or outdated impressions of what individual aging entails. Because our social institutions have responded to aging as a problem, we tend to see only losses and to overlook opportunities in the process of aging.

An important point to remember is that the solutions to yesterday's problems may not prove adequate for the challenges we face today, or for those we'll face in the future. For example, Social Security has proven vital in protecting older Americans from the threat of poverty in old age. But Social Security was never designed to help promote second careers or new forms of productivity among older people. We may need to think in new ways about pensions and retirement in the future. Similarly, Medicare has proved to be an important, although expensive, means of guaranteeing access to medical care for older people, but it was never designed to address the problems of long-term care for older people who need help to remain in their own homes. Finally, as the sheer number of people over age 65 increases, the United States as a society will need to consider which institutions and policies are best able to provide for the needs of this growing population.

Social gerontologist Matilda White Riley pointed out that our failure to think deeply about population aging is a weakness in gerontology as a discipline. Gerontologists know more about individual aging than about opportunity structures over the whole life course. By "opportunity structures," we mean that the way society is organized or structured affects an individual's chance or ability to gain certain rewards or meet certain goals. A good example is the way the life course has been shaped, with transitions from education to work to retirement. These transitions do not seem to prepare us for an aging society in the future. In effect, we have a "cultural lag" in facing the future. We know that in this century, the age of leaving the workforce to retire has been gradually going down, whereas the age for leaving schooling has been going up. Riley pointed out that, if we were to project these trends into the future, sometime in the 21st century, people would leave college at age 38 and immediately enter retirement. This scenario, of course, is not serious. But it does make a serious point. We must not take current trends and simply project them into the future.

Part of the problem is that we have less knowledge than we ought to have about the interaction between individual lives and the wider society. During the 20th century, nearly three decades were added to human life expectancy. Now more than a third of adult life is spent postwork. People over age 65 are healthier and better educated than ever before. Yet opportunity structures are lacking to integrate this older population into major institutions of society such as education or the workplace. We have yet to design a blueprint for what an aging society of the future might look like. And there are important questions to be asking about what such a blueprint might look like. Today, we grow old differently from the way our grandparents did, and in a way differently than will our children, so it does little good to look backward as we move into the 21st century.

The challenge is to change our way of anticipating and planning for the future by thinking critically about our underlying assumptions. This task of critical thinking may actually be more difficult in gerontology than in other fields because of the familiarity and deeply personal nature of aging. Revolutionary changes took place in the 20th century, but most of us tend to assume that aging and the human life course have remained the same. Despite our commonsense perceptions, however, history and social science tell us that the process of aging is not something fixed, but is both changeable and subject to interpretation.

Taking a more critical and thoughtful stance, we know that "stages of life" have been viewed differently by different societies. Even in our own society, the experience of growing older is not uniform, but means different things to individuals depending on their gender, ethnicity, and social class. From this perspective, a familiar practice such as retirement turns out to be less than a century old and now is in the process of being reexamined and redefined. Even in the biology of aging, scientists are engaged in serious debate about whether it is possible to extend the maximum human life span from what we have known in the past.

In short, wherever we look—biology, economics, the social and behavioral sciences, and public policy—we see that "aging," despite its familiarity, cannot be taken as a fixed fact of human life. Both individual aging and population aging are socially and historically constructed, subject to interpretation, and therefore open to controversy, debate, and change.

It is astonishing to realize that more than half of all the human beings who have ever lived beyond age 65 are alive today. What aging will mean in the 21st century is not something we can predict merely by extrapolating from the present and the past. Still less can the study of aging consist of an accumulation of facts to be assimilated, as if knowing these facts could somehow prepare us for the future. The changes are too significant for such an approach.

What we need most of all is to consider facts about individual and population aging in a wider context: to understand that facts and theories are all partial, provisional, and, therefore, subject to interpretation and revision. That is the second major reason that the study of aging in this book is presented in the form of controversy and debate, offering all of us an opportunity to reflect on and construct an old age worthy of "our future selves."

ABOUT THE AUTHORS

Harry R. Moody is a graduate of Yale University and received his PhD in philosophy from Columbia University. He has taught philosophy at Columbia University, Hunter College, New York University, and the University of California at Santa Cruz. For 25 years he was at the Brookdale Center for Healthy Aging and Longevity at Hunter College of the City University of New York, where he served as cofounder and Executive Director. With the National Council on Aging in Washington, DC, he served as Codirector of its National Policy Center. He is the author of more than 100 scholarly articles and several books: *Abundance of Life: Human Development Policies for an Aging Society* (1988), *Ethics in an Aging Society* (1992), and most recently *The Five Stages of the Soul* (1997), a study of spiritual growth in the second half of life. He is known for his work in older adult education and served as Chairman of the Board of Elderhostel. He has also been active in the field of biomedical ethics and edits the newsletter *The Soul of Bioethics*. Dr. Moody is currently the Director of Academic Affairs for AARP in Washington, DC. He also serves as Senior Fellow of Civic Ventures.

Jennifer R. Sasser is Chairperson and Associate Professor in the Department of Human Sciences, and coordinates the gerontology program, at Marylhurst University in Portland, Oregon. She joined the Marylhurst faculty in 1997 and since that time has been involved in the design and implementation of many on-campus and Web-based courses and programs for adult learners, including the graduate and undergraduate certificate programs in gerontology. As an undergraduate she attended Willamette University, in Salem, Oregon, graduating cum laude in psychology and music; her interdisciplinary graduate studies at University of Oregon and Oregon State University focused on the human sciences, with specialization areas in adult development and aging, women's studies, and critical social theory and alternative research methodologies. While conducting her doctoral work at OSU she was a graduate teaching and research fellow, as well as the first recipient of the AARP/Andrus Foundation Graduate Fellowship in Gerontology. Her doctoral dissertation became part of a book published by Routledge in 1996 and coauthored with Dr. Janet Lee, titled *Blood Stories: Menarche and the Politics of the Female Body in Contemporary US Society*. Over the past 20 years Jennifer has been involved in inquiry in the areas of creativity in later life; older women's embodiment; critical gerontological theory; and transformational adult learning practices. Jennifer served on the Oregon Gerontological Association Board of Directors starting in 2005 and was Chairperson of the Board for three years. She is the 2012 recipient of the Association for Gerontology in Higher Education Distinguished Teacher award.

A LIFE COURSE
PERSPECTIVE ON AGING

Multigenerational families provide a vivid illustration of the life course perspective: Aging is a gradual, lifelong process we all experience, not something that happens only in later life.

When we think about "aging," we often call to mind the image of an old person. But the process of aging actually begins much earlier in life. We cannot fully understand what *old age* means unless we understand it as part of the entire course of human life, and this approach is called the **life course perspective** (Settersten, 2003).

1

Often our image of old age is misleading. For example, try to conjure a mental image of a college student. Now imagine a recent retiree, a grandmother, and a first-time father. Hold those images in mind and then consider the following facts:

- Each year, half a million people over age 60 are studying on college campuses.
- Retirees from the military are typically in their 40s or 50s.
- In some inner-city neighborhoods, it is not at all unusual to meet a 35-year-old grandmother whose daughter is a pregnant teenager.
- It is no longer surprising for men in second marriages to become a father for the first time at age 40 or 50.

Did some of those facts contradict the images you conjured, particularly your images about how old people are when they fill certain roles? What this exercise tells us is that roles such as "student," "retiree," "grandmother," and "first-time father" are no longer necessarily linked to chronological age. Today, what we are learning about aging is forcing us to reexamine traditional ideas about what it means to grow old. Both biomedical science and social behavior among older adults depart from stereotypical images of what is "right" or "appropriate" for old age.

Although we tend to think of old age as a stage at the end of life, we recognize that it is shaped by a lifetime of experience. Conditions of living, such as social class, formal education, and occupational experience, are determinants of the individual's experience in old age. In other words, the last stage of life is the result of all the stages that come before it. The implication is that we no longer accept the quality of life in old age, or even the meaning of old age, as a matter of destiny. Rather, we view it as a matter of individual choice and social policy. Whether older people feel satisfaction and meaning may therefore depend on what they do and what social institutions do to give them new purpose in later life.

Recent biological research demonstrates that indeed people do not suddenly become old at the time we have defined as *old age*. Aging is a gradual process, and many human capabilities survive long past the time when Americans are considered of an age to retire. We are learning more every day about how and why people grow old, with the hope that we can make the last stage of life just as meaningful in its own way as earlier stages are.

Age Identification

A central concept in any discussion of aging is the meaning of age itself. Age identification is partly an acknowledgment of chronological age or years since birth, but it is also a powerful social and psychological dimension of our lives.

From early childhood, we are socialized to think about what it means to "act your age," a process described as **age differentiation**. We learn that different roles or behaviors are considered appropriate depending on whether we are a toddler or a teenager, an adult or an older person. **Age grading** refers to the way people are assigned different roles in society depending on their age (Streib & Bourg, 1984). Theorists of **age stratification** emphasize that a person's position in the age structure affects behavior or attitudes.

People also come to define themselves, at least in part, in terms of their age. Consider when you started thinking of yourself as an adult instead of a kid. Did you suddenly lose interest in some of the things that had once fascinated you because you considered them "childish" interests? Do you anticipate that when you become "middle aged" or "old" you will no longer be quite the same person you are now?

People within a culture have widely shared expectations about the "right time" for an event to happen. In Western society, for example, marriage at age 13 or retirement at age 30 would be considered "off time," but graduation from college at 22 or retirement at age 65 would be "on time." In other words, we have a shared **social clock** concerning the appropriate age for life events (Helson, Mitchell, & Moane, 1984). However, the timetable for life events varies somewhat with social class and occupation; the career timetable of a medical student, for instance, is quite different from that of a migrant farmworker. In addition, age norms change over time. For example, Americans today tend to first marry in their mid- to late 20s, but a century ago, people that age (particularly women) would have been considered rather old for a first marriage.

Cultural understandings about what is "age appropriate" are part of a tradition going back to antiquity (Falkner & de Luce, 1992). In the comedy of ancient Rome, for example, older adults are often ridiculed for unseemly behavior, and hostility is expressed toward old men who take young lovers, a theme often repeated in medieval literature (Bertman, 1976).

What do we think is appropriate for "older people" in our culture today? For one answer, we can look to the images in our symbols, rituals, and myths. Storytellers and minstrels have expressed traditional societies' concepts of age, but today in advanced industrial societies, those concepts are frequently transmitted and reinforced by TV and other mass media. As a rule, people on TV are young and good looking; older people are not visible on TV in anything like their proportion in the actual population (Davis & Davis, 1985; Peterson & Sautter, 2003). When they are depicted, they tend to be one step removed from the action. Even when TV advertisers try to appeal to the "gray market" of older consumers, they present idealized images of good health and vigorous activity. It seems sometimes that we are trying to ignore the inevitability of old age. However, some analysts find that older people use TV in ways that connect public concerns with their everyday lives (Riggs, 1998).

Beyond stereotypes, electronic media have a latent effect that is both more subtle and more pervasive. TV occupies a perpetual present dominated by novelty and momentary images (Meyrowitz, 1985). The effect is to weaken any sense of continuity over the life course and to undercut any authority or meaning for old age (Moody, 1988). Traditional cultures tend to prize their older members as links in a historical chain reaching back to the ancestors. But the contemporary culture of TV, like the Internet, tends to put all age groups on an equal footing (Gilleard & Higgs, 2000). The result is the "disappearance of childhood" and perhaps of old age too (Postman, 1982).

Mythic images of aging are of course oversimplified and based on fantasy. But sometimes they provide insight into the deeper meaning of the last stage of life. The Western view of old age tends to be ambivalent. In ancient Hebrew religious literature, for instance, old age is venerated as a reward for righteous living: The Fifth Commandment to honor one's parents contains a promise of long life. In contrast, there is a realistic dread of frailty and a fear that children may reject aged parents (Isenberg, 2000). The Book of Job even questions the assumption that old age brings wisdom and recognizes that the wicked can live just as long as the righteous.

The Greek and Roman views of late life also reflect profound ambivalence. In the first great work of Western literature, Homer's *Iliad,* we find worship of youth in the figure of the young, strong warrior Achilles, but the aged Nestor is revered for his wisdom. In the philosophical tradition, Plato and Aristotle took opposing views on aging. For Plato, later life offered a possibility of rising above the body to attain insight into the eternal nature of reality. In contrast, Aristotle saw middle age as the summit of life, a time when creative intellectual powers were at their peak, but later life as a time of decline.

In our culture today, we explore similar issues, especially in feature-length films. The myths of aging range from the quest for rejuvenation through the fountain of youth (*Cocoon*) to the psychological self-fulfillment of the aged hero returning home (*Wild Strawberries* or *The Trip to Bountiful*). At its best, film can present images of the older person as a genuine hero triumphing over circumstance (*Driving Miss Daisy* or *The World's Fastest Indian*). The images of old age purveyed by mass media have a profound effect on attitudes toward aging in all industrialized societies (Featherstone & Wernick, 1995).

THE STAGES OF LIFE

Since the dawn of civilization, human beings have recognized a progression through the life course, from infancy through old age. The overall progression appears universal, yet the time between birth and death has been organized in distinctive ways by different societies (Boyle & Morriss, 1987). The simplest concept of the life course has been a division into two stages: childhood and adulthood. But as societies become more complex and as longevity increases, they tend to develop a greater number of life stages.

Greek and Roman ideas were influential in shaping how we think today about aging and the life course. One of the greatest Greek tragedies is the three-part *Oedipus* cycle, the last play written when its author, Sophocles, was nearly 90 years old. In this story, Oedipus became king because he solved the famous riddle of the Sphinx: "What creature walks on four legs in the morning, two legs at noon, and three legs in the afternoon?" The answer is the human being at successive life stages: infancy (crawling on four legs), adulthood (walking on two), and old age (using a cane, a third leg, to support the other two). The Greek medical writer Hippocrates described four stages of life, or "ages," corresponding to the four seasons of the year. Similar ideas were put forward by the Roman physician Galen and by the astronomer Ptolemy. Ptolemy developed an idea of seven stages of life, which had great influence during the Middle Ages.

During the Middle Ages, Christian civilization balanced the image of multiple stages with the metaphor of life as a journey or spiritual pilgrimage. From that standpoint, no single stage of life could be viewed as superior to another. Just as the natural life cycle was oriented by the recurrent cycle of the seasons, so the individual soul would be oriented toward the hope of an afterlife (Burrows, 1986). The human life course as both cycle and journey was thereby endowed with transcendent meaning and wholeness (T. Cole, 1992).

With the coming of the Reformation and the Renaissance, ideas about the life course changed into forms we recognize as modern. Writing in this epoch, Shakespeare expressed the traditional idea of the "Seven Ages of Man":

> All the world's a stage
>
> And all the men and women merely players,
>
> They have their exits and entrances;
>
> And one man in his time plays many parts,
>
> His acts being seven ages.

(As You Like It, Act II, Scene 7)

To Shakespeare, the periods of life were merely "roles" acted out on the stage of society, and the role losses of old age appeared as the final act of the play. Thus, a theatrical metaphor replaced the ideal of a cosmic cycle or a spiritual journey.

At the dawn of modern times, a generation after Shakespeare, drawings and engravings began to depict the stages of life in a new way. The traditional image of a completed circle became an image of a rising and falling staircase, where midlife occupied the peak of power. That image promoted the idea of life as a "career," in which individuals could exercise control over later life through, for example, extended education, good health care, and capital accumulated through savings during earlier stages.

During the 16th and 17th centuries, the stages of life began to be demarcated in ways we recognize today. Childhood became a period of life in its own right, separate from adulthood and old age (Aries, 1962). By the 20th century, as the practice of retirement became well established, old age became a distinct phase as well. Some sociologists argue that such stages reflect patterns of socialization tied to dominant institutions such as the school or workplace (Dannefer, 1984); in other words, retirement exists as a separate phase of life partly because society needs to make way in the workplace for younger workers.

Today, a person will spend, on average, at least one fourth or even one third of adulthood in retirement (Kohli, 1987). Partly as a consequence, distinctions are now sometimes made between the **young-old** (ages 65–74), the **old-old** (ages 75–84), and now the **oldest-old** (ages 85 and over). Demarcating a stage of life following the working years is more important to us than ever, yet we have simultaneously become less certain about what it means to grow older or to "act your age" at any point in life. Issues around the potential for new forms of self-expression and contributions to society in later life are discussed later in this book.

THE LIFE COURSE AND AGING

The study of aging as a historical phenomenon reveals a variety of views about the stages of life, about when old age begins, and about what it involves (Minois, 1989). When we read about aging in the Bible or in works by such writers as Shakespeare and Cicero, we might imagine that "old age" is a fixed stage of life, always part of the natural pattern of things, such as birth and death. But now, at the beginning of the 21st century, it has become clear that human aging is far more ambiguous than might have been imagined in earlier epochs. We can most fruitfully understand old age not as a separate period of life, but as part of the total human life course from birth to death.

Increasingly, aging is seen from this life course perspective (Markson & Hollis-Sawyer, 2000). In other words, we look at old age as one phase of the entire course of life and the result of influences that came earlier than old age. We distinguish here between the span of a lifetime, which is the total number of years we live, and the course of life, which refers to the meaningful pattern seen in the passing of time. Gerontology is enriched and broadened by the life course perspective. Instead of merely describing the limited characteristics of old age, which are tied primarily to biophysical changes, we shift the framework to include all phases of life, from childhood, adolescence, and adulthood right up through the last period of old age. We also view the complex interaction of age, social status, cohort effects, and history (M. W. Riley & J. W. Riley, 1994). **Longitudinal research**, which follows individuals over long periods of time, is a key research tool to reinforce the life course perspective because such an approach allows researchers to view developmental changes as they unfold within the same people.

The life course perspective insists that, to make sense of old age, we need to understand the entire life history. As people move through the life course, they are socialized to act in ways appropriate to successive social roles: student, parent, worker, retiree, and so on. But these structural factors only set boundaries; the meaning and experience of aging vary significantly by culture and are influenced by powerful factors such as gender, socioeconomic status (SES), and ethnicity. There is also room for individual variety and freedom of choice as human beings interpret age-related roles in distinctive ways.

Life Transitions

A life course perspective will recognize markers of the passage through life: important life events or transition points, such as graduation from school, first job, marriage, and retirement. In some respects, life transitions have become more predictable than was true earlier in history. For example, today people commonly die in old age, whereas in an earlier era, death was not unusual at any time of life. Thus, an event such as the death of a spouse or a parent is now a more predictable marker of later adulthood than it once was.

At the same time, however, certain transitions are less often tied to a particular age or stage of life than they might have been in earlier times. For example, during the 1950s and 1960s, college students were expected to graduate at the age of about 21. But today's college students graduate at any age from the early 20s to the 30s and beyond, and news photos of gray-haired grandparents wearing a cap and gown are no longer uncommon. Graduation may occur either before marriage, sometime during childbearing age, or well after. Whatever the age or circumstance of the graduate, however, the transition still marks a major role change.

Special events that mark the transition from one role to another—such as a bar mitzvah, confirmation, graduation ceremony, or wedding—are known as **rites of passage** (Van Gennep, 1960). These rituals reinforce shared norms about the meaning of major life events. Some traditional rites of passage, such as the sequestration of adolescents prior to induction into adult society, are no longer commonly observed in our society. However, we continue to observe a great many, including markers of old-age transitions such as retirement parties, 50th wedding anniversaries, and funerals.

How are we to understand the significance of life transitions? As the human life course became an object for scientific study, the stages of life were no longer seen as part of a cosmic order of meaning (Cole & Gadow, 1986; Katz, 1996). Instead, psychology tried to explain change over the course of life as a natural process unfolding through time. The result was the rise of a new field: life span development psychology. Erik Erikson, an influential developmental psychologist, depicted the life course as a series of psychological tasks, each requiring the person to resolve conflicting tendencies (Erikson, 1963). For middle age, Erikson posited a conflict between stagnation and generativity: roughly, being trapped by old habits versus going beyond self-absorption to nurture the next generation (Kotre, 1984). For old age, Erikson saw a conflict between ego integrity and despair—that is, accepting one's life versus feeling hopeless and depressed about the limited time remaining.

Related to Erikson's basic ideas has been the attention on psychological changes during midlife transition, a time when people in middle age confront facts about mortality and the limits of youthful dreams (Jacques, 1965). Psychologist Daniel Levinson (1978) has described life transitions characteristically associated with ages such as 30, 40, and 50. These are times when people at midlife reassess themselves and ask, "Where have I come from, and where am I going?" Many of these psychological "passages" or changes of adult life have been popularized by journalists. However, doubts have been raised about just how universal such "passages" and age-related transitions actually are (P. Braun & Sweet, 1984). Midlife, just like old age, turns out to be a time of life that is different for different people (Brim, Ryff, & Kessler, 2004).

In contrast, many theorists today see personality in terms of continuity or flexible adaptation over the life course. These theories are more optimistic than those that see old age as a time of loss resulting in either passive adjustment or dependency and depression. Today, most gerontologists believe that people bring positive resources to aging, including a personal sense of meaning. Empirical studies show that people generally cope well with life transitions such as retirement, widowhood, and the health problems of later life. When problems come, styles of coping tend to remain intact, and people adapt. Because of this capacity for adaptation, old age is not usually an unhappy time.

Nevertheless, many behavioral or psychological problems come about because of the difficulties of preparing for transitions without the help of widely observed rites of passage and institutional structures. For example, the transition from adolescence to adulthood is typically marked by events such as marriage, parenthood, and employment (Hogan & Astone, 1986). Although schools, job orientation, and marriage counseling help people make transitions to adulthood, few social institutions exist to help people with the transitions of the second half of life.

In addition, we currently have no consensus about how people are supposed to act when in late life they confront events that have traditionally been linked to younger ages (Chudakoff, 1989). How are older widows supposed to go about dating? How much help should older parents expect from their children who are themselves at the point of retirement? When confronted with a 70-year-old newlywed or a 60-year-old "child," we realize how unsettled the norms are relating to many of the transitions in later adulthood (Featherstone & Hepworth, 1993).

THEORIES OF AGING

Modernization Theory

How do we make sense of the contradictory images of aging found in modern culture? One influential account that tries to do so is the **modernization theory of aging** (Cockerham, 1997). According to this theory, the status of older adults declines as societies become more modern. The status of old age was low in hunting-and-gathering societies, but it rose dramatically in stable agricultural societies, in which older people controlled the land. With the coming of industrialization, it is said, modern societies have tended to devalue older people. The modernization theory of aging suggests that the role and status of older adults are inversely related to technological progress. Factors such as urbanization and social mobility tend to disperse families, whereas technological change tends to devalue the wisdom or life experience of elders, leading to a loss of status and power (Cowgill, 1986). Some investigators

Becoming a grandparent can be an important life course transition.

have found that key elements of modernization were, in fact, broadly related to the declining status of older people in different societies (Roger Clark, 1992–1993).

This account strikes a responsive chord because it echoes the "golden age" picture of later life, which depicts the old as honored in preindustrial societies (Stearns, 1982), a version of the "world-we-have-lost" syndrome (Laslett, 1965/1971). But imagining that elders were well treated in "the good old days" is a big mistake, and modernization theory has been widely criticized (C. Haber, 1983; Quadagno, 1982). As we have seen already, in primitive, ancient, and medieval societies, older adults were depicted and treated in contradictory ways: sometimes abandoned, sometimes granted power. The history of old age includes variations according to race, gender, social class, and culture. Modernization has clearly reshaped the meaning of old age, yet the image and reality of old age have never entirely coincided, as the cross-cultural study of aging confirms (E. Holmes & L. Holmes, 1995).

URBAN LEGENDS OF AGING

"Respect for elders was higher in the past."

This is a common myth, debunked by historian Peter Laslett 40 years ago in his classic *The World We Have Lost* (1971). Maybe there's a reason why the Bible contains the Fifth Commandment, "Honor Thy Mother and Thy Father." It's the only Commandment that carries a reward for following it.

At the core of the history of old age, there has always been ambivalence: both resentment and guilt, both honor and oppression. The psychological basis for ambivalence is understandable. Why shouldn't adults feel guilt and dread at the sight of vulnerable old age stretching before them? Why shouldn't we harbor ambivalent feelings toward those who accumulate power and wealth over a long lifetime? We see the same ambivalence today. Older people as a group receive many benefits from the government based on their age, yet they are sometimes depicted, perhaps unfairly, as selfish or unconcerned with other generations. The truth is different from what popular images convey.

A decisive change with industrialization has been a growing rationalization and bureaucratization of the life course—a greater rigidity among the "three boxes of life" of childhood, adulthood, and old age (Bolles, 1981). At the same time, as we have seen, mass media and rapid flux in cultural values have begun to erode any special qualities linked to distinctive life stages. With rising longevity, more people are living to old age, and older adults as a group are becoming a larger proportion of the total population. The power of older people has grown by sheer numbers. Meanwhile, the achievement of old age has been devalued simply by becoming more familiar. Perhaps most important, old age has been stripped of any clear or agreed-on meaning because the entire life course has changed in ways that will have unpredictable effects on what aging may be in the 21st century.

The problem of constructing an overall theory of aging for social gerontology can be compared with a parallel problem in the biology of aging: Is aging truly something inevitable (Olshansky & Carnes, 2002)? The challenge for evolutionary biology begins with a paradox: Why does aging appear at all? From the standpoint of survival of the fittest, there seems to be no reason for organisms to live much past the age of reproduction. Old age, in short, should not exist. Yet human beings do live long past the period of fertility; indeed, human beings are among the longest-living mammals on earth.

Thus, the meaning of old age is a problem even for biology, and biologists have put forward a whole variety of theories to explain it: somatic mutation theory, error catastrophe theory, autoimmune theory, and so on. No single theory has proved decisive, but all have stimulated research enabling us to better understand the biology of aging. Similarly, the changing condition and meaning of old age have provoked a variety of theories in social gerontology. Just as with the biology of aging, there is no clear agreement that a single theory is best. But two early theories of aging are still worth closer examination because they demonstrate just how deeply held values affect all theories of aging and how these theories are related to enduring questions about the meaning of old age.

Disengagement Theory

One of the earliest comprehensive attempts to explain the position of old age in modern society is the **disengagement theory of aging** (Cumming & Henry, 1961). The disengagement theory looks at old age as a time when both the older person and society engage in mutual separation, as in the case of retirement from work. This process of disengagement is understood to be a natural and normal tendency reflecting a basic biological rhythm of life. In other words, the process of disengagement is assumed to be "functional," serving both society and the individual. The disengagement theory is in fact related to modernization theory. It was assumed that the status of older adults must decline as society became more modern and efficient, so it was natural for older adults to disengage.

The disengagement theory grew out of an extensive body of research known as the Kansas City Study of Adult Life, which was a 10-year longitudinal study of the transition from middle age to old age (Williams & Wirths, 1965). The idea of disengagement presented itself not only as an empirical account based on those findings but also as a theory to explain why the facts turned out the way they have. But gerontologists have criticized the theory of disengagement (Hochschild, 1975); some have pointed out that the theory evolved during the 1950s and reflected social conditions quite different from those of today.

Although the original theory is no longer widely accepted, the pattern of disengagement does describe some behavior of *some* older people—for example, the popularity of early retirement. But there are growing numbers of older people whose behavior cannot be well described as withdrawal or disengagement from society. Disengagement as a global pattern of behavior can hardly be called natural or inevitable.

Another problem arises when we describe disengagement as "functional," which is a synonym for "useful." The same process that might be functional or useful for an organization—for instance, compulsory retirement at a predictable age—may not be at all useful for individuals, who might prefer flexible retirement or who may need to continue working because of economic necessity. In fact, it was widespread resentment at being forced to retire at a fixed age that led Congress to end mandatory retirement in 1986.

There is also a lack of clarity about what behavior is actually being described by the concept of disengagement. For example, individuals might partially withdraw from one set of activities, such as those in the workplace, to spend more time on other activities, such as family and leisure pursuits; total withdrawal is quite uncommon. Although advancing age at some point is usually accompanied by losses in health, physical ability, and social networks, those who age most successfully adjust to and compensate for these losses by putting the changes of later life into a wider perspective, an attitude sometimes described as "wisdom." Later life today, at least for those who remain healthy, is often filled with a rich range of activities. The Kansas City Study investigators found that with advancing age, there is, in fact, a trend toward greater *interiority,* meaning increased attention to the inner psychological world (Neugarten, 1964). Individuals appear to reach a peak of interest in activity and achievement in their middle years. As they anticipate later life, they may become more detached, more inclined to "ego transcendence," as if in anticipation of predictable role losses in later life (Peck, 1968).

Understood in this way, "disengagement" need not necessarily describe the outward behavior of individuals, but may refer to an inner attitude toward life. Furthermore, there is no reason to assume that all older people are inclined toward even a psychological stance of disengagement; some may have ambivalence about their own activities and attachments. Perhaps the greatest example in literature of that ambivalence is the tragic fate of Shakespeare's King Lear, who tried to give up his role as king but was not quite able to withdraw from power and prestige. As a result, he brought disaster on his family. King Lear's example suggests that disengagement depends on having some sense of personal meaning that is distinct from the office one holds. The ability to achieve some degree of detachment, at any age, is a matter of wide individual difference. In later life, disengagement is the preferred style for some, whereas continued activity remains attractive for others.

Activity and Continuity Theories of Aging

At the opposite pole from the disengagement theory is the **activity theory of aging**, which argues that the more active people are, the more likely they are to be satisfied with life. Activity theory assumes that how we think of ourselves is based on the roles or activities in which we engage: We are what we do, it might be said. The activity theory recognizes that most people in later life continue with the roles and life activities established earlier because they continue to have the same needs and values.

The **continuity theory of aging** makes a similar point, noting that people who grow older are inclined to maintain as much as they can the same habits, personality, and style of life they developed in earlier years (Costa & McCrae, 1980). According to both the activity theory and the continuity theory, any decreases in social interaction are explained better by poor health or disability than by some functional need of society to "disengage" older people from their previous roles (Havighurst, Neugarten, & Tobin, 1968).

A large body of research seems to support some aspects of activity theory. Continued exercise, social engagement, and productive roles all seem to contribute to mental health and life satisfaction. But other studies indicate that informal activity or even merely perceived social integration may be more important in promoting subjective well-being. In other words, our attitudes and expectations about activity or detachment may be more important than our formal participation patterns (Longino & Kart, 1982). In fact, what counts as "activity" depends partly on how we look at things, not on external behavior alone. This point is emphasized by those who adopt a phenomenological standpoint under-lying the interpretation of aging.

If retirement or age limitations make actual participation impossible, the activity theory suggests that people will find substitutes for roles or activities they have to give up (Atchley, 1985). A great many social activities encouraged by senior centers or long-term-care facilities are inspired by an assumption that if older people are active and involved, then all will be well. This "busy ethic" and its hostility to retirement are expressed in similar terms, and the sentiment seems widely shared (Ekerdt, 1986). For instance, former *Cosmopolitan* magazine editor Helen Gurley Brown (1993), in a self-help book for older women (*The Late Show*), wrote that work is "our chloroform . . . our life . . . our freedom from pain . . . supplier of esteem." Along the same lines, essayist Malcolm Cowley, in his book *The View From 80* (1980), also expressed the ideal of the activity theory of aging when he wrote: "Perhaps in the future our active lives may be lengthened almost to the end of our days on earth; that is the most we can hope for."

But such active involvement may be more feasible for the young-old than for the old-old, and certainly there are differences between individuals as well. Biological limitations cannot easily be overcome by voluntary effort. The ideal of active aging seems more like a prolongation of the values of middle age than something special or positive about the last stage of life. Finally, despite progress in recent years, society still places many obstacles to social engagement in old age. For example, remarriage is more difficult for older women than for older men because the proportion of older women is larger than the proportion of older men in the population at every age after 65, and in the labor market, age discrimination is a real barrier preventing middle-age and older people from taking up a second career. According to the U.S. Department of Labor, anyone over age 40 is officially an "older worker." A more realistic recognition of these facts might allow adults to live out their years with greater dignity instead of trying to stay forever young.

INFLUENCES ON THE LIFE COURSE

Every theory of aging has its limits: None of them fully explains the variety of ways in which individuals experience old age, and many theories seem to reflect social values in uncritical ways—for example, by holding up either activity or disengagement as the ideal goal or social norm for later life. The advantage of thinking in terms of age transitions is that we can see adult development as more open ended than people have tended to see it in the past. As a result, the meaning of old age is less fixed, and the choices are more varied. We can contrast this wider social freedom with stereotypes that still persist about human development in the second half of life.

The most widely pervasive view of adulthood is not based on positive development at all, but assumes continuous deterioration and decline. Consider the message on birthday cards: Aging is a disaster; after youth, it's all downhill (Demos & Jache, 1981). This pessimistic, age-as-decline model gives priority to biological factors and is the basis for the widely shared prejudice called **ageism**. We are better off appreciating how social class, life history, and social institutions and policies create variation in the experience of aging. Although aging is a negative experience for some people, for others, it opens the door to meaningful new roles and activities.

Social Class and Life History

The pathway through life depends on social class, which is strongly related, in turn, to the number of years spent in the educational system. Socioeconomic position predicts not only how long people remain in school but also, for instance, whether they will be "late" parents—that is, with a first child born after age 30. Social class, above all, has a lifelong influence on health status (Hemingway et al., 1997).

Earlier life events have long-lasting effects. The basic rule of accumulated advantage or disadvantage is that "the rich get richer and the poor get poorer." For example, early completion of college and entry into a favorable occupation is converted during middle age into increased wealth in the form of home ownership and pension vesting (Henretta & Campbell, 1976). Women who enter the labor force at the beginning of childbearing usually have to accept career interruptions and tend to have diminished income later in life; thus, gender differences in old-age poverty are explainable partly as the result of life course choices made decades earlier.

History also plays a profound role in shaping lives. For example, a large historical event like the Great Depression can cause a dramatic and unexpected drop in income and status for many people (Elder, 1974). The cohort who were in their prime working years at the time were typically worse off financially in old age than their children, the current generation of retirees. This recognition of the influence of historical events has stimulated new interest in using interviews and oral history to understand how social forces affect people's lives (A. Cole & Knowles, 2001).

Unpredictable or non-normative life events, such as getting divorced or losing a job, also have a significant effect on the life course. A longitudinal study during the 1970s showed that around one third of people experienced an unexpected but significant drop in income due to such non-normative life events (Duncan, 1988). Research has also shown that negative life events such as widowhood or job loss can cause a dramatic downturn in personal health and can profoundly affect an individual's financial status during retirement. Such events induce a psychosomatic response to stress, and negative life events therefore become risk factors that predict the onset of illness (T. H. Holmes & Rahe, 1967). Yet the impact of life events is not a simple process. The same stressful life event—for example, becoming a widow—may have different effects on different people. The impact depends on whether the event was expected or anticipated and also on what kind of personal or family resources are available. Support from family and friends can help older people cope with stress and maintain self-esteem.

Social Institutions and Policies

The structure of the life course in modern times has been shaped by the power of the educational system and the workplace. In the 19th century, the rise of public schools began to lengthen the period of formal education and introduce credential requirements for most types of work. The United States, a self-consciously "modern" nation, took a lead in these progressive developments (Achenbaum, 1978; Fischer, 1977). Early in the 20th century, adolescence was recognized as a distinct phase of life and became more prolonged, whereas middle age became an important period of the life course (Neugarten, 1968).

The Industrial Revolution brought far-reaching demographic and economic changes, as well as new cultural ideas about age-appropriate behavior (Hareven & Adams, 1982). Bureaucratic institutions, from local school systems to the Social Security Administration, always favor rule-governed, predictable procedures, so it is not surprising that with the rise of bureaucracy came an emphasis on defining life stages by chronological age. With falling birthrates in the 20th century, the modernized life course became established.

Today, social policies and institutions still define transitions throughout the life course. The educational system defines the transition from youth to adulthood, just as retirement defines the transition from middle age to old age.

Like progression through the school system, the movement into retirement seems more orderly than midlife transitions because employment policies and pension coverage closely regulate retirement. But the timing of retirement today is becoming less predictable than in the past because of turbulence in the U.S. labor market and because of the disappearance of mandatory retirement. Economic pressures force some to retire early, whereas others are encouraged to go back to school or take on part-time employment. The result is that previously clear boundaries—"student," "retiree," and so on—are becoming blurred.

If societal forces shape the life course, then it is reasonable to think that some of the negative features of old age may be due, at least in part, to institutional patterns that could be changed. A good example is the pattern known as **learned helplessness**, or dependency and depression reinforced by the external environment (Seligman, 1975). It has been suggested that some of the disengagement often seen in old age is not inevitable, but comes from social policies and from practices in institutions that care for dependent older adults (M. M. Baltes & P. B. Baltes, 1986). For instance, nursing home residents often suffer a diminished **locus of control**, in which they lose the ability to control such basic matters as bedtime and meal choices. When residents feel manipulated by forces beyond their personal control, they may become more withdrawn, fail to comply with medical treatment, and become fatalistic and depressed.

Without interventions to reduce dependency, older adults in ill health all too commonly lose hope and self-esteem as they experience declining control (Rodin, Timko, & Harris, 1985). But this downward spiral is not inevitable. The institutional structures responsible for such dependency can be changed. In an experiment with nursing home residents, psychologists offered small opportunities to increase locus of control—for example, allowing residents to choose activities or giving them responsibility for taking care of plants. The result was a dramatic improvement in morale and a decline in mortality rates (Rodin & Langer, 1980).

AGING IN THE 21ST CENTURY

Today, in the early 21st century, we no longer have a shared map for the course of life. The timing of major life events has become less and less predictable at all levels of society. In upper socioeconomic groups, for example, a woman with a graduate degree and career responsibilities may delay having a first child until age 35 or later; in other parts of society, where teenage pregnancy rates have soared, a 35-year-old woman may well be a grandmother. We are no longer so surprised when a 60-year-old retires from one career and takes up a new one, perhaps in consulting if the retiree has been an executive or a professional, or in small-electronics repair if the retiree has been a technician. In many other ways as well, the life course is becoming more "deinstitutionalized," more fragmented, disorderly, and unpredictable (Held, 1986; Hockey & James, 2003). Major events of life are no longer parts of a predictable or natural pattern.

Although the rigidity of the linear life plan has failed to keep up with new demographic realities, it did offer a degree of security. In the new, "postindustrial" life course, we are increasingly each on our own. Familiar social institutions such as marriage and employment can no longer be counted on for security throughout adulthood, and therefore the last stage of life also becomes less predictable.

Society has not yet come to terms with the meaning of "aging" in such unpredictable times. Optimists believe that medical science will soon permit us to delay aging-related decline until later and later in life. Yet economic forces seem to move in the opposite direction. In science and engineering, knowledge becomes obsolete within 5 or 10 years, so life experience counts for less than exposure to the latest technological advances. On the one hand, biology promises to postpone aging, but, on the other hand, social forces such as age discrimination make the impact of aging on individuals more important than ever.

Time and the Life Course

Expectations about time remain a major element in how we think about aging and the life course today (Hendricks & Peters, 1986). Just as industrialization imposed time schedules on workers to improve efficiency in the workplace, so the life course became "scheduled" by differentiated life stages. The factory and the assembly line had their parallel in the linear life plan. But that mode of organization has become outdated. In a postindustrial "information economy," the pace of life is speeding up, and flexible modes of production require a more flexible life course. The volatile economy demands multiple job changes and thereby makes every career unpredictable. Individuals at any age may be called on again and again to rewrite their biographies, although reinventing oneself gets more difficult as the résumé gets longer.

Another example of our contemporary time orientation is the prolonged period of life devoted to education. The knowledge explosion and pressure for specialization put a premium on added years of schooling, and the job market has fewer places for those without advanced skills. Our postindustrial economy is increasingly based on "knowledge industries," where emerging fields, such as computer software and biotechnology, favor cognitive flexibility.

The trend toward cognitive flexibility also poses a distinct challenge for an aging society. Middle-age and older workers, who are perceived to be less creative than younger workers, may be at a disadvantage in the fast-moving labor market. For instance, in some branches of media or advertising, employees are viewed as "old" if they are over the age of 40. But if retirement, the defining institutional feature of old age, is to remain economically feasible, then we will have to develop ways to keep people working as long as they remain productive. Retraining for displaced workers of whatever age is likely to become imperative in the future. These trends underscore the importance of adaptability and lifelong learning.

Parenthood offers still another example of our changing time orientation. Demographers estimate that, in the 1930s, 90% of a woman's years after marriage were spent raising dependent children (Gee, 1987). By the 1950s, that proportion had dropped to 40%, giving rise to what some observers have dubbed the "empty nest syndrome"—an extended postparental period of life that occurs after children have grown up and left home (Lowenthal & Chiriboga, 1972). Because of women's roles and responsibilities in the family, their later lives typically have greater variability than do men's (Rindfuss, Swicegood, & Rosenfeld, 1987). Gender roles are increasingly shaped by the power of culture (Gullette, 2004).

Another change has been the postponement of childbearing. People often spend more of their lives in their roles as adult children of aging parents than as parents themselves (Brubaker, 1985). But what does it mean, in psychological and social terms, when a "child" is 50 or 60 years old or even older? Even to ask these questions shows that the human life course has changed in ways that are still not fully recognized.

The Moral Economy of the Life Course

The changing structure of the life course has profound implications for obligations and expectations across the life course. We can speak about these expectations in terms of the "moral economy" of the life cycle (Minkler & Estes, 1998). The moral economy embodies expectations of what is fair or right: Stay in school and you'll get a good job, become a senior citizen and you'll have a right to retirement income, and so on. But the old moral economy, with its characteristic distribution of work and leisure according to chronological age, is losing its power, and we do not have anything as well defined to replace it.

To overcome limitations of the old map of life, we may need to develop bolder ideas about the positive social contributions that can be made by the old; we also need to think more deeply about the meaning of life's final stage. Cicero (106–43 BC), author of the classic essay "On Old Age," offered a realistic account of both the gains and the losses of aging. Cicero was inspired by the hope that the mind can prevail over the body. Thus, he viewed old age not exclusively as a time of decline or loss but also as an opportunity for cultivating compensatory wisdom. Cicero, in fact, was one of the first and most eloquent proponents of the ideal of "successful aging" (P. B. Baltes & M. M. Baltes, 1990).

Despite Cicero's wise words about later life, we should not sentimentalize the status of old age in the past. But at least in the past, those who had lived a full life span could take for granted shared values and shared experience across the generations simply because the pace of change was slower. With the rapid social changes of the 21st century, we can too easily stereotype those who are older as people who are "behind the times" or lacking in creativity and wisdom.

One role well suited to older people in such an environment might be mentoring, or guiding the next generation in the capacity of teacher, coach, or counselor (Neikrug, 2000). This idea is attractive for several reasons: It encourages intergenerational relationships, and it takes advantage of generativity and wisdom, the virtues to be cultivated in the second half of life, according to Erikson. Mentors, however, still have to develop up-to-date skills and attitudes if their advice is to be respected by younger workers.

There appears to be a mismatch between the flexibility of the individual aging experience and the rigidity of outdated social structures, such as retirement practices (M. W. Riley & J. W. Riley, 1994). Instead of treating the life course as fixed, in the future, we may come to see later life as a period more susceptible to intervention and improvement. Instead of viewing aging only as decline, it is possible to provide incentives that modify the lifestyles and behaviors of older people. The goal would be to move from an age-differentiated society to an age-integrated society, where opportunities in education, work, and leisure are open to people of every age.

THE BIOLOGY OF AGING

The life course perspective on aging offers an optimistic view of possibilities open to older people. That view is sensible, given the prolongation of vigor among older people in our times. But will changes in aging go even further in the future? For instance, the film *Cocoon* (1985) tells the story of older adults who gain access to a drug that can reverse the process of aging and make them young again. In the movie, the audience has the experience of seeing famous older actors Don Ameche and Hume Cronyn grow young before their eyes. The film, of course, is science fiction. But it's only the latest version of a recurrent hope as old as humanity: the search for the fountain of youth (Olshansky & Carnes, 2002). Sometimes the dream takes the shape of the "hyperborean theme": a conviction that people in a remote part of the earth—for example, the Caucasus or the mountains of Peru—live extremely long lives. James Hilton's novel *Lost Horizon* (1933) popularized the idea of an imaginary place called "Shangri-La," which harbored the secret of longevity, and a movie based on the book had wide appeal.

But researchers have never found groups of people who live beyond the normal life span. Scientists who have diligently examined the facts have failed to find any place on earth where people live beyond the maximum human life span of around 120 years. Death remains a biological inevitability, and so far, we have not learned how to overcome the physiological limits that we know as aging.

The biology of aging remains one of the great unsolved mysteries of science. Scientists ask how the same process that leads to decline and death can be intrinsic to life. From an evolutionary point of view, aging poses a puzzle: How can a process of physiological decline—detrimental to the survival of organisms—actually be preserved by natural selection (Hayflick, 1996)? Biologists who study how aging takes place have accumulated a large body of knowledge, and experiments with lower organisms have proved that genetic and environmental manipulations can change life expectancy and maximum life span. Thus, scientists are now beginning to confront the question of whether it is possible to postpone, or even reverse, the process of biological aging.

The New Science of Longevity

Normal aging is not a disease, but denotes a series of progressive changes associated with increasing risk of mortality. But not all age-related changes involve mortality. For example, hair typically turns gray with advancing age, but this change does not diminish survival prospects. By contrast, other progressive changes lead to losses in functional capacity or the ability of biological structures to perform their proper jobs. For example, as blood vessels age, they tend to lose elasticity, a tendency known as arteriosclerosis or hardening of the arteries. Over time, arteriosclerosis can increase the likelihood of blockage and therefore the risk of damage that we describe as stroke or heart attack.

At the biological level, aging seems to result from changes taking place at the level of molecules, cells, tissues, and the whole organism. How do we recognize such changes? The simplest way to study the effects of aging at these levels is to compare younger and older organisms and note the differences. Such studies employ a **cross-sectional methodology**; that is, they look at physical function of people at different chronological ages, but at a single point in time. The general conclusion from such studies of human beings suggests that most physiological functions decline after age 30, with some individual variations.

A purely cross-sectional design is not necessarily the best way, however, to measure the changes presumably brought about by aging. For one thing, it is difficult to be sure we have taken into account all of the possible variables that might contribute to changes with age in the organism. Thus, a contrasting methodological approach, a *longitudinal design*, is sometimes used. The same individuals are followed over a long period to measure changes in physical function, or other abilities, at different ages. This approach also has problems. For instance, with human beings, we need to consider the influences of a changing external environment. Furthermore, carrying out longitudinal studies is expensive and not easy when the subject is a long-lived organism like a human being. But the results can be of great importance.

One of the most important studies of this kind has been the Baltimore Longitudinal Study of Aging, sponsored by the National Institute on Aging (Shock et al., 1984). In the Baltimore study, scientists looked at 24 distinct physiological functions (Sprott & Roth, 1992). These functions are called **biomarkers**, or biological indicators that can identify features of the basic process of aging (Shock, 1962). Some of the most commonly measured biomarkers are diastolic and systolic blood pressure and auditory or visual acuity. Others include the ability of the kidney to excrete urine and the behavior of the immune system. All these tend to decline with chronological age (Warner, 2004), but the rate and amount of decline differ between individuals.

Many age-related changes in physical function have already been documented, some of them familiar. For instance, with increased age, height tends to diminish while weight increases; hair becomes thinner, and skin tends to wrinkle. Another change is the loss in vital capacity, or the maximum breathing capacity of the lungs. With aging, both respiratory and kidney function decrease. But this decline chiefly results in a loss of **reserve capacity**, or the ability of the body to recover from assaults and withstand peak-load demands, as during physical exertion. Diminished reserve capacity may not have any discernible impact on the normal activities of daily living. For instance, not having reserves to run a marathon race is probably irrelevant to most activities of daily life.

A key finding from studies of biological aging is that chronological age alone is not a good predictor of functional capacity or "biological age." In other words, people of the same chronological age may differ dramatically in their **functional age**, which can be measured by biomarkers (Anstey, Lord, & Smith, 1996).

Scientists have not yet identified a single overall mechanism that gradually reduces functional capacity. Increasingly, however, they have come to believe that the process of aging is controlled at the most basic level of organic life. The key to reversing the process of aging may lie in the strands of the molecule called DNA (deoxyribonucleic acid), the basis for heredity in living cells.

For each species, there appears to be a maximum time, or **life span**, for how long a member of that species can survive. By contrast, **life expectancy** from birth is the average number of years an individual may be expected to live. Maximum life span, in other words, is always higher than average life expectancy. Maximum human life span, or longevity, may be determined by biological processes separate and distinct from those that bring about the time-related declines we see as aging. In fact, it might turn out that maximum life span is determined by factors much simpler than whatever degrades functional capacity. At this point, it seems likely that longevity is genetically determined. Some scientists have argued that natural selection may have promoted longevity-assurance genes (Sacher, 1978). In other words, evolution may have arranged for us to live as long as we do, but not necessarily for us to have the signs and symptoms of aging that we do.

Medawar (1952) was one of the first to advance the idea that a species might carry harmful genes whose time of onset was delayed until after the period of reproduction. If those same genes had the positive virtue of promoting reproduction, then such genes would be transmitted to future generations. This idea of a trait beneficial in early life but harmful in later life is known as **antagonistic pleiotropy**. For example, the disease known as sickle-cell anemia, prevalent in Africa, is genetically linked to resistance to malaria. Those people born with the sickle-cell trait are more likely to survive malaria and pass the gene on to their children. This idea helps explain how diseases and senescence could actually be the product of natural selection through evolution (Williams, 1957).

Much remains to be discovered about genetic links between evolution and longevity. Genes with a favorable influence early in life, perhaps by maintaining reproductive capacity for a longer time, could have a harmful influence later on by allowing individuals to pass on linked genes with a negative impact, such as a shorter life span. In contrast, the genes that determine maximum life span could turn out to be linked to genetic factors that forestall the degenerative diseases of late life. Thus, under the most favorable scenario, if we were to discover and intervene in the genetic causes of longevity, we might also find the key to reducing the disabilities and dysfunctions of old age.

Scientists studying genetic influences on aging and longevity have moved in a number of suggestive research directions. From an evolutionary point of view, for example, there seems to be no obvious reason that human beings should live beyond 30 or 40 years, which gives them enough time to reproduce. There seems to be a trade-off between the biological investment made in survival for reproduction and maintaining organs and tissues beyond the end of the reproductive period.

In fact, we see from population studies of animals in the wild that aging rarely exists. The sea anemone, for example, seems to exhibit no physiological losses with chronological age at all. Animals in the wild exhibit survival curves similar to those of human populations; that is, most individuals die during a certain age range, but others die when very young or when very old. What follows from this evolutionary argument is that there is no intrinsic biological necessity for aging, and thus no reason that the extension of maximum life span would be impossible.

According to one optimistic view, most of the decremental changes associated with aging—including potentially preventable diseases, such as Alzheimer's—are not the result of any preprogrammed, built-in requirement for decline, but are the result of environmental causes (R. Cutler, 1983). However, maximum life span seems to be largely shaped by specific genetic endowment, rather than environmental factors. Perhaps, then, aging is a passive or indirect result of biological processes, whereas maximum life span is a positive or direct result of evolution. From this perspective, it follows that both the rate of aging and the maximum life span of a species could change—and change relatively quickly.

Some provocative questions follow: Would it be possible by direct intervention to alter the genetic code and thus delay the onset of age-dependent illnesses and perhaps to retard the rate of aging itself? With deeper biological knowledge, could the maximum life span be extended to 150 or 200 years or beyond? Even to ask these questions shows just how far we have come from a traditional view of the human life course, in which birth, aging, and death were facts simply taken for granted as part of the unalterable nature of things (Aaron & Schwartz, 2004).

MECHANISMS OF PHYSICAL AGING

We sometimes think of aging as a process applying uniformly to the whole organism, yet physiological studies show that different parts of the body age at different rates. For example, white blood cells die and are replaced within 10 days, but red blood cells last 120 days. The stem cells that produce all blood cells reveal no signs of aging at all. Cells in the brain last as long as the body lives; once the brain is fully formed, neurons do not exhibit significant cell division, and, unless damaged by illness, they remain largely intact. But apart from long-living stem cells and brain cells, most parts of the body are constantly subjected to damage and repair. The mechanisms that contribute to this process of aging include wear and tear, the effects of free radicals, and the decline of the immune system.

Wear and Tear

The organic process of life is a delicate balance between forces that wear down structures—forces that lead to cell death, for instance—and those that repair damage at the molecular and cellular levels. The structure and metabolism of each living thing maintain this balance over time. But over time, the balance begins to shift: Damage occurs faster than it can be repaired. Moreover, repair capacity is not unlimited; mechanisms for maintenance and repair can be maintained only at a certain cost. In other words, there are trade-offs involved in longevity. As a result, damage tends to accumulate with age, and the body gradually loses its capacity to repair that damage.

Like other components of the body, DNA in the nucleus of cells is always being damaged and repaired, although not always perfectly. Among mammals, the longer-lived species are the ones that have greater capacity to repair damaged DNA. But as DNA replicates over and over, those small errors, or mutations, progressively alter the organism's genetic code.

Can we conclude that we could possibly control aging by reprogramming our genes? Perhaps, but manipulating genes to retard aging might not increase the human life span. Imagine that an older person starts to exhibit signs of arteriosclerosis, and so we "fix" the individual's DNA. We might prolong one person's life, but we haven't really done anything to prolong the lives of that person's children or successive generations. They will also need DNA fixes when they become older. The problem is that a harmful mutation expressed at an advanced age, only after reproduction, will not be removed from the gene pool. In fact, we may want to leave a mutation that contributes to aging in place for the next generation because the mutated gene could have positive effects at an earlier age. In other words, the same biological processes promoting health and vigor among young organisms can have a negative impact in later life.

Free Radicals

Like the effects of wear and tear, the action of **free radicals** contributes to physical aging. As they engage in metabolism, all cells produce waste products. Among those waste products are free radicals, or molecules of ionized oxygen, which have an extra electron. Those ionized oxygen molecules cause damage because they more readily bond with proteins and other physiological structures. Sometimes the proteins become inactive and unable to carry out their functions. Even oxygen, the essential element required for energy transformation in living organisms, can become a destructive force.

Certain physiological processes can fight the effects of free radicals, but over time, the reduction of functional capacity damages the organism. Free radicals have been implicated in many processes of physical aging (Armstrong et al., 1984).

A similar mechanism of physical aging is **glycosylation**. Among the most universal of all chemical changes in living things are those involving sugar (glucose). Along with oxygen, glucose is the basis for metabolism in all organisms. When foods such as meat and bread are heated, the proteins combine with sugar and turn brown, in a process known as *caramelization.* In our bodies, the sticky by-products of this chemical reaction can literally gum up our cells. Glycosylation is behind much of the damage created in adult-onset diabetes, as well as stiffened joints and blocked arteries.

Is it possible to reverse the symptoms of aging caused by free radicals or glycosylation? Perhaps so, but that intervention is not likely to be simple—at least not as simple, for instance, as taking an "anti-aging pill."

The Immune System

The decline of the immune system is another important mechanism of physical aging. The immune system's job is to defend the body from invaders like viruses, bacteria, and parasites. To perform this job, it sends a variety of cells, which are categorized as T cells, B cells, and accessory cells, coursing through the body. These cells interact in complex ways to destroy or neutralize antigens, the foreign organisms that trigger an immune response.

The cells of the immune system also remove damaged and mutant cells produced within the body, which may become cancers.

With normal aging, the immune system's ability to fight off invaders and mutants gradually declines; it may even begin mistakenly attacking healthy cells. The process begins at puberty, when the levels of a certain hormone begin to decrease. The components of the immune system, particularly the T cells, gradually lose their efficiency.

The aging immune system leaves the body increasingly vulnerable. No longer does it mount the maximum response to very small doses of antigen; below a certain dose, it may no longer even recognize some antigens. It seems to develop an especially sluggish response to some tumor cells. Thus, infection and age-related cancers, such as prostate and colon cancer, become more likely. In addition, the immune system becomes more likely to attack healthy cells, which may lead to an increase in rheumatoid arthritis and other autoimmune diseases. However, most autoimmune diseases do not seem to be a function purely of age, and genetic predisposition may also play a role. In any case, the gradual decline of the immune system leaves the body more and more susceptible to a wide variety of diseases, each of which takes its toll on the functioning and vigor of the organism as a whole.

AGING AND PSYCHOLOGICAL FUNCTIONING

Research continues into ways to forestall physical aging in the hopes that someday we will discover a way to stay young as long as we live. The crusade against senescence and death is particularly appealing to Americans, who idealize success. The sentiment "You're only as old as you think you are!" expresses an optimistic outlook that fits well with our can-do attitude toward life. According to this optimistic picture of later life, both physical and psychological decline can be offset by vigorous exercise and engagement with the world. "Use it or lose it!" seems to be the motto here, a philosophy that has been applied to everything from "sex after 60" to lifelong learning.

No wonder that the strategy called "successful aging" has become the goal of many gerontologists. In part, they wish to reject age-based stereotypes, and, in part, they wish to counter the assumption that aging means a rapid decline into frailty and senility (Rowe & Kahn, 1997). They are certainly right to reject such stereotypes. But the idea of successful aging should never be based on denial of real losses in functioning in the last stage of life. The importance of the idea of successful aging is that it encourages older people to optimize the capacities that remain while compensating for inevitable losses (P. B. Baltes & M. M. Baltes, 1990). The measure of successful aging is life satisfaction and a sense of well-being in the face of decline. Successful aging therefore involves the psychological side of aging, including self-concept, social relationships, and cognitive processes.

Self-Concept and Social Relationships

The way people see themselves has several dimensions, including personality, self-esteem, body image, and social roles. The aspect of self-concept that changes least with age seems to be personality. For instance, an extroverted person, one who enjoys interacting with other people, is likely to remain extroverted from childhood into the final stage of life. An older person who is skeptical or gullible is likely to have been that way all along.

Other aspects of self-concept do tend to change, however. The aging mind is not so much the impetus for change as is the progression of circumstances in which people find themselves. Take body image, for example. Naturally, the image of one's body changes as hair becomes grayer and skin gets more wrinkled. Self-esteem varies throughout life with one's successes, whether they are interpersonal, occupational, intellectual, or otherwise.

Social relationships in old age tend to exhibit the most predictable types of change. For one thing, people's social networks become smaller as time goes on. With retirement, work relationships diminish, if not disappear. An older person may have survived a great many friends and family members. In addition, it is difficult to add new members to the network if one is no longer engaged in work or wider community life.

Social Roles

Another aspect of self-concept that changes with age has to do with the social roles we occupy. Growing up and growing older, we leave behind earlier roles, such as child, student, employee, parent, and eventually, perhaps, spouse and friend. In the process, whether as an adolescent or a recent retiree, it is natural enough to ask: Who am I? Psychologist Carl Jung described the psychological task of the second half of life as "individuation"—that is, becoming more and more our genuine individual self as opposed to carrying out the social roles required of people in midlife (Chinen, 1989).

Gerontologists have spoken about this late-life transition as a matter of **role loss** or role discontinuity. In earlier-life transitions, role losses are typically accompanied by new roles that take their place: Ceasing to be a child in one's family of birth, one grows up and takes on the role of parent. But in old age, some roles, such as those ended by widowhood or retirement, may never be replaced.

From one sociological standpoint, then, old age can be described as a *roleless role* (Blau, 1981). Once defined in this way, it is a natural step to see aging as a "social problem." A different perspective is possible, however. Other sociologists look on old age as a period when individuals maintain informal roles that are individually negotiated and perhaps continually redefined and constructed. In other words, the meaning of age, subjectively experienced, would not be decisively determined by the external roles, such as spouse, employee, and parent, that typically shape behavior earlier in the life course. On the contrary, once freed from conventional roles, the development of the self in later life may become a highly individual matter. From a philosophical point of view, old age can actually appear as an unexpected form of "late freedom" (Rosenmayr, 1984).

The importance of the meaning we bring to situations encountered in life has been underscored by several theories of aging. For example, Hans Thomae has developed his **cognitive theory of aging** based on empirical results of the Bonn Longitudinal Study of Aging (Rudinger & Thomae, 1990). The cognitive theory of aging argues that it is perception of change, rather than actual objective change, that has the most impact on behavior. The same life event—such as retirement—might be perceived by one person as loss and by another as freedom from an oppressive work situation. Cognitive, emotional, and motivational factors shape the way we perceive any event, and adjustment depends on a balance that changes over the life course. Studies of stress and coping in old age reveal individual differences in mastery depending on perception and adaptation.

The subjective experience of meaning is closely related to individual well-being, as Carol Ryff (1989) has argued. She has defined multiple psychological dimensions, including self-acceptance, which may come from reviewing one's life; positive relations with other people; autonomy and self-determination; mastery of the surrounding environment; beliefs that give purpose to life; and a sense of personal growth and development over the life course. Ryff's conceptualization gives a new approach to the definition of *activity* by shifting our attention to the inner dimension of the relationship between ourselves and the world.

But the old theory of disengagement has received some support as well—for example, through the idea of "gerotranscendence," Lars Tornstam (1997) has suggested that people find the deepest meaning in the last stage of life by overcoming self-centeredness and fear of death in favor of a spiritual focus.

Cognitive Functioning

The search for interpretive meaning in later life underscores the importance of cognitive functioning in old age. Contrary to the popular stereotype, we don't "lose a million neurons every day" as we grow older. Most people over age 65 *do not* suffer from memory defects or dementia. Among all those over age 65, there are a significant number—perhaps one in five—who have mild or moderate mental impairment. That means the overwhelming majority of older people have no mental impairment. Memory defects are actually quite limited among the large majority of older people. Nevertheless, some thinking processes do decline or change with age. Cognitive skills such as remembering, solving complex problems, paying attention, and processing language are affected by age- and disease-related changes in the brain.

Cognitive functioning is a critical issue because it is the aspect of psychological functioning most affected by aging. In addition, cognition has a greater effect than the other types of psychological functioning on the ability to perform the **activities of daily living (ADLs)**. Those whose memories fail may not be able to keep up with needed medications or remember to turn off the stove when they've finished cooking. People who lose their judgment or become more impulsive may make extremely foolish decisions about how to spend their money or whom to trust. What's worse, they may lose the ability to recognize their own mental shortcomings.

The picture for most older people, however, is considerably more positive. Although memory, reaction time, and basic information-processing and problem-solving abilities appear to decline with normal aging, other cognitive functions seem to remain stable or even improve. Wisdom and knowledge about the ways of the world, for example, are typical strengths of older people. In addition, training and practice in problem-solving skills, memory techniques, and other cognitive strategies can noticeably improve the abilities of older people.

It is important to remember the influence of life history and context as well. Older people who never learned a foreign language, for example, might have a more difficult time doing so in retirement than would an older person who had been bilingual from childhood. A person who worked as a carpenter in middle age might later be able to solve simple geometry problems faster than someone who worked as a nurse simply because the retired carpenter had been called on to solve geometry problems all his life.

Older people also remain adaptive and learn ways to cope with losses in cognitive functioning. For example, other people may help them to compensate for cognitive losses through a social process dubbed "interactive minds" or "collaborative cognition" (P. B. Baltes & Staudinger, 1996). Imagine that a young relative has asked two older adult sisters about their parents. One may start out with the story of how their parents met but be stymied by the issue of who introduced them. The other sister may say, "Remember, it was their friend from the old neighborhood. What was his name? He always used to bring us butterscotch candies." "Oh, I know who you mean," says the first sister. "The man who played the accordion." The other sister then remembers his name: "Yes, Mr. Catano. His family lived next door to Mother's family, and he was in the band that Dad played in." Similar teamwork helps older people function cognitively much better than they might on their own. That is one of the reasons that losing a spouse, another relative, a close friend, or some other central member of one's social network can be such a problem for the older adults.

Another form of cognitive adaptation is *selective optimization with compensation* (P. B. Baltes & M. M. Baltes, 1990), which is actually one definition of successful aging. The idea of selective optimization is that older people gradually narrow the scope of the capabilities they seek to maintain to those who are most useful, just as we all do throughout life. For instance, a freshman college student may begin studying engineering, but by the time she begins to pursue a career, she may decide to specialize in the microchips that make appliances work and forget about becoming knowledgeable about hydraulics, engines, or other aspects of engineering.

The idea of compensation is that people seek new ways of accomplishing things that become difficult or impossible because of losses in functional capacity. An example of selective optimization with compensation is the case of pianist Arthur Rubinstein. As Rubinstein grew older, he reduced the music he played to those piano pieces he knew best and then practiced just those pieces more often (selective optimization). When it came time to play the faster passages, he would slow down his playing speed just beforehand to maintain the apparent difference between the slower and faster passages (compensation).

Although much research remains to be done on cognitive capacity in old age, particularly among the oldest-old, it is safe to say that the old stereotype of feebleminded seniors is not only counterproductive but also inadequate as a description of cognitive functioning in later life. Older people do lose some thinking abilities, but the losses through normal aging are gradual and for the most part can be accommodated until late in life. What is more, older people often have other cognitive gains to offer. Their experience of living gives them an understanding of the world and an ability to apply its lessons that younger people typically have not had time to develop.

CONCLUSION

As long as there have been old people, there has been ambivalence about old age. The psychological basis for ambivalence is understandable. Why shouldn't people feel uncertain dread at the prospect of vulnerable old age stretching before them? In contrast, why shouldn't they look forward to a time when it seems possible to finally drop the burdens of coping with the complications of life? We see the same ambivalence today, but the truth is different from what popular images often convey. Old age in our day cannot easily be characterized. Especially in early old age, just past retirement, most people remain active and capable despite their removal from economically productive roles. Inevitably, the human body declines and dies, but normally, even in middle and late old age, humans retain more capabilities than they are often given credit for.

Industrialization brought growing rationalization and bureaucratization of the life course, a greater rigidity that took the shape of stronger demarcations between youth and adulthood and adulthood and old age. At the same time, rapid developments in medical science and cultural values have begun to erode the concept of distinctive life stages. With rising longevity, more people are living to old age, and older adults as a group are becoming a larger, more influential proportion of the total population. Today, because the entire life course is changing, the meaning of old age is ambiguous.

The problem of understanding what it means to be old in postindustrial society can be compared with a parallel problem in the biology of aging: Why does physical aging occur? There seems to be no reason for organisms to live much past the age of reproduction. Old age, in short, should not exist. Yet human beings do live long past the period of fertility; indeed, human beings are among the longest-living mammals on earth. But the challenge is not just to discover why we live so long—or even how to allow people to live longer. It is to understand how we can make the final phase of the life course more meaningful—for our elders and, eventually, for our future selves.

Toward a New Map of Life

We can think of the stages of life as a kind of "map" of unknown territory through which we must move. Until recently, some regions of that territory, such as midlife transition, were completely "unmapped" and unacknowledged. Other regions, such as adolescence, have been delineated or cultivated only in the last century, although now they seem familiar and predictable. The symbolism of life stages was once easily understood in societies where a map was thought to depict a common geographic or social "space" that was stable and enduring, the same for each generation. This familiar ideal of life stages reappears in popular forms of life span development psychology, such as the theories of Erikson and Levinson. The ideal seems to correspond to a fundamental and universal fact about human psychology: the need to define the predictability of life. Now, however, some are coming to call into question this whole approach to the life course. Perhaps the metaphor of a map is mistaken.

Today, at the beginning of the second decade of the 21st century, we no longer have confidence in a shared timetable for the course of life. The timing of major life events has become less and less predictable at all levels of society. As a result, we may need a new "map of life" corresponding to the changed conditions of demographic circumstances, economics, and culture in a postindustrial society (Laslett, 1991).

The meaning of aging has changed in contradictory ways. Optimists believe that medicine will soon permit us to displace aging-related disease and declines until later and later in life, a pattern known as **compression of morbidity**. Yet economic forces seem to move in the opposite direction from biology as some individuals accumulate financial assets during a lifetime of working.

To overcome limitations of the old map of life, we need to develop bolder ideas about the positive social contributions that can be made by older people; we also need to think more deeply about the meaning of life's final stage. Without such new understanding, there is a risk that older people may be dismissed as "uncreative" or that people will lose any shared sense of the positive meaning from survival into old age. Successful aging in the future will involve new ways of tapping the creative potential of later life (Adams-Price, 1998).

Creativity and wisdom depend on cognitive development over the life course. Whether our society cultivates such qualities among older people will depend, in the end, on creating more imaginative policies and institutions. The challenge of an aging society in the 21st century is to nurture the special strengths of age in an environment that prizes change, novelty, and flexibility. That challenge is what is ultimately at stake in debates about the meaning of the last stage of life (Bateson, 2010; Roszak, 1998).

Suggested Readings

Butler, Robert N., *The Longevity Revolution: The Benefits and Challenges of Living a Long Life,* PublicAffairs, 2008.

Cutler, Neal E., Whitelaw, Nancy A., and Beattie, Bonita L., *American Perceptions of Aging in the 21st Century: A Myths and Realities of Aging Chartbook,* Washington, DC: National Council on Aging, 2002.

Greenbaum, Stuart (ed.), *Longevity Rules: How to Age Well Into the Future,* Eskaton, 2010.

Does Old Age
Have Meaning?

A human being would certainly not grow to be seventy or eighty years old if this longevity had no meaning for the species. The afternoon of human life must also have a significance of its own and cannot be merely a pitiful appendage to life's morning.

—Carl Jung

The Meaning of Age

Most of the characteristic qualities of old age are uniquely human. For instance, among animal species in the wild, we never see offspring take care of the aging parents who gave birth to them. On the contrary, young animals typically abandon their parents when they themselves reach maturity, like baby birds that leave the nest to fly on their own. It is only the human being who cares for and honors the oldest members of the species, just as only human beings care for and remember their dead. In both cases, we might ask: Why?

The answer is that human beings live in a symbolic world of shared meaning, and the power of meaning can be a matter of life and death. For example, acts of bravery in crisis or wartime prove that people are willing to sacrifice their lives for what outlives the individual self—whether they act on behalf of family, religion, patriotism, or something else. Outliving the self—what Erik Erikson called "generativity"—is not limited to acts of sacrifice (Kotre, 1984). Awareness of a meaning that transcends individual life is a universal human quality. Transcendence and the search for meaning are what make us human.

Human beings contemplate aging and death, and they reach backward and forward in time to pose questions about the meaning of existence. In remembering the dead and in caring for the aged, we express our deepest convictions about the meaning of life. Old age is a time when we are likely to come face-to-face with questions about ultimate meaning. In fact, it was only in the 20th century that a sizable proportion of the population survived to experience old age, and it is therefore natural that, in our time, the meaning of old age has become an issue.

The question about whether old age has meaning is both a personal question and a challenge for social gerontology. The personal question is ultimately a matter of values:

What is it that makes my life worth living into the last stage? Put this way, it may seem like an abstract or philosophical question. But as we see in discussion about end-of-life decisions, this question becomes practical for families and health professionals.

Whether old age has meaning is central to what we understand to be life satisfaction or morale in old age (S. R. Kaufman, 1986). If aging threatens deeply held values—such as the desire to be independent, to have control, or to be socially esteemed—then both society and individuals will seek to avoid age or deny it as much as possible. The denial of aging and the denial of death are central problems for our society (Becker, 1973).

Thus, there are two questions we need to examine: Does old age have a meaning for society? How do individuals actually experience their lives as meaningful in the last stage of life? Both questions are related, and both pose a challenge to social gerontology. A key issue is whether we have a theory of aging that can explain the facts about old age, including the different meanings old age takes on over the course of life and through history. To focus on these questions about meaning and aging, we can begin with two domains—leisure and religion—that express contrasting values of activity and disengagement and thereby offer alternative perspectives on how people find meaning in later life.

LEISURE ACTIVITIES IN LATER LIFE

Leisure is not simply what we do with "leftover time," but a multidimensional quality of life.

Old age is characteristically a time when the work role becomes less constricting. Leisure may take its place as a way of finding meaning in life. We might think of leisure simply as "discretionary time," which becomes more available during the retirement years. But more deeply, leisure can be defined as activity engaged in for its own sake, as an end in itself. Leisure is not simply what we do with "leftover time," but a multidimensional quality of life different from paid employment, household maintenance, or other instrumental activities. Aristotle described leisure as a realm in which human beings gain freedom for self-development when the necessities of life have been taken care of.

Does leisure in retirement actually replace the work role in later life? Does it become a powerful source of meaning in its own right? The answer to these questions depends on the quality of subjective experience during leisure. Leisure may be an end in itself, but moments of leisure also have a developmental structure; they are not complete in themselves. For example, if we play sports, perform music, or read a book, each moment leads to the next in some purposeful developmental pattern. By contrast, other common leisure activities, such as TV viewing, take up a lot of time for older people, but tend to be passive or less demanding. If leisure activity is to be a path to deeper meaning, then it must have some dimension of growth or personal development (McGuire, Boyd, & Tedrick, 2004).

As people get older, they usually engage in the same activities as earlier in life, but with advancing age, there tends to be an overall decline in the level of participation. It is a mistake to think in stereotypes about "old people's" activities, such as shuffleboard, bingo, and singing old-time songs. That stereotype is wrong because age alone does not serve as a good predictor of what people do with their leisure in later life. Old people are not all alike. Variations and individual differences, along with the influence of gender and socioeconomic status, play a big part.

Changing Leisure Participation Patterns

How do patterns of leisure activity change over the life course? Broadly speaking, people over age 65 continue to engage in the same activities with the same people as they did in middle age. Although there is some selective age-related withdrawal, active engagement remains a key to life satisfaction and positive meaning in later life. But subgroups among older adults display markedly different patterns. For instance, the "young-old" can generally be categorized as the "active-old," a group of increasing interest to advertisers and marketers (Furlong, 2007).

Social structures, not age itself, determine the uses of time in later life. According to surveys of time use, as people age, they spend varying proportions of time in paid work, family care, personal care, and free time. Most of the variation comes from a decrease in time spent working, not from any demonstrable effects of aging. People who are still in the labor force after age 65 have time use patterns similar to those of younger people. Retirement frees up time—on average up to 25 hours a week for men and 18 hours a week for women. After taking into account household labor, most of this gain in time is taken up by interacting with media such as TV, radio, and newspapers (Robinson, 1997).

Some leisure activities decline with age, but others remain the same. A study of leisure found that the number of people starting new activities does diminish as we get older (Iso, Jackson, & Dunn, 1994). In addition, certain activities show a marked decline in participation rates: For example, moviegoing drops from 38% in midlife to 17% after age 65. Involvement in indoor fitness shows a decline, and travel diminishes significantly among people over age 75. Other activities, such as outdoor gardening, show only modest declines, and still others, such as TV viewing, watching sports, and engaging in informal discussion, show no age-related decline at all. Church participation and community activities tend to be maintained. Age-related declines appear to come partly from barriers to physical exertion or access. Activities based in the home, such as reading or socializing with familiar people, remain strong until well into advanced old age.

Patterns of late-life leisure have important implications for the economy in an aging society. Americans over the age of 50 offer a huge and growing market for business. They command more than half of all discretionary income and account for 40% of consumer demand. Older consumers are highly heterogeneous, varying by family status, ethnicity, education, geography, and social class. The "gray market" is stratified by age, however. The young-old are much more likely to be interested in travel than the old-old. Old-age leisure is often advertised as a consumption good or a status symbol, but leisure is also a means of affirming one's identity, a vital dimension of our phenomenological "life world" at a time when other roles may be lost (Hendricks & Cutler, 1990). Leisure time activities, then, are an important part of our personal world of meaning and also part of a shared horizon of socioeconomic transactions that shape the meaning of leisure over the entire life course.

Explaining Patterns of Leisure

A study of activity patterns in old age sheds interesting light on different theories of aging, such as activity theory, disengagement theory, and continuity theory. The Ontario Longitudinal Study of Aging found that most people engage in activities that are familiar and that they maintain stable activity patterns, as continuity theory would predict (Singleton, Forbes, & Agwani, 1993). The Ontario study also found that education and income are big factors, however. Retired people who have more choices because they have more resources are also likely to change their patterns of activity more often.

We also find some support for the idea of disengagement as people age, but not as a global generalization or stereotype. Disengagement, in other words, is not a universal pattern, but is highly selective, an example of selective optimization with compensation (P. B. Baltes & M. M. Baltes, 1990). As long as leisure activities remain accessible, people will go on doing what they find worthwhile and meaningful as long as they can. When physical impairments impose obstacles, most people adapt to optimize whatever resources they still have. Most people do not simply disengage altogether from meaningful activities.

Other explanations for the decline in leisure participation can also be found. For a segment of the older population with limited income, travel or cultural activities may be economically out of reach. But those with limited income may pursue activities outside the marketplace: for example, informal socializing with others. Another cause for constricted activity is declining health or age-related declines in vision, which might limit participation in fitness or sporting activities, as well as driving at night. Even among those who remain healthy, loss of companions for leisure activities can be a limiting factor. As a result, decline in leisure, as we might expect, is most severe among the oldest-old.

RELIGION AND SPIRITUALITY

Religion is "very important" in their lives, say three quarters of older Americans in public opinion surveys. Their behavior matches what they say: 70% of Americans reportedly belong to a congregation, and 40% attend services weekly (Ehmann, 1999). On the one hand, it is natural to expect that interest in religion might increase with advancing age given its association of old age with increased mortality. On the other hand, the continuity theory of aging reminds us that, as people age, they tend to maintain earlier patterns of practice and belief.

But religion is more complicated than responses to a poll or attending a worship service might indicate. To understand the role of religion, we need to distinguish formal religious behavior from subjective attitudes toward religion, what we might call an inner attitude of spirituality. Across many different dimensions, religion and spirituality continue to play vital roles in the lives of older adults and helps them find meaning in later life (Atchley, 2009).

Religious Involvement Over the Life Course

Religious involvement in old age displays a pattern that some investigators have called "multidimensional disengagement." What this means is that as people grow older, they may withdraw from some activities, such as attending church, but at the same time show an increase in personal religious practice, such as Bible study or listening to religious TV and radio. The number who report praying "once a day" or "several times a day" increases steadily from age 55 to the highest levels among those over 75. By contrast, other empirical studies show declining frequency of church attendance after age 75, perhaps reflecting frailty and physical limitations among the old-old. Older people seem to disengage from some organized religious roles, but make up for this loss by intensifying their non-organizational religious involvement—for example, personal prayer, meditation, and other forms of spiritual practice.

As they grow older, Americans continue to display patterns of religious identification similar to those among younger age groups: 65% identify themselves as Protestant, 25% as Catholic, and 3% as Jewish. Older women tend to have higher levels of religious participation and belief than do men. Overall, 50% of all older adults attend religious services at least once a week, and attendance tends to be positively related to measures of personal adjustment. When we look at church attendance from a life course perspective, we see the influence of family structure. Parents with young children often get involved in church activities, but after middle age, attendance falls off.

Despite these variations, older people are still more likely to be involved with church or synagogue than with other kinds of community organizations. Among mainstream Protestant and Catholic churches, as well as Jewish synagogues, a large proportion of the congregation is over age 50. Adults over age 65 are twice as likely to attend church regularly as those under 30. But it is a mistake to assume that people simply become more religious as they get older. Today's older generation appears to be more religious, but that effect may be due more to cohort or generational effects than to age. For instance, older adults may have gone to Sunday school or been involved in religious activities throughout life. Such lifelong religious identification explains higher religiosity in old age.

Churches and religious organizations play many roles in the lives of older people: in formal religious programs, through pastoral care programs, and as sponsors or providers of social services. Elders find fulfillment in a variety of church-sponsored volunteer activities, but, ironically, organized religion has often emphasized services and activities for youth. Innovative programs—such as Bible classes geared to older people, intergenerational programs, and new volunteer roles—could change that picture in the future. Some successful national initiatives, such as the Faith in Action program and the Shepherd's Centers of America, remind us that religious organizations represent a great, partly untapped resource for older people to find deeper meaning in life.

Religious Participation and Well-Being

Researchers have been interested in the benefits that religion and spirituality can have for older people. Cross-sectional studies have found a positive correlation between measures of well-being and religious beliefs among the old (MacKinlay & McFadden, 2004). Those with high levels of religious commitment also have higher levels of life satisfaction than those with no such sense of meaning. This relationship holds true even when controlling for age, marital status, education, and perceived health status.

But the significance of these correlations may be less than meets the eye. How do we define or measure what "religiousness" actually means in people's lives? Another difficulty is the partial confounding of religious involvement with measures of functional health status. Does religious engagement actually promote physical health? The answers to these questions are elusive.

Participation in religious activities and spiritual practices has been connected to well-being in later life.

Empirical studies have shown that religion can serve as a means of helping older people cope with stress. For example, the Duke Longitudinal Study found that older persons who used religion as a coping mechanism were more likely to exhibit higher levels of adjustment than others, even during intense life stress, such as bereavement and chronic illness. Nearly half of the respondents in the Duke study reported that religious attitudes or behavior helped them cope with stressful life events. Among those who relied on religion, coping strategies reflect different patterns of disengagement or activity. Private religious beliefs and behaviors, such as trust, faith in God, and prayer, were cited as coping strategies more frequently than church-related and religious social activities.

Investigators theorize that religion helps older adults cope in a variety of ways:

- By reducing the impact of stress in late-life illness
- By providing a sense of order and meaning in life
- By offering social networks tied to religious groups
- By strengthening inner psychological resources, such as self-esteem

URBAN LEGENDS OF AGING

"Religion is good for your health."

It's true that people who attend church tend to live longer, but no one knows why. Some studies suggest that volunteerism, the arts, lifelong learning, or even having a pet will give the same result. It could be that religion has little to do with it; maybe bowling would do the same (but not bowling alone). Here, as so often, correlation is not causation.

Spirituality and the Search for Meaning

Habits of religiosity, like other behaviors, tend to remain stable as people move into later life, but faith can take on new meanings with age. One research team found that, among those who had undergone some distinct change in religious faith, 40% reported experiencing such a change after the age of 50. The researchers concluded that changes in religious faith are not limited to youth, but can occur at any time in the life course (Koenig, 1994).

Often the personal search for meaning leads to deeper understanding of religious faith. James Fowler (1981) developed a framework of "faith stages" describing how people move from simpler, more literalist ideas of religion to levels where they see themselves and their lives in more universal terms, as the greatest saints and mystics have preached. As examples of those who have reached the highest stage of faith, Fowler cites personalities like Dag Hammarskjöld, Abraham Heschel, Thomas Merton, and Mahatma Gandhi.

Theologians who have reflected on the life course tend to view aging not as a problem that calls for a solution, but as an existential condition that can provide an opportunity for personal growth, or what some have called a "spiritual journey" (Bianchi, 1982) that can lead to a contemplative dimension for aging (Tornstam, 1997). In terms of Erik Erikson's developmental theory, older adults struggle with a psychological conflict between ego integrity and despair. Faith can be a way of enhancing ego integrity—an attitude of acceptance toward life and the world that is part of positive mental health. Stressing the importance of religion for mental health, Blazer (1991) has identified six dimensions of **spiritual well-being**: self-determined wisdom, self-transcendence, the discovery of meaning in aging, acceptance of the totality of life, revival of spirituality, and preparation for death. None of these tasks is easy, but the fact that some older people undertake this spiritual journey makes us believe that the effort can yield a profound sense of meaning for the later years.

GLOBAL PERSPECTIVE

The Search for Meaning in Asian Religions

The great civilizations of India, China, and Japan have all paid attention to the search for meaning in later life. Images of positive aging are embodied in these traditional religions of Asia.

Hindu Stages of Life

According to traditional Hinduism, spiritual freedom is the ultimate goal of life, to be attained by introspection and meditation. Aging as part of the total life course was understood to be crucial in ancient Hindu culture, which divided life into four major stages (*ashramas*). The first stage is discipleship, or learning from a guru. The second stage is the householder, based on marriage and family. The third stage is becoming a forest dweller, devoted to study of scripture. The fourth stage is complete renunciation, becoming a sannyasi, which may include teaching others as part of the path of transcendence.

(Continued)

(Continued)

Chinese Religion

According to Chinese Confucianism, filial piety is the primary virtue, including a duty to keep our body healthy to fulfill the demands of justice. We should feel gratitude toward elders and toward all of nature. The other traditional Chinese religion was Taoism, which emphasized health promotion even more. Taoists have even believed that immortality could be possible if only human beings followed the true laws of nature. Taoism, in this way, joins forces with traditional Chinese medicine and its emphasis on attaining a proper balance among the various elements and energies in the human body.

Japanese *Ikigai*

The Japanese word *ikigai* can be translated as "source of value in one's life" or "what makes life worth living." *Ikigai* can have a range of meaning, extended from devotion to one's children up to wider needs for fulfillment, such as personal growth, freedom, and self-actualization. In this respect, the concept of *ikigai* is similar to what contemporary psychologist Abraham Maslow called a "hierarchy of needs." The ideal of *ikigai* has not been relegated to traditional virtues. This ideal has now been adopted by Japan's Ministry of Health, Labour and Welfare in its national health promotion plan encouraging people to remain active beyond age 80.

SOURCES:

Shrinivas Tilak, *Religion and Aging in the Indian Tradition,* State University of New York Press, 1989.

Fumimaso Fukui, "On Perennial Youth and Longevity: A Taoist View on Health of the Elderly," *Journal of Religion and Aging,* 4:3–4 (1988), pp. 119–126.

Noriyuki Nakanishi, "'Ikigai'" in Older Japanese People," *Age and Ageing* (May, 1999).

Social Gerontology and the Meaning of Age

As a branch of the human sciences, gerontology tries to depict the facts about old age as a way of understanding the meaning of aging. But the approach of gerontology sometimes looks at "meaning" from the outside. Perhaps a better place to begin is to ask: What do older people themselves say about what gives meaning to their lives? When a sample of participants at a senior center was asked that question, nearly 90% of respondents described their lives as meaningful (Burbank, 1992). For most of them (57%), the meaning came from human relationships, followed by service to others (12%), religion, and leisure activities. Another study revealed that the most damaging threat to well-being in later life is loss of life purpose and boredom, not fear of absolute destitution or poor health. Responses show that people find purpose or meaning in a variety of ways: work, leisure, grandparenting, and intimate adult relationships. Respondents reported that, unless they were sick or depressed, they "didn't feel old" (Thompson, 1993), which suggests what has been called "the ageless self" (S. R. Kaufman, 1986).

Looking at verbal responses or patterns of behavior is suggestive, but may not get us any closer to understanding meaning in the last stage of life. Questionnaires about life satisfaction tell us only a limited amount about these deeper issues (Windle & Woods, 2004). Inevitably, values and philosophical assumptions reveal themselves in our discourse.

According to one widely shared view, the agenda for social gerontology should be to promote better social integration of the aged (Rosow, 1967) by means of group activities, social involvement, and participatory roles of all kinds. We see that view in the popularity of "productive aging," intergenerational programs, and other strategies. The ideal of an "age-integrated society" is a comprehensive enunciation of the same goal (M. W. Riley & J. W. Riley, 1994). Whether through work, leisure, or attendance at religious services, the aim of social integration is for people to stay engaged throughout life. Workers in senior centers and nursing homes often share this outlook. But if we view role losses of old age as an opportunity for self-development beyond conventional roles, then integration in group activities may no longer seem so compelling. Other values might assume greater importance.

We might still encourage older people to maintain social connections or affiliate with groups, but the form of that engagement would be based on a strategy for individual development, not conformity to social norms or activities. An example of such individual development might be a creative arts program designed to encourage self-expression; another example might be a religious retreat designed to permit individual prayer and meditation. These last kinds of pursuits seem in keeping with the potential for interiority and individuation in later life. Whether individual contemplation or social activity is the more desirable approach still remains debatable, of course, but that is precisely what is at issue in the controversy about whether old age has meaning or offers some special opportunity not readily available at other stages of life. The question is what makes it important for gerontology to look more deeply at what inspires a shared sense of meaning in life's last stage (T. Cole & Gadow, 1986).

The Meaning of Aging in the 21st Century

The life course perspective tends to view "stages of life" as social constructions reflecting broader structural conditions of society. As conditions change, so will our view of how people find meaning at different stages of life. Consider the weakening of age norms and beliefs about what is "appropriate" for different stages of life. In a world where retired people may go back to college or where a woman may have a first child at age 40, it makes less sense to link education or work with strict chronological ages. Indeed, one attractive strategy for an aging society might well be to introduce more flexibility for people of all ages to pursue education, work, and leisure over the entire course of life, rather than link these activities stereotypically to periods of youth, middle life, and old age, as modern societies have done in the past.

It is not clear how the meaning of old age will change in contemporary postindustrial societies. On the one hand, older Americans have achieved gains in income levels, health, and political power. On the other hand, as the stages of life have evolved and become blurred, the entire image of "old age" is giving way to more of an "age-irrelevant" image of the life course (Neugarten, 1983). As an empirical matter, chronological age, by itself, loses predictive value and importance for many purposes.

Does this trend mean that old age, as a distinct stage of life, no longer has any special meaning or significance? Here, we again must distinguish between a meaning that society ascribes to old age and what individuals find meaningful in their own lives. In postmodern culture, it is increasingly difficult to ascribe anything special to the last stage of life. But if nothing special is to be found in later life, we wonder, does it follow that personal meaning in old age must simply be "more of the same," that is, continuing whatever values gave meaning earlier in life (Moody & Carroll, 1997)? Or does lifelong growth imply a constant effort to overcome old habits and change our view of what offers meaning in life? These questions have no easy answers.

Activity or Reflection?

The previous discussion initially looked at two classical theories of aging—disengagement and activity. We saw how both theories implicitly appeal to deeply held values, but point in opposite directions. When we think about the question of whether old age has meaning, we come back, over and over again, to two fundamental alternatives: on the one hand, continuation of midlife values into old age, and, on the other hand, discovering some new or special challenge that belongs to the last stage of life.

The selection by Simone de Beauvoir offers the view of a philosopher who rejects traditional ideals of old age as a time of tranquility or disengagement. On the contrary, she believes that only continued activity on behalf of new goals will give our lives meaning, whether in old age or at any other time of life. Along these lines, Rowe and Kahn's strategy of "successful aging" represents a way of preserving meaning by adapting ourselves to diminished reserve capacity. Rowe and Kahn believe that "success" is best defined by optimizing capacity for continued engagement with the activities of life.

Erik Erikson, one of our most distinguished psychologists, shares this endorsement of engagement, but takes a different approach. Erikson sees each stage of life as a period with its own purpose or psychological task to be achieved. Old age is different from other stages because it offers a kind of culmination to life as a whole. Erikson believes that, through concern for the welfare of future generations, older people find a sense of meaning in later life. In the personal journal of Florida Scott-Maxwell, we find an echo of Carl Jung's belief that advanced age is a time for turning inward for deeper reflection. Her rich reflections prove that even when outer activity is cut off, it is still possible to find deep meaning in the last stage of life (Berman, 1986).

It seems ironic that modernization has made it possible for people to live a greater portion of their lives in old age than ever before in history. At the same time, the distinctive stance of postmodern culture tends to preclude finding any special meaning or purpose for the last stage of life. Whether modernization has reduced the power of the old seems debatable. Public spending for old-age benefits suggests that just the opposite may be true. But there is no doubt that modernization has helped to erode traditional ideas about fixed "stages of life" once based on shared meaning (Gruman, 1978). The result is a sense of openness or uncertainty about the meaning of old age. Such openness to new ideas and to contradictory answers is disconcerting to some and exhilarating to others. However, the future of an aging society will be shaped by all of us because, in the end, the old are simply "our future selves."

FOCUS ON PRACTICE

Reminiscence and Life Review

As people grow older, it is not unusual for them to reminisce about the "good old days." Feelings of both nostalgia and regret are commonly part of this attitude toward the past. A stereotypical response to reminiscence is to assume that older people are interested only in the past or, still worse, to see those who dwell on past memories as showing signs of escapism or even mental impairment. But late-life reminiscence may be a normal form of **life review**, which Robert Butler (1963) defined as a natural, even universal process stimulated by awareness of approaching death. He also wrote:

> The life review is characterized by a progressive return to consciousness of past experience, in particular the resurgence of unresolved conflicts which can now be surveyed and integrated. . . . If unresolved conflicts and fears are successfully reintegrated they can give new significance and meaning to an individual's life. (Butler, 1974, p. 534)

Butler's view is similar to that of Erik Erikson, who sees the psychological task of late life as achieving ego integrity, a reintegration of all aspects of the individual's life. Both Erikson and Butler based their psychological theories on the importance of finding meaning in the last stage of life. But do the facts support their theories? Just how important is reminiscence in old age?

Some studies have shown that older adults actually do not spend much more time daydreaming about the past than do people of other ages (Gambria, 1977), so it may be a mistake to see life review as a universal process. However, regardless of frequency, reminiscence may have adaptive value; that is, it may promote better mental health in old age. One early study of reminiscence found that people who spend time thinking about the past are less likely to suffer depression (McMahon & Rhudick, 1967). Some psychologists who have studied life review feel it may be a psychological defense mechanism that helps some people adjust to memories of an unhappy past. In that sense, reminiscence could be described as an adaptive feature of old age (Coleman, 1974), which is something to be encouraged (Brennan & Steinberg, 1983–1984).

Reminiscence and life review appear to help some older people bolster their self-image. By recalling the past, older adults can improve self-esteem and establish solidarity with others of their own generation. We might interpret older people's interaction with the young as a way to help them maximize perceived power or status, just as the exchange theory of aging predicts. When activity is the preferred style, older people are likely to downplay reminiscence in favor of talking about present or future events. But when disengagement is the preferred style, older people may emphasize past accomplishments.

Some gerontologists recommend that reminiscence and life review can have great value for older people who can no longer remain active (Haight, 1991). For that reason, reminiscence groups have been encouraged as a form of therapy among some nursing home residents and senior center participants. Guided autobiography is a method used as a basis for education in the later years (Birren & Deutschman, 1991). Spiritual autobiography groups have played a similar role in religious congregations.

(Continued)

(Continued)

All these methods can be useful for practitioners who work with older people, but techniques to encourage reminiscence as a form of practice must not divert us from a basic question: Is reminiscence or life review the best way of achieving a sense of meaning in old age? The response to that question cannot be purely scientific, but depends on basic values and philosophy of life. For example, if we follow philosopher Simone de Beauvoir's view, then activity and future orientation are the best approach to finding meaning in old age. She would therefore discourage people from spending time reminiscing about the past, unless past memories can somehow contribute to improving the world. Psychologist Carl Jung, in contrast, would see great value in inwardness or interiority in old age. The purpose or meaning of old age, in his view, is not necessarily to be active, but to know ourselves better and to accept ourselves as individuals. If life review can promote that goal, then Jung would encourage it, and Florida Scott-Maxwell, for example, follows along the lines suggested by Jung.

Does the reminiscence and life review by older people have meaning for people of other ages? Clearly, there is something special about old age precisely because it is the final stage of life. The last stage includes an awareness of finitude and a shortened time perspective (Kastenbaum, 1983). Furthermore, as the pace of social change increases, older people can no longer take for granted that their values will be shared by other cohorts; the 1960s and World War II generations may be quite different, not only from one another, but from Generation X, born during the baby bust after the mid-1960s. The old may be perceived by others or perceive themselves as belonging to "the past," regardless of their own subjective time orientation. Young people may assume that reminiscence is something appropriate only for the old.

In fact, the process of life review or autobiographical consciousness is not limited to old age, but occurs at transitions across the adult life course—for instance, in self-assessment after a job loss or another major life change. The life course perspective helps us appreciate links between subjective and objective time orientations and to see life review in broader terms. The search for meaning in life occurs not only at the end of life, but every time human beings become aware of their limited time on earth. It is perhaps for that reason that in the Bible the Psalms include a prayer for God to help us all to "number our days" and thus to cherish each passing moment, whatever our age may be.

READING 1

The Coming of Age

Simone de Beauvoir

Die early or grow old: there is no other alternative. And yet, as Goethe said, 'Age takes hold of us by surprise.' For himself each man is the sole, unique subject, and we are often astonished when the common fate becomes our own—when we are struck by sickness, a shattered relationship, or bereavement. I remember my own stupefaction when I was seriously ill for the first time in my life and I said to myself, 'This woman they are carrying on a stretcher is me.' Nevertheless, we accept fortuitous accidents readily enough, making them part of our history, because they affect us as unique beings: but old age is the general fate, and when it seizes upon our own personal life we are dumbfounded. 'Why, what has happened?' writes Aragon. 'It is life that has happened, and I am old.' . . . When we are grown up we hardly think about our age anymore: we feel that the notion does not apply to us; for it is one which assumes that we look back towards the past and draw a line under the total, whereas in fact we are reaching out towards the future, gliding on imperceptibly from day to day, from year to year. Old age is particularly difficult to assume because we have always regarded it as something alien, a foreign species: 'Can I have become a different being while I still remain myself?' . . .

Thus, the very quality of the future changes between middle age and the end of one's life. At sixty-five one is not merely twenty years older than one was at forty-five. One has exchanged an indefinite future—and one had a tendency to look upon it as infinite—for a finite future. In earlier days, we could see no boundary mark upon the horizon: now we do see one. 'When I used to dream in former times,' says Chateaubriand, harking back to his remote past, 'my youth lay before me; I could advance towards the unknown that I was looking for. Now I can no longer take a single step without coming up against the boundary-stone.' . . .

A limited future and a frozen past: such is the situation that the elderly have to face up to. In many instances, it paralyzes them. All their plans have either been carried out or abandoned, and their life has closed in about itself; nothing requires their presence; they no longer have anything whatsoever to do. . . .

Clearly, there is one preconceived notion that must be totally set aside—the idea that old age brings serenity. From classical times, the adult world has done its best to see mankind's condition in a hopeful light; it has attributed to ages that are not its own, virtues that they do not possess: innocence to childhood, serenity to old age. It has deliberately chosen to look upon the end of life as a time when all the conflicts that tear it apart are resolved. What is more, this is a convenient illusion: it allows one to suppose, in spite of all the ills and misfortunes that are known to overwhelm them, that the old are happy and that they can be left to their fate. . . .

Why should an old person be better than the adult or child he was? It is quite hard enough to remain a human being when everything—health, memory, possessions, standing, and authority—has been taken from you. The old

---

SOURCE: Excerpts from *The Coming of Age* by Simone de Beauvoir. Copyright © 1972 by Andre Deutsch. Reprinted by permission of the Putnam Publishing Group.

person's struggle to do so has pitiable or ludicrous sides to it, and his fads, his meanness, and his deceitful ways may irritate one or make one smile; but in reality it is a very moving struggle. It is the refusal to sink below the human level, a refusal to become the insect, the inert object to which the adult world wishes to reduce the aged. There is something heroic in desiring to preserve a minimum of dignity in the midst of such total deprivation. . . .

On the intellectual plane, old age may also bring liberation: it sets one free from false notions. The clarity of mind that comes with it is accompanied by an often bitter disillusionment. In childhood and youth, life is experienced as a continual rise; and in favourable cases—either because of professional advancement or because bringing up one's children is a source of happiness, or because one's standard of living rises, or because of a greater wealth of knowledge—the notion of upward progress may persist in middle age. Then all at once a man discovers that he is no longer going anywhere, that his path leads him only to the grave. He has climbed to a peak, and from a peak there can be a fall. 'Life is a long preparation for something that never happens,' said Yeats. There comes a moment when one knows that one is no longer getting ready for anything and one understands that the idea of advancing towards a goal was a delusion. Our personal history had assumed that it possessed an end, and now it finds, beyond any sort of doubt, that this finality has been taken from it. At the same time, its character of a 'useless passion' becomes evident. A discovery of this kind, says Schopenhauer, strips us of our will to live. 'Nothing left of those illusions that gave life its charm and that spurred on our activity. It is only at the age of sixty that one thoroughly understands the first verse of Ecclesiastes.' . . .

If *all* were vanity or deceit, there would indeed be nothing left but to wait for death. But admitting that life does not contain its own end does not mean that it is incapable of devoting itself to ends of some kind. There are pursuits that are useful to mankind, and between men there are relationships in which they reach one another in full truthfulness. Once illusions have been swept away, these relationships, in which neither alienation nor myth form any part, and these pursuits remain. We may go on hoping to communicate with others by writing even when childish images of fame have vanished. By a curious paradox, it is often at the very moment that the aged man, having become old, has doubts about the value of his entire work that he carries it to its highest point of perfection. This was so with Rembrandt, Michelangelo, Verdi, and Monet. It may be that these doubts themselves help to enrich it. And then again it is often a question of coincidence: Age brings technical mastery and freedom while at the same time it also brings a questioning, challenging state of mind. . . .

Freedom and clarity of mind are not of much use if no goal beckons us anymore: but they are of great value if one is still full of projects. The greatest good fortune, even greater than health, for the old person is to have his world still inhabited by projects: then, busy and useful, he escapes from both boredom and decay. The times in which he lives remain his own, and he is not compelled to adopt the defensive or aggressive forms of behavior that are so often characteristic of the final years. . . .

There is only one solution if old age is not to be an absurd parody of our former life, and that is to go on pursuing ends that give our existence a meaning—devotion to individuals, to groups, or to causes—social, political, intellectual, or creative work. In spite of the moralists' opinion to the contrary, in old age, we should wish still to have passions strong enough to prevent us from turning in upon ourselves. One's life has value so long as one attributes value to the life of others, by means of love, friendship, indignation, compassion. When this is so, then there are still valid reasons for activity or speech. People are often advised to 'prepare' for old age. But if this merely applies to setting aside money, choosing the place for retirement, and laying on hobbies, we shall not be much the better for it when the day comes. It is far better not to think about it too much, but to live a fairly committed, fairly justified life so that one may go on in the same path even when all illusions have vanished and one's zeal for life has died away.

READING 2

Vital Involvement in Old Age

Erik H. Erikson, Joan M. Erikson, and Helen Q. Kivnick

Elders have both less and more. Unlike the infant, the elder has a reservoir of strength in the wellsprings of history and storytelling. As collectors of time and preservers of memory, those healthy elders who have survived into a reasonably fit old age have time on their side— time that is to be dispensed wisely and creatively, usually in the form of stories, to those younger ones who will one day follow in their footsteps. Telling these stories, and telling them well, marks a certain capacity for one generation to entrust itself to the next, by passing on a certain shared and collective identity to the survivors of the next generation: the future. Trust, as we have stated earlier, is one of the constant human values or virtues, universally acknowledged as basic for all relationships. Hope is yet another basic foundation for all community living and for survival itself, from infancy to old age. The question of old age, and perhaps of life, is how—with the trust and competency accumulated in old age—one adapts to and makes peace with the inevitable physical disintegration of aging.

After years of collaboration, elders should be able to know and trust, and know when to mistrust, not only their own senses and physical capacities, but also their accumulated knowledge of the world around them. It is important to listen to the authoritative and objective voices of professionals with an open mind, but one's own judgment, after all those years of intimate relations with the body and with others, is decisive. The ultimate capacities of the aging person are not yet determined. The future may well bring surprises.

Elders, of course, know well their own strengths. They should keep all of these strengths in use and involved in whatever their environment offers or makes possible. And they should not underestimate the possibility of developing strengths that are still dormant. Taking part in needed and useful work is appropriate for both elders and their relationship to the community.

With aging, there are inevitably constant losses—losses of those very close, and friends near and far. Those who have been rich in intimacy also have the most to lose. Recollection is one form of adaptation, but the effort skillfully to form new relationships is adaptive and more rewarding. Old age is necessarily a time of relinquishing—of giving up old friends, old roles, earlier work that was once meaningful, and even possessions that belong to a previous stage of life and are now an impediment to the resiliency and freedom that seem to be requisite for adapting to the unknown challenges that determine the final stage of life.

Trust in interdependence. Give and accept help when it is needed. Old Oedipus well knew that the aged sometimes need three legs; pride can be an asset but not a cane.

When frailty takes over, dependence is appropriate, and one has no choice but to trust in the compassion of others and be consistently surprised at how faithful some caretakers can be.

Much living, however, can teach us only how little is known. Accept that essential "not-knowingness" of childhood and with it also that playful curiosity. Growing old can be an interesting adventure and is certainly full of surprises.

One is reminded here of the image Hindu philosophy uses to describe the final letting go—that of merely being. The mother cat picks up in her mouth the kitten, which completely collapses every tension and hangs limp and infinitely trusting in the maternal benevolence. The kitten responds instinctively. We human beings require at least a whole lifetime of practice to do this. The religious traditions of the world reflect these concerns and provide them with substance and form.

THE POTENTIAL ROLE OF ELDERS IN OUR SOCIETY

Our society confronts the challenge of drawing a large population of healthy elders into the social order in a way that productively uses their capacities. Our task will be to envision what influences such a large contingent of elders will have on our society as healthy old people seek and even demand more vital involvement. Some attributes of the accrued wisdom of old age are fairly generally acknowledged and respected. If recognized and given scope for expression, they could have an important impact on our social order. We suggest the following possibilities.

Older people are, by nature, conservationists. Long memories and wider perspectives lend urgency to the maintenance of our natural world. Old people, quite understandably, seem to feel more keenly the obstruction of open waterfronts, the cutting of age-old stands of trees, the paving of vast stretches of fertile countryside, and the pollution of once clear streams and lakes. Their longer memories recall the beauty of their surroundings in earlier years. We need those memories and those voices.

With aging, men and women in many ways become less differentiated in their masculine and feminine predilections. This in no way suggests a loss of sexual drive and interest between the sexes. Men, it seems, become more capable of accepting the interdependence that women have more easily practiced. Many elder women today, in their turn, become more vigorously active and involved in those affairs that have been the dominant province of men. Some women come to these new roles by virtue of their propensity to outlive the men who have been their partners. Many younger women have made a similar transition by becoming professional members of the workforce. These women seem capable of managing parenting and householding along with their jobs, particularly if they have partners who learn cooperation in these matters as an essential component of the marriage contract.

Our subjects demonstrate a tolerance and capacity for weighing more than one side of a question that is an attribute of the possible wisdom of aging. They should be well suited to serve as arbiters in a great variety of disputes. Much experience should be a precursor of long-range vision and clear judgment.

The aged have had a good deal of experience as societal witnesses to the effects of devastation and aggression. They have lived through wars and seen the disintegration of peace settlements. They know that violence breeds hatred and destroys the interconnectedness of life here on our earth and that now our capacity for destruction is such that violence is no longer a viable solution for human conflict.

Ideally, elders in any given modern society should be those who, having developed a marked degree of tolerance and appreciation for otherness, which includes "foreigners" and "foreign ways," might become advocates of a new international understanding that no longer tolerates the vicious name-calling, depreciation, and distrustfulness typical of international relations.

It is also possible to imagine a large, mature segment of the aging population, freed from the tension of keeping pace with competitors in the workplace, able to pursue vigorously art activities of all varieties. This would bring an extraordinary liveliness and artfulness to ordinary life. Only a limited portion of our adult population now has either the time or the money to be involved in activities of art expression or as appreciative supporters of the performing arts. Widespread participation in the arts is possible only if children are encouraged to develop those roots of imaginative play that arise from stimulating sensory experience. Elders learn this as they undertake to open these new doors of experience and could promote the inclusion of the arts in the educational system. The arts offer a common language, and the learning of that language in childhood could contribute to an interconnection among the world's societies.

The development of a new class of elders requires a continued upgrading of all facilities for the health care and education of people at all stages of life, from infancy to old age. Organisms that are to function for a hundred years need careful early nurturing and training. Education must prepare the individual not only for the tasks of early and middle age, but for those of old age as well. Training is mandatory for both productive work and the understanding and care of the senses and the body as a whole. Participation in activities that can enrich an entire lifetime must be promoted and made readily available. In fact, a more general acceptance of the developmental principle of the life cycle could alert people to plan their entire lives more realistically, especially to provide for the long years of aging.

Having started our "joint reflections" with some investigation of the traditional themes of "age" and "stages," a closing word should deal with the modern changes in our conception of the length and the role of old age in the total life experience. As we have described, modern statistics predict for our time and the immediate future a much longer life expectancy for the majority of old individuals rather than for a select few. This amounts to such a radical change in our concept of the human life cycle that we question whether we should not review all the earlier stages in the light of this development. Actually, we have already faced the question of whether a universal old age of significantly greater duration suggests the addition to our cycle of a ninth stage of development with its own quality of experience, including, perhaps, some sense or premonition of immortality. A decisive fact, however, has remained unchanged for all the earlier stages, namely, that they are all significantly evoked by biological and evolutionary development necessary for any organism and its psychosocial matrix. This also means that each stage, in turn, must surrender its dominance to the next stage, when its time has come. Thus, the developmental ages for the pre-adult life stages decisively remain the same, although the interrelation of all the stages depends somewhat on the emerging personality and the psycho-social identity of each individual in a given historical setting and time perspective.

Similarly, it must be emphasized that each stage, once given, is woven into the fates of all. Generativity, for example, dramatically precedes the last stage, that of old age, establishing the contrast between the dominant images of generativity and of death: one cares for what one has generated in this existence while simultaneously preexperiencing the end of it all in death.

It is essential to establish in the experience of the stages a psychosocial identity, but no matter how long one's life expectancy is, one must face oneself as one who shares an all-human existential identity, as creatively given form in the world religions. This final "arrangement" must convince us that we are meant as "grandparents," to share the responsibility of the generations for each other. When we finally retire from familial and generational involvement, we must, where and when possible, bond with other old-age groups in different parts of the world, learning to talk and to listen with a growing sense of all-human mutuality.

READING 3

———————————————— Successful Aging ————————————————

John Rowe and Robert Kahn

Satchel Paige, baseball's legendary, indestructible African-American pitcher, was as famous for his fast answers as for his fastball. He began pitching at the age of seventeen and was for many years restricted to what was then called the Negro Baseball League. Born near the turn of the century, he was already a veteran at the pitcher's mound when the racial barrier was relaxed. However, the decades rolled by, and he continued to pitch. As he did so, Paige became purposefully vague about his age, a subject of increasing speculation among sportswriters. When one of them put the question bluntly—"How old *are* you?"—Paige gave him a classic answer: "How old would *you* be if you didn't know how old you was?" The question—and Paige's answer—have as much to do with society's definitions and expectations of aging, and successful aging, as with Paige's own personal experience. By physical measures, at least, Paige was certainly aging successfully. But his wariness about coming clean with a hard number speaks volumes about our society's skepticism about competence in old age. What, after all, does it mean to "age successfully"? Does America think of aging per se as a bad thing, even when good things continue to develop—or emerge for the first time—with age? What, actually, is "success"? . . .

SUCCESSFUL AGING OR THE IMITATION OF YOUTH?

Modern society, perhaps especially American society, seems to regard aging as something to be denied or concealed. Women are freed, happily, from the corsets and similar instruments of torture that fashion once decreed. But a massive and inventive cosmetics industry does its best to persuade middle-aged and elderly women—and, increasingly, men—that they will lead happier lives if they change their hair color from gray to some improbable shade of blonde or red, camouflage their hair loss, and cover, erase, or abrade their wrinkles.

Photographs that advertise the products in question show people who are invariably young in appearance; photographer and makeup artist collaborate to send the incessant message of youth. And what cosmetics and computer-enhanced photography cannot do, plastic surgery offers to accomplish. The implication of all this information and misinformation is that the ultimate form of successful aging would be no aging at all. A psychologist might be tempted to say that underlying this denial of the aging process is a more deep-seated denial: refusal to acknowledge the fact of human mortality and the inevitability of death.

Our view of successful aging is not built on the search for immortality and the fountain of youth. George Bernard Shaw, when he was in his nineties, was asked whether he had any advice for younger people. He did. "Do not try to live forever," said Shaw, "you will not succeed." Or, as psychologist Carol Ryff put it in a thoughtful article, "Ponce de León missed the point."

In short, successful aging means just what it says—aging well, which is very different from not aging at all. The three main components of successful aging—avoiding disease and disability, maintaining mental and physical function, and continuing engagement with life—are important throughout life, but their realization in old age differs from that at earlier life stages. . . .

Old age has been called a "roleless role," a time when it is no longer clear what is expected of the elderly person or where he or she can find the resources that will make old age successful.

For earlier life stages, the expectations are clearer. Children are expected to attend school; in fact, they are legally required to do so. Able-bodied adults are expected to be employed or to be actively seeking paid employment. Parents of young children are expected to care for them.

None of these societal expectations generates perfect compliance, but all of them are felt and most of them are backed by law.

The years after child-rearing and employment present a sharp contrast to these expectational patterns and arrangements for their fulfillment. Almost nothing is expected of the elderly. The spoken advice from youth to age is "take it easy," which means do nothing or amuse yourself. The unspoken message is "find your own way and keep out of ours."

Many older men and women do better than that. . . . They find new friends, partially replace paid employment with useful voluntary activity, maintain some form of regular exercise, and enjoy a measure of increased leisure. But many others do much less and age less well.

READING 4

The Measure of My Days

Florida Scott-Maxwell

Age puzzles me. I thought it was a quiet time. My seventies were interesting and fairly serene, but my eighties are passionate. I grow more intense as I age. To my own surprise, I burst out with hot conviction. Only a few years ago, I enjoyed my tranquility; now I am so disturbed by the outer world and by human quality in general that I want to put things right, as though I still owed a debt to life. I must calm down. I am far too frail to indulge in moral fervor.

Old people are not protected from life by engagements, pleasures, or duties; we are open to our own sentience; we cannot get away from it, and it is too much. We should ward off the problematic and, above all, the insoluble. These are far, far too much, but it is just these that

attract us. Our one safety is to draw in and enjoy the simple and immediate. We should rest within our own confines. It may be dull and restricted, but it can be satisfying within our own walls. I feel most real when alone, even most alive when alone. . . .

Age is truly a time of heroic helplessness. One is confronted by one's own incorrigibility. I am always saying to myself, "Look at you, and after a lifetime of trying." I still have the vices that I have known and struggled with—well it seems like since birth. Many of them are modified, but not much. I can neither order nor command the hubbub of my mind. Or is it my nervous sensibility? This is not the effect of age; age only defines one's boundaries. Life has changed me greatly, it has improved me greatly, but it has also left me practically the same. I cannot spell, and I am overcritical, egocentric, and vulnerable. I cannot be simple. In my effort to be clear, I become complicated. I know my faults so well that I pay them small heed. They are stronger than I am. They are me. . . .

Another day to be filled, to be lived silently, watching the sky and the lights on the wall. No one will come probably. I have no duties except to myself. That is not true. I have a duty to all who care for me—not to be a problem, not to be a burden. I must carry my age lightly for all our sakes, and thank God I still can. Oh that I may to the end. Each day, then, must be filled with my first duty, I must be "all right." But is this assurance not the gift we all give to each other daily, hourly? . . .

Another secret we carry is that, although drab outside—wreckage to the eye mirrors a mortification—inside we flame with a wild life that is almost incommunicable. In silent, hot rebellion, we cry silently—"I have lived my life haven't I? What more is expected of me?" Have we got to pretend out of noblesse oblige that age is nothing, in order to encourage the others? This we do with a certain haughtiness, realizing now that we have reached the place beyond resignation, a place I had no idea existed until I had arrived here.

It is a place of fierce energy. Perhaps passion would be a better word than energy, for the sad fact is this vivid life cannot be used. If I try to transpose it into action, I am soon spent. It has to be accepted as passionate life, perhaps the life I never lived, never guessed I had it in me to live. It feels other and more than that. It feels like the far side of precept and aim. It is just life, the natural intensity of life, and when old we have it for our reward and undoing. It can—at moments—feel as though we had it for our glory. Some of it must go beyond good and bad, for at times—although this comes rarely, unexpectedly—it is a swelling clarity as though all was resolved. It has no content, it seems to expand us, it does not derive from the body, and then it is gone. It may be a degree of consciousness which lies outside

activity and which when young we are too busy to experience. . . .

It has taken me all the time I've had to become myself, yet now that I am old, there are times when I feel I am barely here, no room for me at all. I remember that in the last months of my pregnancies, the child seemed to claim almost all my body, my strength, my breath, and I held on wondering if my burden was my enemy, uncertain as to whether my life was at all mine. Is life a pregnancy? That would make death a birth.

Easter Day. I am in that rare frame of mind when everything seems simple—when I have no doubt that the aim and solution of life is the acceptance of God. It is impossible, imperative, and clear. To open to such unimaginable greatness affrights my smallness. I do not know what I seek, cannot know, but I am where the mystery is the certainty.

My long life has hardly given me time—I cannot say to understand—but to be able to imagine that God speaks to me, says simply—"I keep calling to you, and you do not come," and I answer quite naturally—"I couldn't, until I knew there was nowhere else to go." . . .

I am uncertain whether it is a sad thing or a solace to be past change. One can improve one's character to the very end, and no one is too young in these days to put the old right. The late clarities will be put down to our credit I feel sure.

It was something other than this that had caught my attention. In fact, it was the exact opposite. It was the comfortable number of things about which we need no longer bother. I know I am thinking two ways at once, justified and possible in a notebook. Goals and efforts of a lifetime can at last be abandoned. What a comfort. One's conscience? Toss the fussy thing aside. Rest, rest. So much over, so much hopeless, some delight remaining.

One's appearance, a lifetime of effort put into improving that, most of it ill judged. Only neatness is vital now, and one can finally live like a humble but watchful ghost. You need not plan holidays because you can't take them. You are past all action, all decision. In very truth, the old are almost free, and if it is another way of saying that our lives are empty, well—there are days when emptiness is spacious and non-existence elevating. When old, one has only one's soul as company. There are times when you can feel it crying, you do not ask why. Your eyes are dry, but heavy, hot tears drop on your heart. There is nothing to do but wait and listen to the emptiness which is sometimes gentle. You and the day are quiet, and you have no comment to make. . . .

I don't like to write this down, yet it is much in the minds of the old. We wonder how much older we have to become and what degree of decay we may have to endure. We keep whispering to ourselves, "Is this age yet? How far must I go?" For age can be dreaded more than death. "How many years of vacuity? To what degree of deterioration must I advance?" Some want death now as a release from old age; some say they will accept death willingly, but in a few years. I feel the solemnity of death and the possibility of some form of continuity. Death feels a friend because it will release us from the deterioration of which we cannot see the end. It is waiting for death that wears us down and the distaste for what we may become.

These thoughts are with us always, and in our hearts we know ignominy as well as dignity. We are people to whom something important is about to happen. But before then, these endless years before the end, we can summon enough merit to warrant a place for ourselves. We go into the future not knowing the answer to our question.

But we also find that as we age we are more alive than seems likely, convenient, or even bearable. Too often our problem is the fervor of life within us. My dear fellow octogenarians, how are we to carry so much life, and what are we to do with it?

Let no one say it is "unlived life" with any of the simpler psychological certitudes. No one lives all the life of which he was capable. The unlived life in each of us must be the future of humanity. When truly old, too frail to use the vigor that pulses in us, and weary, sometimes even scornful, of what can seem the pointless activity of mankind, we may sink down to some deeper level and find a new supply of life that amazes us.

All is uncharted and uncertain; we seem to lead the way into the unknown. It can feel as though all our lives we have been caught in absurdly small personalities, circumstances, and beliefs. Our accustomed shell cracks here, cracks there, and that tiresomely rigid person we supposed to be ourselves stretches, expands, and, with all inhibitions, is gone. We realize that age is neither failure nor disgrace, although mortifying we did not invent it. Age forces us to deal with idleness, emptiness, not being needed, not able to do, helplessness just ahead perhaps. All this is true, but one has had one's life, one could be full to the brim. Yet it is the end of our procession through time, and our steps are uncertain.

Here we come to a new place of which I knew nothing. We come to where age is boring, one's interest in it by-passed; further on, go further on, one finds that one has arrived at a larger place still, the place of release. There one says,

> Age can seem a debacle, a rout of all one most needs, but that is not the whole truth. What of the part of us, the nameless, boundless part who experienced the rout, the witness who saw so much go, who remains undaunted and knows with clear conviction that there is more to us than age? Part of that which is outside age has been created by age, so there is gain as well as loss. If we have suffered defeat we are somewhere, somehow beyond the battle. . . .

A long life makes me feel nearer truth, yet it won't go into words, so how can I convey it? I can't, and I want to. I want to tell people approaching and perhaps fearing age that it is a time of discovery. If they say, "Of what?," I can only answer, "We must each find out for ourselves, otherwise it won't be discovery." I want to say, "If at the end of your life you have only yourself, it is much. Look, you will find."

FOCUS ON THE FUTURE

Conscious Aging

In recent years, there has been a surge of public interest in spiritual topics, as shown by the popularity of best sellers like *The Road Less Traveled* and *The Care of the Soul*. This interest in things spiritual takes different forms, ranging from an interest in exotic New Age phenomena to a revival of traditional mystical teachings from Judaism and Christianity.

Some recent research suggests that mystical experience is becoming more common, with broad implications for an aging society. For example, Jeffrey Levin (1993) looked at age differences in reports of extrasensory perception, spiritualism, and numinous experience, which he defined as being "close to a powerful, spiritual force that seemed to lift you out of yourself." Using data from a representative cross-sectional population survey, Levin found that between 1973 and 1988, composite mysticism scores increased with successive age cohorts. Private and subjective religiosity is positively related to overall mystical experience, but organizational religiosity is inversely related, suggesting that those pursuing spiritual growth may find it in places other than church on Sunday. In light of Levin's findings, it is not surprising that large proportions of older Americans are already making use of so-called alternative therapies, including meditation, as part of their health practices (McMahan & Lutz, 2004).

Compared with European societies, the United States has historically been more religiously oriented, but spiritual revival today goes beyond mainstream religion. *Individual growth* is the new watchword. In keeping with that trend, one of the most fascinating developments today is the rise of "conscious aging," an idea based on an assumption that late life can be a period for positive spiritual growth. Zalman Schachter-Shalomi, a pioneer of the Jewish Renewal Movement, and Ram Dass, once a Harvard psychology professor and later a spiritual teacher, emerged as national leaders of the conscious-aging movement. Holistic health care, life review, and mystical religion are all important elements in conscious aging (Schachter-Shalomi & Miller, 1995).

One of the earliest initiatives on behalf of conscious aging was a self-help project known as Senior Actualization and Growth Explorations (SAGE), developed in California in 1974 to promote holistic health activities among older adults. In that project, weekly group activities drew on methods of humanistic and transpersonal psychology, such as meditation, guided discussions, yoga, massage, dream analysis, and exercises built around nutrition and holistic health care. SAGE was founded by Gay Luce, who was inspired to share her enthusiasm for meditation and biofeedback techniques with her 71-year-old mother (Luce, 1979). Exercises in visualization, meditation, and relaxation helped participants move toward self-discovery while confronting fears of dying.

A central practice of conscious aging is personal meditation (Goleman, 1988), whether it takes the form of yoga, Zen, and other Eastern disciplines or the form of contemplative prayer, which has a long history in the Christian church. Meditation as a spiritual discipline is a way of looking at ourselves as beings with depths beyond the conscious mind or ego. The same outlook permeates the work of Jungian psychiatrist Allan Chinen, who has opened up new vistas for the interpretation of fairy tales about the second half of life (Chinen, 1989). Conscious aging represents a coming together of religion and psychology so that each can enrich the other.

Conscious aging goes beyond conventional assumptions about adaptation or personality development over the life course. An early proponent of this view was Abraham Maslow, founder of humanistic psychology. Maslow believed that most people use only a small part of human potential, a potential demonstrated in what Maslow called "peak experiences." At these high points in our life, we have a chance to move toward self-actualization, that is, to become more fulfilled as a human being. Maslow himself believed that most people who are self-actualized are to be found among those who are mature in years—middle aged or older.

Mainstream psychology has, for the most part, not looked closely at the higher reaches of human potential, whether in young people or in old. One result of that limitation may be the "decline-and-fall" view of aging criticized by researchers who have looked at the emergence of wisdom in later life (P. B. Baltes, 1993). But some life span developmental psychologists go further. They argue that mature thought in adulthood entails a dimension of "transcendence" (Miller & Cook-Greuter, 1994), the province of transpersonal psychology (Walsh & Vaughan, 1993). Transpersonal psychology includes elements such as attention training, emotional transformation, refining awareness, and the achievement of wisdom through detachment and integration.

The conscious-aging perspective may have something to contribute to gerontology on matters such as health care, intergenerational relations, and adult education. For example, research over the past two decades has documented the tangible benefits of meditation for physical and mental health. What happens in meditation has long been familiar to medical and psychological researchers under the name of "autogenic training," or self-induced modification of lower brain centers. More than two decades ago, Dr. Herbert Benson of Harvard Medical School published his groundbreaking article on "the relaxation response," which explained altered states of consciousness in yoga and Zen in terms of the central nervous system. Since then, extensive research on biofeedback and alpha waves in the brain has confirmed the feasibility of studying consciousness.

There has also been some interesting experimental confirmation of strategies of conscious aging as a means of overcoming what psychologist Robert Kastenbaum (1984) calls **habituation**. In Kastenbaum's view, the essence of aging is a process of becoming gradually deadened or more mechanical in our response to life because of the power of habits. By contrast, meditation can be viewed as a progressive growth in powers of attention to overcome habituation in old stimulus–response patterns.

Conscious aging is a struggle to establish new cognitive structures, new ways of looking at the world. Researcher Arthur Deikman (1966/1990) has described how the process of deautomization can come from practicing meditative disciplines such as yoga or Zen. Deikman, for instance, conducted a procedure of "experimental meditation," after which subjects reported sensory experience that was more vivid and luminous. Deikman's work and other experiments like it suggest that deliberate concentration and meditation can modify the selectivity of sensory input to the brain.

These findings could have implications for an aging society. For example, a controlled study in a geriatric population found that meditation-relaxation techniques can have a major impact in reducing anxiety and depression, an impact superior to conventional cognitive-behavioral techniques (DeBerry, Davis, & Reinhard, 1989). Another study, funded by the National Institute of Mental Health, looked at the impact of Transcendental Meditation to see whether it can have benefits beyond simple relaxation. That study confirmed the point that cultivation of "mindfulness," a state of consciousness free of content but alert, does have measurable consequences for learning, cognitive flexibility, and overall mental health. These positive results remained with the individuals years later (Alexander, Chandler, Langer, Newman, & Davies, 1989).

Conscious aging is trying to apply these lessons from research and practice to a growing older population. Interest in health promotion, productive aging, and lifelong learning is likely to make conscious aging a subject of continuing importance as America becomes an aging society in the 21st century. It may prove an intriguing glimpse of things to come.

Questions for Writing, Reflection, and Debate

1. Some critics have argued that disengagement theory may have accurately characterized the behavior of the older population in the 1950s, but that it was a mistake to infer that this pattern was universal. According to these critics, activity theory or continuity theory might well be a better description of how older people actually live today. If the critics' views are correct, does it mean that any theories of aging simply express the way aging appears at a certain time in history? If so, how would it be possible to develop an account that is more general and not limited to a certain time and place?

2. America as a society tends to place a high value on success and achievement. Does that fact suggest that the goal of "successful aging" is an appropriate approach to thinking about growing old in America? Are there aspects of growing older that could present a problem for the goal of successful aging?

3. Psychologist Carl Jung believed that the psychological goal of later life is to become more and more oneself as an individual. What does this goal mean in practice? What drawbacks to this idea can you think of? If we adopt Jung's approach, how would we evaluate older people who remain very much as they have always been, in contrast to other people who dramatically change their lives, say, after the point of retirement or widowhood?

4. Imagine that you are now 80 years old and have discovered that you may not have long to live. Your grandchildren have asked you to write a statement about what you've learned about the meaning of life, especially in the last few years. In your statement, contrast what you believe now (as a future 80-year-old) with what you believed in the past (at what is your present age).

5. Assume that you are the activities director of a church-affiliated nursing home that prides itself on promoting the quality of life of residents. Write a memorandum for the nursing home director outlining a range of activities that would help enhance the residents' sense of the meaning of life in the long-term facility.

6. Is the idea of "meaning" in life something purely personal and private, or does it have some wider social importance? Does discussing the question of meaning give us an understanding of older people's behavior, or is it simply confusing? In addressing this question, consider other issues discussed in this book, such as assisted suicide, work and leisure, and the allocation of health care resources for life prolongation. How would the idea of a meaning for old age affect one's view of these questions?

7. Visit the website devoted to "religion, aging, and old age" at http://www.trinity.edu/~mkearl/ger-relg .html. Based on what you see at this site, what recommendations would you make to a reporter whose editor has given an assignment to write a general article on "religion and old age"?

8. Consider carefully Lars Tornstam's concept of "gerotranscendence" as this idea is expressed and developed on his website: http://www.soc.uu.se/ research/gerontology/gerotrans.html. Now try the following exercise. Using only the simplest and most everyday language, try to give an explanation of "gerotranscendence" to a friend or relative who knows nothing about gerontology and is not particularly sympathetic to religion.

Suggested Readings

Atchley, Robert, *Spirituality and Aging*, Baltimore: Johns Hopkins University Press, 2009.

Bateson, M. C., *Composing a Further Life: The Age of Active Wisdom,* New York: Knopf, 2010.

Frankl, Viktor, *Man's Search for Meaning: An Introduction to Logo Therapy* (trans. Ilse Lasch), New York: Pocket Books, 1973.

Haight, Barbara K. and Haight, Barrett S., *The Handbook of Structured Life Review*, Health Professions Press, 2007.

Kimble, M. A., McFadden, S. H., Ellor, J. W., and Seeber, J. J. (eds.), *Aging, Spirituality and Religion: A Handbook*, Minneapolis: Fortress, 2003.

Student Study Site

Visit the Student Study Site at **www.sagepub.com/moody7e** for these additional learning tools:

- Flashcards
- Web quizzes
- Chapter outlines

- SAGE journal articles
- Web resources
- Video and audio resources

Controversy 2

WHY DO OUR
BODIES GROW OLD?

Oliver Wendell Holmes (1858/1891), in his poem "The Wonderful One-Hoss Shay," invokes a memorable image of longevity and mortality, the example of a wooden horse cart or shay that was designed to be long-lasting:

> Have you heard of the wonderful one-hoss shay,
>
> That was built in such a logical way,
>
> It ran a hundred years to a day . . . ?

This *wonderful* "one-hoss shay," we learn, was carefully built so that every part of it "aged" at the same rate and didn't wear out until the whole thing fell apart all at once. Exactly a century after the carriage was produced, the village parson was driving this marvelous machine down the street, when

> What do you think the parson found,
>
> When he got up and stared around?
>
> The poor old chaise in a heap or mound,
>
> As if it had been to the mill and ground!
>
> You see, of course, if you're not a dunce,
>
> How it went to pieces all at once,
>
> All at once, and nothing first,
>
> Just as bubbles do when they burst.

The wonderful one-horse shay is the perfect image of an optimistic hope about aging: a long, healthy existence followed by an abrupt end of life, with no decline. The one-horse shay image also suggests that life has a built-in "warranty expiration" date. But where does this limit on longevity come from? Is it possible to extend life beyond what we know? The living organism with the longest individual life span is the bristlecone pine tree found in California, more than 4,500 years old, with no end in sight.

The maximum human life span appears to be around 120 years. In fact, we have no valid records of anyone living much beyond that length. There have been claims of people living to the advanced age of 150 or even longer. Some claims have persuaded the *National Enquirer,* and others even convinced a scientist at Harvard Medical School. But whatever the *Enquirer* or Harvard wanted to believe, there has never been proof of such longevity. Quite the contrary. Despite the fact that we have millions upon millions of verified birth records in the 20th century, until recently, there were no proven cases at all of any human being living beyond age 120. Then, in 1995, a Frenchwoman named Jeanne Louise Calment reached the proven age of 122 (Robine, 1998). Madame Calment actually remembered seeing Vincent van Gogh as a child!

Some scientists argue that even the idea of maximum life span is based only on empirical observation. With biological breakthroughs in the future, might we someday surpass that limit? Indeed, optimists ask, why settle for the one-horse shay?

On the face of it, prolonging the human life span sounds good. But is it feasible? Will it make our lives better? One cartoon in *The New Yorker* shows a middle-aged man at a bar complaining to his companion: "See, the problem with doing things to prolong your life is that all the extra years come at the end, when you're old" (Mankoff, 1994). Another cartoon depicts two nursing home residents in wheelchairs confiding to each other: "Just think. If we hadn't given up smoking, we'd have missed all this."

These cartoons point to the fact that often the consequences of biophysical aging appear well before reaching maximum life span. Some believe that the proper aim of medicine should therefore be to intervene, perhaps even to slow down the rate of aging, so that more and more of us can remain healthy up to the very end of life. At that point, the body would simply "fall apart" all at once, like the wonderful one-horse shay (Avorn, 1986). This view, mentioned earlier, is known as the *compression of morbidity*, an idea developed and promoted by James Fries (1988).

The compression-of-morbidity hypothesis looks forward to greater numbers of people who postpone the age of onset of chronic infirmity (Brooks, 1996). In other words, we would aim for a healthy old age, followed by rapid decline and death. Sickness or morbidity would be compressed into the last few years or months of life. But things don't always work out that way. We may succeed in postponing deaths from heart disease, cancer, or stroke. But what happens if we live long enough to get other diseases? The same preventive measures can have, as an unintended result, increased rates for chronic conditions such as dementia, diabetes, hip fracture, and arthritis (Roush, 1996). Many observers worry that as increasing numbers of people live to advanced ages, the challenge of compressing morbidity will become more and more difficult. One study of mortality did find some evidence that survival curves became more rectangular—that is, deaths became concentrated around a point later in life (Nusselder & Mackenbach, 1996), but other studies have found the opposite (G. Kaplan, 1991).

It is important to distinguish here between life expectancy and maximum life span. Life expectancy, or expected years of life from birth, has risen, but life span, which is defined as the maximum possible length of life, has evidently not changed at all. To the best of our knowledge, no human being has ever lived beyond 120 years or so. The causes of maximum life span and of aging itself still remain unknown. Biological evidence suggests that maximum life span is genetically determined, and therefore fixed, for each species.

With this concept of limit in mind, compression of morbidity is attractive because delaying dysfunction would enhance the quality of life, extend life expectancy, and reduce health care costs (Butler, 1995). A compression-of-morbidity strategy would move life expectancy closer to the hypothetical upper bound of maximum life span. Instead of expecting to live only to age 85, people who reached 65 could expect to become centenarians, yet in good health nearly to the end of life.

Such a gain in active life expectancy would have dramatic consequences for our society (Seltzer, 1995). To judge whether this strategy for compression of morbidity is actually feasible, we need to examine in more detail what is involved in normal aging. We need to understand what is known about the biology of aging and what may be discovered in the future.

THE PROCESS OF BIOLOGICAL AGING

Normal aging can be defined as an underlying time-dependent biological process that, although not itself a disease, involves functional loss and susceptibility to disease and death. One way to measure susceptibility to death is to look at death rates. For contemporary humans, these rates double every 8 years. This pattern is known as **Gompertz law** (Kowald, 2002). In other words, a 38-year-old is about twice as likely to die as a 30-year-old, a 46-year-old is four times more likely to die than a 30-year-old, and so on. At any given age, there is an important gender difference: Men are about twice as likely to die as women. Although men and women age at the same rate, women at every age are less biologically fragile than men—just the contrary to what our cultural stereotypes might suggest.

Studies of different species of organisms show that aging is almost universal, but the causes of aging are complex. For instance, among animals whose body mass and metabolism are comparable, the rate of aging varies greatly. Consider the differences in maximum life span among some familiar animal species shown in Exhibit 1.

The rate of aging can be correlated, in a general way, with the amount of time it takes for the mortality rate of a species to double. The doubling time is around 8 years for humans today, but only 10 days for a fruit fly and 3 months for a mouse. In rough terms, we can say that a mouse ages at around 25 times the rate of a human being.

What accounts for these clear differences in aging and life span across species? Comparative anatomy—the study of the structure of different species—generates some insights into this question. For example, among mammals and other vertebrates, an increase in relative brain size is positively related to an increased life span. Other factors correlated with life span are lifetime metabolic activity, body size, body temperature, and the rate of energy use. For example, a tiny hummingbird has a rapid heartbeat and a high rate of energy metabolism; it also lives a comparatively short time, as if it were more quickly using up its total lifetime energy or action potential (Sacher, 1978).

Biologists have discovered intriguing relationships among life span, body size, relative brain size, and metabolic intensity. For example, a chipmunk has a maximum life span of 8 years, but an elephant can achieve 78 years. These facts suggest a more general idea known as the **rate-of-living concept**: roughly, the concept that metabolism and life expectancy are closely correlated. Smaller organisms, which tend to have a more rapid metabolism for each unit of body mass, also tend to have shorter life spans. A short-lived mouse and a long-lived elephant both have approximately the same temperature, but the mouse produces more heat per unit of mass. At the other extreme, slow-moving turtles are likely to have life spans longer than the more active mammals. Another fascinating fact is that no matter their total body mass, mammals have approximately the same number of heartbeats in a lifetime. Still,

Exhibit 1   Some Organisms' Maximum Life Span

Organism	Maximum Life Span (in Years)
Tortoise	177
Human being	120
Horse	62
African elephant	50
Golden eagle	50
Chimpanzee	37
Dog	34
American buffalo	26
Domestic cat	21
Kangaroo	16
Domestic rabbit	12
House mouse	3
Fruit fly	25 days

SOURCE: Data from Walford (1983) and Encyclopædia Britannica online.

despite these tantalizing correlations, the rate-of-living theory has largely been rejected by biologists, along with the notion that biological aging is somehow necessary for the good of the species (Austad, 1997).

In comparison with other species of mammals, the human being has the longest life span and also expends more energy per body weight over the total life span than any other mammal. Energy metabolism per body weight across the life span in humans is about four times greater than that for most other species of mammals. Human beings have an average life expectancy and a maximum life span about twice as great as those of any other primate.

Compare the chimpanzee and the human being. The maximum human life span appears to be around 110 to 120 years; the chimpanzee's is close to 50 years. But when we look at DNA from both species, we find that their DNA is more than 98% identical. These figures suggest that the rate of aging may be determined by a relatively limited part of the genetic mechanism. Calculations suggest that if a cell is determined by around 100,000 genes, then perhaps no more than a few hundred alterations in the genetic code are needed to change the rate of aging.

Scientists have judged that a large increase in maximum human life span occurred fairly recently—probably within the last 100,000 years. The speed of this development suggests that only a tiny portion of the human genome, representing less than 1% of the genetic code, was likely to be involved. If so few genetic mechanisms determine aging, then we can perhaps hope to intervene to delay the process of aging (Finch, 1990).

BIOLOGICAL THEORIES OF AGING

The facts about aging and maximum life span have led many biologists to believe that biophysical aging, or senescence, may have a single fundamental cause. In their efforts to find such a single primary process to explain those time-dependent changes that we recognize as biophysical aging, they have developed many different ideas. Biologist Zhores Medvedev (1972) enumerated more than 300 biological theories of aging. At present, no single theory of aging explains all the complex processes that occur in cells and body systems, but ongoing research is under way that is leading to new insights into why we grow old.

Broadly speaking, we can distinguish between two kinds of theories of aging (Finch & Kirkwood, 2000):

- *Chance.* Some theories see aging as the result of external events, such as accumulated random negative factors that damage cells or body systems over time. For example, these factors might be mutation or damage to the organism from wear and tear.
- *Fate.* Some theories see aging as the result of an internal necessity, such as a built-in genetic program that proceeds inevitably to senescence and death.

In either case, the question remains open: Is it possible to intervene to correct damage to the aging body or modify the genetic program? The most likely interventions are those that would make sense depending on which theory best explains the facts about aging (Ludwig, 1991).

Wear-and-Tear Theory

The **wear-and-tear theory of aging** sees aging as the result of chance. The human body, like all multicellular organisms, is constantly wearing out and being repaired. Each day, thousands of cells die and are replaced, and damaged cell parts are repaired. Like components of an aging car, parts of the body wear out from repeated use, so the wear-and-tear theory seems plausible.

The wear-and-tear theory is a good explanation for some aspects of aging—for example, the fact that joints in our hips, fingers, and knees tend to become damaged over the course of time. A case in point is the disease of osteoarthritis, in which cartilage in joints disintegrates. Another is cataracts, in which degeneration causes vision loss. Our hearts beat several billion times over a lifetime, so with advancing age, the elasticity of blood vessels gradually weakens, causing normal blood pressure to rise and athletic performance to decline.

The wear-and-tear theory of aging goes back to Aristotle, but in its current form was expounded by one of the founding fathers of modern biogerontology, August Weismann (1834–1914). He distinguished between the two types of cells in the body: germ plasm cells, such as the sperm and egg, which are capable of reproducing and are in some sense "immortal," and somatic cells comprising the rest of the body, which die. Weismann (1889), in his famous address "On the Duration of Life," argued that aging takes place because somatic cells cannot renew themselves, and so living things succumb to the wear and tear of existence.

What we see as aging, then, is the cumulative, statistical result of "wear and tear." Consider the case of glassware in a restaurant, which follows a curve similar to that for human populations. Over time, fewer and fewer glasses are left unbroken, until finally all are gone. The "life expectancy" or survival curve of the glassware follows a linear path over time, but the result for each individual glass comes about because of chance. Nothing decrees in advance that a specific glass will break at a fixed time. Glasses are just inherently breakable, so normal wear and tear in a restaurant will have its inevitable result. Like everyone born in a certain year (e.g., 1880), the "glasses" disappear one by one until none are left.

Some modern biological theories of aging are more sophisticated versions of this original wear-and-tear theory. For example, the **somatic mutation theory of aging** notes that cells can be damaged by radiation and, as a result, mutate or experience genetic changes (Szilard, 1959). The somatic mutation hypothesis would seem to predict higher cancer rates with age, yet survivors of the atomic bomb at Hiroshima showed higher rates of cancer, but no acceleration of the aging process.

Even without actual mutation, over time, cells might lose their ability to function as a consequence of dynamic changes in DNA. According to the so-called error accumulation theory of aging, or error catastrophe theory, decremental changes of senescence are essentially the result of chance or random changes that degrade the genetic code (Medvedev, 1972). The process is similar to what would happen if we were to use a photocopy to make another copy. Over time, small errors accumulate. The errors eventually make the copies unreadable. Similarly, the error catastrophe theory suggests that damaged proteins eventually bring on what we know as aging through dysfunction in enzyme production.

The **accumulative waste theory of aging** points to the buildup in the cells of waste products and other harmful substances. The accumulation of waste products eventually interferes with cell metabolism and leads to death. Although waste products do accumulate, there is little evidence of harm to the organism. The key to longevity may be the extent to which cells retain the capacity to repair damage done to DNA. In fact, DNA repair capacity is correlated with the metabolic rate and life span of different species. Some studies suggest that DNA damage in excess of repair capacity may be linked to age-related diseases such as cancer.

Autoimmune Theory

The immune system is the body's defense against foreign invaders such as bacteria. The immune system protects and preserves the body's integrity, and it does this by developing antibodies to attack hostile invaders. We know that the immune system begins to decline after adolescence, and the weakening of immune function is linked to age-related vulnerability. According to the **autoimmune theory of aging**, the system may eventually become defective and no longer distinguish the body's own tissues from foreign tissues. The body may then begin to attack itself, as suggested by the rising incidence of autoimmune diseases, such as rheumatoid arthritis, with advancing age (Kay & Makinodan, 1981).

Aging-Clock Theory

According to the **aging-clock theory of aging**, aging is programmed into our bodies like a clock ticking away from the moment of conception. One of the best examples of an aging clock in humans is the menstrual cycle, which begins in adolescence and ends with menopause. The aging-clock theory is part of programmed aging, in which aging is seen as a normal part of a sequence leading from conception through development to senescence and finally to death.

One version of the aging-clock theory emphasizes the roles of the nervous and endocrine systems. This version postulates that aging is timed by a gland, perhaps the hypothalamus, the thymus, or the pituitary gland. Such a gland acts like an orchestra conductor or a pacemaker to regulate the sequence of physiological changes that occur over time. Some support for this idea comes from observations that the hormone dehydroepiandrosterone (DHEA) is found in higher levels among younger people. Experimenters have also discovered that DHEA supplements help laboratory rats live longer.

The aging-clock theory has encouraged research on the role of hormones secreted by the thyroid, pituitary, and thymus glands (Lamberts, van den Beld, & van der Lely, 1997). These include human growth hormone, which can now be manufactured in quantity through genetic engineering. In experiments, volunteers injected with growth hormone lost flabby tissue and grew back muscle, essentially reversing some manifestations of the aging process for a time. Other investigators are interested in hormones produced by the pineal gland, which may help regulate the "biological clock" that keeps time for the body.

Hormones and the endocrine system clearly play a major role in the process of aging. Hormones control growth, development, and reproduction in plants and animals. Biologists recognize a phenomenon here called *semelparity.* The best example is the Pacific salmon, which swims upstream to lay its eggs and then dies. So-called annual plants also exhibit semelparity: The tomato plant flourishes, produces fruit, and then dies away as the autumn leaves begin to fall.

But we find no comparable biological process in humans. We do recognize the profound age-related hormonal change of menopause, which comes with the loss of cells in the ovary that produce estrogen. Female mammals are born with a finite number of egg cells, so menopause is an example of a "preprogrammed" life event linked to age. Menopause is tied to health problems of aging because the loss of estrogen often weakens bone-mineral metabolism, resulting in thinner bone structure—a condition known as *osteoporosis.* Thin bones can lead to fractures, which in turn may compromise an older person's ability to live independently.

Cross-Linkage Theory

Connective tissue in the body, such as the skin or the lens of the eye, loses elasticity with advancing age. We recognize the result as wrinkling of skin and cataracts. The explanation for this change lies in a substance known as **collagen**, a natural protein found in skin, bones, and tendons. According to the **cross-linkage theory of aging**, the changes we see result from the accumulation of cross-linking compounds in the collagen, which gradually become stiff. As in the waste accumulation theory, the piling up of harmful molecules is thought to eventually impair cell function. Some of this cross-linking may be caused by free radicals, which are cited in several different theories of aging.

Free Radicals

Free radicals are unstable organic molecules that appear as a by-product of oxygen metabolism in cells (Armstrong, Sohal, Cutler, & Slater, 1984). Free radicals are highly reactive and toxic when they come in contact with other cell structures, thus generating biologically abnormal molecules. The result may be mutations, damage to cell membranes, or damage by cross-linkage in collagen.

Free-radical damage has been related to many syndromes linked with aging, such as Alzheimer's disease, Parkinson's disease, cancer, stroke, heart disease, and arthritis. According to the **free-radical theory of aging**, damage created by free radicals eventually gives rise to the symptoms we recognize as aging.

An important point about this theory is the fact that the body itself produces so-called **antioxidant** substances as a protection against free radicals. These antioxidants "scavenge" or destroy free radicals and thus prevent some of the damage to cell structures. The production of antioxidants is, in fact, correlated with the life span of many mammals.

Free-radical theory has prompted some to believe that consuming antioxidant substances, such as vitamin E, might retard the process of aging. Genetic engineering techniques can now be used to produce antioxidants in vast quantities, but antioxidants are also supplied by the normal food we eat. Vitamins A, C, and E, as well as less familiar enzymes, play a role as antioxidants. Animal studies to date, however, show that consumption of antioxidants produces only minimal effects on aging.

Cellular Theory

A major finding from cell biology is that normal body cells have a finite potential to replicate and maintain their functional capacity. This potential appears to be intrinsic and preprogrammed, part of the genetic code. The **cellular theory of aging** argues that aging ultimately results from this progressive weakening of capacity for cell division, perhaps through exhaustion of the genetic material. That cellular limit, in turn, may be related to the maximum life span of species (Stanley, Pye, & MacGregor, 1975).

One of the major milestones in the contemporary biology of aging was the discovery that cells in laboratory culture have a fixed life span. In 1961, Leonard Hayflick (1965) and associates found that normal human cells in tissue culture go through a finite number of cell divisions and then stop. This maximum number of divisions is known as the **Hayflick limit**. Hayflick found that cells replicate themselves around 100 times if they are taken from fetal tissue. But if taken from a 70-year-old, they reach their limit of "aging" after 20 or 30 divisions.

Cells taken from older organisms divide proportionately fewer times than those taken from younger ones. Normal human cells that are frozen at a specific point in their process of replication and later thawed seem to "remember" the level of replication at which they were frozen. Furthermore, normal cells from a donor animal that are transplanted will not survive indefinitely in the new host.

Cell division in the laboratory sheds light on an interesting question: Can human bodies become immortal? The answer is yes, but there's a catch. We have to get cancer to do it. The classic instance is the case of so-called "HeLa" cells—an immortal remnant of a terminally ill young woman named Henrietta Lacks (HeLa), who died in Baltimore in 1951 (Skloot, 2010). Before she died, a few cancerous cells were removed from her body and put into tissue culture: essentially, put down on a glass lab dish and supplied with cell nutrients. Scientists were surprised to find that these HeLa cells just kept dividing and growing. In the years since, the cells haven't stopped growing. So we might say that a little piece of Henrietta Lacks has achieved immortality in a laboratory dish.

By contrast, in normal cell differentiation, cells divide and become more specialized, and their ability to live indefinitely simultaneously declines. The Hayflick limit may not so much be an intrinsic limit on living cells as a limit on when cells begin to differentiate, as in development of the embryo. When cells approach a limiting point, a genetic program normally shuts down the capacity for further division. If the genetic program doesn't work, the result is uncontrolled multiplication, or cancer. The Hayflick limit doesn't keep all cells from dividing—after all, germ cells such as eggs and sperm continue to divide—but it may give a clue about why aging brings an increase in cancer and a weakening of the immune system.

Studying aging by examining cells in a test tube raises some questions. For instance, the nutrient medium in a cell culture does not contain all the nutrients and hormones that a cell would normally receive. In addition, cells in the body become differentiated tissues and organs and remain in equilibrium in ways quite different from the way cells replicate in a test tube.

Fundamentally, the cellular theory of aging sees aging as somehow "programmed" directly into the organism at the genetic level. In this view, it is just as "natural" for the body to grow old as it is for the embryo or the young organism to develop to maturity, as we see in annual plants or the Pacific salmon. Does the program theory of aging therefore apply to higher organisms such as mammals and, specifically, human beings? Perhaps, but it does not apply as obviously as it does in organisms in which rapid aging is tied to reproduction.

One of the most intriguing points in favor of the cellular approach to aging is the discovery that tiny tips at the ends of chromosomes—structures known as *telomeres*—become shorter each time a cell divides. Telomeres, it seems, comprise a biological clock marking the unique age of a cell as it divides. Studies are under way to explore the link between aging at the cellular level and what we recognize as aging in complete organisms (Bodnar et al., 1998).

Is Aging Inevitable?

The biological aging process may not be the result of a rigid genetic program; it may simply be the complex and indirect result of multiple traits in the organism tied to normal development. In other words, the body may not be preprogrammed to acquire gray hair, wrinkles, or diminished metabolic functions. Rather, these supposed signs of aging may simply be telltale side effects of activities of the organism.

Consider the analogy of an aging car. Suppose a distinctive "species" of automobile were designed to burn fuel at a fixed temperature with an efficient rate of combustion. That specific rate of combustion is required for appropriate acceleration, cruising speed, fuel mileage, and so on. But, alas, when the car performs this way, it also inevitably produces certain emission by-products. Over time, these by-products clog the cylinders, reduce efficiency, and lead to the breakdown and final collapse of the machine.

In the case of the human "car," burning oxygen in normal metabolism generates harmful by-products—namely, free radicals that prove toxic to the organism. The trade-off is that oxygen is essential for life yet harmful to our long-term well-being. Although the human "car" is not intentionally designed to accumulate toxic emissions in order to collapse, the car cannot function at optimum levels without creating destructive by-products.

Now suppose we could find some special fuel additive that eliminates toxic emissions. Would we then have an "immortal" car? Probably not. Changing the fuel in your car won't prevent accidents, nor will any fuel additive prevent rusting or the wearing down of springs and shock absorbers.

The "human car" analogy has its limits because an organism, unlike a manufactured object, has a capacity for repair and self-regeneration, at least up to a certain point; unlike an automobile, human beings have consciousness and can make choices about how to live out their life span. Nevertheless, to find out how we might modify or retard biological aging, we must find out why capacity for self-repair seems unable to keep up with the damage rate—in short, why aging and death appear to be universal.

URBAN LEGENDS OF AGING

"Anti-aging medicine today is making rapid progress."

Actually, no progress is being made at all; no intervention has ever been shown to slow the biological process of aging, other than caloric restriction (eating drastically less). Herbal supplements sold in health food stores are totally unregulated; many are dangerous. None, including antioxidants, has ever been proven effective in slowing aging.

WAYS TO PROLONG THE LIFE SPAN

Biological aging is inevitable, but diet and exercise may promote healthier aging.

Most theories of aging depict biological aging as an inevitable process, like a disease to which we must all eventually fall victim. Some theories look on the organism as succumbing to chance events, whereas others see it as driven by a built-in biological clock. Yet whether aging is thought to occur by chance or by fate, most theories seem to reach a pessimistic conclusion about the inevitability of aging.

But aging is not a disease; rather, it is a process of change, part of which may make us vulnerable to disease. Instead of being driven by a single primary process timed through a single biological clock, aging is driven by many different clocks, each on a different schedule and unfolding in parallel developmental patterns.

Biological theories of aging could have enormous importance for an aging society. For example, the compression-of-morbidity idea assumes that there is a definite human life span, roughly 85 years, with a broad range from 70 to 100 years. There are thousands who live beyond age 100, but the maximum number of years any human being has lived is 122 years. An age around that level is often assumed to be the maximum life span possible. But today, basic research in the biology of aging is challenging assumptions about a fixed maximum life span and the inevitability of aging as a biological process. Two approaches have been found that could extend the maximum life span for a species: one based on environmental intervention through diet, the other on a genetic approach.

URBAN LEGENDS OF AGING

"Aging is not a disease."

Most gerontologists agree that aging is not a disease. Yet a growing number of biologists reject this proposition, and serious work on slowing the process of aging is now under way in the laboratory. In other words, they treat aging as if it's a curable pathological condition. Of course, no one has ever defined exactly what a "disease" is, so it's hard to prove the point one way or another (Moody & Hayflick, 2003).

Environmental Approach

For more than 60 years, scientists have known of only one environmental intervention—restricting food intake—that extends life span in mammals. Dietary restriction produces gains in longevity in laboratory animals, raising maximum life span by up to 40% and reducing diseases of aging.

As long ago as the 1930s, scientists discovered that the life span of rats can be extended by restricting food intake after weaning. Caloric restriction in mice has similar effects even when it is begun in midlife. Rodents live longer if they eat a diet with 40% fewer calories than normal, as long as their diet remains otherwise nutritionally sound. When caloric intake is restricted, age-related deterioration slows down, and age-related diseases, such as kidney problems and autoimmune syndromes, are diminished (Bronson & Lipman, 1991). The rats' condition does not deteriorate until late in a long life. Under such a diet, both average life expectancy and maximum life span increase by 30%. Apparently, the rate of acceleration of aging has been reduced.

What accounts for this dramatic, well-established impact of dietary restriction in enhancing longevity? The longevity gain is achieved not through reduction in any specific component of the diet, but simply because of fewer total calories consumed. One possible explanation is that caloric reduction slows metabolism, or the rate at which food is transformed into energy (Demetrius, 2004). With caloric reduction, the basic biological clock slows down. But we cannot be sure of the explanation because caloric restriction is consistent with many different mechanisms of biological aging, including DNA, free radicals, and a stronger immune system. The results are clear enough for rodents, and experiments with primates have begun to confirm that caloric restriction is effective there as well (Couzin, 1998).

In human terms, caloric reduction would mean surviving on a diet of 1,400 calories a day, but, in return, it would mean, in theory, gaining 30 extra years of life. To achieve this goal, Roy Walford (1986), one of the premier investigators of the biology of aging, has proposed a so-called high-low diet that incorporates high nutritional value with low calories.

A similar approach is suggested by *cryobiology,* or the study of organisms at low temperatures. Lowering internal body temperature can increase life span in fruit flies as well as vertebrates, such as the fence lizard, an animal that lives twice as long in New England as its cousins do in sunny Florida. Experiments with fish demonstrate that with lower temperature, life span is prolonged in the second half of life. Lower temperature can significantly reduce DNA damage. We don't yet know whether cryobiological processes apply to warm-blooded animals like humans. However, calorie restriction also seems to lower body temperature a small amount. Calorie-restricted mice have a lower average body temperature, and the temperature changes according to biorhythm.

Caloric restriction somehow protects genes from damage by the environment and perhaps serves to strengthen the immune system. Caloric restriction also reduces the incidence of cancer. The experimental findings on caloric reduction converge with what is known about indirect regulation of genetic expression that controls the aging process. Research is under way to explore the implications of caloric restriction for increasing human longevity.

Genetic Approach

Many lines of evidence point toward the central role of genetics in fixing the longevity for each species, although for any individual, length of life will be the result of both genetic and environmental factors. We often think of genetic inheritance as the element that is fixed and unalterable, but some genetic studies have shown a dramatic ability to improve maximum life span over generations.

For example, studies have been conducted on bread mold, fruit flies, mice, and nematode worms. In all these species, genetic manipulation has been shown to modify maximum life span. For example, some mutated forms of nematode worms have exhibited substantial increases. Among mice, large differences in average life expectancy and maximum life span exist among different strains because of hereditary differences. In the fruit fly, scientists have achieved an increase in average as well as maximum life span by using artificial selection as a breeding technique.

Some recent genetic experiments have produced astonishing gains in longevity. For example, Michael Rose, a population geneticist, used artificial selection to produce fruit flies with a life span of 50 days—double the normal average of 25 days; the equivalent would be a human being living to 240 years of age. Rose, in effect, has in the laboratory mimicked an increase in the evolutionary rate of change. As a result, successive generations of fruit flies passed along genes favoring prolonged youth and longevity (Rose, 1994).

Thomas Johnson, a behavioral geneticist, went further and altered a single gene (known as Clock-1) out of the roundworm's 10,000 genes. He also achieved a doubling of the worm's 3-week life span (T. Johnson, 1990). Still other recent studies suggest that in some fruit fly populations, the risk of mortality may actually decrease with advancing age, a finding that challenges previous assumptions about maximum life span (Barinaga, 1992). These dramatic successes, through breeding or direct genetic manipulation, point to the way that genetic change may have come about rapidly through natural selection.

Whether any of these findings can be applied to humans is, again, unknown, but we can draw some conclusions about the genetics of aging. For instance, in at least several of the animal studies cited here, the genes involved governed antioxidant enzymes and mechanisms for repair of damage to DNA, which have been at the center of several theories about the biology of aging. Second, in the species benefiting from genetic change, a small number of genes have been involved in determining longevity. Thus, these results could possibly be applied to higher animal species.

New horizons for genetic application are already visible. Scientists have found a way to double the life of skin cells by switching off the gene that regulates production of a specific protein responsible for manifestations of aging. A similar method of genetic engineering has been used with tomatoes, permitting them to be stored and shipped without decay. The key here is the so-called mortality genes, which determine the number of times that cells divide. Thus, this intervention addresses the Hayflick limit, which remains central to aging at the cellular level. Even without affecting maximum life span, this sort of gene therapy could have major applications in the future, perhaps leading to a cure for age-related diseases such as Parkinson's, Alzheimer's, and cancer (Anderson, 1992; Freeman, Whartenby, & Abraham, 1992).

GLOBAL PERSPECTIVE

Blue Zones for Longer Life

A Japanese grandmother carries her young grandson.

When we think of Italy, we often think of pizza or the ancient city of Rome. But if you're thinking about longevity, think instead about the Italian island of Sardinia. Demographers have there identified its mountain slopes as a distinctive *Blue Zone,* a region of high longevity. In fact, the proportion of centenarians in Sardinia is more than twice as high as in the rest of Italy. That prompts a question: Why do people there live so long? Lifestyle and genetics may both play a part, but in what proportion? The isolated, mountain-dwelling Sardinians tend to be descendants of settlers dating back to the Bronze Age. Sardinians have also been known for eating a *Mediterranean diet* and for maintaining a traditional, family-oriented way of life. So, gerontologists wonder, what makes Sardinia such a standout as a Blue Zone for extreme longevity?

Some answers can be found on the opposite side of the globe, in Okinawa, an island portion of the longest-living country, Japan. Okinawans have an average life expectancy of more than 82 years and also enjoy an old age largely free of disabilities. Rates of heart disease, cancer, and dementia are lower than among Americans. Again, we wonder, what's the reason? Some observers point to the Japanese word *ikigai*, which means "purpose for living." A traditional Okinawan diet of vegetables, tofu, and a small amount of fish is also part of the picture. Finally, Okinawans have strong social ties among family, friends, and neighbors. Gerontologists have confirmed that the power of such social networks on longevity is comparable to giving up smoking a pack of cigarettes each day.

Blue Zones around the world are *natural laboratories* for the study of longevity. Lessons learned from these regions can help give guidance for a healthier and happier old age closer to home.

SOURCE: Buettner, Dan, "The Secrets of Long Life" (cover story), *National Geographic* (November 2005), 208(5): 2–27.

The recent Human Genome Project has produced a comprehensive map of the entire sequence of genes on the human chromosome. Genetic engineering could draw on that knowledge in ways that might dramatically change what we have thought of as the process of aging and even our assumptions about the maximum human life span (Watson, 1992). Such speculations, however, belong to the future.

Compression or Prolongation of Morbidity?

Biology has not yet succeeded in unraveling the mystery of aging, so it is not surprising that medical science has produced no technology or method for raising the maximum life span of human beings. Caloric reduction and genetic methods have worked with lower organisms, but human beings are more complex organisms, and the research studying humans has yet to provide conclusive evidence on this point. To extend life expectancy and promote healthy aging, we may need to identify genes responsible for harmful mutations, whether expressed early or late in life. A parallel approach would be to identify those environmental agents (e.g., diet, sunshine, and smoking) that have a cumulative impact on sickness and survival. Health promotion might then succeed in postponing chronic illness, thereby making more possible the idea of the one-horse shay.

Progress in these directions depends on answering the question of why we age. In the reading titled "Why Do We Live as Long as We Do?" Leonard Hayflick highlights some basic facts about the biology of aging that are relevant to our hopes for compressing or extending our longevity.

In the other readings that follow, we hear different voices in the compression-of-morbidity debate. On one side, James Fries and Lawrence Crapo take the optimistic position that improving life expectancy will also lead to compressed morbidity: People will live longer and not be sick until the very end of their natural life span. Fries and Crapo believe that successful aging involves optimizing life expectancy while reducing physical, psychological, and social morbidity. Their "sunny" view of aging is paradoxical, in a way, because it presumes that the maximum life span remains fixed, a limitation other biologists might reject. In support of their view, we can note that some postponement of morbidity has already occurred: Declining death rates from heart disease and stroke reflect improvements in health due to lifestyle, diet, hypertension detection, and so on.

But not everyone is persuaded by Fries and Crapo's interpretation of the evidence on morbidity and death rates. Researchers and demographers disagree about whether compression of morbidity is actually occurring and whether maximum human life span is really finite, as Fries believes. Mor notes that, although it seems that morbidity rates and functional decline have decreased in the industrialized world, because of population aging, there will be more older people than ever before suffering from chronic and disabling health conditions. Still other conditions, such as depression and sensory losses, are not linked to causes of improved life expectancy at all, so we remain haunted by the fear that longer life might mean only prolongation of morbidity (Olshansky, Carnes, & Cassel, 1990; Verbrugge, Lepkowski, & Imanaka, 1989).

Finally, as we look further in the 21st century, we might consider possibilities beyond the range of current science and medicine. Parker and Thorslund suggest that, in tracking changes in compression of morbidity and health among older adults, we need to be thinking about the implications of these trends for providing services and resources. In more visionary terms, biologist Aubrey de Grey believes, contrary to most gerontologists, that aging is a disease, a condition to be "cured." S. Jay Olshansky, by contrast, reminds us that a "cure for aging" is a fantasy that has deluded seekers for biological immortality down through the ages. Olshansky does favor research on the biology of aging, but he distrusts any claim that raising the maximum life span is right around the corner.

The debate over why we grow old shows that scientific "facts" are rarely as simple as we imagine. The meaning of the facts depends on our theories and interpretations, and our own hopes about the aging experience, and it is therefore subject to debate and construction in different ways. Different views of the facts about illness and survival in old age today are leading us to new ways of thinking about mortality and morbidity among older adults. Indeed, the debate about compression of morbidity is rooted in biology, but it has implications for health care economics in an aging society, as well as how individuals experience later life. What can we expect in the future if medical technology succeeds in prolonging life still more? How much emphasis should we give to health promotion as opposed to curing diseases in old age? Whatever our view, the compression-of-morbidity theory stands out as an important reminder of how critical biological research will be for the future of an aging society.

FOCUS ON PRACTICE

Health Promotion

Can we take steps now to control our own longevity? The consumer market for "anti-aging" products is growing. Magazines on the subject can be found on every newsstand. But most claims for life-extending products are not proved by science. For example, melatonin, antioxidants, human growth hormone, and DHEA have all been hailed as anti-aging breakthroughs, but proof has not lived up to the promise. There are no diets, hormone injections, or vitamin or mineral supplements that have so far been proven to slow down the process of aging (Butler et al., 2002). However, it is possible that a breakthrough in our knowledge of the biology of aging could give us ways to slow down aging in the 21st century.

When we think about the prospect of slowing the process of aging or dramatically extending maximum life span, many questions present themselves. Would people really want to triple their life spans? Would they want to hold the same job or be married to the same person for 150 years? What would society be like if people lived for centuries instead of decades (Post & Binstock, 2004)?

These questions are still in the realm of science fiction, but many interventions already have been shown to promote health and longevity in ways that can benefit people today (D. Haber, 1999). For example, the death rate from cardiovascular disease has been cut in half in the last two decades chiefly because of a reduction in high-risk behaviors such as smoking. Changes in diet or exercise patterns could provide further gains in adult life expectancy.

The secrets of keeping aging's effects at bay are actually well known (J. Brody, 2001). Most of the causes of lost years of life today are related to lifestyle choices: alcohol, tobacco, exercise, and diet (Arking, 1991). Dr. Herbert de Vries, a highly regarded exercise physiologist from the University of Southern California, has estimated that regular exercise could give a huge boost to the life expectancy of most people. Millions of Americans have already started eating a low-fat,

(Continued)

(Continued)

high-fiber diet, just as they have given up smoking. Others go even further and seek to minimize free radical damage to cells by including more antioxidant carotenes in their diets (Walford, 1986).

The topic of health promotion and aging engenders a familiar argument between "optimists" and "pessimists." On the one hand, Hayflick argues that calorie-restricted, long-living mice are merely living out their fixed natural life spans. In the end, our genetic program prevails, and environmental interventions, such as diet, can accomplish only a limited amount. If Hayflick is right, then Walford (1986), like Juan Ponce de León, has embarked on a vain search for the fountain of youth.

But the optimists hold a different view. According to one scenario for the future, as a result of prudent nutrition and more exercise, during the 21st century, the average life span could well rise from 76 to beyond 80 years. Then, early in the next century, through hormone replacement and genetic engineering, the maximum life span could push well beyond the current limit of 110 or 120 years. Optimists believe that lifestyle enhancement and new technologies could combine to delay or even reverse aging, thus extending youthfulness and pushing the limits of the life span itself (Hall, 2003).

Steps to improve longevity are already becoming part of the popular culture. Changes in diet and exercise, reductions in smoking, and health-promotion activities of many kinds are now far more common than they were two decades ago. As baby boomers experience middle age, these activities are likely to spread and have an impact on longevity.

In thinking about these scenarios for the future, we should retain a measure of skepticism. We should also focus on practical steps that are proven and feasible right now. Health promotion has to be based on science, not on conjecture, fear of frailty and mortality, or hopes for the future.

Health promotion seems clearly to be a desirable trend, but it also raises some difficult questions about personal and social responsibility (Centers for Disease Control and Prevention, 2003). What should we do about groups in our society who cannot or will not change their unhealthy behaviors? Are harmful behaviors ultimately a matter of free choice, or do environmental and social factors also shape behavior? The cost of Medicare depends a great deal on the cost of chronic illnesses. If we embrace an ethic of personal responsibility for health care, might we be less willing to support public funding for medical care? Should health promotion take account of inequality in income, education, and access to health care? How do we motivate people in favor of health promotion when the results of "bad choices"—such as smoking, poor diet, lack of exercise, or use of alcohol—don't show up until decades later? These questions will remain both personal and societal issues for years to come.

URBAN LEGENDS OF AGING

"Drinking red wine will make you live longer."

A lot of people believe this one based on a TV story on *60 Minutes*. There is a substance, resveratrol, found in red wine and grapes that has been shown by some laboratory studies to promote longevity in mice (Bauer, 2006). But you would have to drink amounts of wine far beyond what is humanly possible in order to have any of the hypothetical effects. Studies of resveratrol on human longevity continue, but alcoholics need to find a different excuse for drinking more wine.

READING 5

——————— ## Why Do We Live as Long as We Do? ———————

Leonard Hayflick

The premise upon which the following ideas rest is that the survival of a species depends upon a sufficient number of its members reaching sexual maturation and producing enough progeny that reach independence to guarantee the continuation of the species. Natural selection, guided by beneficial mutations, has molded the biology and the survival strategies of all living things to achieve this fundamental goal. As previously indicated, the best strategy to guarantee that an animal or human will survive long enough to mature sexually is to provide it with more than the minimum required capacity in its vital organs. In this way, if damage or pathology occurs in an essential system before sexual maturation, there is a greater likelihood that the animal will still survive to reproduce and pass on to its progeny its superior physiological capacity. This general strategy, essential for the survival of all species, has evolved in different ways for various life forms. Energy and purpose are concentrated to achieve reproductive success which assures the immortality of the genes. The continuation of the germ line is the driving force of natural selection. Longevity of individual animals is of secondary importance.

Animals are selected through evolution for having physiological reserves greater than the minimum necessary to reach sexual maturation and rear progeny to independence, but once this critical goal has been attained, they have sufficient excess reserve capacity to "coast" for a period of time, the remainder of which we call their life span. This time period, then, is *indirectly* determined genetically. During the coasting period the animal functions on its excess capacity. This physiological reserve of energy and functional capacity does not renew at the same rate that it incurs losses, so molecular disorder—entropy—increases. Random changes or errors appear in previously well-ordered molecules, resulting in the normal physiological losses that we call age changes. These changes increase the vulnerability of the animal or human to predation, accidents, or disease (Holiday, 2004).

What happens after reproductive success and raising progeny to independence is not important for the survival of a species. What happens next, of course, is aging and, ultimately, death. Wild animals, because they rarely live long enough, do not experience aging. The entire scenario is analogous to the ticking on of a cheap watch after the guarantee period has ended. The watch's guarantee period corresponds to the time spent by animals to reach sexual maturation and to finish rearing progeny. After the warranty period ends, the watch does not simply "die" because it would be prohibitively expensive to put a mechanism in a cheap watch that would cause it to self-destruct on the day after the guarantee expires. Likewise, it would cost too much of energy to make a system in an animal that would cause it to die precisely on the day that its progeny become independent. What happens after the guarantee period expires in watches and after the reproductive period in animals is aging, which inexorably leads to failure in watches and death in animals.

In this way of thinking, survival to sexual maturation is accomplished by postponing until after reproductive maturity the effects of genes that perform well in youth but become mischief-makers later. When these once good, now harmful genes eventually do switch on, they provide the blueprint for age changes. . . .

Until now we have almost always thought about aging by asking, "Why do we age?" And biogerontologists have designed their experiments to attempt to answer this question. The results have not been impressive. With the exception of the discovery that age changes occur within individual cells, we do not know much more today about the fundamental cause of aging than we did a century ago. Most of what we have learned is descriptive: we know much more about *what* happens than we did before but very little about *why* it happens. Biogerontologists have described changes that occur as we age from the molecular level up to the level of the whole animal. However, these descriptive observations add little to our understanding of the basic process.

It is for this reason that George Sacher proposed that we have been asking the wrong question. Instead of asking "Why do we age?" we should ask "Why do we live as long as we do?" By asking that question we might reorder our thinking and be able to design experiments to obtain more fundamental information. I think this is a useful new approach and I hope that more biogerontologists will come to appreciate the subtle but important reason for asking this better question.

Implicit in the question "Why do we live as long as we do?" is the idea that our longevity has increased and may be capable of increasing further. That appears to be true, since the human life span is known to have increased since prehistoric times. If our life span has increased, then it is likely that the start of the aging process has changed within the new time frame. Based on this reasoning, we may conclude that the aging process is malleable, that we can understand how it occurs, and that perhaps we can tamper with it. . . .

I do not believe that we have a sufficient understanding of either the aging process or the determinants of life span to expect to significantly manipulate either during our lifetime. A more important issue, however, is whether it would be desirable to manipulate either process. The capacity to halt or slow the aging process, or to extend longevity, would have consequences unlike most other biomedical breakthroughs. Virtually all other biomedical goals have an indisputably positive value. It is not at all clear whether or not the ability to tamper with the processes that age us or determine our life span would be an unmixed blessing. As pointed out earlier, resolution of all disease and other causes of death would result in a life expectation of about one hundred years. I am apprehensive about extending average life expectation beyond age one hundred once the leading killers are resolved because the result would be disease-free but nonetheless functionally weaker, still inexorably aging people. . . .

Virtually all biomedical research has the implicit goal of eliminating disease in all of its forms. It is logical to ask what will happen if we are successful. The answer seems to be that if we are successful, our life expectation will be increased but we will eventually die from the basic aging processes that lead to failure in some vital system.

READING 6

Vitality and Aging

Implications of the Rectangular Curve

James F. Fries and Lawrence Crapo

Why do we age? Why do we die? How can we live longer? How can we preserve our youth? Questions about life, aging, and death are fundamental to human thought, and human beings have speculated about the answers to these questions for centuries. Our own age values the methods of science—the methods of gathering evidence, of observation, of experiment—above the musings of philosophy. Yet, philosophical speculation and scientific theory may interact and enhance each other. The scientific theories of Copernicus and the conception of a sun-centered solar system, of Newton and an orderly universe, of Einstein and the relationship between matter, energy, and spacetime, of Darwin and the evolution of species have influenced our notions of who we are, where we are, how we came to be here, and the meaning of life itself. Similarly, the study of health and aging may contribute a new philosophical perspective to these age-old questions about life and death.

The implications of new scientific discoveries are often not widely appreciated for many years. Scientific knowledge develops by small increments within a relatively cloistered scientific community, whose members are sometimes more interested in the basic ideas than in their social implications. . . .

So it is with the study of human aging. The ancient philosophical questions have largely fallen to those who search for the biological mechanisms that affect our vitality and that cause our death. The study of aging as a separate scientific discipline is relatively new and is not yet the province of any single science. Independent observations have been made in medicine, in psychology, in molecular biology, in sociology, in anthropology, in actuarial science, and in other fields. There are remarkable parallels in the ideas that have emerged from these independent fields of research. It is our intention to review these parallel developments and to present a synthesis of scientific ideas about human aging that will offer insights into the fundamental questions about the nature and meaning of the life process, aging, and death.

THE INCOMPLETE PARADIGM

The growth of scientific knowledge historically has been impeded by thought systems (paradigms) that worked well for a time but that increasingly failed to explain new observations. For the study of aging, the contemporary paradigm is often called the medical model. The medical model defines health as the absence of disease and seeks to improve health by understanding and eradicating disease. This model of life and health, while useful, has obscured a larger perspective. There are four prevalent beliefs in the medical model that have proved to be limiting (see box). Certainly, few

SOURCE: Excerpts from *Vitality and Aging: Implications of the Rectangular Curve* by James F. Fries and Lawrence Crapo. New York: W. H. Freeman, 1981. Reprinted by permission of the authors.

present scholars hold these beliefs literally, but these ideas nonetheless have largely defined contemporary opinion about the aging process.

The Limiting Premises

1. The human life span is increasing.

2. Death is the result of disease.

3. Disease is best treated by medication.

4. Aging is controlled by the brain and the genes.

These four premises seem to imply the following conclusions. If the human life span is increasing, then our scientific goal can be the achievement of immortality. If death results from disease, our objective must be the elimination of disease. If disease is best treated with medication, our strategy is to seek the perfect drug or surgical procedure. With regard to aging, the medical model suggests that we should perform basic research to understand the genetic, neurologic, or hormonal mechanisms that control the process, and then learn to modify them.

Historically, these premises, objectives, and strategies have been useful. They are still worthy and deserving of study and hope. But they are certainly incomplete, and, taken literally, they are misleading. The human life span is not increasing; it has been fixed for a period of at least 100,000 years. The popular misconception of an increasing life span has arisen because the average *life expectancy* has increased; the *life span* appears to be a fixed biological constant. Three terms must be understood. The maximum life potential (MLP) is the age at death of the longest-lived member of the species—for human beings, 115 years. The life span is the age at which the average individual would die if there were no disease or accidents for human beings, about 85 years and constant for centuries. The life expectancy is the expected age at

death of the average individual, granting current mortality rates from disease and accident. In the United States, this age is 78 years and rising.

Death does not require disease or accident. If all disease and all trauma were eliminated, death would still occur, at an average age not much older than at present. If premature death were eliminated, and it may be in large part, we would still face the prospect of a natural death.

Medical treatment is not the best way to approach current national health problems. The major chronic diseases (atherosclerosis, cancer, emphysema, diabetes, osteoarthritis, and cirrhosis) represent the major present health threats. They are deserving of continued medical research, and further advances are to be expected. But abundant evidence points to personal health habits as the major risk factors for these diseases. Preventive approaches now hold far more promise than do therapeutic approaches for improving human health.

Aging does not appear to be under direct control of the central nervous system or the genes. Rather, the aging process occurs in cells and in organs. The aging process is most likely an essential characteristic of biological mechanisms. The process of aging, or *senescence,* is an accumulation in cells and organs of deteriorating functions that begins early in adult life. Aging may result from error-prone biological processes similar to those that have led to the evolution of species.

So the prevailing ideas about aging are incomplete. An increasing body of new scientific information requires revision and extension of these ideas. The time for a new synthesis has arrived, heralded by a number of new discoveries that do not fit well into the old paradigm but that as yet lack a coherent paradigm of their own.

COMPETING THEMES

Changes in our ideas about health and aging are now being reflected in our social institutions and lifestyles. Change in a prevalent system of thought is often turbulent, and such turbulence is now manifest in health by a set of new movements. Within the medical community, there has been increasing recognition of the importance of preventive medical approaches. Such technical strategies as mass screening have been promoted. New departments of preventive medicine have been developed within medical schools; previously, such efforts were largely carried out within schools of public health. These developments are not entirely successful (screening efforts have proved disappointing, and some departments of preventive medicine have not thrived), but their very creation acknowledges the ferment of new approaches to health care.

The public has asked for more active involvement in consumer choices and for more accurate information on which to base such choices. In response, a self-care movement in health has developed, which now represents a considerable social force. At its best, this movement encourages critical consumption of medical services and increased autonomy from professional dominance. At its worst, the self-care movement takes an adversary stance and would replace professional medical treatment with idiosyncratic folk remedies. Still, the growth of these movements indicates discontent with the prevailing medical orthodoxy.

Recent changes in personal lifestyles have been even more significant. Joggers organize footraces in which tens of thousands compete, and cocktail party conversations concern the number of miles run per week. The number of militant antismokers has grown, and the nonbelievers are being packed into smaller and smaller spaces in the back of the airplane. Such spontaneous social changes are very likely to have constructive effects on health, and we applaud them, but the point is that the phenomenon itself represents a profound changing of the public consciousness.

Within professional medicine, new themes are evident. There is an increased interest in long-term patient outcome as a goal and less interest in correcting the trivial laboratory abnormality that does not materially affect the patient. Benefit-cost studies are sometimes advocated as a solution to the astronomical increases in the cost of medical care. Many observers have pointed out that orthodox medical approaches have reached the area of diminishing returns. The quality of life, rather than its duration, has received increasing emphasis.

Both psychologists and physicians have recently described strong relationships between psychological factors and health, and theories explaining such relationships have been developed that emphasize life crises, helplessness, loss of personal autonomy, depression, and other psychological factors. Correction of some psychological problems, it is implied, will improve health, and indeed the circumstantial evidence that this may be true is quite convincing. Again these approaches are outside the orthodoxy of the medical model.

Two new research areas have recently been emphasized—chronic disease and human aging. Increasingly, researchers recognize the central roles that aging and chronic disease play in our current health problems. The study of aging and chronic disease is oriented toward long-term outcomes, is interdisciplinary, requires preventive strategies, seeks to demonstrate the relevance of psychological factors, and uses lifestyle modification as a major tactic. The student of aging and the student of the diseases of the aged now have a unique opportunity to harmonize the incomplete old orthodoxy and the emerging new themes.

A NEW SYLLOGISM

Using new knowledge of human aging and of chronic disease, we attempt here to provide a model that harmonizes these competing and chaotic themes, one that points toward new strategies of research and of health attainment. Our theoretical structure allows predictions to be made, and the predictions are strikingly different from those traditionally expected.

These curves are correct. They converge at the same maximum age, thereby demonstrating that the maximum age of survival has been fixed over this period of observation.

Figure 1 shows the actual data. Quite . . . startling conclusions follow from these data. The number of extremely old persons will not increase. The percentage of a typical life spent in dependency will decrease. The period of adult vigor will be prolonged. The need for intensive medical care will decrease. The cost of medical care will decrease, and the quality of life, in a near disease-free society, will be much improved.

Figure 1   Human Survival Curves for 1900, 1920, 1940, 1960, and 1980

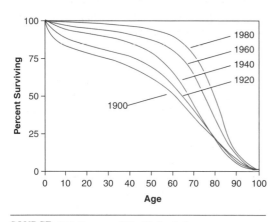

SOURCE: U.S. Bureau of Health Statistics.

Adult life may be conveniently divided into two periods, although the dividing line is indistinct. First, there is a period of independence and vigor. Second, for those not dying suddenly or prematurely, there is a period of dependence, diminished capacity, and often lingering disease. This period of infirmity is the problem; it is feared, by many, more than death itself. The new syllogism does not offer hope for the indefinite prolongation of life expectancy, but it does point to a prolongation of vitality and a decrease in the period of diminished capacity.

There are two premises to the syllogism; if they are accepted, then it follows that there will be a reversal of the present trend toward increasing infirmity of our population and increased costs of support of dependency. . . . The first premise is almost certain; the second is very probable. If, after careful evaluation of the supporting data, one accepts the premises of this syllogism, then one must accept the conclusion and the implications of the conclusion.

SOME QUESTIONS OF SEMANTICS

Nuances of meaning may mask the substance of a subject, and slight changes in emphasis may allow a new perspective to be better appreciated. There are problems with several of the terms often used to describe health, medical care, and aging. Among these are *cure, prevention, chronic, premature death,* and *natural death.* We will use these terms in slightly different senses than is usual.

Cure is a term with application to few disease processes other than infections. The major diseases of our time are not likely to be cured, and we have tried to avoid this term. *Prevention* is better but is unfortunately vague; this term, as we shall see, is sometimes misleading. We prefer the term *postponement* with regard to the chronic diseases of human aging, since prevention in the literal sense is difficult or impossible. *Chronic* is a term usually used to denote illnesses that last for a long period of time. It serves as a general but imprecise way of distinguishing the diseases that may be susceptible to cure (such as smallpox) from those better approached by postponement (as with emphysema). Regrettably, this important distinction cannot be based solely on the duration of the illness, since some diseases that last a long time both are not chronic conditions and might eventually be treatable for cure (such as rheumatoid arthritis and ulcerative colitis). We limit our use of the term *chronic* to those conditions that are nearly universal processes, that begin early in adult life, that represent insidious loss of organ function, and that are irreversible. Such diseases (atherosclerosis, emphysema, cancer, diabetes, osteoarthritis, cirrhosis) now dominate human illness in developed countries. We have defined *premature death* simply as death that occurs before it must, and we have used *natural death* to describe those deaths that occur at the end of the natural life span of the individual. . . .

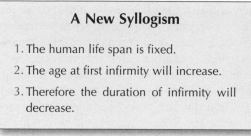

A New Syllogism

1. The human life span is fixed.

2. The age at first infirmity will increase.

3. Therefore the duration of infirmity will decrease.

THE RECTANGULAR CURVE

Survival curves for animals show a similar pattern of rectangularization with domestication or better care. Old age in wild animals is very rare, as it probably was for prehistoric man living in a dangerous environment. In uncivilized environments, accidental deaths and violent deaths account for a greater proportion of deaths than the biologically determined life-span limit. For the great majority of wild animal species, there is a very high neonatal mortality, followed by an adult mortality rate that is almost as high and is nearly independent of age. In such environments, death occurs mostly as a result of accidents and attacks by predators. One day is about as dangerous as the next.

By contrast, animals in captivity begin to show survival curves much more rectangular in shape. Such animals are removed from most threats by accident or predator, and for them the second term of the equation, that of the species' life span, begins to dominate. Figure 2 shows

theoretical calculations of this phenomenon after Sacher (1977). Such rectangularization has been documented for many animals, including dogs, horses, birds, voles, rats, and flies. . . .

Figure 2   Theoretical Survival Curves for an Animal Become Progressively More Rectangular as the Environment Progresses From Wild to Domestic

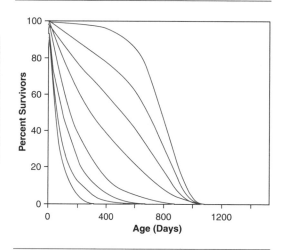

SOURCE: Redrawn from G. A. Sacher, "Life Table Modification and Life Prolongation," in C. E. Finch and L. Hayflick, eds., *Handbook of Aging,* Van Nostrand Reinhold, 1977. Redrawn with permission.

Figure 3 is drawn from the data Shock developed in 1960, and it is modified only slightly from what has been called "the most frequently shown data in the field of gerontology." The data show that many important physiological functions decline with age, and the decline is quite close to being a straight line. It is important to emphasize that these data were obtained from healthy human subjects in whom no disease could be identified that was related to the function being measured. Thus, the observed decline does not depend on disease.

Figure 3 is a major oversimplification of complex data. . . . The lines are not actually as

Figure 3 The Linear Decline of Organ Function With Age

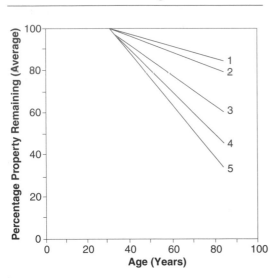

SOURCE: U.S. Bureau of Health Statistics. Redrawn from N. Shock, "Discussion on Mortality and Measurement of Aging," in B. L. Strehler, S. D. Ebert, H. B. Glass, and N. W. Shock, eds., *The Biology of Aging: A Symposium.* Copyright © 1960, American Institute of Biological Sciences. Redrawn with permission.

straight as portrayed, and some of the data have been contested. The point is that a considerable body of research supports a gradual, nearly linear decrease in organ function with age.

Normal, healthy organisms maintain an excess organ reserve beyond immediate functional needs. We have four to ten times as much reserve function as we need in the resting state. The heart during exercise can increase its output sixfold or more. The kidneys can still excrete waste products adequately if five-sixths of the functional units, the nephrons, are destroyed. Surgeons can remove one entire lung, and sometimes part of the second, and still have an operative success. Three-fourths of the liver can be removed, under some circumstances, and life is still maintained.

However, the mean level of reserve in many of our organs declines as we grow older. We seldom notice this gradual loss of our organ reserve. Only in the circumstances of exceptional stress do we need all that excess function anyway. Shock and others suggest that the decline may be plotted as a straight line.

HOMEOSTASIS AND ORGAN RESERVE

The human body may be viewed as a remarkable assembly of components functioning at various levels of organization. Systems of molecules, cells, and organs are all marvelously integrated to preserve life. The eminent nineteenth-century physiologist Claude Bernard emphasized that these integrated components act to maintain a constant internal environment despite variable external conditions. Bernard saw life as a conflict between external threats and the ability of the organism to maintain the internal milieu.

These fundamental observations have stood well the test of time. Indeed, the human organism cannot survive if the body temperature is more than a few degrees from normal, if acid-base balance is disturbed by a single pH unit, or if more than 20% of the body water is lost. Body chemicals are regulated closely, often to within 2% or 3% of an average value. A change in one direction in body constituent is often followed by a complicated set of responses that act to restore equilibrium.

Bernard also noted that living beings change from a period of development to a period of senescence or decline. He stated that "this characteristic of a determined development, of a beginning and an end, of continuous progress in one direction within a fixed term, belongs inherently to living beings."

The regulation of bodily functions within precise limits was termed *homeostasis* by Cannon (1932). Living organisms under threat from an extraordinary array of destructive sources maintain their internal milieu despite the perturbations, using what Cannon called the "wisdom of the body." Dubos (1965) has pointed out that this "wisdom" is not infallible. Homeostasis is only an ideal concept; regulatory mechanisms do not always return bodily functions to their original state, and they can sometimes be misdirected. Dubos sees disease as a "manifestation of such inadequate responses." Health corresponds to the situation in which the organism responds adaptively and restores its original integrity.

The ability of the body to maintain homeostasis declines inevitably with decreasing organ reserve. Figure 3 shows the decline for lungs, kidneys, heart, and nerves. The decline is not the same for all individuals, nor is the decline the same for all organs. For example, nerve conduction declines more slowly than does maximal breathing capacity. And some organs, such as the liver, intestinal lining cells, and bone marrow red cells, seem to show even less decline with age.

The important point, however, is that with age there is a decline in the ability to respond to perturbations. With the decline in organ reserve, the protective envelope within which a disturbance may be restored becomes smaller. A young person might survive a major injury or a bacterial pneumonia; an older person may succumb to a fractured hip or to influenza. If homeostasis cannot be maintained, life is over. The declining straight lines of Figure 3 clearly mandate a finite life span; death must inevitably result when organ function declines below the level necessary to sustain life. . . .

IMPLICATIONS OF THE RECTANGULAR CURVE

The rectangular curve is a critical concept, and its implications affect each of our lives. The rectangular curve is not a rectangle in the absolute sense, nor will it ever be. The changing shape of the curve results from both biological and environmental factors. Many biological phenomena describe what is often called a normal distribution. This is the familiar bell-shaped or Gaussian curve. If one studies the ages at death in a well-cared-for and relatively disease-free animal population, one finds that their ages at death are distributed on both sides of the average age of death, with the number of

individuals becoming less frequent in both directions as one moves farther from the average age at death. A theoretical distribution of ages at death taking the shape of such a curve in humans is shown in Figure 4. This simple bell-shaped curve, with a mean of 85 years and a standard deviation of 4 years, might exemplify the age at death of an ideal disease-free, violence-free human society. The sharp downslope of the bell-shaped survival curve is analogous to the sharp downslope of the rectangular curve. In Figure 5, the first part of the curve becomes ever flatter, reflecting lower rates of infant mortality. Several factors prevent the total elimination of infant mortality and thus prevent the curve from becoming perfectly horizontal. These premature deaths are the result of birth of defective babies, premature disease, and violent death. Improvements in medicine can lower but never eliminate the birth of defective babies and premature disease. It seems likely that the ever dominant proportion of violent deaths during early life will prove recalcitrant to change and will form an ever larger fraction of total premature deaths.

So, the rectangular curve has an initial brief, steep downturn because of deaths shortly after birth, a very slow rate of decline through the middle years, a relatively abrupt turn to a very steep downslope as one nears the age of death of the ideal Gaussian curve, and a final flattening of the curve as the normal biological distribution of deaths results in a tail after the age of 90. . . .

Thus, two profound characteristics of the mortality of man, the elimination of premature disease and the development of the sharp downslope representing natural death, have remained far from the public consciousness. These data have been available for many years. The first solid comments about rectangularization of the human survival curve can be found in prophetic statements in the 1920s. Many statisticians and actuaries working with national health data since that time have noted the increasingly rectangular shape of the curve, and many have speculated that it represents a natural species life limit. Entire theories of the aging process . . .

Figure 4 Sequential Survival Curves in the United States

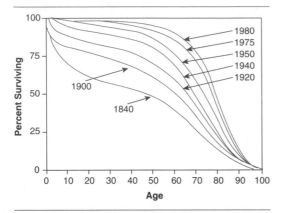

SOURCE: U.S. Bureau of Health Statistics.

Figure 5 Ideal Mortality Curve in the Absence of Premature Death

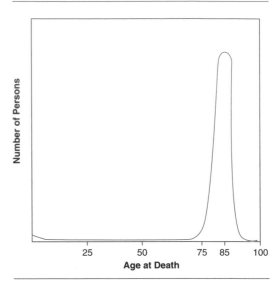

SOURCE: U.S. Bureau of Health Statistics.

have been built around the observed fact of a natural life span in man and animals. Yet, the public has remained largely ignorant of these developments.

A society in which life expectancy is believed to increase at every age and in which one becomes increasingly feeble as one grows older is a society heading for trouble. A society moving according to the curves of Figure 5, as our society is, is a society moving toward a world in which there is little or no disease, and individuals live out their natural life span fully and vigorously, with a brief terminal period of infirmity. . . . Dramatic changes in mortality patterns result in equally dramatic social changes.

REFERENCES

Cannon, W. B., *The Wisdom of the Body,* New York: Norton, 1932.

Dubos, R., *Man Adapting,* New Haven, CT: Yale University Press, 1965.

Sacher, G. A., "Life Table Modification and Life Prolongation," in J. Birren and C. Finch, eds., *Handbook of the Biology of Aging,* New York: Van Nostrand Reinhold, 1977, 582–638.

READING 7

—— The Compression of Morbidity Hypothesis ——

A Review of Research and Prospects for the Future

Vincent Mor

Cross-national evidence for the validity of the compression of morbidity hypothesis originally proposed by Fries is generally accepted. Generational improvements in education and the increased availability of adaptive technologies and even medical treatments that enhance quality of life have facilitated continued independence of older persons in the industrialized world. Whether this trend continues may depend upon the effect of the obesity epidemic on the next generation of older people.

For more than 2 decades, gerontologists have been debating the implications of the progressive reductions in old-age mortality and increasing survival of the very old, with some noting that lengthening life necessarily extends the duration of functional dependency in an aging population. It has been hypothesized that increasing survival does not necessarily mean that the added years of life accruing to older individuals would be spent sick and disabled.[1] This compression of morbidity hypothesis stated that better health care, an active lifestyle, and greater preventive health behavior would preserve health even in the face of increasing survival. Shortly thereafter, other researchers[2] were able to quantify "active life expectancy," setting the stage for the application of sophisticated demographic techniques to test the hypothesis that the duration of morbidity and disability would not increase or might even be reduced in the population, even as mortality was decreasing. . . .

There are several important clarifications that should be made to better understand the dispute surrounding the compression of morbidity hypothesis. First, using morbidity and disability interchangeably ignores the evidence that the presence of different diseases may have quite different effects on mortality, hospitalization (a health services use-based marker of real morbidity), disability, and functional impairment.[9] For example, although cardiovascular disease mortality declined, partially because of improved treatment, outreach efforts also led to earlier identification of more individuals with early-stage disease. Earlier detection of disease (morbidity) is one reason why increases in the prevalence of chronic illness have not translated into increases in disability and impaired function.[10, 11] . . .

Efforts to understand what has caused the reduced rate of functional decline in the aged population have focused on the improved education of the newer cohorts of elderly, improvements in the built environment (e.g., barrier-free housing, elevators) and material amenities, and improvements in function-enhancing medical interventions.[14] First, the average 75-year-old in developed countries of the 1990s is less likely to be constrained by stairs and more likely to have an automobile and to live in housing that is architechnally barrier free. Second, that same older person has his/her Social Security check directly deposited, meals warmed in a microwave, and groceries ordered over the telephone. Third, the disability of cataracts has been virtually eliminated with new surgical techniques, disabling arthritic hips and knees are routinely replaced, and improvements in the medical management of heart disease, for example, clearly facilitate retained functioning and independence. The relative contributions of each of these major classes of technological innovations to improvements in population functioning is not known, but it is likely that these, as well as other significant shifts in the lives of older persons in the industrialized world, have improved their quality of life and functional independence.[6, 15] Indeed, it may be that the education advantage observed in most studies that find reduced functional decline is partially achieved by older persons being able to manipulate the environment without exertion.[11]

Nevertheless, before we celebrate and ignore the pending explosion of the aged population in all industrialized countries in the world, it is critical to understand that, even if the rate of functional decline has dropped several percentage points over the last decades, the sheer numerical increase in the size of the aged population over the next 30 years will mean that the number of older persons who are dependent, disabled, and suffering the functional consequences of multiple chronic conditions will be larger than it has ever been, far larger than most countries are prepared to manage. Healthy life expectancy (another expression of "active life expectancy") is increased via reductions in mortality and morbidity, but disease prevalence is increasing, so functional independence must be maintained in the face of advancing age and comorbidity.[11]

The emerging epidemic of obesity among the middle-aged population, particularly in the United States, is another factor that may temper the optimism some have expressed about being able to compress the duration of functional morbidity.[6, 15, 16] Recent evidence of the rising prevalence of obesity in the middle aged and the consequences of obesity for independence and for the ability to function and fill social roles suggests that we may be in for a reversal of the hard-fought gains of functional decline.[12, 17–19] Indeed, these findings reinforce the importance of an active lifestyle and low-risk health habits such as avoiding obesity in maintaining functional independence into the advanced years. Although technology, the built environment, and medical care advances may have yielded benefits in function and quality of life for the "greatest" generation, unless the health habits of the baby boomers change dramatically, future researchers may be trying to explain the cohort effect that found a short-lived reduction in the duration of age-related functional impairment.

REFERENCES

1. Fries J. Aging, natural death and the compression of morbidity. N Engl J Med 1980; 303:130–133.

2. Katz S, Branch LG, Branson MU, et al. Active life expectancy. N Engl J Med 1983;309:1218–1224.

3. Manton KG, Carder J, Stallard E. Chronic disability trends in elderly United States populations: 1982–1994. Proc Natl Acad Sci USA 1997;94:2593–2598.

4. Manton KG, Stallard E, Corder L. Changes in the age dependence of mortality and disability: Cohort and other determinants. Demography 1997;34:135–157.

5. Manton KG, Gu X. Changes in the prevalence of chronic disability in the United States black and nonblack population above age 65 from 1982 to 1999. Proc Natl Acad Sci USA 2001;98:6354–6359.

Cutler D. Declining disability among the elderly. Health Aff 2001;20:11–27.

6. Freedman VA, Martin LG, Schoeni RF. Recent trends in disability and functioning among older adults in the United States: A systematic review. JAMA 2002;288:3137–3146.

7. Hubert HB, Bloch DA, Oehlert JW, et al. Lifestyle habits and compression of morbidity. J Gerontol A Biol Sci Med Sci 2002;57(6):M347–M351.

8. Verbrugge LM, Patrick DL. Seven chronic conditions: Their impact on US adults' activity levels and use of medical services. Am J Public Health 1995;85:173–182.

9. Freedman VA, Martin LG. Contribution of chronic conditions to aggregate change in old-age functioning. Am J Public Health 2000;90:1755–1760.

10. Crimmins EM, Saito Y. Trends in healthy life expectancy in the United States, 1970–1990: Gender, racial, and educational differences. Soc Sci Med 2001;52:1629–1641.

11. Crimmins EM, Saito Y. Change in the prevalence of diseases among older Americans: 1984–1994. J Dem Rsrch.

12. Vita AJ, Terry RB, Hubert HP, et al. Aging, health risks, and cumulative disability. N Engl J Med 1998;338:1035–1041.

14. Cutler DM. The reduction in disability among the elderly. Proc Natl Acad Sci USA 2003;98:6546–6547.

15. Freedman VA, Martin MG, Schoeni RF. Recent trends in disability and functioning among older adults in the United States. JAMA 2002;288:3137–3146.

16. Fries J. Measuring and monitoring success in compressing morbidity. Ann Intern Med 2003;139:455–459.

17. Lakdawalla DV, Bhattacharya J, Goldman DP. Are the young becoming more disabled? Health Aff 2004;23:168–176.

18. Peeters A, Barendregt JJ, Willekens P, et al. Obesity in adulthood and its consequences for life expectancy: A life-table analysis. Ann Intern Med 2003;138:24–32.

19. Sturm R, Ringel J, Andreyeva T. Increasing obesity rates and disability trends. Health Aff 2004;23:199–205.

READING 8

——————— Health Trends in the Elderly Population ———————

Getting Better and Getting Worse

Marti G. Parker and Mats Thorslund

Health trends in the fastest growing sector of the population, the oldest old, have received much attention during the past decade because of the rising costs of medical and long-term care. Many studies have suggested a compression of morbidity in this sector, implying that the future care needs of elderly people will not follow the demographic prognoses. Most of these studies have used health indicators based on disability, a concept that is contextually embedded. We have taken a closer look at health-trend surveys with a focus on the health indicator used. Our findings reveal that although disability measures often show improvement, there is a simultaneous increase in chronic disease and functional impairments—health components that require care resources. That is, an expansion of other health problems may accompany a compression of disability. Therefore, a concept of general morbidity is not sufficient when discussing health trends and the need for care services in the elderly population. Because different indicators do not show the same trends over time, we suggest a more refined discussion that distinguishes between different health components. In addition, different components have different implications for the amount and kind of care resources needed. If the current positive trends in disability continue, future need for social services and long-term care may not parallel demographic projections. Trends in disease and functional limitations seem to have taken a different direction, suggesting a parallel or increased need for resources in medical care,

rehabilitation, and compensatory interventions such as assistive technology.

The 20th century was incredibly successful in regard to aging. Although change first appeared in infant mortality, mortality also decreased in elderly age groups due to improvements in living conditions, better control of infectious diseases, and medical advancements (Centers for Disease Control and Prevention, 1999). In the second half of the century, cardiovascular disease replaced infectious diseases as the major killer, but survival here also has increased dramatically. Even elderly people who were once considered to have a very high mortality risk now seem to be surviving longer (Crimmins & Saito, 2000; Rosén & Haglund, 2005). . . .

The prevalence of health problems increases sharply with age with associated costs for medical care, social services, and long-term care. Therefore, health trends in the oldest sector of the population are of particular interest when estimating need for future care resources. As the average expected life span increases, an important issue is whether the years added to life are characterized by good health and independence or by health problems and the need for care.

Bearing in mind the complex interplay for mortality and morbidity, we pose the following question: How should researchers best measure the health of the elderly population to reflect need for care? Different population surveys utilize different kinds of measures, all of which are related to health and are therefore often loosely referred to as health indicators. When used in surveys of the oldest

SOURCE: "Health Trends in the Elderly Population: Getting Better and Getting Worse" by M. G. Parker and M. Thorslund in *The Gerontologist, 47*(2), 150–158. Copyright © 2007 by The Gerontological Society of America.

sectors of the population, indicators need to span the entire spectrum of health. Representative samples will include healthy and independent people as well as people who are bedridden and dependent on extensive social and medical services. Thus it is difficult to construct health indicators that avoid floor and ceiling effects. Disease, one of the most common measures of ill health, usually reflects a need for medical care, but without clinical information about severity, disease may say little about the need for the most expensive service, long-term care. For example, people who report that they have heart failure or Parkinson's disease could be fully independent or institutionalized. Consequently, most surveys use measures of function or disability (i.e., measures that reflect the cumulative consequences of disease and other living conditions). There is no consensus about how to define these concepts or which are the best health or function indicators for population surveys. . . .

The idea of compressed morbidity among the oldest sector of the population has received wide publicity because of its optimistic implications for future resource need. It suggests that, although the population is aging, future elderly cohorts may not need as many care resources as do current cohorts. . . . In this discussion, however, it is important to remember that even the most optimistic prognoses foresee an absolute increase in resource need. . . .

Commonly Used Health Indicators

Global Self-Rated Health

This item asks respondents to rate their own general health on a 3- to 5-point scale. Self-rated health reflects the total picture of health as perceived by the individual. As such, it probably reflects dimensions of health that are most meaningful to each individual. . . .

Specific Self-Reported Health Items

Many surveys include items that ask about specific health problems, either diseases or symptoms. When posed in survey interviews, questions about diseases often necessitate that the respondent be diagnosed, be informed of the diagnosis, remember the diagnosis, and report it during the interview. . . .

Functional Impairment

Many surveys include instruments or items that refer to specific functions (e.g., walking, rising from a chair, lifting a heavy object, or seeing and hearing). . . .

Mobility is one of the most commonly studied functions because of its importance in independent living. . . .

Disability

One of the most commonly used indicators of health trends in the elderly population is disability. It is particularly useful because of its close correlation with need for social services. Most often researchers measure it with some form of primary activities of daily living (ADLs; e.g., ability to dress, use the toilet, eat, bathe) and secondary instrumental ADLs (IADLs; e.g., ability to clean house, prepare food, shop for groceries). . . .

Tests of Function

Several surveys have incorporated simple tests of function in their batteries. Tests provide more objective measures that are less susceptible to individual interpretations or expectations. They are also less affected by environmental factors. . . .

Conflicting Evidence

Table 1 presents the trends for the indicators mentioned previously as a summary of health-trend studies of the elderly population over the past two decades. . . .

Results of trend studies using self-rated health and measures of self-rated function have shown mixed results: Some have shown improvement, and others have shown worsening. Results concerning specific diseases and symptoms have leaned overwhelmingly toward increased prevalence.

Results for ADL limitations (severe disability) have been mixed, although there is much evidence for improvement or no change. Results

Table 1 Health Trends in Elderly Populations According to Various Health Indicators

Indicator	Trend
Self-rated health	Mixed results
Specific health items: diseases and symptoms (self-ratings, costs, medical journals)	Increased prevalence
Self-rated physical function	Mixed results
Self-rated activities of daily living	Mixed, mostly improvement
Self-rated instrumental activities of daily living	Improvement or no change
Tests of functional ability	Increased prevalence or no change

for IADL limitations (moderate disability) have leaned heavily toward improvement. We found no study showing increased prevalence of IADL limitations.

DISCUSSION

From this overview of international studies, it is clear that research results diverge, and even conflict, in regard to health trends among elderly populations. Investigators can expect to see different trends in different countries due to different demographic and mortality patterns. . . .

Explaining Change in Prevalence Rates

One can explain some change in prevalence over time as a result of changes in reporting, particularly for symptoms and diseases. Most population surveys are based on self-reports from respondents or proxies. Many factors other than the pathological condition of the respondent can influence these reports. . . .

Awareness of problems such as depression or hypertension among elderly people has increased in the medical profession, leading to more frequent diagnosing. Physicians may also be more likely today than they were in previous years to tell their patients about the diagnoses. In general, it has also become more socially acceptable to talk about certain problems, such as depression

and incontinence; respondents may therefore be more willing to report these problems now than their counterparts would have been years ago.

Nonetheless, there also seems to be substantial evidence for increases in disease prevalence among older sectors of the population. . . .

How can investigators explain the increased prevalence of health problems, both reported disease and symptoms as well as tested functional ability? Most population studies, understandably, look at prevalence at a particular time. As the mortality rate decreases, more people—even those with diseases—survive with their problems. In particular, survival among even very old people with stroke and cardiac infarct has improved (Rosén & Haglund, 2005). Many of these people survive, but they often have chronic health problems. . . .

Despite the widely divergent results, the general tendency seems to be that older sectors of the population report more diseases and health problems at the same time that they seem to be coping better with many of the activities necessary for independent living (Crimmins, 2004; Parker et al., 2005; Spillman, 2004). Decreases in disability levels do not seem to be the result of less disease or fewer symptoms. . . .

The fact that most studies have found that IADL disability seems to be improving could well reflect the many environmental changes that can facilitate these activities: improved accessibility, wheeled walkers with baskets, ready-made meals, and microwave ovens. There are fewer technological interventions available to facilitate primary ADLs such as maintaining personal hygiene, dressing, and eating. . . .

CONCLUSIONS

The study of health patterns over time will lead to better understanding of contextual factors that may be correlated to health. Experts may also use health trends in the oldest sectors of the population to make projections of possible future resource need. However, most studies aimed at estimating future need do not specify which kinds of resources will be needed. Studies that use only disability measures give misleading results in regard to the total resource need that can be expected in the future. Elder care includes a wide variety of services, from highly specialized medical care to long-term-care facilities to simple but essential home services. The resources, in terms of cost and competence, vary accordingly. Therefore, if the study of health trends is going to be of any use in planning resource distribution in the future, investigators must examine the different components of health separately. This entails using a variety of measures as indicators.

Nebulous concepts of morbidity have clouded discussion on, and research about, health trends. Studies have shown that during a single time period there are different trends for different components of health in the elderly population and that the correlations between different components also change. This review suggests that the prevalence of symptoms, disease, and functional limitations is expanding at the same time that disability is being compressed, or at least postponed. . . .

Functional limitations imply rehabilitative and compensatory measures, whereas disability among elderly people often entails need for social services and/or long-term care. . . .

In summary, trend studies using disability measures give a skewed picture of overall health development in the elderly population.

Furthermore, the implication that future need for care may be lower for future cohorts of elderly people is dangerously deceptive. To adequately study health trends among the very old, surveys need to include multiple health indicators. . . .

REFERENCES

Crimmins, E. M. (2004). Trends in the health of the elderly. *Annual Review of Public Health,* 25, 79–98.

Crimmins, E. M., & Saito, Y. (2000). Change in the prevalence of diseases among older Americans: 1984–1994. *Demographic Research,* 3(9), 1–20.

Freedman, V. A., Crimmins, E., Schoemi, R. F., Spillman, B. C., Aykan, H., Kramarow, E., et al. (2004). Resolving inconsistencies in trends in old-age disability: Report from a technical working group. *Demography,* 41, 417–441.

Freedman, V. A., Martin, L. G., & Schoemi, R. F. (2002). Recent trends in disability and functioning among older adults in the United States: A systematic review. *Journal of the American Medical Association,* 288, 3137–3146.

Idler, E. L., & Benyamini, Y. (1997, March). Self-rated health and mortality: A review of 27 community studies, *Journal of Health and Behavior,* 38, 21–37.

Manton, K. G., Stallard, E., & Corder, L. S. (1998). The dynamics of dimensions of age-related disability 1982 to 1994 in the U.S. elderly population. *Journal of Gerontology: Biological Sciences,* 53A, B59–B70.

Parker, M. G., Abacic, K., & Thorsland, M. (2005). Health changes among Swedish oldest old: Prevalence rates from 1992 and 2002 show increasing health problems. *Journal of Gerontology: Medical Sciences,* 60A, 1351–1355.

Rosén, M., & Haglund, B. (2005). From healthy survivors to sick survivors: Implications for the 21st century. *Scandinavian Journal of Public Health,* 33(2), 151–155.

Spillman, B. C. (2004). Changes in elderly disability rates and the implications for health care utilization and cost. *Milbank Quarterly,* 82, 157–194.

READING 9

We Will Be Able to Live to 1,000

Aubrey de Grey

Life expectancy is increasing in the developed world. But Cambridge University geneticist Aubrey de Grey believes it will soon extend dramatically to 1,000. Here, he explains why.

Ageing is a physical phenomenon happening to our bodies, so at some point in the future, as medicine becomes more and more powerful, we will inevitably be able to address ageing just as effectively as we address many diseases today.

I claim that we are close to that point because of the SENS (Strategies for Engineered Negligible Senescence) project to prevent and cure ageing.

It is not just an idea: it's a very detailed plan to repair all the types of molecular and cellular damage that happen to us over time.

And each method to do this is either already working in a preliminary form (in clinical trials) or is based on technologies that already exist and just need to be combined.

The Alternative View

Nothing in gerontology even comes close to fulfilling the promise of dramatically extended lifespan.

—S. Jay Olshansky

This means that all parts of the project should be fully working in mice within just 10 years and we might take only another 10 years to get them all working in humans.

When we get these therapies, we will no longer all get frail and decrepit and dependent as we get older, and eventually succumb to the innumerable ghastly progressive diseases of old age.

We will still die, of course—from crossing the road carelessly, being bitten by snakes, catching a new flu variant, etcetera—but not in the drawn-out way in which most of us die at present.

I think the first person to live to 1,000 might be 60 already.

So, will this happen in time for some people alive today? Probably. Since these therapies repair accumulated damage, they are applicable to people in middle age or older who have a fair amount of that damage.

It is very complicated, because ageing is. There are seven major types of molecular and cellular damage that eventually become bad for us—including cells being lost without replacement and mutations in our chromosomes.

Each of these things is potentially fixable by technology that either already exists or is in active development.

'YOUTHFUL NOT FRAIL'

The length of life will be much more variable than now, when most people die at a narrow range of ages (65 to 90 or so), because people won't be getting frailer as time passes.

SOURCE: "We Will Be Able to Live to 1,000" by Dr. Aubrey de Grey is reprinted with permission from BBC News at bbcnews.co.uk.

> There is no difference between saving lives and extending lives, because in both cases we are giving people the chance of more life.

The average age will be in the region of a few thousand years. These numbers are guesses, of course, but they're guided by the rate at which the young die these days.

If you are a reasonably risk-aware teenager today in an affluent, non-violent neighbourhood, you have a risk of dying in the next year of well under one in 1,000, which means that if you stayed that way forever you would have a 50/50 chance of living to over 1,000.

And remember, none of that time would be lived in frailty and debility and dependence—you would be youthful, both physically and mentally, right up to the day you mis-time the speed of that oncoming lorry.

SHOULD WE CURE AGEING?

Curing ageing will change society in innumerable ways. Some people are so scared of this that they think we should accept ageing as it is.

I think that is diabolical—it says we should deny people the right to life.

The right to choose to live or to die is the most fundamental right there is; conversely, the duty to give others that opportunity to the best of our ability is the most fundamental duty there is.

There is no difference between saving lives and extending lives, because in both cases we are giving people the chance of more life. To say that we shouldn't cure ageing is ageism, saying that old people are unworthy of medical care.

PLAYING GOD?

People also say we will get terribly bored but I say we will have the resources to improve everyone's ability to get the most out of life.

People with a good education and the time to use it never get bored today and can't imagine ever running out of new things they'd like to do.

And finally some people are worried that it would mean playing God and going against nature. But it's unnatural for us to accept the world as we find it.

Ever since we invented fire and the wheel, we've been demonstrating both our ability and our inherent desire to fix things that we don't like about ourselves and our environment.

We would be going against that most fundamental aspect of what it is to be human if we decided that something so horrible as everyone getting frail and decrepit and dependent was something we should live with forever.

If changing our world is playing God, it is just one more way in which God made us in His image.

READING 10

Don't Fall for the Cult of Immortality

S. Jay Olshansky

Some 1,700 years ago the famous Chinese alchemist, Ko Hung, became the prophet of his day by resurrecting an even more ancient but always popular cult, Hsien, devoted to the idea that physical immortality is within our grasp.

Ko Hung believed that animals could be changed from one species to another (the origin of evolutionary thought), that lead could be transformed into gold (the origin of alchemy), and that mortal humans can achieve physical immortality by adopting dietary practices not far different from today's ever-popular life-extending practice of caloric restriction.

The Alternative View

I think the first person to live to 1,000 might be 60 already.

—Aubrey de Grey

He found arrogant and dogmatic the prevailing attitude that death was inevitable and immortality impossible.

Ko Hung died at the age of 60 in 343 AD, which was a ripe old age for his time, but Hsien apparently didn't work well for him.

The famous 13th Century English philosopher and scientist, Roger Bacon, also believed there was no fixed limit to life and that physical immortality could be achieved by adopting the "Secret Arts of The Past." Let's refer to Bacon's theory as SATP.

According to Bacon, declines in the human lifespan occurred since the time of the ancient patriarchs because of the acquisition of increasingly more decadent and unhealthy lifestyles.

What do the ancient purveyors of physical immortality all have in common? They are all dead.

—S. Jay Olshansky

All that was needed to reacquire physical immortality, or at least much longer lives, was to adopt SATP—which at the time was a life-style based on moderation and the ingestion of substances such as gold, pearl, and coral—all thought to replenish the innate moisture or vital substance alleged to be associated with aging and death.

Bacon died in 1292 in Oxford at the age of 78, which was a ripe old age for his time, but SATP apparently didn't work well for him either.

Physical immortality is seductive. The ancient Hindus sought it, the Greek physician Galen from the 2nd Century AD and the Arabic philosopher/physician Avicenna from the 11th Century AD believed in it.

Alexander the Great roamed the world searching for it, Ponce de Leon discovered Florida in his quest for the fountain of youth, and countless stories of immortality have permeated the literature, including the image of Shangri-La portrayed in James Hilton's book *Lost Horizon,* or in the quest for the holy grail in the movie *Indiana Jones and the Last Crusade.*

What do the ancient purveyors of physical immortality all have in common? They are all dead.

PROPHETS OF IMMORTALITY

I was doing a BBC radio interview in 2001 following a scientific session I had organised on the question of how long humans can live, and sitting next to me was a young scientist, with obviously no sense of history, who was asked the question: "how long will it be before we find the cure for ageing?"

Without hesitation he said that with enough effort and financial resources, the first major breakthrough will occur in the next 5–10 years.

My guess is that when all of the prophets of immortality have been asked this question throughout history, the answer is always the same.

The modern notion of physical immortality once again being dangled before us is based on a premise of "scientific" bridges to the future that I read in a recently published book entitled *Fantastic Voyage* by the techno-guru Ray Kurzweil and physician Terry Grossman.

They claim unabashedly that the science of radical life extension is already here, and that all we have to do is "live long enough to live forever."

What Kurzweil and others are now doing is weaving once again the seductive web of immortality, tantalising us with the tale that we all so desperately want to hear, and have heard for thousands of years—live life without frailty and debility and dependence and be forever youthful, both physically and mentally.

The seduction will no doubt last longer than its proponents.

'FALSE PROMISES'

To be fair, the science of ageing has progressed by leaps and bounds in recent decades, and I have little doubt that gerontologists will eventually find a way to avoid, or more likely delay, the unpleasantries of extended life that some say are about to disappear, but which as anyone with their eyes open realises is occurring with increasing frequency.

There is no need to exaggerate or overstate the case by promising that we are all about to live hundreds or even thousands of years.

The fact is that nothing in gerontology even comes close to fulfilling the promise of dramatically extended lifespan, in spite of bold claims to the contrary that by now should sound familiar.

What is needed now is not exaggeration or false promises, but rather, a scientific pathway to improved physical health and mental functioning.

If we happen to live longer as a result, then we should consider that a bonus.

FOCUS ON THE FUTURE

"I Dated a Cyborg!"

Dateline: 2030. As Tony walked back to the college dormitory, his feelings were confused. He needed to talk to his roommate.

"You know, I really like her," Tony began. "I mean I really fell for her. And now . . . I just don't know . . ." Tony's voice trailed off.

"What's the problem?" Tony's roommate asked.

"Well, you know that girl I've been dating—Cynthia? It turns out she's a lot older than I thought she was."

"So. How much older?" asked his roommate.

"Hey, she remembers the assassination of President Kennedy, which happened when she was 10. That makes Cynthia 77 years old. She's 55 years older than I am! Can you believe that?"

Tony's roommate was aghast. He'd seen Cynthia. He figured she was around 30, not much more. Tony was pleased about going out with an "older" woman. But neither Tony nor his roommate had guessed just how much older she really was.

"I don't believe it! I mean, how could she be so old?" stammered Tony's roommate.

"Well, I found out she's had skin grafts and plastic surgery on her face; that's why there are no wrinkles. And of course her hair is dyed so there's no gray at all. But it's the rest of her that's . . . I don't know how to say it . . . that's all been replaced. It's weird. It's like Cynthia's body is artificial, the way it is with a cyborg.

"To begin with, she's got silicone breast implants. OK, not so unusual. But inside she's artificial, too: all plastic valves in her heart, a liver transplant, hip replacements, and a lot of artificial bones. She's been on estrogen replacement for years and on other anti-aging hormones, too. That's why she looks so young.

"Cynthia never talked much about things that happened before the turn of the century, and now I see why. I never suspected that she was born in the early 1950s. She admitted it to me last night. I came home and suddenly realized I've been dating a cyborg!"

Science fiction stories have had titles such as "I Married a Martian," and a Star Trek film, *First Contact,* featured a female "Borg" (for cyborg) as a leading character. Star Trek fans remember that the alien species known as the Borg are creatures that are part human and part machine. Like Tony, Captain Picard found himself in a relationship with a Borg and faced perplexing questions. Is the experience of Tony or Captain Picard a glimpse of things to come?

Cyborgs are not outside the realm of possibility (Gray, 1995). In fact, the era of modern bioethics may be said to have started in 1967, when Louis Washkansky received a heart transplant from Dr. Christiaan Barnard. Tissue transplants have long become a standard part of modern medicine. Some tissues, such as cartilage and the cornea of an eye, are transplanted easily. With proper safeguards, blood can be safely transfused. Modern medicine has also shown success in transplanting skin, bone, kidneys, and, more recently, lungs, livers, and hearts. The development of monoclonal antibodies, which help suppress rejection of transplanted tissues, has opened up a vast field of surgery to replace organs diseased or worn out with age.

At the same time, biomedical scientists are developing artificial tissues and organs that have been successfully inserted into the human body. Bioengineering has already made possible a variety of "replacement parts":

- *Skin:* Skin tissue has been successfully grown in the laboratory, and biotechnology companies are now producing it in quantity for use with burn victims.
- *Cartilage:* One of the most common effects of aging is the wear and tear on cartilage. Surgeons can now use cartilage grown in the lab to treat joint injuries.
- *Bone:* Hip replacements have long been a staple of geriatric medicine—even Elizabeth Taylor had one. Today, biotechnology companies are selling bone substitutes manufactured from artificial substances. Companies are working on grafts that would enable the body to replace living tissue with artificial bone.

More exciting innovations are on the horizon:

- *Breast Tissue Regeneration:* Breast implants made of silicone have long been in use, but the results have been controversial. Tissue engineers are working on new techniques to stimulate women's bodies to grow new breast tissue. Already, plastic surgery has become enormously popular. Tissue engineering and "body sculpting" are likely to become even more important in years to come.
- *Artificial Vision:* In *Star Trek: The Next Generation,* the character Geordi (played by LeVar Burton) is able to see by using a "VISOR"—an artificial vision device worn over the eyes. Today, older adults are the age group most likely to be afflicted with impaired vision or total blindness. But in the future, electronic devices may replace lost visual capacity.

(Continued)

(Continued)

- *Heart Valves:* Cardiovascular disease is the biggest cause of death among older Americans. Researchers have long been at work on a totally implantable artificial heart. Today, heart valves from pigs have been transplanted into humans. Researchers have discovered how to grow valves from blood vessel cells in the laboratory, and these lab-grown valves work well in lambs. In the future, thousands of people could benefit from artificially grown heart valves.
- *Bladder:* Urinary incontinence is one of the most troubling afflictions for older adults, and it is a factor in nursing home placement. But scientists are working on producing molded lab-grown cartilage that could function as a valve to keep urine flowing in the proper direction.
- *Pancreas:* Late-life diabetes is one of the most serious diseases of old age, entailing complications such as blindness, amputation, and heart failure. Diabetes results from basic organ failure. The pancreas doesn't produce enough insulin to metabolize sugar properly. Bioengineers are now working on implants made of pig islet cells, which could produce insulin without injections for people who develop diabetes.
- *Brain:* No one expects medical science to produce anything like *Donovan's Brain*, a tissue-culture brain that was the centerpiece of a 1950s science fiction movie. But drugs to stimulate nerve growth are under investigation today, and techniques may soon be available to implant cells or introduce growth factors that would reverse damage to the central nervous system.

So far, cyborgs, like *Star Trek,* are just science fiction. But bioengineering work on transplants and artificial organs is not fictional. Moreover, other scenarios are possible. For instance, Bruce Sterling's (1996) novel *Holy Fire* has as its heroine a wealthy 94-year-old woman who gets total cellular rejuvenation based on new genetic material added to chromosomes in her body. The result is an organism constructed from "designer genes," which is different from Cynthia and her replacement parts. Stay tuned as the 21st century progresses as biomedical technology reshapes our vision of what human aging is all about.

Questions for Writing, Reflection, and Debate

1. What are the arguments for and against the view that aging in and of itself is actually a disease? Pick one side of this issue and then try listing the points that can rebut the opposing point of view.

2. What do James Fries and Lawrence Crapo mean by *natural death?* What is the relationship between *natural death* and the *natural life span?* Should we consider the natural life span to be identical to the maximum life span?

3. Swedish data have turned up the surprising fact that death rates for the oldest-old (85+) have actually been going down. Some scientific studies suggest an ever-increasing life expectancy is quite possible. These findings sound like good news. Do we have any reasons to believe that these findings are not good news? What would Fries and Crapo's response be to these claims?

4. The Human Genome Project has now produced a complete map of all human chromosomes. Considering the different theories of aging, what are some of the ways in which new genetic knowledge might change how we think about the causes of biological aging? What are the social and ethical implications of that knowledge?

5. Write a science fiction scenario or imaginary picture of how the United States might look in the year 2030 if dramatic breakthroughs in the genetics of aging occur. In developing this picture, be sure to state the year you expect the key discoveries or inventions to occur, and describe the likely social consequences of those discoveries or inventions.

6. What is the best scientific evidence in favor of, or against, the compression-of-morbidity thesis?

Visit one of these websites on this subject—http://www.healthinaging.org/agingintheknow/chapters_ch_trial.asp?ch=2 or http://www.ncbi.nlm.nih.gov/pmc/articles/PMC20133—where you can review disability trends among older adults. What questions are left open by this website's report—for example, what exactly is "disability" as measured across different societies?

Suggested Readings

Austad, Steven N., *Why We Age: What Science Is Discovering About the Body's Journey Through Life,* New York: Wiley, 1997.

Hayflick, Leonard, *How and Why We Age,* New York: Ballantine Books, 1994.

Masoror, Edward J., *Challenges of Biological Aging,* New York: Springer, 1999.

Ricklefs, Robert E., and Finch, Caleb E., *Aging: A Natural History,* New York: Scientific American Library, 1995.

Williams, T. F., Sprott, R., and Warner, H., "Biology of Aging," *Generations* (Fall 1992), 16(4) [entire issue].

Student Study Site

Visit the Student Study Site at **www.sagepub.com/moody7e** for these additional learning tools:

- Flashcards
- Web quizzes
- Chapter outlines

- SAGE journal articles
- Web resources
- Video and audio resources

DOES INTELLECTUAL FUNCTIONING DECLINE WITH AGE?

We shall not cease exploring

And the end of all our exploring

Will be to arrive where we started

And know the place for the first time.

—T. S. Eliot, *The Four Quartets*

The view that intelligence and creativity decline with age is widely shared. Albert Einstein won a Nobel Prize for his contribution to quantum theory, a creative breakthrough that appeared in published form when he was only 26 years old. He later remarked that "a person who has not made his great contribution to science before the age of 30 will never do so." Was Einstein right?

The question of age and intellectual ability is an important one for individuals who worry about becoming irrelevant in a fast-paced world. The question is also important for society. The French demographer Alfred Sauvy (1976) feared that an aging society would result in a "population of old people ruminating over old ideas in old houses." In the coming decades, the U.S. population will become older. The workforce will be aging during a historical period when workplaces are being pushed to adopt new methods to improve competitive performance. Can we expect middle-aged and older workers to exercise creativity and initiative, or can we expect them to resist new ideas? What will happen to American inventiveness and scientific creativity as the average age of scientists goes up (Stephan & Levin, 1992)? These questions are disturbing for those who see in an aging America the "specter of decline" (Moody, 1988).

Some of these fears have foundation in fact. For instance, there is a common stereotype that older people take longer to learn new things, and this is a stereotype that turns out to be true. Compared with younger people, older people *do* tend to proceed more slowly in new learning situations, but slower speed is partly explained by lack of practice, differences in learning style, or motivation. In addition, reaction time tends to slow down with age—probably the result of "hardware" limits in the nervous system. By itself, chronological age doesn't explain much about learning ability. In any case, slower speed or reaction time usually isn't a factor in everyday performance. For example, short-term memory may weaken, but it usually doesn't seriously affect daily life.

Along with the stereotype of low creativity, there is a common assumption that older people overall are just plain bored, yet the Duke Longitudinal Study of Aging found that nearly 9 out of 10 respondents said they had never been bored in the previous week (Palmore, 1981). Another stereotype suggests that older people cannot adapt to change, yet a little reflection shows this stereotype to be wrong. Consider only the enormous changes that most people are likely to face in their later years, such as retirement, widowhood, adapting to chronic illnesses, and so on.

The debate about age, intelligence, and creativity is important for America's future. A number of gerontologists, perhaps with one eye on their own advancing years and the other on a changing society, have tried to determine what can be expected regarding cognitive changes that come with aging, especially whether creativity declines with age. They have faced a number of practical obstacles in their research, the most basic being an acceptable definition of *creativity.* Other types of cognitive function, notably intelligence and memory, have proved easier to pinpoint, although definitions of these cognitive functions are not without debate.

ELEMENTS OF COGNITIVE FUNCTION

Creativity has been related to intelligence, specifically, **fluid intelligence**, which is intelligence applied to new tasks or the ability to come up with novel or creative solutions to unforeseen problems (Horn, 1982). Some believe the key to fluid intelligence is divergent thinking, which is the ability to come up with lots of different ideas in response to a problem-solving challenge.

The other side of the coin is **crystallized intelligence**, which reflects accumulated past experience and the effects of socialization (Horn, 1982). Whereas fluid intelligence denotes a capacity for abstract creativity, crystallized intelligence may signify the acquisition of practical expertise in everyday life—in short, wisdom. Some components of wisdom have long been familiar. Philosophers going back to Socrates have argued that wisdom lies in a balanced attitude toward what we think we know: knowing what one does not know but, at the same time, refusing to be paralyzed by doubt (Meacham, 1990). Another key feature of wisdom would seem to be the ability to transcend bias or personal needs that may distort one's perception of a given situation (Orwoll & Perlmutter, 1990). Wisdom, then, involves more than cognitive development alone; it requires a degree of detachment and freedom from self-centeredness that has been described as *ego transcendence* (Peck, 1968).

Older people, if they develop a degree of detachment, might be in a position to achieve such wisdom. But, of course, no one has suggested that wisdom is a universal or inevitable result of chronological age alone. Something more is required than merely living a certain number of years, but psychologists do not agree about what that "something more" might be.

Some psychologists have wondered whether there is a trade-off between creativity and wisdom, with one declining while the other increases with advancing age. In this view, wisdom and creativity are seen in opposition to one another. Other psychologists argue that the cognitive processes involved in wisdom, intelligence, and creativity are all basically the same, but are put to different uses by different kinds of people. Wise people, we might say, have a high tolerance for ambiguity because they appreciate how difficult it is to make reliable judgments. They see the world "in depth." By contrast, the creative person seeks to go beyond whatever is given in the immediate environment to create something new.

Yet genuine creativity need not be identified with novelty for its own sake, as contemporary Western societies often do. In some societies of the East—for example, India, China, or Japan—old age is viewed as an appropriate time for spiritual exploration and artistic development. Late-life disengagement is balanced by opportunities for personal growth and creativity. "A Confucian in office, a Taoist in retirement," went the Chinese proverb, so retirement roles might include meditation or traditional landscape painting. In the Hindu doctrine of life stages, as well, later life was a period culminating in spiritual insight and wisdom.

What happens when a creative artist grows older and also develops a measure of wisdom applied to the creative process? Part of the answer may be found by looking at those creative artists who continued to be productive in old age. One of the greatest examples was the Dutch painter Rembrandt, whose style changed and deepened as he grew older. The aged Rembrandt practiced looser brushwork and became more preoccupied with the inner world of the people he painted. Another example is the impressionist Monet, who continued to paint his famous water lilies even after he was confined to his home in his 70s. Frail health also plagued the aging Matisse, who was forced to give up painting in favor of creating colored cardboard "cutouts" that distilled a lifetime of artistic experience into simple, powerful designs. It is as if the older artist is able to discard mere technical achievement in favor of some essential and elemental quality of art. We see a similar development of "late style" among poets such as Goethe and W. B. Yeats. All these examples suggest that, in the last stage of life, many of the greatest creative minds experience a change or a deepening of their creative style that could be attributed to an accumulation of wisdom.

The sources of creativity and productivity in later life are complex and result from many different factors. For example, most productive individuals produce both successes and failures; they have more successes than do less productive individuals partly because they have more failures as well. There is no law of fate that decrees that creativity must decline with age. Late-life creativity is unquestionably real, but it is far from universal, and it takes unpredictable forms. For example, it is well known that so-called late bloomers—such as the painter Grandma Moses—may attain the peak of their careers much later in life than others. What is known about creativity in later life suggests that individual differences in creative potential are so substantial that they largely go beyond the effect of aging (Simonton, 1998). Finally, we should note that many examples of creativity in later life focus on extraordinary creative older people—a composer like Verdi or a choreographer like Martha Graham. But there is also creativity and intelligence manifest in everyday life. So-called ordinary people can exhibit capacity for new modes of thinking and acting with innovation and creativity.

Self-Portrait by Rembrandt Harmensz. van Rijn, 17th Century

THE CLASSIC AGING PATTERN

Creativity in itself is difficult to define or measure, but psychologists have had long experience in measuring human intelligence. The **Wechsler Adult Intelligence Scale (WAIS)** is the most influential measure of global or general intelligence in use today. The WAIS includes verbal and performance scales, which are combined to assess IQ. The verbal part focuses on learned knowledge, including comprehension, arithmetic, and vocabulary; the performance part measures ability to solve puzzles involving blocks or pictures. As people grow older, their verbal scores on the WAIS tend to remain stable, but their performance scores tend to decline (Sattler, 1982). This persistent difference in performance on measures of the two components has been found so often that it is called the **classic aging pattern**.

Some leading researchers have cautioned against taking the classic aging pattern too seriously. They question what is actually being measured by IQ tests. In other words, they challenge the very validity of IQ tests as a measure of the "real" intelligence of older adults. Perhaps test performance should not be equated with real differences in intelligence at all, they say. This controversy has a familiar ring. It is the same kind of challenge that has been heard about the use of IQ tests and Scholastic Aptitude Tests when those tests show poorer scores for some minority groups. Critics argue that "intelligence" is a more complex, multi-dimensional capacity than the tests measure (Gardner, 1985).

The evidence certainly indicates that age and intelligence have a complex relationship. In a test of basic memory skills of young and older adults, the average 70-year-old will take three or four times longer than a 20-year-old to identify a mental picture linking a word and a location and will tend to make more mistakes. Things are completely different when we test people for knowledge transmitted across generations through culture, however. Older people do well on language skills as well as knowledge about how to handle life's ups and downs. For example, when presented with a difficult hypothetical dilemma, older adults score much better than younger adults. How, then, do we develop a valid measure of late-life intellectual ability?

MEASURES OF LATE-LIFE INTELLIGENCE

Interest in the validity problem, or the problem of measuring "real" intelligence, has helped stimulate psychologists to ask whether any positive cognitive developments come with age. The long debate has at least confirmed that conventional methods of measuring intellectual abilities have not always acknowledged the skills used by adults in coping with the demands of everyday life. As a result, some psychologists have become interested in devising new approaches and methods, such as an age-relevant intelligence test.

Tests to measure the relation of wisdom or creativity to age are seeking to capture something elusive. Everyday intelligence is a multidimensional capacity involving more than logic or information processing alone. Everyday intelligence—what we sometimes call "common sense"—involves pragmatic or social judgment, which is more than abstract reasoning (Cornelius, 1990). What is involved is something akin to *everyday problem solving* (Cornelius & Caspi, 1987) or *expertise in life planning*. Some of these same cognitive capacities are evident in what we call "wisdom." The wisdom of later life probably includes several distinct but interrelated attributes—reflective judgment in the face of uncertainty, "problem finding" (as opposed to solving an already given problem), integrated thought about one's life, and intuition (i.e., the empathic ability to understand a concrete situation). These qualities are obviously difficult to measure on a test.

Paul Baltes, perhaps the leading psychologist investigating wisdom today, has tried to develop a psychological test to measure wisdom. Baltes and his associates presented adult test subjects with questions such as this one: "A 14-year-old girl is pregnant. What should she, what should one, consider and do?" In scoring the test, Baltes was not looking for any specific answer, but instead was trying to measure how wise people go about dealing with difficult questions. Not all older people are wise, but more than half of the top responses on Baltes's (1992) "wisdom test" came from people beyond 60 years of age.

Baltes went on to define wisdom as an expert knowledge system derived from experience and capable of dealing with pragmatic problems (Baltes & Staudinger, 1996). That definition is similar to the commonsense understanding of wisdom as consisting of good judgment in response to uncertain problems of living. If we follow this approach, we can understand why wisdom, potentially at least, might increase with age. The reason goes back to the distinction between fluid intelligence, which operates by the mechanics of information processing, and the content-rich, pragmatic knowledge of crystallized intelligence that is honed through long-life experience.

Exhibit 2   Percent of People 55 Years and Over by Educational Attainment, Age, and Sex: 2002

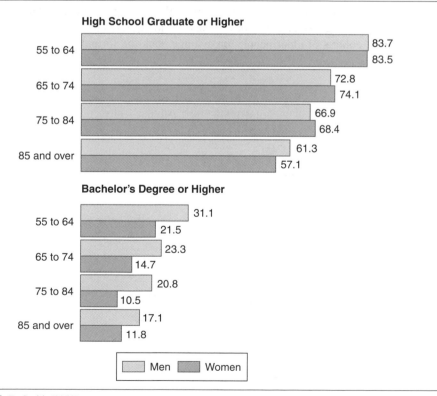

Steps toward defining or measuring wisdom are still in the early stages, but the effort holds promise. Research on the aging mind has moved from a simple view of growth versus decline to a more complex assessment of potentialities and limits. The cognitive mechanics of the computer—information processing—can be compared with fluid intelligence, which is biologically based and tends to decline with age. In contrast, cognitive pragmatics—factual knowledge and problem solving—can grow with age and can compensate for losses in processing power. We could say that with advancing age hardware declines, whereas software becomes enriched (Baltes, 1993).

STUDIES OF AGE AND COGNITIVE FUNCTION

Different research methodologies have been used to measure cognitive changes associated with aging. Cross-sectional studies look at groups of young and old people at a single point in time, and longitudinal studies follow subjects over many years. Optimists on the subject of creativity and age point out that cross-sectional studies of intelligence may be revealing differences that come not from age but from characteristics and experiences of different cohorts (Dixon, Backman, & Nilsson, 2004).

For instance, young people taking IQ tests tend to be quite familiar with test taking from recent experience in school. As a group, they show far less test anxiety than do older people (Whitbourne, 1976). Furthermore, many older people have internalized ageist attitudes and believe that with advancing age, intelligence—and especially memory ability—inevitably declines. Older people also tend to be more cautious than younger people, and thus they may be more reluctant to guess at the right answers on an IQ test (Birkhill & Schaie, 1975). Finally, current cohorts of older people, on average, lack the formal schooling enjoyed by younger age groups.

Exhibit 3   Education Participation Among Older Adults: 2005

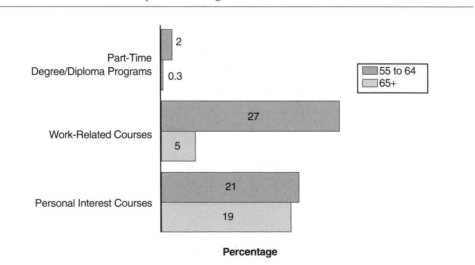

SOURCE: U.S. Department of Education, National Center on Education Statistics, 2007.

Given the tendency of cross-sectional studies to overestimate the impact of chronological age, longitudinal studies make sense. One of the most extensive sources of knowledge about intelligence and aging comes from the Seattle Longitudinal Study, which followed individuals ranging in age from 25 to 81 years over two decades (Schaie, 1996). That investigation and others have found that the steepest average intellectual declines come after age 60. Averages conceal large differences among individuals, but even on longitudinal studies, the classic aging pattern emerges. Still, research findings do challenge the idea of inevitable, global intellectual decline for all individuals. Even more important, intellectual decline in older people may be halted or reversed by specific interventions, such as training and education (Boron et al., 2007). These findings suggest that intellectual decline in later life is by no means irreversible or inevitable.

Indeed, longitudinal studies show that successive cohorts of older people are, in fact, improving their performance on intelligence tests, perhaps reflecting higher educational attainment. In addition, anywhere from 60% to 85% of those tested maintain their scores over time or even improve specific abilities. Among those over age 80, only between 30% and 40% of participants in the Seattle study had declining scores.

These studies indicate that few people show any global decline in intelligence as they age, suggesting that people can optimize their cognitive functioning by drawing on their strengths or compensating for losses. Perhaps most important, even in their 80s and 90s, people tend to remain quite competent in familiar everyday situations. Both cross-sectional and longitudinal studies, however, do show the classic aging pattern, with uniform decline among subjects beyond their 70s.

Studies of creativity, as opposed to cognitive function in general, have been more difficult to conduct. Again, the problem is defining creativity.

Both cross-sectional and longitudinal studies, using many different kinds of tests, have shown that divergent thinking does decline with advancing age, and the decline is not attributable simply to reductions in speed of response (McCrae, Arenberg, & Costa, 1987). The way that creativity is measured in this study is not completely satisfactory. However, in his classic study of creativity and aging, Harvey Lehman (1962) used a public consensus approach instead. First, he recognized public consensus about products that clearly demonstrate superior creativity—for example, Mozart's symphonies, Newton's theory of gravity, or Edison's invention of new electrical devices. Lehman found that the curves of publicly acknowledged creativity followed exactly the curves of fluid intelligence: They both peaked after age 30 and declined with each subsequent decade.

Wayne Dennis (1966), a critic of Lehman's work, looked at different data and found that for most people, the decades of the 40s and 50s were the most productive period. Dennis's conclusions were based on quantitative measures of productivity, however (e.g., how many publications), not on qualitative measures (how important the contribution was). Therefore, Dennis's results do not actually refute Lehman's findings.

Still other investigators measuring scientific creativity found that productivity among scientists peaked in the early 40s—later than Lehman said—and then declined slowly after age 50 (S. Cole, 1979; A. Diamond, 1986). A longitudinal study of creativity among mathematicians found that those who published a great deal when young did continue to publish as they became older, at least through middle age. An important question about these findings has to do with whether there are specific styles of creative production in different professions or fields, such as scientific versus artistic fields, and, thus, whether professional creativity is a uniform phenomenon that can be studied one way.

The evidence thus shows that age does not necessarily mean loss of cognitive function. Nevertheless, performance on intelligence tests does decline. Psychologists speculating about the reasons cite strong evidence that declining speed with advancing age does have a negative effect on performance on intelligence tests, but the precise reasons remain unclear (Salthouse, 1985a). Aging is, in fact, accompanied by a clear loss in **cognitive reserve capacity**—that is, the degree of unused potential for learning that exists at any given time. Studies of reaction time in training also show that the speed of information processing definitely declines with age. Older adults, for instance, do not reach the same peak of performance in reaction time as do younger adults (Salthouse, 1985b), nor do older people achieve comparable performance when trained in memory skills (P. B. Baltes & M. M. Baltes, 1990). Still other psychologists distinguish between maintained and vulnerable cognitive function, once again underscoring the importance of the concept of reserve capacity (A. Kaufman, 2001).

Optimists counter that, although fluid intelligence abilities decline with age, crystallized abilities tend to increase. In addition, declines in cognitive ability among older people can often be compensated for by the expertise acquired with aging, a phenomenon that has been called *decrement with compensation.* In other words, wisdom and pragmatic knowledge compensate for declines in speed or fluid intelligence. For instance, despite declines in typing speed, some older typists demonstrate superior typing productivity. They apparently compensate for loss of speed by reading farther ahead in the manuscript they are typing, which is a pragmatic response demonstrating knowledge of how to type more effectively (Salthouse, 1984).

CORRELATES OF COGNITIVE STABILITY

The debate about the causes and meaning of the measurable decline in IQ scores with age comes down to a difference between those who think of themselves as "realists" and those who take a more optimistic view. On the optimistic side, some psychologists speak of the "myth of the twilight years." They suggest that intelligence actually need not decline in later life at all (P. B. Baltes & Schaie, 1974). But other, equally expert psychologists bitterly reject this conclusion (Horn & Donaldson, 1977). These realists contend that declines in fluid intelligence in the classic aging pattern are empirical facts to be accepted, no matter how unpleasant. Although we might find individuals who do not exhibit the pattern, the realists insist, such cases do not refute an overall decline in average performance.

Taking another tack, the optimists have explanations other than chronological age for the classic aging pattern. One possible factor could be ill health, which does become more frequent with aging, although not universally so. Studies reveal consistent differences in IQ test performance depending on even modest declines in health status. Poor health and disability also tend to cause retirement and therefore probably weaken learning opportunities. Note, then, that both biological changes, such as health status, and social changes, such as retirement, may be responsible for changing cognitive abilities. It may be possible to change these biosocial factors to such a degree that the classic aging pattern no longer holds true.

The ability to adapt or compensate for decrements in cognitive function is probably related to cognitive style or personality. According to some psychologists studying personality, basic personality dispositions include traits such as being neurotic, extroverted, open to experience, and conscientious. These dispositions predict how people adapt to changing life circumstances. Surprisingly, according to this model, basic personality and temperament change little after the age of 30 (Costa & McCrae, 1980; McCrae & Costa, 1990). Longitudinal studies show that personality is stable throughout adulthood, even in response to health problems, economic setbacks, and bereavement (Costa, Metter, & McCrae, 1994).

However, psychological characteristics over the life span do not emerge entirely from the isolated individual. Behavior often reflects social conditions and socially structured transitions in the life course (Schooler & Schaie, 1987). For example, retirement may boost the cognitive performance of people who retire from routine or boring jobs, but accelerate cognitive decline for those who have held complex jobs. In addition, some psychological traits can be intensified by life course transitions. For instance, middle-aged people with flexible attitudes are less likely to experience a decline in psychological competence as they grow older than are those who could be described as cognitively rigid (Schaie, 1984).

We should thus be skeptical of any broad generalizations or unqualified claims about either the decline or the stability of intelligence with aging. Experiments in training have shown that declines in intellectual functioning among older people can be reversed. In the Seattle Longitudinal Study, investigators found that 40% of participants who showed a decline in mental abilities benefited from training; following training, they achieved intelligence scores at least as high as those measured at the beginning of the 14-year study (Cunningham & Torner, 1990). Critics question, however, whether the reversal reflects practice or a genuine reversal of changes induced by aging.

Despite the criticism, psychological studies with older people have demonstrated that intelligence, defined as the ability to think and learn new things, has a great measure of plasticity or potential for growth even at advanced ages. Results from studies of groups of healthy people between ages 60 and 80 demonstrate that they benefit from practice and show performance gains just as younger people do. One series of studies showed that older adults could even be trained to become much more skilled than before at short-term memory tasks: In effect, they could become *memory experts* (P. B. Baltes & M. M. Baltes, 1990). When older people are stimulated and intellectually challenged, this capacity for learning and remembering is impressive.

URBAN LEGENDS OF AGING

"We lose a million neurons every day."

This familiar nugget of ageism is now long disproven by neuroscience. It's true that we're always losing, and gaining, neural connections each day, and it's also true that the brain shrinks with age. But research by scientists such as Marian Diamond and others demonstrated the remarkable neuroplasticity of the human brain into advanced age. Even if you were losing a million neurons a day, it would take centuries to lose your mind.

CREATIVITY IN AN AGING POPULATION

The experiments discussed suggest that the debate about the effect of aging on creativity and intelligence is by no means settled. The readings that follow represent the classic positions in this debate. The selection by Harvey Lehman gives some of the data from his public consensus studies and provides Lehman's major conclusions. Wayne Dennis, one of Lehman's strongest critics, attacks the claim that creativity declines with age. More recent and perhaps more balanced perspectives on aging and creativity are provided by the readings from Carstensen and Cohen, as well as Levy and Langer.

These discussions of wisdom and aging should remind us of how little we know about what is possible in old age. It was during the 20th century that we first saw gains in longevity on a massive scale. Only in recent decades have substantial numbers of people experienced old age in relatively good health and with high levels of education. Therefore, studies of older people in previous decades may not be a good basis for judging what older people are capable of today or in the future.

We are left to take hope from examples of individual achievement in the past. For example, a number of well-known creative artists have made outstanding contributions in their old age. At age 71, Michelangelo was named chief architect of St. Peter's in Rome. Titian painted some of his greatest works in his 80s, and Picasso produced drawings and paintings into his 90s. Martha Graham continued her choreography into her 80s, and Jessica Tandy won an Oscar at age 80. These are not isolated examples. In fact, recent research has confirmed that creativity can continue into the later years, and that, for some at least, higher-order mental abilities such as creativity and wisdom can flourish late into life (Lindauer, 2003).

With improving opportunities to practice the arts and pursue lifelong learning, tomorrow's elders could take up the challenge of creativity and wisdom in the later years in ways unimagined today. The old age once reserved for elites could become an opportunity for all to grow in later life. As art critic Ananda Coomaraswamy put it, it is not that the artist is a special kind of person, but rather it is that each person is a special kind of artist. Viewed in those terms, the real debate about age and creativity has barely begun.

Georgia O'Keeffe, an original and distinctively American painter, continued painting until the end of her long life.

FOCUS ON PRACTICE

Older-Adult Education

Increasingly, education is not limited to the first stage of life, but is instead extended over the life course. One obstacle to late-life education, however, is a stereotype that, by later life, people are too old to learn. Sometimes older people accept the stereotype, but we have seen that continued involvement in learning helps to maintain the ability to learn.

Today's opportunities for late-life learning are more plentiful than ever before (Findsen, 2005). Along with organized educational programs, many informal opportunities for older people also abound. One example of a successful program is Elderhostel (now called Road Scholar), founded in 1975 as a summer residential college program for people over age 55. It offers noncredit courses in the liberal arts and now attracts more than 200,000 participants each year at 1,000 campuses around the United States and in 70 countries overseas. Elderhostel involves no homework, papers, or grades. But it does offer an opportunity for low-cost travel and an intellectual challenge for those interested in learning.

For those who do not want to travel to another community, tuition-free space-available courses are offered at most public universities. In addition, a national survey of community colleges showed that up to a quarter of two-year institutions provide some offerings for older adults, mostly in the areas of personal financial planning, health, and life enrichment (e.g., arts and humanities, exercise, and nutrition) and contemporary civic or political issues (Ventura-Merkel & Doucette, 1993).

Still another approach is the local Institute for Learning in Retirement, where retired people with special skills or knowledge teach courses to one another. This mutual-aid model has been replicated in more than 300 communities around the United States and is now sponsored by Elderhostel. In the Scandinavian countries, France, Spain, and other countries, older people have created similar "Universities of the Third Age" affiliated with institutions of higher education.

In the future, we can expect that older-adult education will increase substantially. One reason is the rising level of prior education among successive cohorts of older people. Previous education is the best predictor of interest in life-long learning. The median level of education for people over age 65 in the year 1900 was only 8 years, whereas by the 1980s it had risen to 12 years. Between 1970 and 1994, the proportion completing high school rose from 28% to 62%. In the past, younger people had comparatively higher levels of education, but today Americans over age 65 have approximately as many years of schooling as the general adult population.

The overall market for adult education has grown substantially in recent decades, and this growth has included older people. For example, the percentage of those ages 66 to 74 who took at least one adult education class in the previous year more than doubled—from 8.4% in 1991 to 19.9% in 1999, according to data from the National Household Education Survey. The same survey found that in 2001, 22% of those ages 66 or above participated in some kind of formal adult education during the past 12 months. The number of older people participating in adult education courses is growing rapidly, as are the numbers of people over 50 participating in higher education seeking degrees (American Council on Education, 2008). The expanding population of educated adults should make lifelong learning even more appealing over the entire life course.

GLOBAL PERSPECTIVE

Universities of the Third Age

Universities of the Third Age are local mutual-aid learning groups where retired people pursue lifelong learning activities. The University of the Third Age (U3A) idea was originally based at the University of Toulouse in France, but the original academic model has since spread to many countries in Europe and beyond. By the early 1980s, the U3A model appeared in Britain, where it operates more in the spirit of mutual self-help. U3As have now appeared in countries as distant as Canada, New Zealand, and India. In the United States, a comparable program exists in the form of Institutes for Learning in Retirement and Osher Lifelong Learning Institutes.

Since its beginning, the U3A has become a global phenomenon and comprises, literally, thousands of locally developed educational programs of remarkable variety. Following the British approach, learning groups in the U3A are conducted by leaders drawn from the ranks of retirees with a lifetime of experience and specialized expertise. Learning groups are largely run by volunteers and may have only a loose affiliation with a local university. Groups may focus on liberal arts subjects but are also organized around leisure-time activities such as travel and the creative arts.

At one level, U3As may appear to be simply clubs for culturally oriented older people, but others have seen the U3A as a distinctive contribution to our idea of later-life learning: in particular, the idea of learning from life experience in contrast to what some have termed a "banking model" of education as the transfer of skills or information. The renowned British historian Peter Laslett in his own later years became a leader of the U3A movement and presented some of these ideas in his book, *A Fresh Map of Life: The Emergence of the Third Age* (1991).

SOURCES: Midwinter, E., *Mutual Aid Universities: A Self-Help Approach to Educating Older People*, Routledge, 1984; Moody, H. R., "Structure and Agency in Late-Life Learning," in E. Tulle (ed.), *Old Age and Agency*, Nova, 2004.

READING 11

Age and Achievement

Harvey Lehman

What are man's most creative years? At what ages are men likely to do their most outstanding work? In 1921, Professor Robert S. Woodworth, of Columbia University, published this statement in his book, *Psychology: A Study of Mental Life:* "Seldom does a very old person get outside the limits of his previous habits. Few great inventions, artistic or practical, have emanated from really old persons, and comparatively few even from the middle-aged. . . . The period from twenty years up to forty seems to be the most favorable for inventiveness" (p. 519). . . .

Assuming that the method by which one arrives at a conclusion is no less important than is the conclusion itself, let us see what is found when the inductive method is employed in the study of man's most creative years. Let us first examine the field of creative chemistry and attempt to answer the question whether chemists display more creative thinking at some chronological age levels than at others.

In his book, *A Concise History of Chemistry* . . . , Professor T. P. Hilditch, of the University of Liverpool, presents the names of several hundred noted chemists and the dates on which these chemists made their outstanding contributions to the science of chemistry. . . .

When the birth dates of the chemists listed by Hilditch were ascertained, insofar as data were available, it was possible to determine the ages at which the world's most renowned chemists made their most significant contributions, both theoretical and experimental, to the science of chemistry. A sample of the findings is set forth graphically in Figure 1.

Figure 1   Average Number of Contributions by Chemists During Each Five-Year Interval of Their Lives

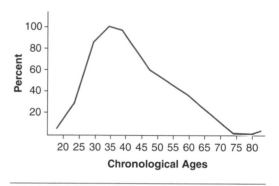

NOTE: Based on 993 significant contributions by 244 chemists now deceased.

Figure 1 presents, by five-year intervals, the chronological ages at which 244 chemists (now deceased) made 993 significant contributions to the science of chemistry. In studying Figure 1 it should be borne in mind that it sets forth the average number of chemical contributions per five-year intervals. Full and adequate allowance is thus made for the larger number of youthful workers. . . .

Figure 2 presents the ages at which 554 notable inventions were made by 402 well-known inventors. . . . When Figure 2 was displayed to interested friends and colleagues, several persons immediately said, "What about Edison?" It is, of course, well-known that Thomas A. Edison was very active as an

SOURCE: Excerpted from *Age and Achievement* by Harvey Lehman. Princeton, NJ: Princeton University Press, 1953. Reprinted by permission of the American Philosophical Society.

inventor throughout his entire life. Figure 3 reveals, however, that 35 was Mr. Edison's most productive age. Moreover, during the four-year interval from 33 to 36, Edison took out a total of 312 United States patents. This was more than a fourth (28 per cent) of all the United States patents taken out by him during an inventive career that lasted for more than 60 years. . . .

Figure 2 Average Number of Practical Inventions During Each Five-Year Interval of the Inventors' Lives

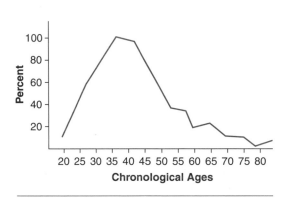

NOTE: Based on 554 inventions from 402 inventors now deceased.

Figure 3 Age Versus Inventions Patented in the U.S.A.

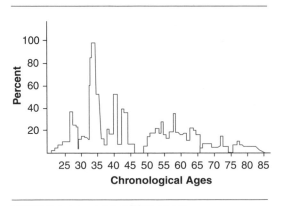

NOTE: Based on a total of 1,086 patents.

The shape of a performance age-curve varies with a number of things: (1) the type of performance, (2) the excellence of the performance, and (3) the kind of measurement employed. This last fact can perhaps best be illustrated by use of an analogy. Thus, one might construct an age-curve setting forth the average ability of individuals within each of the several age-groups to do the ordinary high jump. At almost every age level some persons would be found who are more or less able to perform this feat. One might, therefore, test out large numbers at each age level and with the resultant data it would be quite possible to construct age-curves disclosing the *average* height that could be attained by the members of each age group.

But there are several other possible procedures which might be employed for comparing the several age groups. Thus, within each of the age groups, one might ascertain the per cent of individuals able to high-jump six feet, the per cent able to high-jump five feet, etc. With the obtained data it would then be possible to construct one curve that would show for each age group the per cent of individuals able to do six feet, another curve showing the per cent able to do five feet, and so on. If a number of these curves were to be constructed, it seems obvious that that curve which set forth age differences in the ability to do six feet would start its rise later and would fall off both earlier and much more rapidly than would another curve showing age differences in the ability to do, say, two feet. It is evident that very superior high jumping is likely to occur during a narrower age-range than would be found for a much lower degree of ability.

If we think in terms of actual performance, the foregoing situation seems to exist in such diverse fields of endeavor as athletics, mathematics, invention, science, chess, the composition of enduring music, and the writing of great books. For each of these types of behavior, very superior achievement seems most likely to occur during a relatively narrow age-range, and the more noteworthy the performance, the more rapidly does the resultant age-curve descend after it has attained its peak. The findings with . . . reference to sculptured works, oil paintings, and etchings suggest similarly that there is an optimal chronological age level for superlatively great success within these particular fields also. . . .

The work of the genius in his old age may still be far superior to the best work that the average man is able to do in his prime. Therefore, for the study of age differences in creativity, it is not valid merely to compare the achievements of the aged genius with the more youthful accomplishments of the average person. If one wishes to ascertain when men of genius have done their very best work, it is necessary to compare the earlier works of men of genius with their own later works. . . .

Sculpture. Effort was made to ascertain the ages at which the most noted sculptors of early Greece executed their most famous works, but this information could not be obtained. Data for Figure 4 were found in Lorado Taft's *The History of American Sculpture* . . . , which attempts to list the best works of the most famous American sculptors. It seems safe to assume that Taft's list contains no age bias. From his book the dates of execution were found for 262 sculptured works by 63 sculptors now deceased. For these 262 works, Figure 4 sets forth the average number executed during each five-year interval of the artists' lives. . . .

Figure 4   Age Versus Famous Sculpture

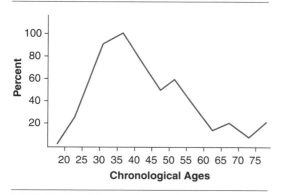

NOTE: Based on 262 works by 63 sculptors.

By means of statistical distributions and graphs [we] show the ages (1) at which outstanding thinkers have most frequently made (or first published) their momentous creative contributions, [and] (2) at which leaders have most often attained important positions of leadership. . . .

The most notable creative works of scientists and mathematicians were identified by experts in the various specialized fields of endeavor. For such fields as oil painting, education, philosophy, and literature, a consensus of the experts was obtained by a study of their published writings. In each field listed below the maximum average rate of highly superior production was found to occur not later than during the specified range of ages. For example, item 1 of this list, chemistry, 26–30, is to be interpreted as follows: in proportion to the number of chemists that were alive at each successive age level, very superior contributions to the field of chemistry were made at the greatest average rate when the chemists were not more than 26–30. The remaining items here and those in the tabular lists that follow are to be interpreted in similar manner.

Physical Sciences, Mathematics, and Inventions:

1. Chemistry, 26–30

2. Mathematics, 30–34

3. Physics, 30–34

4. Electronics, 30–34

5. Practical Inventions, 30–34

6. Surgical Techniques, 30–39

7. Geology, 35–39

8. Astronomy, 35–39

Biological Sciences:

9. Botany, 30–34

10. Classical Descriptions of Disease, 30–34

11. Genetics, 30–39

12. Entomology, 30–39

13. Psychology, 30–39

14. Bacteriology, 35–39

15. Physiology, 35–39

16. Pathology, 35–39

17. Medical Discoveries, 35–39 . . .

For most types of superior music, the maximum average rate of good production is likely to occur in the thirties. Here are the maxima.

18. Instrumental Selections, 25–29

19. Vocal Solos, 30–34

20. Symphonies, 30–34

21. Chamber Music, 35–39

22. Orchestral Music, 35–39

23. Grand Opera, 35–39

24. Cantatas, 40–44

25. Light Opera and Musical Comedy, 40–44

For the study of literary creativity, fifty well-known histories of English literature were canvassed. The works most often cited by the fifty literary historians were assumed to be superior to those cited infrequently. Best-liked short stories were identified similarly by use of 102 source books, and "best books" were ascertained by study of a collation of fifty "best book" lists. As is revealed by the following tabulation, literary works that are good and permanently great are produced at the highest average rate by persons who are not over 45 years old. It is clear also that most types of poetry show maxima 10 to 15 years earlier than most prose writings other than short stories.

26. German Composers of Noteworthy Lyrics and Ballads, 22–26

27. Odes, 24–28

28. Elegies, 25–29

29. Pastoral Poetry, 25–29

30. Narrative Poetry, 25–29

31. Sonnets, 26–31

32. Lyric Poetry, 26–31

33. Satiric Poetry, 30–34

34. Short Stories, 30–34

35. Religious Poetry (Hymns), 32–36

36. Comedies, 32–36

37. Tragedies, 34–38

38. "Most Influential Books," 35–39

39. Hymns by Women, 36–38

40. Novels, 40–44

41. "Best Books," 40–44

42. Best Sellers, 40–44

43. Miscellaneous Prose Writings, 41–45 . . .

Although the maximum average rate of output of the most important philosophical books occurred at 35–39, the total range for best production extended from 22 [to] 80, and for mere quantity of output—good, bad, and indifferent—the production rate was almost constant from 30 [to] 70. . . .

A very large proportion of the most renowned men of science and the humanities did their first important work before 25, and . . . in general the earlier starters contributed better work and were more prolific than were the slow starters. . . .

For most types of creative work the following generalizations have been derived. Within any given field of creative endeavor: (1) the maximum production rate for output of highest quality usually occurs at an earlier age than the maximum rate for less distinguished works by the same individuals; (2) the rate of good production usually does not change much in the middle years and the decline, when it comes, is gradual at all the older ages—much more gradual than its onset in the late teens or early twenties; (3) production of highest quality tends to fall off not only at an earlier age but also at a more rapid rate than does output of lesser merit. . . .

Item 62 in the following tabulation shows that, in proportion to the number of men who were still alive at each successive age level, presidents of American colleges and universities have served most often at 50–54. The other items in this tabulation are to be interpreted similarly.

62. Presidents of American Colleges and Universities, 50–54

63. Presidents of the U.S. Prior to Truman, 55–59

64. U.S. Ambassadors to Foreign Countries From 1875 to 1900, 60–64

65. U.S. Senators in 1925, 60–64

66. Men in Charge of the U.S. Army From 1925 to 1945, 60–64

67. Justices of the U.S. Supreme Court From 1900 to 1925, 70–74

68. Speakers of the U.S. House of Representatives From 1900 to 1940, 70–74

69. Popes, 82–92

An analysis of age data for the most highly successful athletes reveals that their modal ages differ less from the norms for intellectual proficiency than is commonly supposed. The following comparisons are illustrative.

70. Professional Football Players, 22–26

71. Professional Prizefighters, 25–26

72. Professional Ice Hockey Players, 26

73. Professional Baseball Players, 27–28

74. Professional Tennis Players, 25–29

75. Automobile Racers, 26–30

76. Leading Contestants at Chess, 29–33

77. Professional Golfers, 31–36

78. Breakers of World Billiards Records, 31–36

79. Winners at Rifle and Pistol Shooting, 31–36

80. Winners of Important Bowling Championships, 31–36 . . .

When seven groups of earlier-born athletic champions were compared with seven groups of those more recently born, the field of sport being kept constant in each comparison, the later-born were found to be older than the earlier-born. The changes that have taken place in the modal ages of creative thinkers, leaders, and athletes all evidence the fact that these modal ages are not due solely to genetic factors. Whether the modal ages will continue to change and whether they can be subjected to some kind of human control are quite different questions.

A mere increase in man's longevity should not change greatly the modal ages at which man exhibits his greatest creative proficiency since, both for long-lived and for short-lived groups, the modal age occurs in the thirties. . . .

POSSIBLE CAUSES FOR THE EARLY MAXIMA IN CREATIVITY

At present we are in no position to explain these curves of creativity that rise rapidly in early maturity and then decline slowly after attaining an earlier maximum. Undoubtedly multiple causation operates in these complex behaviors and no discovered contributing condition is likely to be of itself a sufficient or necessary cause. Nevertheless, it is profitable here to list sixteen of the factors which have been suggested as contributing to these representative functions with their early maxima, for such factors indicate possible lines for further research. Here is the list.

(1) A decline occurs prior to 40 in physical vigor, energy, and resistance to fatigue. This decline is probably far more important than such normal age changes as may occur in adult intelligence prior to outright senility.

(2) A diminution in sensory capacity and motor precision also takes place with advance in age. For example, impaired vision and hearing handicap the older individual in many cumulative ways, and writing by hand also becomes more difficult with advance in age.

(3) Serious illness, poor health, and various bodily infirmities more often influence adversely the production rates of older than of younger age groups.

(4) Glandular changes continue throughout life. It is conceivable that hormone research may some day reveal a partial explanation for the changes and especially for the early maxima.

(5) In some instances unhappy marriages and maladjustment in the sex life, growing worse with advance in age, may have interfered with creative work.

(6) The older age groups, more often than the younger, may have become indifferent toward creativity because of the death of a child, a mate, or some other dear one.

(7) As compared with younger persons, older ones are apt to be more preoccupied with the practical concerns of life, with earning a living, and with getting ahead.

(8) Less favorable conditions for concentrated work sometimes come with success, promotion, enhanced prestige, and responsibility.

(9) In some cases the youthful worker's primary ambition may not have been to discover the unknown or to create something new but to get renown. Having acquired prestige and recognition, such workers may try less hard for achievement.

(10) Too easy, too great, or too early fame may conceivably breed complacency and induce one to rest on his previously won laurels before he has done his best possible creative work.

(11) Some older persons may have become apathetic because they have experienced more often the deadening effect of non-recognition and of destructive criticism.

(12) As a result of negative transfer, the old generally are more inflexible than the young. This inflexibility may be a handicap to creative thinking, even though it is dependent on erudition.

(13) Perhaps in part because of the foregoing factors, some older persons experience a decrease in motivation which leads to a weaker intellectual interest and curiosity.

(14) Younger persons tend to have had a better formal education than their elders, they have grown to maturity in a more stimulating social and cultural milieu, and they have had less time to forget what they have learned.

(15) In some few cases outright psychosis has clouded what was previously a brilliant mind. Psychoses occur more often in the latter half of the normal life span.

(16) In other extreme cases, the individual's normal productive powers may have been sapped by alcohol, narcotics, and other kinds of dissipation. Here, as elsewhere, it is difficult to separate cause from effect. . . .

Upon the basis of all these statistics what is one to conclude? Whatever the causes of growth and decline, it remains clear that the genius does not function equally well throughout the years of adulthood. Superior creativity rises relatively rapidly to a maximum which occurs usually in the thirties and then falls off slowly. Almost as soon as he becomes fully mature, man is confronted with a gerontic paradox that may be expressed in terms of positive and negative transfer. Old people probably have more transfer, both positive and negative, than do young ones. As a result of positive transfer the old usually possess greater wisdom and erudition. These are invaluable assets. But when a situation requires a new way of looking at things, the acquisition of new techniques or even new vocabularies, the old seem stereotyped and rigid. To learn the new they often have to unlearn the old and that is twice as hard as learning without unlearning. But when a situation requires a store of past knowledge then the old find their advantage over the young.

Possibly every human behavior has its period of prime. No behavior can develop before the groundwork for it has been prepared, but in general it appears that the conditions essential for creativity and originality, which can be displayed in private achievement, come earlier than those social skills which contribute to leadership and eminence and which inevitably must wait, not upon the insight of the leader himself, but upon the insight of society about him.

READING 12

—————————— Age and Achievement ——————————

A Critique

Wayne Dennis

The recent book by Lehman,[1] *Age and Achievement,* seems to indicate that in many fields relatively little creative work of importance is done by persons past 45 or 50 years of age. This generalization does not hold in all fields of creativity, but the preceding sentence expresses Lehman's most striking finding.

That the production of first-rate work in poetry, art, science, and other creative areas decreases markedly with age is a matter of prime importance. If correct, it suggests that the creative worker in many fields should plan for early superannuation. If the conclusion drawn by Lehman is erroneous, the impression which it has created should be corrected with dispatch, for a conviction that early deterioration is inevitable may itself have deleterious consequences. Clearly the relationship of age to achievement is a topic in regard to which conclusions should be drawn with extreme care.

It is the thesis of this [essay] that much of the apparent decline in creative achievement revealed by Lehman's tables and graphs is due to factors other than age. We believe Lehman's data give a spurious appearance of age decrement in creativity.

Let us note first that the studies presented by Lehman are so numerous and so varied that it is difficult to do justice to them in a brief recapitulation. However, it is not incorrect to say that Lehman has been interested primarily in determining the 5 or 10 year age-period in which important creative works have most often been produced. The first step in this procedure, typically, consists in identifying important works in some field. To avoid introducing a bias of his own, he always uses a list of works drawn up by some other person. Lehman then determines the age at which each item was produced. He has done this for many creative fields, including mathematical discoveries, contributions to chemistry, lyric poems, and operas, to mention only a few. The first six chapters of his book are devoted to presenting the results of these analyses.

The graphs in these chapters almost all indicate that the production of outstanding works rises to a peak relatively early in the adult years and then declines. The age at which the peak of productivity is reached varies from field to field. It is as early as ages 22–26 for lyrics, ballads, and odes, and as late as 40–45 for novels, metaphysics, and miscellaneous prose writings. However, for a considerable number of fields the top rates for the production of outstanding works occur between ages 30 and 39.

Many aspects of these curves are worthy of attention, but we are concerned chiefly with the decrements which follow the peaks. In most instances, as presented by Lehman, the decrements are very striking. For example, . . . at ages 40–45 chemists produce, per man [*sic*], only one

SOURCE: "Age and Achievement: A Critique" by Wayne Dennis, *Journal of Gerontology*, 2(3): 331–333, 1956. Copyright © 1956 The Gerontological Society of America. Reprinted by permission. Publication of this article was supported by a grant from the Forest Park Foundation to the Journal of Gerontology.

half as many significant contributions as they produce between ages 30–35. By ages 60–65 their rate of production is only 20 per cent of their peak rate. Other graphs give very similar data for other sciences. The fine arts also show a severe decrement. For example, . . . by ages 45–50 the production of orchestral music judged to be of highest merit is only about 10 per cent as great as it was 10 years earlier. By ages 55–60 the composition of orchestral music of high quality decreases to 20 per cent of the maximum rate.

Examination of such findings, page after page, creates an impression of inevitable decline. If these charts are taken at their face value, we must conclude that in most kinds of creative work the output of work of first-rate quality is greatly reduced after the thirties.

But should these charts be taken at their face value? Let us consider this question.

A major methodologic weakness in Lehman's treatment of data lies in the fact that in most instances a table or graph combines information pertaining to men of different degrees of longevity. Thus a table usually presents data for men nearly all of whom reached age 30, but only part of whom attained age 40, and still fewer of whom completed half a century of life. To equate for differences in numbers of subjects at different ages, Lehman found the mean number of important contributions per person for persons surviving each decade. We shall attempt to show that this method of treating data acts in part to produce the productivity differentials which Lehman discovers.

Let it be noted that each man whose record is used by Lehman is required to produce only one important work in order to qualify for inclusion. In most lists of outstanding works used by Lehman, each individual contributes one, or only a few items. In his collections of data, the mean number of contributions per man is often only two or three. Furthermore, the mean number of "significant" contributions per man is only slightly greater for the men who lived to age 70 than it is for the men who died relatively early.

In order to be included the short-lived man must have produced a significant work at an early age. To qualify for inclusions, a long-lived man was required to produce one significant work but this could have been done either early or late. In other words, in order to achieve a certain degree of eminence, the short-lived man must have fulfilled in a few years what the longer-lived achieved in a more leisurely fashion. We shall show that the consequence of combining data for men of different longevities is a higher average productivity in the early decades.

In this connection Lehman says, "Adequate allowance for the unequal numbers of individuals alive at successive age levels was made. . . ." It seems to us that no adequate allowance can be made for the fact that all of the significant contributions of short-lived people occur in the early decades, whereas the long-lived can contribute both early and late. In tabulating entries for different decades, the twenties or thirties receive a score for each short-lived person. On the other hand, the later decades, such as the sixties, contain no entries for short-lived persons and only part of the entries for the septuagenarians. When data from men of different degrees of longevity are included in the same table, the early decades have an inevitable loading which is not shared by the later decades. To give the later decades a similar loading, it would be necessary to adopt the rule of including a long-lived person only if he made a significant contribution in his later years, because, conversely, the short-lived person is included only if he made a significant contribution in the early decades of life. This is a somewhat subtle point, but one which is essential to the correct evaluation of Lehman's data.

From the point of view of the consideration presented above, a very interesting table is presented by Lehman in his penultimate chapter. . . . This table represents 1,540 notable contributions to various sciences. In this case, the data for persons of different longevity are treated separately. For this reason, the criticism presented above does not apply.

The table shows that for each group the decade . . . of the thirties is most productive but the differences between the thirties and the forties are not large. The largest difference between the thirties and forties occurs among those dying in the forties. In this group ill-health may have contributed to the decrement. For longer-lived groups, even the decrements in the fifties, compared to the thirties, are not dramatic. No group in the fifties drops to the extent which is found when persons are not segregated according to longevity. In other words, this table shows that the combining of data for men of unequal longevity in other tables seems to have exaggerated the apparent age decrement. Nevertheless, even when data refer to men of equal life-spans some age decrement is still found.

This table is so significant in regard to age decrement that it is surprising that Lehman makes no reference to it when discussing the striking decrements reported in his earlier chapters. Nor are its findings adequately reflected in the summary chapter of his book. For these reasons it seems necessary here to emphasize the importance of the data which it contains.

We believe that much of this residual decrement is the product of other deficiencies in methodology. For one thing, it seems likely that the very high peaks of productivity which Lehman reports in his early chapters may be due to errors in sampling and to choosing age-intervals in such a way as to maximize the effects of sampling errors.

Many, but not all, of the curves presented by Lehman are based upon a relatively small number of entries. Thus figure 14 is the result of only 52 entries, figure 16, 30, figure 51, 53, figure 53, 67, and figure 56, 40. These entries are divided among age-intervals, usually 5-year periods, extending from age 20 to age 70 and beyond. With small numbers of entries divided among 10 or more age intervals, one would expect that, even though no true age differences are present, high values in some age-intervals would frequently be obtained through the operation of sampling errors. This fact is important because the highest age score in any body of data is taken as the peak from which decrement is measured. Therefore any exaggeration of the peak naturally results in finding exaggerated decrements.

This factor is further aggravated by the fact that Lehman did not limit himself to a fixed set of age-intervals, but apparently altered them in order to determine the particular "peak years" which seem to characterize a particular set of data. Thus, as the final chapter indicates, the step-intervals for peak years for different activities are variously reported as 22–26, 24–28, 25–29, 26–31, 30–34, 32–36, etc. The modification of age-intervals in order to find "ages of maximum productivity" would be legitimate if the findings were cross-validated against new data, but this was seldom done. Hence the extent to which "peak years" are affected by random errors of sampling is unknown.

There can be little doubt that some part of the decrements reported by Lehman are to be explained by the considerations just presented. The reader of Lehman's book will note that decrements are less precipitous in the graphs which are based upon numerous data and in the construction of which the step intervals follow the decimal system instead of being varied to maximize the peaks.

The preceding arguments have been of a mathematical or statistical sort. Those which follow are of a different kind, but, we think, no less cogent.

Lehman used as a criterion for inclusion of a work as a "significant contribution" the appearance of the work in histories of the appropriate area, or its appearance in lists of "best" books, "best" operas, etc. Perhaps, no better indices of importance are available, but it should be pointed out that these criteria may have certain weaknesses from the point of view of the study of age differences. It is possible that biographies, histories, and lists of best works contain systematic errors somewhat favoring a man's early work at the expense of his later products, and Lehman's findings may reflect these biases. For example, the art historian may be more likely to mention a painter's first significant contribution than he is to mention his last important piece of work. Likewise, an historian of science may be more likely to mention a young man's pioneering research which opened a new vista than he is to describe the subsequent painstaking investigations which were necessary to develop and validate the promise of the pioneering study. It is difficult to know to what extent an apparent age decrement may be due to the proclivities of anthologists and historians rather than to age itself.

In this connection, the possibility of a bias against the evaluation of recent contributions should be considered. It is our impression that critics and historians tend to consider the evaluation of recent contributions to be more difficult than the evaluation of more remote works. They may, therefore, suspend judgment in connection with recent contributions. Now a considerable number of Lehman's subjects were born after 1800. . . . Their later works were recent works at the time of the preparation of the source books from which Lehman obtained his data. Unwillingness, on the part of historians and editors, to evaluate recent works would therefore lessen the number of significant works recorded for the later years of some of Lehman's subjects. Consonant with this interpretation is Lehman's report that in former centuries the decrement with age in several fields seems not to have been as great as in recent times. . . . A century or more ago the apparent decline of creativity with age was slight.

Let us note, too, that the assessment of the relative excellence of work done early and late in a man's career is made exceedingly difficult, if indeed not impossible, by the changes in standards which occur during a man's lifetime. For example, the situation in biology in 1880, when Darwin was 71, was extremely different from what it was when "The Origin of Species" appeared in 1859 when Darwin was 50. In fact, the difference was due in large part to Darwin's own work. It seems relatively meaningless to compare biologic contributions made before and after the publication of the theory of evolution. This argument, of course, is not limited to biology. Changing standards characterize all fields, whereas judgments of quality in regard to works separated by several decades seem to imply absolute standards.

Standards for the judgment of quality are further complicated by the great increase in the number of creative workers in most fields which has taken place in recent times. Thus the best psychologist in America in 1900 was the best in a group of approximately 100. The best psychologist today, if he were ascertained, would have to be judged the best among 13,000. A psychologist living in 1900 and still living today, had 99 competitors for distinction in his youth and has 12,999 rivals (or thereabouts) in his later years. Similar, if perhaps less striking, increases in personnel have taken place in other fields. Curves for age changes in number of significant contributions do not, and probably cannot, correct for changes in standards of evaluation which occur during a lifetime.

In summary, we have presented several reasons for skepticism in regard to accepting the view that there is a decrement with age in the production of creative works of high level. We have not attempted to be exhaustive in this treatment. We submit, however, that there is a reasonable doubt that the curves presented by Lehman depict an age decline. Quality of creative work *may* decrease with age, but data presently available do not offer satisfactory evidence.

We would like to be able to suggest a method by which valid conclusions concerning changes in the quality of creative contributions with age could be reached, but we are unable to do so. All sources of data, and all methods of evaluation which we have considered seem to suffer from one or more of the difficulties discussed above. Nevertheless, it has been noted that as the methodologic difficulties in Lehman's work are reduced, the apparent decline with age becomes smaller. Whether ideal data would show no decline prior to extreme old age it is at present impossible to say, but this possibility should not be ignored.

NOTE

1. Lehman, H. C.: *Age and Achievement,* Princeton University Press, 1953.

Growing Old or Living Long

Take Your Pick

Laura L. Carstensen

PSYCHOLOGICAL SCIENCE AND LONGEVITY

In psychology, as in most of the biological and social sciences, research on aging has focused mostly on decline. And it has found it. The aging mind is slower and more prone to error when processing information. It is less adept at considering old information in novel ways. Memory suffers. In particular, working memory–the ability to keep multiple pieces of information in mind while acting on them–declines with age. . . .

These changes begin in a person's 20's and 30's and continue at a steady rate across the adult years. They occur in virtually everyone, regardless of sex, race, or educational background. In all likelihood, these effects are accounted for by age-related changes in the efficiency of neurotransmission.

Despite these changes in cognitive processing, the subjective experience of normal aging is largely positive. By experiential and objective measures, most older people remain active and involved in families and communities. The majority of people over 90 live independently. . . . Research shows that in areas of expertise, age-related decline is minimal until very advanced ages.

Arguably even more interesting and important is growing evidence that performance–even on basic processes such as semantic or general memory–improves under certain conditions. One

of the first such studies was reported by Paul Baltes and Reinhold Kliegl in 1992. They demonstrated rather striking improvement in memory with practice. Baltes and Kliegl first enlisted younger and older people's participation in a study of memory training. They assessed the participants' baseline performance and, as expected, younger participants outperformed older participants. However, after this initial assessment, participants attended a series of training sessions in which they were taught memory strategies such as mnemonics. They found that older people's memory performance benefited from practice so much that after only a few practice sessions, older people performed as well as younger people had before they had practiced. Younger people's performance also improved with training, of course, so at no point in the study did older people outperform younger people at the same point in training. But the fact that older people improved to the equivalent of untrained younger people speaks to the potential for improvement. . . .

More recently, scientists have begun to investigate social conditions that also may affect performance. Tammy Rahhal and her colleagues reasoned that because there are widespread beliefs in the culture that memory declines with age, tests that explicitly feature memory may invoke performance deficits in older people. They compared memory performance in younger and older people under two experimental conditions.

In one, the instructions stressed the fact that memory was the focus of the study. The experimenter repeatedly stated that participants were to "remember" as many statements from a list as they could and the "memory" was the key. In the second condition, experimental instructions were identical except that the instructions emphasized learning instead of memory. Participants were instructed to "learn" as many statements as they could. Once again, rather remarkable effects were observed. Age differences in memory were found when the instructions emphasized memory, but no age differences were observed in the condition that instead emphasized learning. . . .

Thus, although there is ample evidence for cognitive deficits with age, the story about aging is not a simple story of decline. Rather, it is a qualified and more nuanced story than the one often told. Even in areas where there is decline, there is also growing evidence that performance can be improved in relatively simple ways. This poses a challenge to psychology to identify conditions where learning is well maintained, to find ways to frame information in ways best absorbed, and ultimately to improve cognitive and behavioral functioning by drawing on strengths and minimizing weaknesses.

My students, colleagues, and I had been studying age-related changes in motivation for several years. We began to wonder whether changes in motivation would affect performance on cognitive tasks, and we set out to explore what we call socioemotional selectivity theory (SST), a life-span theory of motivation.

MOTIVATION MATTERS

SST was initially developed to address an apparent paradox in the aging literature. Despite losses in many areas, emotional well-being is as good if not better in older people as in their younger counterparts. . . . Older people are more satisfied with their social relationships than are younger people, especially regarding relationships with their children and younger

relatives. Fredda Blanchard-Fields and her colleagues find that older people solve heated interpersonal problems more effectively than do younger adults. Many social scientists refer to such findings as "paradox of aging." How could it be that aging, given inherent losses in critical capabilities, is associated with an improved sense of well-being? . . .

When time is perceived as open-ended, as it typically is in youth, people are strongly motivated to pursue information. They attempt to expand their horizons, gain knowledge, and pursue new relationships. Information is gathered relentlessly. In the face of a long and nebulous future, even information that is not immediately relevant may become so somewhere down the line.

In contrast, when time is perceived as constrained, as it typically is in later life, people are motivated to pursue emotional satisfaction. They are more likely to invest in sure things, deepen existing relationships, and savor life. Under these conditions, people are less interested in banking information and instead invest personal resources in the regulation of emotion. . . .

One key tenet of SST is that perceived time horizons, not chronological age, account for age differences in goals preferences. Our research team has examined this theoretical postulate in a variety of ways in a number of studies. We hypothesized that older people would prefer emotionally meaningful goals over informational goals but that these preferences would change systematically when time horizons were manipulated experimentally. In several studies, we showed that younger people display preferences similar to those of the old when their time horizons are shortened, and older people show preferences similar to those of the young when their time horizons are expanded. Importantly, similar changes occur when natural events, such as personal illnesses, epidemics, political upheavals, or terrorism, create a sense of shortened time horizons. Under such circumstances, the preferences of the young resemble those of older people. In other words, when conditions create a sense of the fragility of life, younger as well as older people prefer to pursue emotionally meaningful experiences and goals. . . .

The human brain does not operate like a computer. It does not process all information evenly. Rather, motivation directs our attention to goal-relevant information and away from irrelevant information. We see what matters to us. Imagine walking around a city block with the goal of finding a friend. You see very different things than you would see if you took the same walk while trying to find a particular species of bird. Indeed, in the latter scenario you might walk right by your friend without notice. In the former, you would surely miss the bird.

In an initial study, my former student Helene Fung and I reasoned that because older people prefer emotional goals, they may remember emotional information better than emotionally neutral information. This was an important idea to test because the standard practice in psychological science is to avoid emotional stimuli in tests of memory in order to minimize contamination of "pure" cognitive processes. We wondered if by doing so, experimenters were inadvertently handicapping the performance of older adults. A substantial literature on memory and persuasion shows that people are more likely to remember and be persuaded by messages that are relevant to their goals. Thus, we reasoned that marketing messages that promised emotionally meaningful rewards may be more effective with older people than those that promise to increase knowledge or expand horizons. . . .

In one study, older people remembered the emotional slogans and the products they touted better than did younger people. . . .

THE POSITIVITY EFFECT

Findings from this initial study suggested that in older people, memory of emotional information was superior to memory of other types of information. . . .

A substantial literature in social psychology, albeit based exclusively on young adults, shows

superior memory of negative information. Negative information is also widely believed to be weighted more heavily than positive information in impression formation and in decision making. The burning question was whether such findings, long presumed to represent "human" preferences, actually represented preferences of young people.

We conducted a study in which young, middle-aged, and older adults viewed positive, negative, and neutral images on a computer screen and were then tested for their memory of the images. We found an age-related pattern in which the ratio of positive to negative material recalled increased with age. Younger people recalled equal numbers of positive and negative images. Middle-aged people showed a small but significant preference in memory for positive images. In older people, the preference for positive was striking. Older people remembered nearly twice as many positive images as negative or neutral images. . . .

At that point, we began to think that attention and memory can operate in the service of emotion regulation. That is, focusing on positive memories and images makes people feel good. . . .

To summarize, whereas younger adults favor negative information as much or more than positive information, by middle age this preference appears to have shifted to a preference for positive information. Older adults show a decided preference in memory and attention for positive information. Although longitudinal studies are needed before conclusions about change over time can be drawn, cross-sectional comparisons suggest that the effect may emerge across adulthood. This "positivity effect" has been demonstrated in a range of experimental tasks that assess even the most vulnerable of aspects of cognitive processing, such as working memory. Theoretically, we argue that the pattern represents a shift in goals from those aimed at gathering information and preparing for the future to those aimed at regulating emotional experience and savoring the present. . . .

We maintain that in general a focus on positive information benefits well-being. However, there are probably conditions when a chronic tendency to focus on positive material is maladaptive. . . .

Human need is the basis for virtually all of science. If we rise to the challenge of an aging population by systematically applying science and technology to questions that improve quality of life in adulthood and old age, longer-lived populations will inspire breakthroughs in the social, physical, and biological sciences that will improve the quality of life at all ages. Longevity science will reveal ways to improve learning from birth to advanced ages and to deter age-related slowing in cognitive processing.

READING 14

The Mature Mind

The Positive Power of the Aging Brain

Gene Cohen

"Over the hill."

"Out to pasture."

"Twilight years."

"Retired."

These words reflect a stubborn myth—that aging is a negative experience and that "successful aging" amounts to nothing more than slowing the inevitable decline of body and mind. Rubbish. Some of life's most precious gifts can only be acquired with age: wisdom, for example, and mastery in hundreds of different spheres of human experience that requires decades of learning. Growing old can be filled with positive experiences and "successful" aging means harnessing and manifesting the enormous positive potential that each one of us has for growth, love, and happiness.

Of course, aging brings challenges and losses. As actress Bette Davis once famously quipped, "Getting old isn't for sissies." Sight may blur, hearing may dull, friends may die or become disabled. All of this is true, but it's not the whole truth. Historically, both science and culture in Western societies have focused exclusively on the negative sides of aging and ignored the positive. It's time for a better, truer, and more motivating paradigm—not a rosy, everything-is-wonderful perspective, but a clear-eyed view that acknowledges the hard realities of growing old while at the same time celebrating its benefits, pleasures, and rewards. . . .

The latest research findings are encouraging and important. Denying or trivializing the positive potential of aging prevents people from realizing the full spectrum of their talents, intelligence, and emotions. But when we come instead to expect positive growth with age, such growth can be nurtured. We are still a long way from fully realizing this shift in perspective, but I hope this book will be a forceful catalyst for change in that direction.

SOURCE: Excerpts from *The Mature Mind: The Positive Power of the Aging Brain* by Gene D. Cohen (pp. 3–4, 35, 38, 95, 134, 182). Copyright © 2006 by Gene D. Cohen. Reprinted by permission of Basic Books, a member of Perseus Books Group.

NEW SCIENCE, NEW HORIZONS

Some of the most exciting research supporting the concept of positive aging comes from recent studies of the brain and mind. Much of aging research conducted during the twentieth century emphasized improving the health of the aging body. As a result of this research, life expectancy and overall health did in fact improve dramatically. Aging research at the beginning of the twenty-first century, in contrast, has expanded with a strong focus on improving the health of the aging mind. Dozens of new findings are overturning the notion that "you can't teach old dogs new tricks." It turns out that not only can old dogs learn well, they are actually better at many types of intellectual tasks than young dogs.

The big news is that the brain is far more flexible and adaptable than once thought. Not only does the brain retain its capacity to form new memories, which entails making new connections between brain cells, but it can grow entirely new brain cells—a stunning finding filled with potential. We've also learned that older brains can process information in a dramatically different way than younger brains. Older people can use both sides of their brains for tasks that younger people use only one side to accomplish. A great deal of scientific work has also confirmed the "use it or lose it" adage: the mind grows stronger from use and from being challenged in the same way that muscles grow stronger from exercise.

But the brain isn't the only part of ourselves with more potential than we thought. Our personalities, creativity, and psychological "selves" continue to develop throughout life. This might sound obvious, but for many decades scientists who study human behavior did not share this view. In fact, until late in the twentieth century, psychological development in the second half of life attracted little scientific attention, and when attention was paid, often the wrong conclusions were drawn. For example, Sigmund Freud, whose influence on psychological theory was profound, had this to say about older adults: "About the age of fifty, the elasticity of the mental processes on which treatment depends is, as

a rule, lacking. Old people are no longer educable."

Ironically Freud wrote this statement in 1907, when he was fifty-one, and he wrote some of his greatest works after the age of sixty-five. Furthermore, Sophocles's *Oedipus Rex*, the masterpiece on which Freud based his pioneering psychoanalytic theory, was written when the Greek playwright was seventy-one years old.

Freud wasn't the only pioneer to get things wrong when it came to aging. Jean Piaget, who made an extraordinary contribution to our understanding of cognitive development, ended his description of intellectual development with what he called "formal operations," the kind of abstract thinking that matures during the teenage years. As far as Piaget was concerned, development stopped in young adulthood and then began a slow erosion. . . .

DEVELOPMENTAL INTELLIGENCE

In this book I introduce a novel concept, developmental intelligence, which I see as the greatest benefit of the aging brain/mind. Developmental intelligence is the degree to which a person has manifested his or her unique neurological, emotional, intellectual, and psychological capacities. It is also the process by which these elements become optimally integrated in the mature brain. More specifically, developmental intelligence reflects the maturing synergy of cognition, emotional intelligence, judgment, social skills, life experience, and consciousness. We are all developmentally intelligent to one degree or another, and, as with all intelligence, we can actively promote its growth. As we mature, developmental intelligence is expressed in deepening wisdom, judgment, perspective, and vision. Advanced developmental intelligence is characterized by three types of thinking and reasoning that develop later than Piaget's "formal operations" and hence are referred to as "postformal operations": relativistic thinking (recognizing that knowledge may be relative and not absolute); dualistic thinking (the ability to uncover and resolve contradictions in opposing and seemingly incompatible views); and systematic thinking (being able to see the large picture, to distinguish between the forest and the trees).

These three types of thinking are "advanced" in the sense that they do not come naturally in youth; we prefer our answers black or white, right or wrong. And we usually prefer any answer to none at all. It takes time, experience, and effort to develop more flexible and subtle thinking. Our capacity to accept uncertainty, to admit that answers are often relative, and to suspend judgment for a more careful evaluation of opposing claims is a true measure of our developmental intelligence. . . .

Contrary to societal myths, creativity is hardly the exclusive province of youth. It can blossom at any age—and in fact it can bloom with more depth and richness in older adults because it is informed by their vast stores of knowledge and experience. . . .

It's been said that the mind is what the brain does. The mind is often described as "software" running on the "hardware" of the brain. But this analogy is too simple. The brain is far more malleable and flexible than any computer chip. And the mind, although it seems almost ghostlike, can powerfully influence the brain and, by extension, the body. Mind and brain are really two sides of a single coin—mind/brain.

You may have learned the following "facts" about the brain:

The brain cannot grow new brain cells.

Older adults can't learn as well as young people.

Connections between neurons are relatively fixed throughout life.

Intelligence is a matter of how many neurons you have and how fast those neurons work.

All these "facts" are wrong, as we will see. And that's good news for all of us. The brain is more resilient, adaptable, and capable than we long thought. Research in the past two decades has established four key attributes of the brain that lay the foundation for an optimistic view of human potential in the second half of life.

The brain is continually resculpting itself in response to experience and learning.

New brain cells do form throughout life.

The brain's emotional circuitry matures and becomes more balanced with age.

The brain's two hemispheres are more equally used by older adults.

Now let's be clear. I am not suggesting that the brain is immune to age-related changes. The brain is made of cells, like every other part of the body, and cells can and do "wear out" with age. Certain aspects of brain function do decline with age, such as the raw speed with which complicated math problems are solved, reaction times, and the efficiency of short-term memory storage. But these "negatives" are by no means the whole—or even the most important—story about the aging brain. Unfortunately because much brain research has focused on age-related problems, negative aspects of aging have been emphasized and the positive implications of research have been overlooked. . . . Healthy older brains are often as good as or better than younger brains in a wide variety of tasks. . . .

Developmental intelligence is defined as the maturing of cognition, emotional intelligence, judgment, social skills, life experience, and consciousness and their integration and synergy. With aging, each of these individual components of developmental intelligence continues to mature, as does the process of integrating each with the others. This is why many older adults continue functioning at very high intellectual levels and display the age-dependent quality of wisdom.

As I have emphasized from the start, there is no denying that problems can accompany aging—and research to date has focused mostly on such problems, typically in individual components of our total mental superstructure. Less attention has been paid to how gains and losses can occur at the same time. For example, older adults often experience more trouble with word finding—the "tip-of-the-tongue" experience—but at the same time, the total number of words they use—their vocabulary—continues to increase. If you look only at selected functions, such as certain aspects of memory or mathematical ability, you miss the larger picture of how functions become more integrated, often improving overall performance. This is the heart of developmental intelligence. . . .

It takes time, experience, and learning to develop the capacities for relativistic, dualistic, and systematic thinking. It can be difficult to challenge existing beliefs that offer comforting, but dubious, answers to life's problems. It's sometimes hard to say, "I don't know" instead of, "The answer is . . ." But our capacity to accept some uncertainty, to admit that answers are often relative, and to suspend quick judgment for a more measured evaluation of opposing claims is a real measure of postformal thought and developmental intelligence. In fact, "wisdom" is in some ways a synonym for "developmental intelligence." Wisdom is how developmental intelligence reveals itself. . . .

WISDOM AND POSTFORMAL THINKING

What exactly is wisdom, and how does it develop? One standard definition is that wisdom consists of "making the best use of available knowledge." This rather utilitarian approach implies that wisdom requires specific

knowledge as well as a broad understanding of the context in which that knowledge can be put to use. But this definition isn't completely satisfying. For most people, wisdom also connotes a perspective that supports the long-term common good over the short-term good for an individual. Insights and acts that many people agree are wise tend to be grounded in past experience or history and yet can anticipate likely future consequences. Wise acts, in other words, look both backward and forward. Wisdom is also generally understood to be informed by multiple forms of intelligence—reason, intuition, heart, and spirit. It is fundamentally the manifestation of developmental intelligence—a mature integration of thinking skills, emotional intelligence, judgment, social skills, and life experience. . . .

Social intelligence, memory, and wisdom are closely related fruits that age alone can ripen. The aging brain has greater potential than most people think, and development never stops. Our capacity for social involvement and interpersonal relations remains as strong as ever in later years and is a vital wellspring of both physical and mental health. . . .

We can, if we want to, learn, grow, love, and experience profound happiness in our later years. We need not succumb to difficulties, nor need we accept the myths that still exist about growing older.

READING 15

Aging and Creativity

Becca Levy and Ellen Langer

According to the Peak and Decline Model, creativity increases in adulthood until the late 30's and then begins to decline. According to

the Life Span Developmental Model, creativity does not increase or decline, but rather different types of creativity are expressed in different stages of the life span. We will first define creativity and discuss the conditions which promote it. Then we will outline the two models, and argue that the second model is best supported by existing research on creativity. . . .

Creativity in this article is defined as the ability to transcend traditional ways of thinking by generating ideas, methods, and forms that are meaningful and new to others. It exists on a continuum both within and among individuals. That is, creativity not only differs over time as individuals develop, it also differs between individuals due to differences in personality and how they interact with their environment.

As a person ages both the nature and the degree of creativity vary. Young children at play tend to be very creative. They have not yet internalized many of the societally transmitted rules that can limit creativity. Thus, they can easily apply their imagination to their surroundings. . . .

Those who remain creative in later life have already internalized many social norms. Therefore, unlike children, they must deliberately reject some of these norms. It may be easier to be creative after completing the relevant training in one's discipline. At this point one knows the principles to reject and change. . . .

A. Openness to New Ideas

One characteristic that helps promote creativity is an openness to new ideas, which includes an ability to question surroundings and a tolerance for uncertainty. Essentially uncertainty leads to choice, and choice fosters mindfulness, which paves the way for creativity. Certainty makes individuals believe they know all there is to know and thus feel complacent. This state is at odds with the motivation to explore the target of uncertainty and to create something new. . . .

B. Assertiveness and Focusing Attention

To generate creative ideas it helps if individuals are open to new ideas. On the other hand, to translate their creative ideas into products,

whether it be an elegant mathematical equation or a dramatic sculpture, it is necessary for these individuals to focus their attention. This often requires that individuals be assertive when it comes to guarding their time. . . .

In a study of 91 exceptionally creative people (almost all of whom were over the age of 60 years), Mihaly Csikszentmihalyi found that a majority of the people he studied showed an ability to become single-minded, specialized, and guarded with their time. For example, Albert Einstein insisted that his wife serve him his meals in his home office so he would not be distracted by her or their children. . . .

C. Supportive Environment for Creativity

The third way that creativity can be promoted is through societal expectations and institutions. One needs the proper environment to create. . . .

A stereotype persists that senility increases and creative potential declines in old age. This stereotype may contribute to the decline in creative productivity with age. Research by Becca Levy and Ellen Langer has demonstrated that stereotypes of old age as a time of loss can worsen memory performance and self-efficacy of older adults. A survey found that many older artists doubt their abilities. This may lead to a drop in motivation and an increase in obstacles for creative productivity.

PEAK AND DECLINE MODEL

The Peak and Decline Model fits into the general belief that aging is a time of decline and loss. According to this model, named by Martin Lindauer, creativity increases in early adulthood and then starts to decline starting in one's 30's. This assumes that creativity is the same construct across the life span. Any changes that take place are thought to be due to the quantity, not the quality, of creativity. Therefore, studies within this tradition operationalize creativity consistently throughout the life span. The studies supporting this model have operationalized creativity in two ways.

The first way creativity is operationalized is with psychometric tests. These tests were originally given as paper-and-pencil tests. More recently, psychologists have developed computer programs. Psychometric creativity tests are designed to tap divergent thinking, which is defined as the ability to come up with many different associations. An example of a test item that might appear on a creativity test would be to list as many uses for a brick as you can. Divergent thinking is believed to differ from intelligence, which tends to be based on convergent thinking or the ability to come up with one correct response to a question. . . .

The second way that studies supporting this model have operationalized creativity is by productivity measures. According to the reasoning behind these measures, creativity can be assessed by the number of creative products.

The underlying assumption of the Peak and Decline Model can be seen operating in a variety of decisions and policies. For example, the committee that gives the most prestigious mathematical honor, the Field Award, has decided to only consider mathematicians under the age of 40 years. Many of the academic institutions that employ those who make a profession of creative endeavors have traditionally imposed retirement in the late 60's. Also, Sigmund Freud argued that psychoanalysts should not give therapy to patients over the age of 50 because they tend to lack both personal insight and the ability to make meaningful changes. This was his belief despite the fact that he was over the age of 50 when he made this point and he considered one of the greatest plays to be *Oedipus at Colonus,* which was written by Sophocles at the age of 89.

A. Critique of the Peak and Decline Model: Psychometric Tests

There are numerous problems with this model. We will divide our comments into the two types of data used as evidence for this model:

psychometric test and productivity data. In terms of the psychometric tests the studies that show decline tend to be cross sectional and do not take into account cohort changes, such as the fact that older individuals tended to receive less formal education than those born decades after them. One might argue that declines in tests of divergent thinking are due not to creativity, but to the confound of education with age. Second, although these tests are highly reliable, they tend to have little construct validity. That is, people judged as highly creative by society often perform poorly on tests. Also, people who score highly on creativity tests tend to not display high degrees of creativity in other areas of their lives. . . .

In youth it may be easier to come up with numerous responses for a question or it just may be more desirable for that age group to do so. As individuals get older, research suggests that often there is a shift to a more mature way of thinking where the benefits come from contrasting and integrated ideas in light of one's own experiences. Thus, studies which contrast older and younger adults' scores on psychometric tests may be inappropriately comparing two very different content areas and not creative ability at all.

B. Critique of the Peak and Decline Model: Productivity Tests

Although the studies that chart a loss in the productivity of prominent creative individuals avoid many of the validity problems raised by the psychometric measures of creativity, they have their own problems. First, productivity is not equivalent to quality. Although there are individuals, such as Shakespeare and Beethoven, who have managed to be both extremely prolific and creative, these constructs are not equivalent. . . .

Furthermore, productivity may decline with age for a number of reasons not related to the quality of creativity. These reasons are both internal and external to the creator. As individuals get older, their income and social networks frequently decline, which could limit their ability to produce creative works. As careers progress, there is often an increase in professional obligations, such as committee work, teaching, and public lectures. After achieving success in a field, it is not uncommon for some of the initial passion and motivation to lessen. . . .

It is the earlier work that makes us take notice of an artist's creative ability. After the initial attention the artist receives, it may take a big change in style for the audience to become equally impressed. When artists enter a field it is easy for the public to notice the ways in which their art is distinct since their work is compared to other artists.' Subsequently, artists' later works are compared to their own earlier works. This within person comparison that tends to take place at the end of an artist's career may also make the young appreciate the work of older artists less. By drawing fewer distinctions they are less likely to notice the subtle ways in which the artists have grown.

C. Simonton's Model of Creative Careers

According to a model developed by Dean K. Simonton, age does not predict one's creative productivity. Instead, the predictors he uses are when the creative career begins; the process of coming up with creative ideas; the process of transforming one's ideas into products; and the domain of creativity. In creative domains in which one deals with a finite array of concepts, such as math, creativity tends to peak and decline early, whereas in domains that deal with complex and associative-rich concepts, such as history, the creative productivity tends to peak and decline later. Simonton believes the timing of the creative peak is due to the fact that creative careers tend to be much shorter when there is little delay between coming up with ideas and elaborating them into creative products. This leads to an early consumption of creative potential.

Simonton's model suggests that those with a later start date should have a career peak later than their peers. His model also suggests that individuals may experience multiple career peaks if they find ways to relaunch their career by taking on a new problem, medium, or discipline. . . .

LIFE SPAN DEVELOPMENTAL MODEL

The Life Span Developmental Model, in contrast to the Peak and Decline Model, assumes creativity and productivity are not equivalent. It also assumes that creativity changes with development as a result of the underlying cognitive processes that change with one's life stage, as well as one's experiences. The studies that support this model tend to operationalize creativity as the products judged by society to be novel and significant. For example, in painting, researchers have identified the most creative artists by looking for the number of references to their works that appear in major art history books. . . .

It is particularly instructive to track artists who approached the same subject matter at different times in their lives. Thus, they serve as their own controls. For example, Michelangelo sculpted a Pieta at age 22 and then again at age 90. Similarly, Francis Bacon produced a series of paintings in which he tried to depict a cry. He painted the first image when he was 35 and continued to paint cries until the age of 79. As Gene Cohen points out, Bacon believes that it was not until the last painting that he finally got it right.

A. Why Creativity Changes in Later Life

In this section we will first discuss why creativity may change in later life, and then we will examine the way in which creativity changes. There are a number of theories about late-life development that seek to explain the evolution of creativity within individuals. These include the psychodynamic theories. For example, Erik and Joan Erikson felt that old age is a time of psychodynamic development that could change the quality of creativity in old age; individuals can undergo dramatic changes as they try to resolve conflicts of previous stages as well as new conflicts raised by trying to maintain wisdom despite the factors that can lead to despair, such as trying to come to terms with their approaching death.

Some believe that as individuals approach death they try to find a way to make a lasting mark on the world. This can lead to a surge of creative energy, sometimes called "the swan song." . . .

There are also cognitive changes in old age that could contribute to a change in creativity style. Whereas some cognitive qualities that may be associated with creativity do not seem to change, such as the ability to use imagery, others seem to change. For example, whereas fluid intelligence or the abstract capacity for problem solving may decline, crystallized intelligence or the acquisition of knowledge from experience, such as vocabulary, may increase.

In addition, changes in physical functioning may lead to a new perspective that can fuel creativity. Paul and Margaret Baltes have developed a theory called Optimization with Compensation which suggests that those elderly who face physical or cognitive decline must find ways to change their style to compensate. For example, when Degas began to lose his eyesight he changed his medium from oil paints to the more tactile wax and oily chalk. In a different creative realm, physicist Hans Bethe explained that although he made more mistakes in old age, he became more alert at catching mistakes.

B. How Creativity Changes in Later Life

Art historians have identified an Old Age Style of creativity that is also referred to as "Altersstil."

Its elements include an increased sense of drama; a more profound interpretation of human nature; a more instinctual, less studied approach; looser, freer brushwork; more amorphous corporeal forms; a compression of space such that figures loom close to the picture plane; a lessening of emphasis on setting and background details; a theme of death; and an emphasis on unity and integration. . . .

In this study of seven creative people that he thinks helped change the direction of this century, Howard Gardner describes late life changes in styles of creativity. Albert Einstein in later life turned from a focus on theoretical formulas to public policy. Sigmund Freud also switched his style in later life from writing about medical case studies to broader ideas about civilization and culture. At the end of her life Martha Graham made a dramatic switch. In her younger life she choreographed dances for herself. Then at the age of 73 she reemerged as director of her own troupe, and started to tour the country giving lecture demonstrations to educate the masses about her vision of modern dance. . . .

INFLUENCE OF CREATIVITY ON HEALTH AND LONGEVITY IN OLD AGE

We have discussed how aging influences creativity. Several studies suggest that the reverse also occurs: creativity influences aging and longevity. For example, Lindauer discovered that many of the great artists lived longer than the general population. Another study found that individuals who attend more creative events (vicarious creativity) have extended longevity. This study conducted in Sweden of over 12,000 individuals found that those who were more culturally active tended to outlive those who were less cultural. The authors measured cultural activity by counting reports of attending events such as plays and concerts, and visiting institutions such as museums and art galleries.

FOCUS ON THE FUTURE

Late-Life Learning in the Information Society

Dateline: May 1, 2025, Washington, DC. Associated Press.

Today, President Martha Jefferson welcomed 30,000 delegates to the White House Conference on Aging. At the same time, on a specially dedicated website, she announced the beginning of Older Americans Month.

President Jefferson noted that this White House Conference on Aging was the first to be broadly representative of the American people: There were 10 times as many official delegates—far more than any hotel in Washington could accommodate. In fact, conference "delegates" didn't meet face-to-face at all but "convened" in cyberspace. They used high-speed fiber-optic connections made possible for Internet III CyberSystem.

President Jefferson also took special note of the more than 1,000 older people at the conference who had earned an advanced degree through distance learning under Internet III or its predecessors. She noted that students over the age of 55 are now the fastest growing segment in U.S. higher education.

The likelihood of this scenario all depends on how quickly new computer and telecommunications technologies achieve acceptance and widespread use among the aging population. Technology is advancing rapidly, and signs of late-life learning in an information society are already evident at the dawn of the 21st century. Today, many older people still have anxiety about using a computer. But technophobia is a stereotype, and their anxiety can be overcome.

Studies have shown that computer communication can be an aid to independence for older adults. For instance, one study looked at a sample of women aged 55 to 95 in a Florida community, a group with no prior experience with computers. Participants in the study were at first given a simplified electronic mail and text-editor system, and their software was later upgraded to offer news, weather, movie reviews, health information, and entertainment news. A follow-up survey showed that participants easily learned to use the system and came to value it as a means of social interaction (Czaja et al., 1993).

One of the leaders in the "seniors in cyberspace" movement is SeniorNet, a nonprofit organization founded in San Francisco in 1986 to teach computer skills to older persons (Furlong & Lipson, 1996). SeniorNet has grown rapidly as a membership organization with more than 70 learning centers around the country supporting more than 15,000 individual members. SeniorNet publishes its own educational materials, holds annual conferences, and operates its own online network. SeniorNet Learning Centers, run by senior volunteers, are found in community centers, in senior centers, at schools and on college campuses, in libraries, and at health care facilities. Through America Online, SeniorNet also offers classes and discussion forums, live chats, and file downloading.

SeniorNet is not the only service for elders in cyberspace. The Cleveland Free-Net has become a nationally recognized example of how an entire city can be "wired" to promote maximum access by all groups, with prominent participation from older adults and individuals with disabilities. For example, the local Cleveland Alzheimer's support groups are plugged in to the Free-Net, providing a combination of "high tech" and "high touch." In this way, the "virtual community" of cyberspace becomes a means of reinforcing and extending face-to-face mutual support networks.

The key to lifelong learning in an information society will be to perceive older adults as active users of new technologies rather than as passive recipients (Czaja & Barr, 1989). Two-way interactive TV can address loneliness and isolation among older adults. For instance, a two-way television system in Reading, Pennsylvania, has been programmed, operated, and financed by senior citizens. The Leisure World retirement community in California has long operated its own cable TV station and generated local programming. Interactive and self-directed activities using new technologies can enhance knowledge, skills, and adaptability—a high-tech/high-touch world with great promise for older people in years to come.

Older adults learning to use new computer technology challenge the stereotype that older people can't learn new things.

Questions for Writing, Reflection, and Debate

1. Harvey Lehman's data about the peak years of creativity for different fields are derived from creative people who lived in the past. Would it be reasonable to argue that his conclusions don't apply to older people today because health and life expectancy in recent decades have increased rapidly? Does Wayne Dennis succeed in refuting Lehman's argument that age generally means declining creative power? What are Dennis's strongest points in his criticism of Lehman?

2. Lehman assumes that in judging late-life creativity, we should measure how many "masterpieces" or "breakthroughs" are produced by older people. Do you think this standard is the right one for judging late-life creativity? Would other standards or definitions of creativity be more appropriate?

3. Imagine that you are writing a long obituary for "Louise Bachelard" (an imaginary name), who

died recently at age 78. She was a famous painter whose style changed dramatically in her later years. In the obituary, describe the ways in which the painter's creativity changed as she grew older, and connect this with what you have learned about the psychology of aging.

4. Paul Baltes and his colleagues define wisdom as accumulated expertise, but this definition makes no reference to character or the ethical behavior exhibited by a wise person. Could an infamous bank robber, like Willie Sutton, be judged to have "wisdom" if he showed skillful judgment in crime based on long experience?

5. Pick an example of an older person who seems to you to have developed some of the traits of wisdom, whether in general or in some specific field of activity. Write to a stranger explaining why this wise older person is someone whose advice should be taken seriously.

6. If we were designing classes or educational programs for older adults based on what we know about older adult intelligence and cognition, how should we organize the learning activities? How would such an older adult educational program differ from what is offered in schools and colleges today?

7. Visit three websites: one for Road Scholar (Edlerhostel) (http://www.roadscholar.org), one for Third Age (www.thirdage.com), and one for Senior Net (www.seniornet.org). What similarities do you see in these three sites concerning age-appropriate behavior for older adults? What issues do you *not* see reflected in these sites that seem important for successful living in later life?

Suggested Readings

American Council on Education. *Mapping new directions: Higher education for older adults.* Washington, DC: Author, 2008.

Cohen, Gene, *The Creative Age: Awakening Human Potential in the Second Half of Life,* New York: Harper Paperbacks, 2001.

Hall, Stephen S., *Wisdom: From Philosophy to Neuroscience,* New York: Vintage, 2011.

Simonton, Dean K., *Genius, Creativity, and Leadership: Historiometric Inquiries,* Cambridge, MA: Harvard University Press, 1984.

Sternberg, Robert (ed.), *The Nature of Creativity: Contemporary Psychological Perspectives,* New York: Cambridge University Press, 1988.

Student Study Site

Visit the Student Study Site at **www.sagepub.com/moody7e** for these additional learning tools:

- Flashcards
- Web quizzes
- Chapter outlines

- SAGE journal articles
- Web resources
- Video and audio resources

Aging, Health Care, and Society

Five hundred years ago, the Spanish explorer Juan Ponce de León embarked on a journey to the New World in search of the fountain of youth. He never found it. Instead, he discovered what is today Florida, the state with the largest percentage of older adults. Ponce de León might have smiled at the irony of how his discovery turned out. But discoveries often have a way of turning out differently from what we expect. When we think about medical advances in our time, these also have turned out unexpectedly. For instance, as we discussed in Controversy 2, people are living longer today, but is the prolongation of life into old age always a benefit? Or have recent gains in longevity instead been a prolongation of decrepitude and frailty? Will further medical advances only make matters worse? This question was raised nearly three centuries ago by Jonathan Swift in his satirical novel *Gulliver's Travels* (1726).

THE CHALLENGE OF LONGEVITY

The Case of the Struldbrugs

Swift described a voyage to the fictional country of Luggnagg, where his hero, Lemuel Gulliver, meets a strange group of beings, the "Struldbrugs," who are a race condemned to immortality. It turns out that for the Struldbrugs, unlimited life span has not proved the blessing it promised to be. Longevity has come, but without good health. Their existence is a dismal prolongation of decline and decay, a nightmare like unlimited existence in a nursing home, as Swift describes them:

> They were the most mortifying sight I ever beheld. . . . Besides the usual deformities in extreme old age, they acquired an additional ghastliness in proportion to their number of years, which is not to be described . . .
>
> The diseases they were subject to still continue without increasing or diminishing. In talking they forget the common appellation of things, and the names of persons, even of those who are their nearest friends and relations. . . . The least miserable among them appear to be those who turn to dotage, and entirely lose their memories.

In describing the Struldbrugs, Swift raised a question that is still of compelling interest:

The question therefore [is] not whether a man would choose to be always in the prime of youth, attended with prosperity and health, but how he would pass a perpetual life under all the usual disadvantages which old age brings along with it.

No doubt Swift exaggerated to make his point. To speak of the usual disadvantages of old age misses the positive aspects of aging. Today, we see countless examples of older people who are not debilitated or dependent, but who maintain health and vigor into their later years. Yet Swift's vision does raise profound questions about our values: Are the old less valued than the young? Where will we find the resources to take care of frail older people? Could medical breakthroughs have unforeseen consequences for society, either for good or for ill? These questions have no easy answers. Indeed, they are at the center of the controversies examined in this book.

As a beginning, however, we examine several major challenges that people face as they grow older. The first is the challenge of coping with an aging body. Medical advances that help people live longer may seem beneficial, but a longer period of physical and mental decline has implications for individuals and for society. The second challenge is that of maintaining a valued place in society while aging. Older people are often stereotyped as marginal members of society. However, as the average age in the United States steadily increases, we are beginning to confront questions of when people cross from capable old age to dependency. Finally, as individuals grow older, they do so in the wider context of an entire society that is undergoing a transition to population aging.

Biomedical Advances

There are those who believe that biology will save us from the problem. They argue that biomedical researchers can meet the challenge of longevity by developing techniques for delaying the onset of debilitating conditions in old age. In effect, they hope to postpone sickness until a final, brief period of life and so eliminate prolonged dependency, as discussed regarding compression of morbidity in Controversy 2. Other biologists believe that we can make good on Ponce de León's dream and discover a fountain of youth by altering the fundamental biological mechanism that makes us grow old. Whether by delaying illness or actually preventing biological aging, the scientific optimists believe that the "Struldbrug" problem can eventually be solved.

Rationing Health Care

Their optimism is not shared by all. Others believe that hard choices are called for, and they doubt that technological innovations will save us from making those choices. We do better, it is said, to acknowledge the biological limits, rather than hope for a technological fix for the problems that often come with aging. In this spirit, ethicist Daniel Callahan (as evidenced in the readings later in this chapter) wants to reject high-tech medical care used to prolong life for the very old. Instead, he believes, we do better to ration health care on the basis of age. He recommends forgoing life-extending treatment once older people have lived out a full and natural course of life.

Providing Long-Term Care

If more and more members of the population live into advanced old age, we will see growing numbers of frail, chronically ill older adults in need of long-term care, at home or in institutions. The term *long-term care* covers health care and social services needed by those who have lost the capacity to care for themselves because of a chronic illness or condition (Koff, 1982). It is expected that growing numbers of older people will suffer from chronic disorders that keep them from living independently. In that case, long-term care will loom even larger in the future than it does today. Opinions differ about who should bear the cost of that care, but paying the bill for longevity is already a serious challenge to society.

Self-Determined Death

Neither prolonged debilitation nor rationing of health care is popular with most Americans. But growing numbers today do feel that decline and a diminished quality of life might be sufficient reasons for ending one's own life. Those who hold this view usually reject the idea of society setting limits, but would instead leave the choice about dying up to the individual. Advocates of this idea believe that deliberate termination of treatment must be more openly recognized by law and should be actively supported by health care services.

So here we have four answers to the Struldbrug dilemma: hoping for a medical breakthrough, making tough cost-cutting decisions, providing long-term care, and permitting individuals to end their lives. All are ways of coping with the prospect of a prolonged period of frailty and dependency at the end of life. The options considered here are not mutually exclusive. But each raises profound questions about our values: Are the old less valued than the young? Where will we find the resources to take care of frail older adults? Could scientific breakthroughs in the biology of aging have unforeseen consequences for society, either for good or for ill?

These questions have no easy answers. Indeed, they are at the center of the major debates examined in this book. The biology of longevity, the economics of health care, and the right to die are all related. By appreciating some key facts about biology, economics, and death and dying, we can better approach the debates surrounding these critical issues.

A difficulty arises from the fact that contemporary medical practice in the United States is based on a strategy of curing disease, not promoting health. This familiar strategy has led to the conquest of many killer diseases, such as smallpox and polio, thus permitting a greater portion of the population to reach old age. Since the 1960s, death rates from cardiovascular disease, on an age-adjusted basis, have dropped by 50% (Centers for Disease Control and Prevention, National Center for Health Statistics, 1995a). The net effect of all these interventions has been to raise average life expectancy in the United States from 47 years in 1900 to an all-time high of 77.9, according to the National Center for Health Statistics.

URBAN LEGENDS OF AGING

"We're living much longer today."

We often hear that people are living vastly longer, but is it really true? Life expectancy, for those who reach age 65, has increased by less than 4 years since the mid-20th century (Friedman & Martin, 2011). That's an increase, but hardly staggering. It's true that life expectancy *at birth* went up around 30 years (from 49 to 79) during the 20th century, and that was the greatest gain in history. But the demographic dividend was largely the result of public health interventions early in life, much less from medical breakthroughs enabling us to live longer lives after age 65.

Normal Aging

In a broad sense, one might say that aging begins at birth, but we normally identify aging with changes that come after maturity. Gerontologists often use the term **normal aging** to describe this underlying irreversible process that is characteristic of each species. Aging can be defined as a time-dependent series of cumulative, progressive, intrinsic, and harmful changes that begin to manifest themselves at reproductive maturity and eventually end in death (Arking, 1998). Primary aging would describe those changes that occur over time independent of any specific disease or trauma to the body, whereas secondary aging would describe disabilities resulting from forces such as disease (Blumenthal, 2003).

The idea of normal aging is important because health care professionals see mainly sick people; as a result, it is easy to develop negative stereotypes about older people. One common stereotype depicts older people as frail and sick. But in fact, the majority of people over age 65 are healthy enough to engage in most *activities of daily living (ADLs)*, such as bathing, dressing, and preparing meals. More than four out of five report no limitations on such everyday activities of life, though the probability of limitations increases with increasing age.

Longevity and Disease

Steps toward health promotion, such as improved diet or increased exercise, can reduce the likelihood of illness and thus increase life expectancy. These steps may also reduce **morbidity** in later life—that is, illness or disease—but not invariably so. It is clear that a decline in the mortality rate need not be matched by a decline in morbidity. Data drawn from ongoing surveys conducted by the U.S. Department of Health and Human Services show that between 1969 and 1986 there was little significant improvement in self-reported health among the segment of the U.S. population 65 and over, but self-perceptions of health status have improved over the past two decades. Whether morbidity will be diminished remains an open question about which there are differing opinions. For instance, an older person with a strong cardiovascular system but with dementia could live for many years in a dismal state resembling the Struldbrugs. Hopes for delaying disease by health promotion alone may not be realistic or appropriate. Moreover, a rising curve of survival into old age does nothing to alter maximum life span, the "natural death" for which the Struldbrugs longed.

Scientists have pursued basic research on the biology of aging in the hope of avoiding the Struldbrug problem—namely, having enormous numbers of frail, sick, and dependent older people whose lives are prolonged in a desperate condition. But do we really need to understand the biology of aging to address this challenge? Couldn't we simply concentrate research attention on eliminating the big "killer diseases" that prevent people from living out a full life span? For example, if the most prevalent diseases of later life, the big killers such as stroke, heart disease, and cancer, were eliminated, wouldn't we all live to be over 100? Unfortunately, the answer is no. Curing all these diseases would give us, on average, only a decade or so more years before some other disease would kill us.

URBAN LEGENDS OF AGING

"Prevention and health promotion are the way to save money in health care."

Sounds appealing, but it's probably not true. That's what Professor Louise Russell concluded more than two decades ago in her massive economic study, *Is Prevention Better Than Cure?* The nonpartisan Congressional Budget Office in 2009 agreed, finding that health promotion measures would in fact *not* save money in health care reform. Prevention might be a good idea, but it won't necessarily save money.

What if we could eliminate all diseases? Would immortality then be at hand? Alas, the answer is no. Time and chance take their toll in the form of accidents. Unless we turn our attention to the underlying vulnerability of human biology, we may change life expectancy, but not maximum life span. Still worse, we might succeed in creating more and more long-living "Struldbrugs." It is quite possible that future declines in death rates will actually have a small effect on average life expectancy, but create much larger numbers of very sick old people. The fear, then, according to critics, would be a Struldbrug scenario: an expansion of morbidity.

This trend will take place, pessimists believe, because medical technology is improving survival prospects for patients with disabling conditions associated with fatal disease—Alzheimer's would be a good example. But the basic progression of the disease remains unchanged. The length of life lived with disability for this part of the population would increase, what Olshansky and Carnes (2002) refer to as *manufactured time*. A second reason for the expansion of morbidity is the increasing role of nonfatal diseases of aging, such as arthritis and some forms of stroke (Olshansky et al., 1991).

But optimists take a different view. Analysis of data from the National Long Term Care Survey by Kenneth Manton and colleagues (2006) showed a significant decline in chronic disability in the older adult population between 1984 and 1989. The proportion of older adults with disabilities actually became *lower* in this period, reflecting improved treatments and lifestyle modifications. For instance, the number of those over age 65 with high blood pressure dropped from 46% in 1982 to 39% in 1989; the percentage of Americans with emphysema went down from 8.9% to 6.4%. The research team concluded that there is reason to expect further progress in the future as successive generations of older people show gains in income and education. On the negative side, they pointed to conditions requiring special attention, such as musculoskeletal problems (e.g., arthritis) and dementia (Manton et al., 2006). However, more recent research on morbidity suggests that the opposite trend may be occurring: an increase in longevity and morbidity over the past two decades (Crimmins & Beltran-Sanchez, 2010). It remains to be seen what these patterns will look like as we move farther into the 21st century.

Basic research may find answers to the common diseases of old age. But beyond curing specific diseases, researchers are also looking at interventions that could delay or actually reverse the process of aging. Here, we confront far-reaching questions about the impact of research on the biology of aging. Are we talking about moving the average life expectancy closer to the upper limit of the maximum life span—say, closer to age 120? Or are we talking about pushing that upper limit itself—say, up to age 150 or 200? Or, are we concerned with enhancing to the fullest extent possible however many years one has to live? In either event, successful anti-aging interventions would have large consequences for human society. But until such research yields practical results, society will have to cope with the consequences of having more long-living individuals, and one of those consequences is vulnerability to disability and disease.

EPIDEMIOLOGY OF AGING

Although aging is not in and of itself a disease, with increasing age comes increasing susceptibility to disease. The vulnerabilities of later life are the subject of **geriatrics**, or the medical specialty that focuses on aging issues. Much has been learned about the major diseases of later life, and this subject is important for debates about aging, health care, and society (Blumenthal, 1983).

Basic to our understanding of diseases in society is the discipline of **epidemiology**, which originally acquired its name from the scientific study of epidemics. Today, epidemiology is more broadly understood as the use of statistical techniques to study the distribution of diseases in human populations. A basic goal for the epidemiology of aging is to understand what diseases are most common among older people and to assess their impact (White et al., 1986). An example of how epidemiological data are organized is given in Exhibit 4, indicating selected chronic conditions among older people, which are the leading causes of death.

Major Diseases in Old Age

There are characteristic diseases of old age (Blumenthal, 1983). For example, today, three quarters of all deaths among persons over age 65 come from just three diseases: heart disease, cancer, and stroke. Death rates for heart disease and stroke have declined in recent decades, but they still remain the leading causes of death. If heart disease were completely eliminated as a cause of death, the average life expectancy for someone 65 years old would increase by 7 years, ignoring the likelihood of death from one of the other leading causes. Although often not listed separately as a cause of death in vital statistics, Alzheimer's disease is probably the fourth leading cause of death chiefly afflicting people over age 65 (Hebert et al., 2003).

Along with diseases causing death, we also need to consider **chronic conditions** that persist for a long period, regardless of whether they cause death. Chronic illness is much more common among the old than among the young. Rates of chronic illness are 46% for those over age 65, compared with only 12% for those younger than that age. Currently, 80% of adults 65 and older living in the United States have one chronic condition, and 50% have at least two (Centers for Disease Control and Prevention, 2007). Exhibit 5 shows the prevalence of selected chronic conditions for people over age 65.

Exhibit 4   Chronic Diseases That Are the Leading Causes of Death for U.S. Adults 65+

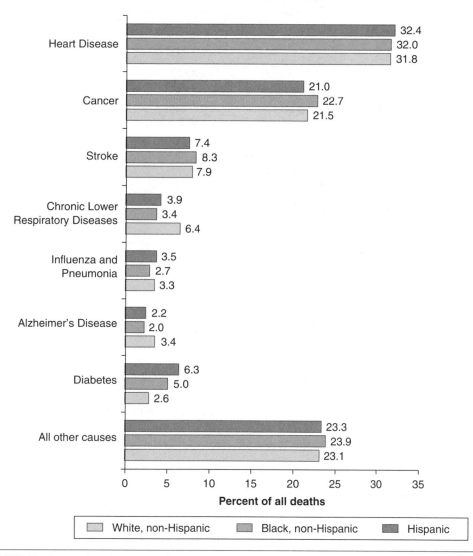

SOURCE: Centers for Disease Control and Prevention, National Center for Health Statistics, National Vital Statistics System, 2006.

Arthritis

Arthritis is the most familiar and one of the most prevalent chronic diseases of later life; it afflicts nearly half of all persons over age 65. Arthritis is basically an inflammation of the joints, also commonly known as "rheumatism," and it is the most important cause of physical disability in the United States. Symptoms include pain and red, swollen joints and muscles. Like cancer, arthritis is actually the name of a group of as many as 100 syndromes, all slightly different. Rheumatoid arthritis can occur at any age, but osteoarthritis is distinctly related to old age and is aggravated by degeneration caused by wear and tear of the joints.

Exhibit 5 Prevalence of Chronic Conditions Among Adults Aged 65 Years or Older by Race/Ethnicity in 2002–2003

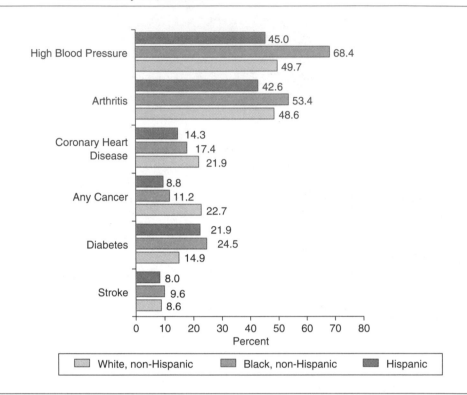

SOURCES: Centers for Disease Control and Prevention, National Center for Health Statistics, National Health Interview Survey, 2006.

NOTE: Rates apply to the noninstitutionalized population. People may have multiple chronic conditions.

Degenerative joint disease in some variety is almost certain to occur in people over the age of 70, but the effect of such disease on ADLs varies tremendously, and most people live full and active lives with it. The cause of arthritis is not known, and there is no cure, but treatment of the disease to reduce symptoms can be effective. Painkilling drugs are not costly, but, for the very serious cases, joint surgery—for example, hip replacement—can be expensive (Moskowitz & Haug, 1985).

Osteoporosis

Osteoporosis is a condition involving the deterioration or disappearance of bone tissue leading to loss of strength and, often, to fracture. The disease is most prevalent in women (four times more common than in men), especially beyond the age of menopause, which occurs around age 51. About one in four White women over the age of 65 will develop osteoporosis. When weakened by osteoporosis, bones are more likely to break, with serious consequences. It is estimated that 1.5 million fractures occur each year as a result of osteoporosis. A hip fracture, often related to a fall, is one of the most common events precipitating admission to a nursing home. About half of those who survive fractures will require some form of long-term care. It is estimated that more than 12 million people in the United States have osteoporosis, and the annual cost of fractures resulting from the disorder is in the range of $7 billion to $10 billion.

Parkinson's Disease

Parkinson's disease is a degenerative neurological disorder characterized by a loss of control over bodily movement. It afflicts about half a million people in the United States, chiefly older people. Symptoms include tremors or shaking of the head and hands, leading to progressive loss of muscle control and the ability to walk unaided. Parkinson's disease is an age-related syndrome, and its incidence increases steadily after midlife. For reasons not clear, dementia is quite prevalent among persons with Parkinson's, and depression is common as well. Parkinson's appears to be caused by lack of dopamine production in brain cells, but there is no treatment that slows the progression of the disorder. Drug treatment, such as L-Dopa, however, can relieve symptoms of the disease (McGoon, 1990).

Cancer

Recent research has focused on aging and changes in the immune system of the body. The strength of the immune function begins to decline after puberty, and with advanced age comes a propensity to develop autoimmune disorders, such as arthritis, as well as higher rates of cancer. In fact, cancer is overwhelmingly a disease of old age, with half of all cancers occurring in people over age 65. The incidence of malignant disease rises progressively with age, so cancer today is the second leading cause of death for Americans over age 65, accounting for 22% of deaths among older people (Centers for Disease Control and Prevention, 2007).

Different forms of cancer seem related to age, but actually may be the result of longer exposure to cancer-causing chemical substances known as carcinogens (e.g., asbestos or tobacco). Because of successful medical interventions, older people who have cancer are living much longer than in the past, so cancer can often become a chronic disease. A person diagnosed with slow-growing or controllable forms of cancer may live many years, thus increasing the cost of medical care over a longer period of time. But it is also possible to prolong the period of dying for those with incurable cancer, raising questions not only about the ethics of termination of treatment but also about the cost of life prolongation.

Cardiovascular Disease

The leading cause of death for people over age 65 remains cardiovascular disease, which includes stroke and heart disease (Kaiser, Morley, & Coe, 1997). Heart disease alone accounts for 32% of all deaths, whereas stroke accounts for another 8% of those deaths (Centers for Disease Control and Prevention, 2007). In the past two decades, there has been a decline of almost 30% in deaths from heart disease, and the cardiovascular condition of older people shows wide variations. According to physiological studies, the heart of a healthy 80-year-old man performs as well as that of someone in his 20s within the normal range of everyday activities. But, unfortunately, about two thirds of men in their 70s have clear evidence of coronary heart disease, so death rates remain high. The economic cost of heart disease is staggering: more than $444 billion in 2010, according to figures from the Centers for Disease Control and Prevention (2011).

Stroke refers to a neurological deficit in the brain arising from a sudden disturbance in the blood supply. A stroke often results in some degree of paralysis, often on one side of the body, or loss of other functions, such as speech, and it can result in coma or death. Although one stroke in three leads to immediate death, another one in three causes permanent disability.

Dementia and Alzheimer's Disease

Dementia is an organic mental disorder involving progressive loss of the capacity to think and remember. It is not characteristic of normal aging, but is the result of a specific disease process. Dementia is characterized by confusion and memory impairment and may manifest itself in a wide range of symptoms, such as wandering or losing things. Dementia can have various causes, but Alzheimer's disease is an important one (Katzman & Bick, 2000). Patients with Alzheimer's may retain social skills and conceal their impairment to some degree. Alzheimer's disease is often hard to diagnose and separate from other cognitive impairments, such as multi-infarct dementia, a condition caused by a series of small strokes affecting the brain.

Senile dementia of the Alzheimer's type (SDAT), or Alzheimer's disease, is the most common cause of irreversible dementia of old age, accounting for two thirds of all dementia conditions. The proportion of people with Alzheimer's disease rises dramatically with each decade of age over 65, doubling every 5 years. It strikes 1 out of 12 persons older than age 65, but the figure rises to 1 out of 3 among those over age 80, at least in some community studies. Between 2 million and 4 million Americans may now be afflicted with the disease. About half the residents of nursing homes have some form of dementia, usually Alzheimer's, but sometimes multi-infarct dementia, which comes from accumulated damage to blood vessels in the brain.

Alzheimer's is a disease caused by deterioration of brain cells, characterized by plaques and tangles. The disorder typically progresses through stages from mild memory loss, through significant cognitive impairment, to very serious confusion and the loss of ability to handle dressing, bathing, or other ADLs (Reisberg, 1983). By the end stage of the disease, there may be incontinence, loss of speech, and inability to walk. A definitive diagnosis of Alzheimer's is difficult, and confirmation usually can be made only upon autopsy. But a mental status examination, such as the Folstein Mini-Mental State Exam, can assess functional cognitive losses produced by the disease (Folstein, Folstein, & McHugh, 1975).

Alzheimer's disease is irreversible and generally foreseeable in its course. In advanced stages, taking care of persons with the disease living in their own homes usually becomes impossible. The result is often placement in a skilled nursing home, sometimes lasting many years. Even when a person's quality of life has severely declined, it is feasible to use modern medical techniques to cure his or her physical illnesses, such as pneumonia or kidney failure, and thus prolong his or her life, resulting in great expense.

In terms of the health care rationing debate, it is worth noting that acute care medical intervention can actually be less costly than long-term care over a period of many years for persons with Alzheimer's (Cassel, Rudberg, & Olshansky, 1992). The National Institute on Aging projects that, unless a cure for Alzheimer's is found, by the middle of the next century, there could be 14 million people with the disorder, costing billions of dollars a year in caregiving costs.

Alzheimer's appears to be one of the most common diseases of late adulthood, and genetic factors clearly contribute to cases of Alzheimer's disease with an early onset. One indication of genetic influence is the association between Down's syndrome and Alzheimer's. Genes found on chromosomes 21 and 14 are known to cause early-onset Alzheimer's, whereas another gene on chromosome 19 seems linked to late onset. The lifetime incidence among relatives of patients with Alzheimer's is estimated at around 20%, or three to four times the risk among comparable groups. If Alzheimer's were purely a genetic disease, however, then it would be expected that identical twins would always come down with the disease. But they do not, thus proving that environmental factors must also play a role in expression of Alzheimer's disease. The classic "Nun Study" showed that individuals with the classic physiological markers of Alzheimer's may not show symptoms of the disease at all (Snowden, 2002). Some neurologists, therefore, point out that it remains very difficult to separate Alzheimer's from other processes of brain aging (Whitehouse & George, 2008).

Although Alzheimer's disease is a major problem, its prevalence among older people should not be exaggerated. Most people over age 65 *do not* suffer from memory defects or dementia. Among all those over 65, perhaps one in five people have a mild or moderate mental impairment. This means the overwhelming majority of older people have no mental impairment at all. Memory defects are quite limited among the large majority of older people, and the capacity for learning and growth in later life remains impressive.

Responses to the Geriatric Diseases

Interventions to eliminate specific diseases, such as cancer and stroke, can increase life expectancy, but they do not raise the maximum life span of individuals. Furthermore, curing a life-threatening illness does not prevent other nonfatal diseases that may bring chronic disability. One of the big questions about aging, health, and society is whether our health care system is capable of dealing with a growing aging population. Many critics charge that it is not. Medicine in the United States has often neglected the dimensions of caring for and coping with people who have illnesses that cannot be cured, such as Parkinson's and Alzheimer's. That neglect is a matter of special concern for geriatric medicine.

The approach of clinical medicine in most advanced countries, and certainly in the United States, focuses almost entirely on discrete causes of disease and their cures. Intrinsic causes within the organism—in other words, vulnerabilities of aging—are not well understood and are not the focus of attention. The paradox here is that, because survivorship has been increasing, the aged have become an increasing proportion of society, and the remaining fatal diseases, whether cancer or Alzheimer's, are linked to the process of aging.

Will a breakthrough in understanding the biology of aging solve this problem? There are reasons for doubt. For example, there is a whole class of age-related changes not likely to be affected by improved DNA repair, a favored mechanism for explaining biological aging. Many physical changes of old age are in the wear-and-tear category and include the decalcification of bones, uric acid encrustation in cartilage of joints, and cholesterol accumulation in blood vessels. It might be possible for geriatric medicine to develop strategies to control causes at the tissue level and to introduce rehabilitative methods that improve the clinical picture. The problem is that many of today's dramatic medical techniques—such as kidney transplants and bypass surgery—do nothing to affect the underlying process of aging. We can keep patients alive, but we can do little to improve their quality of life.

An overview of geriatric epidemiology gives a concrete picture of what the "Struldbrug" problem might look like in the future. Success in curing some forms of cancer or heart disease could raise life expectancy but leave larger numbers of people living with the burden of chronic diseases such as stroke, arthritis, and osteoporosis. A pragmatic approach to geriatric medicine might favor interventions designed to reduce the burden of age-related diseases on individuals as well as society.

Advances in medical technology and adoption of health promotion measures could bring average life expectancy closer to the theoretical upper limit of the maximum life span. But would we then be inadvertently multiplying the Struldbrug problem? Those in favor of age-based health care rationing would cut funding from expensive life-sustaining interventions for the very old and redirect those resources toward quality-of-life interventions for age-related diseases. But there are serious questions about whether paying for extended long-term care is actually cheaper than any alternative we can imagine. Those questions involve the economics of health care.

Adoption of health promotion activities such as exercise could bring average life expectancy closer to the theoretical upper limit of the maximum life span.

Economics of Health Care

The emergence of the Struldbrug problem in the United States has had an important public consequence—namely, rising health care expenditures for the very old. In 2010, older adults comprised 13% of the population. In addition, each older adult consumed significantly more total health care expenditures—more than five times higher than spending per child. This increase has taken place against a background of escalating costs for health care in general. The proportion of the gross national product for health care today is twice what it was in 1965 when Medicare was first enacted, and Medicare remains at the center of the economics of health care for aging (Medicare Payment Advisory Commission, 2004). According to the Centers for Disease Control and Prevention (2007), Medicare spending has grown in the past two decades from $37 billion in 1980 to $336 billion in 2005.

As a nation, the United States has gone from spending approximately 9% of the gross national product on health care in 1980 to spending more than 16% in 2007. Health care is now the second largest item in the federal budget, consuming 20 cents of every dollar spent. Health care spending is growing faster than the general rate of overall inflation, and it remains a concern for the future.

Reimbursement Systems

Medicare is the chief federal government program that pays for health care for 35 million Americans over age 65 and another 5 million people with disabilities of all ages (Centers for Medicare and Medicaid Services, 2004). Medicare has serious limitations: It doesn't pay for the first day of hospitalization; it doesn't cover hearing aids, eyeglasses, or dental care. (See Exhibit 6 for the incidence rates of sensory impairments for older adults.) It also excludes long-term care coverage, except for limited periods after hospital discharge. However, Medicare now covers prescription drugs, as a result of the Medicare Modernization Act (MMA) of 2003. Like most insurance plans, Medicare has deductibles and copayments and covers only 80% of physician expenses. Medicare is available primarily on the basis of age, in contrast to Medicaid, a health program funded by both the states and the federal government, which is available to those below the poverty line and pays for a substantial portion of nursing home care.

Medicare was created in 1965 as part of the Social Security Act. Before Medicare, as many as half of people over age 65 were without health insurance, whereas today almost all people are covered. Much has changed in the Medicare population in more than three decades. Since 1965, life expectancy has risen from 70 to 77, and the 65+ population grew from 9% to 13% of the total U.S. population. Medicare has had a major impact on the health of the older adult population: Since 1965, half as many Americans die of heart attacks and a third as many die of strokes, and this is a tremendous accomplishment.

Like Social Security, Medicare is funded from payroll taxes, with additional funding from general revenues and premiums from beneficiaries. Unlike Social Security, whose problems lie many decades into the future, Medicare faces short-term financing problems. Overall, Medicare spending has risen much faster than the cost of living, and thus it presents government policy makers with a serious problem of cost control.

Medicare actually comprises two distinct programs: Part A, or hospital insurance, and Part B, supplementary medical insurance, covering nonhospital care, which primarily includes physicians' services along with limited home and outpatient services. Medicare Part A is financed by a compulsory payroll tax administered as part of the Social Security tax levied on all wages up to a specified limit. Part B covers 80% of doctors' bills as long as Medicare beneficiaries with incomes less than $85,000 per year pay a $96.40 monthly premium, deducted from their Social Security checks. The monthly premium for beneficiaries with incomes above that range are higher and may increase (see http://questions.medicare.gov). Exhibit 7 shows where money from Medicare goes.

Exhibit 6 Percentage of People Ages 65 and Older Who Reported Having Any Trouble Hearing or Any Trouble Seeing by Sex, 2004

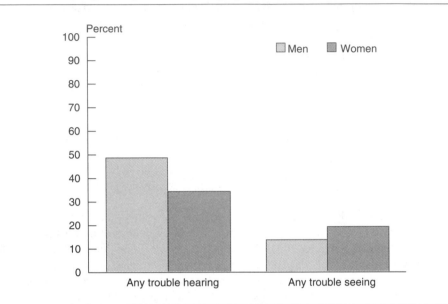

SOURCE: Centers for Disease Control and Prevention, National Center for Health Statistics, National Health Interview Survey, 2004.

NOTE: Rates apply to the noninstitutionalized population. People may have multiple chronic conditions.

Exhibit 7 Where the Medicare Dollar for Older Adults Goes: 2005

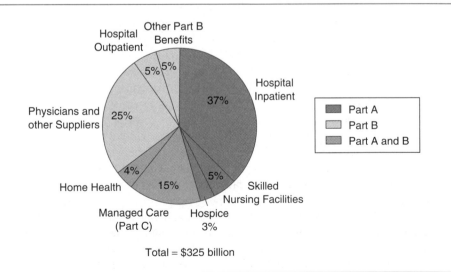

Total = $325 billion

SOURCE: Congressional Budget Office, Medicare Baseline, March 2005.

NOTE: Does not include administrative expenses, such as spending for implementation of the Medicare drug benefit and the Medicare Advantage program. Excludes low-income subsidy payments and items not assigned to particular services.

In 1965, when it was first enacted, Medicare spent a little more than $3 billion. Today, it spends more than $502 billion (Centers for Medicare and Medicaid Services, 2009). Nearly two thirds of that total goes to hospitals, where acute and often high-technology care is provided. If health care rationing on the grounds of age were ever to be introduced, it would probably take place in the Medicare program and would show up in the large sector of Medicare concentrated on hospitals.

Although Medicare expenditures have climbed dramatically, Medicare still covers only about half of the **out-of-pocket medical expenses** of older people: roughly the same percentage as when the Medicare program was first enacted in 1965. Part of the reason is that Medicare Part B reimburses 80% of physicians' "reasonable charges." In fact, the amount reimbursed may or may not reflect actual charges in a specific geographic area. In practice, many physicians in the past have charged much more than the officially allowed Medicare rate, with the patient paying the difference. But that practice has now begun to change. Since 1993, physicians participating in Medicare are limited by law to charging no more than 15% above the rate set for Medicare reimbursement. That law was passed because fewer than half of physicians were willing to accept the official Medicare reimbursement as full payment because the rate was too low. Because of limits on what Medicare will pay, around 30% of Medicare beneficiaries also have private **Medigap insurance** policies to cover the remainder of their medical bills.

Our experience so far with both the Medicare and Medicaid programs gives cause for concern about what might happen if cost-containment measures cut down on physician reimbursement from government insurance programs. Officials of the American Medical Association have rejected the idea of the government setting limits on the fees of doctors, and they have argued that such fee limits will inevitably bring about de facto "rationing" of health care.

Similar fears erupted after 1983, when Congress passed a law limiting payments to hospitals under Medicare. In 1983, Congress responded to the high hospital costs of Medicare Part A by introducing a **prospective payment system**: a new way of reimbursing hospitals for the cost of treating Medicare patients. Under prospective payment, hospitals receive a fixed amount for a specific diagnosis given to a patient no matter how long the hospital stay or the type of service required. Over the past decade, the new prospective payment system has held down hospital costs below what they would have been without these cost controls. But critics charge that the system has resulted in higher outpatient costs and in displacing costs onto families of patients who are discharged "quicker and sicker" (see Exhibit 8).

The system has created hundreds of diagnostic categories, or **diagnosis-related groups (DRGs),** that determine how much a hospital will be reimbursed for patient care. The system, in effect, gives an incentive to hospitals to keep their costs down and discharge patients as early as medically feasible. Despite protests and concerns about the new reimbursement system, DRGs have become an accepted fact of life in American hospitals.

In the 1980s, it was widely feared, and sometimes charged, that these cost-containment measures would lead to "patient dumping" by hospitals, along with widespread deterioration of patient care. Such widespread deterioration did not occur, but the 1983 law did have its intended effect in holding down Medicare Part A spending from where it would have been otherwise. Cost containment for hospital spending proved effective, but during the 1980s, Medicare Part B spending for physicians tripled in size, and outpatient costs—for example, home health care spending—have increased dramatically in recent years.

Exhibit 8   Short-Stay Hospitals: Discharges and Length of Stay for All Payers: 1980–1999

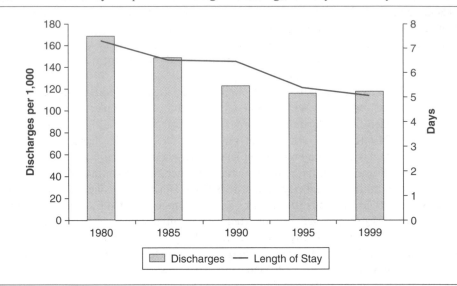

SOURCE: Centers for Disease Control and Prevention, National Center for Health Statistics, Vital and Health Statistics, Series 13 (2005).

NOTE: This chart captures discharges and length of stay for all patients of all payers in nonfederal hospitals with average lengths of stay of less than 30 days.

In part because of the success of DRGs, Congress acted to try to control costs under Medicare Part B. In 1989, Congress passed another law revising the Medicare reimbursement formula for physicians in different medical specialties. The new legislation introduced a so-called Resource-Based Relative Value Scale in the national Medicare program. This Relative Value Scale means that primary health care providers, such as internists, geriatricians, and family practitioners, will be paid more for their services, whereas other specialists, such as some surgeons, will be paid less than they were before.

This reimbursement scheme is an effort to give more incentive to medical specialties involving prevention, health promotion, and quality of life, in contrast to the expensive technologies of life prolongation. Doctors who spend more time with patients but do not use "high-tech" procedures are to be paid more than they were paid previously. The aim of the new measures is to provide a more equitable system of payments reflecting skill, time, and intensity of work.

Despite the ongoing debate about the particulars of Medicare, it commands strong public support as a universal public insurance program for physical illness. By contrast, no consensus has been mobilized to make Medicare a universal public program for long-term care, mental health treatment, or early detection of illness, which might be beneficial in the long run. Medicare will not pay for regular physical examinations or for dental care. Older adults make use of mental health services at only about half the rate of younger people partly because of lower rates of mental illness, but also because today's older generation is likely to be more resistant to using formal services.

Despite recent changes in Medicare, preventive care and health promotion remain low priorities. Critics of this bias note that a great deal of money is spent on acute conditions such as heart disease and cataracts. An expensive procedure such as coronary bypass surgery remains fully covered by Medicare, but a physical exam to detect hypertension or recommend preventive diet or medication is not. Medicare reflects the same priorities favored by the health care system for the nonaging population. The emphasis on technology is in some ways perplexing. Contrary to popular belief, it was not medical technology, but largely social interventions—such as improved sanitation, diet, and public health measures—that accounted for the large drop in mortality in the 20th century. Perhaps further efforts to make lifestyles healthier could help control health care expenditures for our aging population.

The federal government has subsidized some research into the health effects of lifestyle improvements (see, for example, Centers for Disease Control and Prevention, 2007). It has also subsidized research and development in medical technology; in fact, expenditures for biomedical science have increased from $3 million after World War II to more than $26 billion today (Moses et al., 2005). Yet in contrast to private industry, in which investment in research and development leads to lower costs, advances in medical technology have actually led to higher costs for health care. With each new technique for life prolongation, we increase the numbers of those who are very old and very sick.

Prospects for the Future

The escalating cost of health care has become a major problem for older adults and for other groups in society. For example, according to the trustees of the Medicare program, Medicare spending as a share of gross domestic product is projected to grow from under 3% of gross domestic product to more than 13% by 2075. Will biomedical technology help solve the problem or only make it worse?

Some trends in place give cause for concern. First, in health care, new technologies can introduce new services and higher costs. Second, health care costs, even after adjusting for inflation, have continued to rise faster than inflation. Third, the aging of the U.S. population will add to these expenses because incidence of illness and disability is higher among the old. Those over age 65 spend about four times as much money on health care as people below that age. In terms of overall spending for health care, expenditures for those over age 65, who comprise only about 13% of the total population, now amount to more than a third of all health care spending.

It is difficult to predict future levels of use of health care by an aging population. In the past, there were gross underestimates of expenses. In 1965, planners projected the cost of supplemental medical insurance under Medicare. But in 1970, only five years later, there had been a fivefold increase in the cost of that program. Between 1967 and 1975, the rate of use in both parts of Medicare had gone up from 367 per 1,000 enrollees to 528 per 1,000. Recently, Medicare has been growing at a rate three times the rate of inflation.

In light of these huge and rising costs, it is not surprising that there is widespread concern about the prospect of an aging population in the future. Based on U.S. Census Bureau middle-range population forecasts, it is estimated that the Medicare costs for the oldest-old (85+) could increase sixfold by the year 2040 (Schneider & Guralnik, 1990). In light of these trends, as we shall see, there is serious discussion about the rationing of health care in the future.

LONG-TERM CARE

Dramatic end-of-life decisions often attract public attention in debates about the economics of health care. But a far more widespread phenomenon is taking place away from the hospital intensive-care ward for those needing long-term care. People in need of long-term care may live in many different environments, ranging from a nursing home or assisted-living facilities to a single-family residence. Whether in the community or in an institution, people with severe chronic conditions often need help with ADLs, and supportive services can be costly.

How will we provide these needed services? The problem cannot be left for the future. Growing numbers of frail, chronically ill older adults are already in need of long-term care, at home or in institutions. Instead of expecting old people to die early or hoping to find the biomedical fountain of youth, we face the practical problem of how to pay for long-term care, whether furnished by families or in institutions. Opinions differ about who should bear the cost of that care.

Consider the hypothetical case of George and Martha Walton. They never expected to live into their 80s, but they're glad to be alive and glad still to be in their own home in Middletown, USA. Maintaining their home, however, has gotten more difficult since George had his first stroke. Martha finds herself exhausted, and her arthritis prevents her from getting around the way she used to. They can't afford to hire help to come into their home. They've looked into alternative housing arrangements, but the thing George fears most of all is that his condition will deteriorate and he'll end up in a nursing home. They wonder, where will they turn next?

Housing for Older Adults

George and Martha Walton are struggling with long-term care issues regardless of whether they even use or recognize the term *long-term care*. George and Martha like living in their own home, and they don't want to go into a separate residential facility. Their situation, which is typical, shows why the distinction between long-term care services and housing for the aging is not clear-cut. Housing for older adults was long conceived as a bricks-and-mortar affair; that is, it was mainly a matter of financing or subsidizing shelter dedicated to the aged. But increasingly, it is recognized that social as well as physical concerns must be taken into account in planning for housing for the aging population (Newcomer, Lawton, & Byerts, 1986).

Today, around 90% of the older population in the United States lives in conventional housing, made up of mostly single-family houses or apartments. Only 5% of the population over age 65 is in nursing homes, whereas another 5% resides in some form of housing that provides congregate facilities or services. Even among the oldest-old (85+), only about a quarter of the population lives in specialized or supportive housing. But health care for an aging population inevitably brings consideration of housing needs as well.

Housing in the early 21st century of an aging U.S. population may produce greater demand for low-cost housing and coordination of services. Building affordable housing for an aging population is a challenge as funding from the federal government for senior housing continues to shrink. Community-based services, such as home health and adult day care, are likely to be important in the future as cost containment pushes providers to look for alternatives to expensive medicalized facilities such as the nursing home.

The term *nursing home* can refer to any residential facility giving some degree of nursing care to older adults or people with disabilities (Johnson & Grant, 1986). In the United States, about 80% of these facilities are proprietary, that is, operated as commercial, for-profit organizations. Most of the rest are voluntary or nonprofit, with a few run by municipal governments. Among these facilities, it is useful to identify the skilled nursing facility, which is an institution offering medical care, such as a hospital, as well as everyday personal care services to older adults or people with disabilities. An *intermediate-care facility*, in contrast, gives health-related care to patients needing a lower level of support. An *extended-care facility* offers short-term convalescent help to patients coming from hospitals for an extended period of time.

At the same time, interest in new approaches to senior housing is also growing. Today, many public and private sector strategies for planned senior housing are being discussed, including a wide range of options: naturally occurring retirement communities; leisure-oriented and continuing-care retirement communities; board and care homes; adult day care and respite services; and home sharing, assisted living, and medical care in residential settings.

What are the "alternative housing arrangements" that George and Martha Walton might want to consider? In the past, a home for the aged might have been an option. A home for the aged is a facility typically sponsored by a church or fraternal organization and dedicated to helping impoverished or dependent older adults. These residential facilities are less common today, but commercially developed retirement communities have been attractive to more affluent older adults (Hunt, 1983).

Also to be noted is a newer type of facility that has recently seen rapid growth: the **continuing-care retirement community (CCRC)**; Sherwood, Morris, Ruchlin, & Sherwood, 1997). These offer a combination of housing and health care and typically provide a level of social support for those who find it difficult to live on their own. Originally known as "life care communities," CCRCs promise residents the opportunity to "age in place" by combining different levels of health care with housing, nutrition, social supports, and physical security. CCRCs integrate these services under a comprehensive insurance contract that may involve a form of managed care.

Some analysts believe that, at their best, CCRCs can offer a nearly ideal model of health care for older adults because of the guaranteed commitment and integrated approach to housing and long-term care needs (Somers & Spears, 1992). But there are drawbacks. CCRCs are often expensive. George and Martha probably would not qualify. A distinguishing feature of the life care community is that residents are committed to remain there for the rest of their lives: They pay a large entry fee, which can be above $100,000, in return for guaranteed support as they grow older and more frail.

If CCRCs represent the high-income end of the housing continuum, it is important to note the prevalence of **domiciliary care facilities and board-and-care homes** at the lower end (Morgan, Eckert, & Lyon, 1995). These are homes that provide mainly custodial or personal care for older adults and people with disabilities who don't need the intensive medical supervision of a nursing home but who do need help with ADLs.

Another approach is shown by **assisted-living facilities** that offer residents and their families a homelike environment with personal but limited supportive care (Zimmerman et al., 2001). The atmosphere of assisted living promotes a maximum degree of autonomy, independence, and privacy. But assisted-living complexes can also cover the entire continuum of care: from those that provide only minimal help with ADLs to those allowing residents complete nursing care. Assisted living is much more attractive than a nursing home. As hospitals have been pressured to discharge patients earlier and as nursing homes have become facilities for very sick people, assisted living has grown rapidly.

A great advantage of assisted living is that, in contrast to separate retirement communities, assisted-living providers expect to integrate themselves into a surrounding service network, including adult day care, Meals on Wheels, or other social services. Payment for assisted living today is mostly private out of pocket, but insurance and public financing seem likely to grow in the future. Already, 800,000 people are living in 33,000 such facilities around the United States, and it is a fast-growing industry. Still, some questions about assisted living remain unanswered: What happens when residents begin to get sick or seriously impaired? In contrast to skilled nursing facilities, state regulation of assisted living is not consistent (Mollica, 2000).

The federal government subsidizes rental housing through the **Section 202 and Section 8 housing programs** for low-income older adults. But housing programs have often looked only at "bricks and mortar" and failed to take into account the social support needs of older people, which tend to increase with advancing age. Those needs are better taken into account through **congregate housing**: a residential facility providing nutrition, housekeeping, and supportive services for the marginally independent older adult (Chellis, Seagle, & Seagle, 1982). Along the same lines, there has been interest in **shared housing**, an alternative housing arrangement involving either group residence with shared common areas or a homeowner who rents out unused rooms (McConnell & Usher, 1980; Streib, Folts, & Hilker, 1984). These options, including subsidized housing, have mostly been targeted at low-income older adults.

However, for those who can afford it, middle-class and more affluent groups will want to consider leisure-oriented retirement communities, which are different from CCRCs and other supportive living arrangements because they lack a formalized network of social support services. Residents are mostly "on their own" and are expected to live quite independently. Leisure-oriented communities have a prominent focus on recreational activities: By both image and reality, they cater to a healthy, young-old population who aim to enjoy the positive lifestyle offered by such communities.

A question for the future is whether these leisure communities can maintain their recreational identity as the population begins "aging in" and a demand for increasingly intensive support services develops (Folts & Streib, 1994). Still another question, discussed in later chapters of this book, is whether it is socially desirable for housing for older generations to be segregated from housing for younger generations.

All of these options are important, but they probably won't help George and Martha Walton, who just want to remain in their own home. Much of the effort at improved housing for the aging has been planned housing initiated by either the government or the private marketplace. But the overwhelming majority of older Americans live in unplanned housing, typically in the same home and neighborhood they have lived in before, just like George and Martha.

Chronic Care in Old Age

An explosion in demand for long-term care is found in all advanced industrialized countries as a larger and larger proportion of the population survives into old age (Feder, Komisar, & Niefeld, 2000). Compared with the general population, older people on average show twice as many days in which activities are restricted because of chronic conditions. The most important of these conditions are arthritis, rheumatism, and heart conditions. But there are sharp differences in the impact of such conditions among the population over age 65. Apart from people in nursing homes, the young-old group (ages 65–74) have only a small proportion—5.7%—who say they need help with everyday tasks such as household chores, dressing, and going shopping. By contrast, among the oldest-old (ages 85+), the percentage of those needing help jumps to 40% (see Exhibit 9).

Exhibit 9   Percent of Persons With Limitations in Activities of Daily Living by Age
Group: 2007

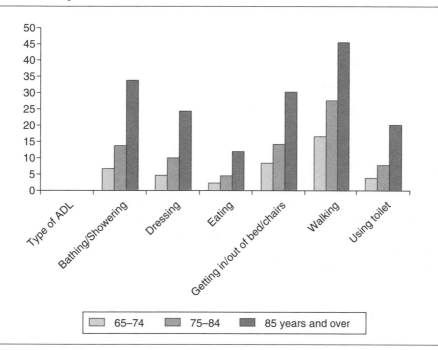

SOURCE: Administration on Aging, 2010.

Long-term care is fundamentally different from acute health care. Acute care is appropriate for conditions that result from a single cause that can be treated by medical intervention. By contrast, the chronic conditions requiring long-term care last a long time and may have varied causes. Examples of such disorders are Alzheimer's disease and other dementias and stroke leading to permanent disability. The result is an inability to perform ADLs (Katz, 1963).

What does this mean in concrete terms? Consider the case of George Walton, who has reached this point. A series of small strokes have affected him profoundly. His condition has deteriorated to the point where he needs help getting to the bathroom and even feeding himself. Martha has done the best she can, but their children, Carol and Robert, have now convinced them that the only alternative is for George to go into the local Middletown nursing home, where he can get the round-the-clock care he needs. George and Martha are afraid to do it; they don't like the idea at all.

In light of George Walton's deterioration, he is likely to enter a skilled nursing facility at some point. As seen in Exhibit 10, in 2006, 1.8 million older Americans live in nursing homes, more than all those in hospitals at any one time, but still less than 5% of the older population. The growth of today's nursing home population is partly a tribute to medical technology and the success of the longevity revolution. But it may also reflect the fact that

Exhibit 10   Nursing Facilities by the Numbers

- Residents: **1.8 million**
- Female: **69.2%**
- Male: **30.8%**
- Median age: **83.2**
- At least one disability: **97.3%**
- Per capita income: **$12,251**

SOURCE: U.S. Census Bureau and Institute for Research on Poverty (2006 figures).

U.S. society has failed to provide accessible and affordable alternatives to living in a nursing home—namely, long-term care based in the home or community. A sizable number of people in nursing homes don't need to be there and could probably live in community settings if appropriate services were available. Estimates of the proportion of the nursing home population in this situation range from 10% to as high as 40%.

Why are George and Martha so fearful about entering a nursing home? Are they right to be afraid? The nursing home has been called a *total institution*, a term used to describe organizations such as prisons, mental hospitals, and boarding schools—that is, facilities that treat people like "inmates" rather than as individuals (Goffman, 1961). In a nursing home, the daily regimen is carefully organized and scheduled, so residents may lose any sense of control over their environment and easily become depressed.

A lot of criticism of nursing homes finds support in careful observational studies of life in these facilities (Gubrium, 1975), and there have been devastating journalistic stories that expose poor conditions in some institutions. Responsible studies have shown how the poor quality of nursing homes arose out of repeated failures in public policy to guarantee good-quality long-term care (Vladeck, 1980). In light of these facts, it is understandable that so many older people today fear institutionalization.

Nevertheless, it is important to remember that, just like schools or hospitals, the quality of long-term care facilities varies widely. The stereotyped view that "all nursing homes are bad" does a disservice to older adults who actually need skilled nursing care, not to mention to the untold numbers of devoted nursing home employees. Government monitoring and regulation have meant that nursing homes today are much better than in the past, and improvements continue (R. A. Kane & R. L. Kane, 1987). Moreover, there is a common misconception that, once someone is admitted to a nursing home, residence there is inevitably a life sentence. In fact, 32% of those in nursing homes stay less than a month; many return home.

How likely is it for older people to anticipate entering a nursing home? Among all people over age 65, only about 5% are in nursing homes at any given time. In other words, it is a mistake to imagine that most or even many older people are in nursing homes. But this low 5% figure may understate the importance of nursing homes in the lives of the very old. It turns out that the percentage of those who will spend some time in a nursing home before they die is much larger: up to 40% of people ages 65 and older. The lower 5% figure comes from citing the percentage of people in a nursing home only at a single point in time. Note that this difference between these two figures—5% in a nursing home at a single point in time versus 40% over the course of a lifetime—shows the dramatic difference in how statistics can be presented. These two figures correspond to a cross-sectional versus a longitudinal view of nursing home residence.

The need for chronic care varies significantly among subgroups of older adults. For those between the ages of 65 and 74, the chance of entering a nursing home is small—only 1 in 100—but for those over age 85, the chance goes up to nearly 1 in 5. Specific risk factors that increase the chances for nursing home placement include mental impairment, chronic disability, advanced age, and spending time in a hospital or another health facility.

Functional Assessment

A key step in determining what kind of help people need is professional long-term care assessment. This determination often serves a "gatekeeping" role in deciding what services will be provided. A **multidimensional functional assessment** takes place when a geriatric professional, such as a doctor or nurse, conducts a full examination of an older adult's physical, mental, and social conditions. This test is more than a physical examination because it covers ADLs, as well as physical and mental health. Among the most important of these activities are feeding, toileting, transferring out of a bed or chair, dressing, and bathing (Katz & Akpom, 1976). A comprehensive functional assessment also looks at social and economic resources, as well as elements such as the physical environment and even strain on caregivers. All these elements play a part in determining the kind of service an older adult may need (see Exhibit 11).

Does a failing score on an assessment test mean that it's time to enter a nursing home? Not necessarily. The key to interpreting an assessment lies in the functional emphasis—that is, asking how an impairment actually affects performance of daily tasks such as shopping, doing housework, handling personal finances, and preparing meals. A comprehensive approach to functional assessment is important because someone with, for instance, mild memory impairment or limited physical mobility may be able to live quite satisfactorily alone in an apartment as long as the environment remains safe and a neighbor or relative comes by regularly to help out. For the same reason, a physical assessment looks not only at biological organ systems but at medications being taken and the impact of sensory impairment on ADLs.

Gerontologists have developed specialized instruments or questionnaires designed to carry out functional assessments (Gresham & Labi, 1984; R. L. Kane & R. A. Kane, 2000). A classic example is the Older Americans Resources and Services (OARS), one of several widely used assessment instruments in the United States today (Duke University, Center for the Study of Aging and Human Development, 1978). The OARS questionnaire gathers information on topics such as mental status, self-assessed well-being, social contact, and help from family. A second part of the instrument looks at the use of services ranging from physical therapy and meal preparation to employment training and transportation. By carefully assessing ADLs in this way, professionals can identify the exact type of help that a client needs (e.g., a walker device for people at risk of falling, a homemaker–home health aide for someone who can't prepare meals, and other kinds of help that might enable people to remain safely in their own homes).

Exhibit 11   Top Functional Problems of Care Recipients

Functional Problem	% Reporting Problem in Previous Week
1. Requires supervision of care tasks	75
2. Taking medications	75
3. Managing money or finances	72
4. Staying alone	70
5. Bathing/showering	69
6. Preparing meals	68
7. Performing household chores	67
8. Dressing	65
9. Grooming	55
10. Mobility	55
11. Using the telephone	52
12. Incontinent	48
13. Using the toilet	45
14. Transferring	43
15. Eating	37
16. Wandering	14
Mean number of functional problems:	9

SOURCE: California Caregiver Resource Centers Uniform Assessment Database (2001).

NOTE: $N = 3,476$

The Continuum of Care

A 65-year-old today can expect to live, on average, for 19 more years. During those years, it is likely that health status and service needs for any individual will change, so provision for long-term care will have to reflect changes over time. Why shouldn't long-term-care services take those changes into account? The idea of a **continuum of care** is based on the goal of offering a range of options responsive to changing individual needs, whether from less intense to more intense, whether at home or in an institution (Brickner et al., 1987).

The ideal of a continuum of care expresses the aim of keeping older people as long as possible out of nursing homes—the most expensive and service-intensive setting. The aim instead is to maintain people in the home, in independent living, or in the least restrictive alternative. If we were to take seriously the ideal of a continuum of care, it would mean spending more money to enlarge the availability of community-based long-term-care services. Such a goal, however, would serve the purpose of promoting maximum independence and personal control and might also help minimize public expense. The reasons for promoting a continuum of care include both choice and economics, but it is rare to find a full continuum of care in most communities in the United States. There are many gaps, and the long-term-care service system remains fragmented and confusing (Binstock, Cluff, & Mering, 1996).

Health care is important, but we should not forget the importance of social care and contact for people like George and Martha. What happens to Martha when she is left all alone after George has entered the nursing home? Who will watch out for her and her needs? If George and Martha were lucky, Middletown, USA, would have a full range of services to help them out, as a few communities already do. The kinds of formal support services delivered to the home that are shown in Exhibit 12 can play a key role in enabling frail older adults to remain in their homes as long as possible (Quinn, Segal, Raisz, & Johnson, 1982).

All of these formal support systems provide a degree of companionship, monitoring, and concrete services for older adults who are frail and isolated. They also can shore up the social network of family, friends, and neighbors—that is, the totality of informal helping relationships that maintain integrity and well-being. Gerontologists have documented the crucial role that these natural support systems play in providing social care and their enormous role in the lives of older adults (Cantor, 1980).

If George Walton had not needed round-the-clock care, there might have been alternatives for him other than going into a nursing home. For instance, why not provide some nursing services on a daytime basis while he remains at home? That, in essence, is the strategy of adult day care, which is usually offered five days a week. Clients visit a community facility, where they are given needed medical and social services as a group during the day, and then return to their homes at the end of the day.

Another alternative is home health care, in which home care aides provide health-related tasks such as rehabilitation exercises or toileting and transferring patients who are bed-bound (Portnow, 1987). Visiting nurses who can dispense medication and perform skilled nursing functions also play a critical role. Home health services have expanded dramatically in recent years as an alternative to institutionalization and as a means of ensuring speedier discharge from hospitals.

These forms of community-based long-term care can sometimes be more cost-effective than a residential nursing home because housing costs are not involved. Most important, they offer an opportunity for those who can to remain relatively independent. The experience of other countries, such as Canada and Great Britain, suggests that adult day care, along with other varieties of community-based long-term care, will have to play a larger role in the United States than it has in the past (R. A. Kane & R. L. Kane, 1987).

Exhibit 12 Support Systems Across the Continuum of Care

Senior centers and congregate housing	Senior citizens' centers offer social and recreational opportunities. Lunches provided for older adults at neighborhood sites, such as senior centers and churches.
Telephone reassurance	Usually performed by peer volunteers. Daily phone calls, typically shortly after wake-up time, to provide support and monitor status. If telephone is not answered, someone goes to the home to check on the client.
Friendly visitor	Volunteer visits, talks with, or reads to a frail homebound older adult.
Chore service or handyman	Visiting person performs outdoor tasks, such as lawn care or snow removal, for the older adult; also may make small repairs and perform minor maintenance.
Homemaker	Visiting person performs light housekeeping (cleaning, washing dishes, vacuuming, laundry, meal preparation, etc.) and food shopping. Services are performed in the home, but do not include services that involve touching the client.
Meals on Wheels	Home delivery of meals supported under the Older Americans Act.
Personal care	Visiting person performs trained but not professional work for the older adult, such as bathing, dressing, and assistance with grooming. Services include touching the client, but not health care services
Home health care	Performed by a trained professional, such as a registered nurse or licensed practical nurse. Administration of medications, measurement of blood pressure, changing of dressings, and so on.
Mental health services	Provision of counseling, psychotherapy, and psychological support services. Practitioners may be psychiatrists, psychologists, nurses, and social workers.
Outpatient medical care	Provision of a range of services, from checkups and diagnostic monitoring through therapeutic procedures short of hospital admission.
Adult day care	Supervision of dependent older adults by professionals or paraprofessionals, offering respite to family caregivers.
Board and care	Residential placement. Meals are provided, housekeeping is performed, and medication reminders are available.
Intermediate-care nursing home	Placement in a facility with (less than 24-hour) supervision and nursing care provided.
Skilled nursing facility	Placement in a nursing home with 24-hour services provided by registered nurses.
Inpatient hospital care	Admission as an inpatient to an acute care facility.

SOURCE: Krain (1995).

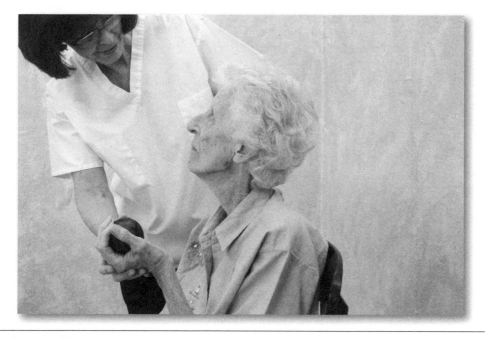

In-home health care and physical therapy services are an example of the continuum of care. The transition into an institutional setting can often be delayed by bringing appropriate services to older adults in their own homes.

Paying for Long-Term Care: An American Dilemma

The costs of long-term care are projected to increase rapidly and dramatically into the 21st century (see Exhibit 13). In the past 10 years, the annual growth rate for nursing home care has been more than 12%. Expenditures now stand at more than $40 billion and are still climbing. Few individuals can afford to pay the complete cost of long-term care in a nursing home. Usually, Medicaid pays part of the bill. Future projections of long-term-care expenditures suggest that private (out-of-pocket) and Medicaid sources will continue to be the biggest source of payment for nursing homes (O'Brien & Elias, 2004).

Advocates for home or other community-based care believe that staying at home costs less than entering a nursing home, just as George and Martha want. But home care is not always cheaper than institutional care. Cost estimates for home care typically fail to include the real value of housing or the value of unpaid family caregiving. Moreover, there is sharp debate about whether we should pay family caregivers to do what is normally done by family members for one another.

The experience of Medicaid payment for nursing home care suggests that some frail older people may end up being placed in nursing homes because institutional care, not home or community-based care, is the only form of long-term care paid for under the U.S. system. When advocates for older adults propose large increases in long-term care, the question arises of who will pay for the expansion (Rivlin & Wiener, 1988).

Should families provide for their own, or should the cost of expanded long-term care be covered by government? Paying for long-term care remains an American dilemma.

Exhibit 13 Projected Nursing Home Expenditures for People Ages 65 and Older by Source of Payment: 1990–2020

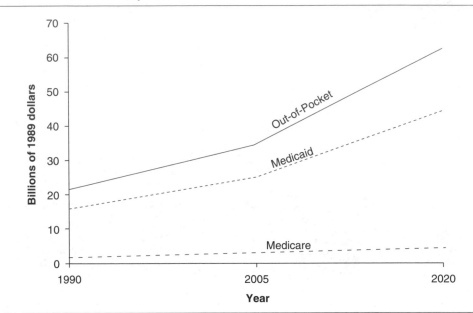

SOURCE: Brookings/ICF, "Long Term Care Financing Model." Washington, DC, 1990.

SELF-DETERMINED DEATH

Our society so far has not been prepared to explicitly ration health care on the grounds of age. Nor do we seem willing to face up to the public policy problem of paying for long-term care. But at some point, decisions become unavoidable, and therefore we turn to our last option: self-determined death. Modern biomedical technology not only enables larger numbers of people to survive into old age, but it has also forced care providers to make explicit decisions about the end of life. The result has been a continuing debate about the so-called right to die, which involves choices from forgoing life-sustaining treatment all the way to assisted suicide (Glick, 1992; Rosenfeld, 2004). In 2005, the right-to-die debate focused on the case of Terri Schiavo, a relatively young woman. But in reality, the debate will increasingly be about end-of-life care for older people.

Today, this debate is taking new forms as the cost of health care rises and the oldest-old population increases in numbers. In the future, termination-of-treatment decisions may unavoidably become intertwined with cost-containment pressures. Instead of individuals claiming a "right to die," we may even see health care providers or policy makers suggesting that some people have a "duty to die" to stop "futile" medical treatment that uses up scarce resources.

This prospect is not just hypothetical. A case in point is the story of Helga Wanglie, who at age 86 broke her hip and was admitted to a nursing home. As a result of complications, Mrs. Wanglie ended up on a respirator and suffered brain damage. The hospital staff felt that, due to her medical condition and advanced age, Mrs. Wanglie should not receive further life support. Her family, however, insisted that treatment be maintained, so the case wound up in court, which agreed with the family. In many other cases, providers have taken a different view and insisted on treating patients while the family asked to end medical treatment.

Another case in which financial considerations became mixed up with termination of treatment was the case of *Grace Plaza of Great Neck, Inc., v. Elbaum* (1993). In this instance, Mrs. Jean Elbaum was in a persistent vegetative state (coma) and was being kept alive by tube feeding. Mrs. Elbaum had made it clear that she would not want to be kept alive under such circumstances, but the nursing home refused to honor the family's wishes. Instead, the facility provided treatment and then sued the family for payment of care provided against their wishes.

In recent decades in the United States, discussion about the right to die has developed along legal and ethical lines focused entirely on individual rights and decisions; it has not focused on resource-allocation issues. But both the Elbaum and Wanglie cases, in different ways, show how end-of-life decisions may now become entangled in considerations about who will pay the bill and whether institutions should expend resources on care that is "medically futile."

The question of medical futility will involve values and will depend on the different treatments involved. One study looked at several different kinds of treatment that might be withheld from older adults and explored the differences among them (U.S. Office of Technology Assessment, 1987). Antibiotics, respirators, cardiopulmonary resuscitation, and kidney dialysis are all different forms of medical technology. A patient's personal decision about one kind of intervention may not hold for another kind. Similarly, a decision may be made in one way at home and differently in a nursing home or a hospital. The setting could make a significant difference in how health care personnel act and what families can expect. Perhaps the most important new developments in the right-to-die debate will center on the question of whether the American health care system can devise practices and forms of treatment that are both respectful of patients' wishes and attentive to the uncertainties involved in end-of-life decisions.

Another question that arises is whether it is actually in the best interest of depressed or debilitated patients to have life-sustaining care terminated because of poor quality of life. The topic is controversial because the patient's best interest may or may not coincide with the interest of the family or of health care providers. When subjective well-being declines and patients want to end their lives, should geriatric health care professionals treat this as a matter of self-determination or a case of suicide prevention?

Most people are uncomfortable when economic considerations become involved with end-of-life decisions. But increasing pressure for cost containment in health care may make it difficult to keep the two matters separate. In 1990, Congress passed the Patient Self-Determination Act to uphold patients' rights. But analysts quickly noted that the law is expected to decrease costs for health care by ending unwanted care. As financial concerns become intertwined with right-to-die considerations, we may wonder whether backdoor rationing of health care could make it more difficult for older patients to assert their rights. It is always cheaper to say no to treatment than to say yes.

Debates about costs and self-determination take place against a background of hopes and fears centered on end-of-life decisions. Our hopes are symbolized, in the poem by Oliver

Wendell Holmes, by the "wonderful one-hoss shay" or carriage, which lasted 100 years and a day and then fell apart all at once, as we saw earlier in this book. Our common hope, in other words, is to live a long life and "fall apart" all at once without decay. But our fears are symbolized by the horrifying image of Gulliver's Struldbrugs, mentioned earlier in this chapter, the same people who today might be wandering in dementia or hooked up to feeding tubes. For increasing numbers of older Americans, self-determined death seems a way to resolve this struggle between hope and fear at the end of life.

Late-Life Suicide

Self-determined death can mean many things, ranging from termination of treatment to active euthanasia or assisted suicide. Those who favor self-determination for end-of-life decisions generally assume that it is possible to make a rational decision to end one's life (e.g., to refuse further treatment and simply permit death to occur). That, at least, is the premise involved in the court decisions that uphold the right to self-determination.

But are these decisions always rational? If they aren't, does that fact mean that end-of-life decisions cannot be left to individual choice? The question is a difficult one. It is not possible to consider the arguments about end-of-life decisions for older people without taking into account mental health issues: specifically, depression, which is a primary cause of old-age suicide. Suicide is now one of the leading causes of death among the old. The suicide rate for the general population is 12 per 100,000, whereas the suicide rate for those over age 65 is 17 per 100,000: nearly 50% higher (see Exhibit 14 for another view of these data).

How can we understand old-age suicide and its causes? The first great sociological investigator of suicide, Émile Durkheim, distinguished several types of suicide. He described "altruistic suicide," or self-sacrifice for the sake of the group or society (Durkheim,

Exhibit 14   Death Rates for Suicide Among People Aged 65 and Over by Race and Sex, 2000

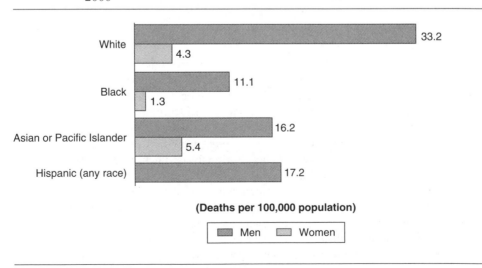

(Deaths per 100,000 population)

SOURCE: National Center for Health Statistics (2003).

1897/1951). A soldier giving up his life on the battlefield to save comrades would be an example of such self-sacrifice. This pattern could describe the voluntary death of some older persons in preindustrial societies facing conditions of economic scarcity. The same pattern might also apply to end-of-life decisions among older people today who fear becoming a burden on their families.

Durkheim also described a form he called "anomic suicide," derived from his sociological concept of anomie, or a condition in which individuals feel hopeless and cut off from any sense of meaning in life. This condition is relevant to thinking about the position of old age in contemporary society. Today, older people commonly experience role loss when they give up previous roles upon retirement, the death of a spouse, or the loss of other social positions. Rosow (1974) described old age in contemporary society as a *role-less role*—that is, a status with no clearly defined purpose or rules of behavior (see also Blau, 1981). A final type of suicide described by Durkheim is "egoistic suicide," where an individual may not be closely integrated into wider society—for example, among the oldest-old, who have outlived most close relatives. In such cases, it might seem perfectly rational for people to end their lives.

As a general rule, the rate of suicide tends to go up with age and to hit a peak after age 65 in the United States as in other advanced industrialized countries. Estimates of suicide remain uncertain because there are 100 suicide attempts for every completed suicide. Among the older population, however, 80% who threaten suicide actually follow through. Furthermore, among older adults who are ill, there is no way to estimate those who end their lives by noncompliance with medical treatment or other forms of self-neglect.

There are pronounced differences in suicide rates among subgroups of older adults, as Exhibit 15 indicates. Among ethnic groups, Blacks have a suicide rate only about 60% of the average for Whites, and, unlike Whites, the rate does not increase in old age. For all age groups, men are much more likely to commit suicide than women, and the difference between the sexes widens with advancing age. For example, according to 1980 data, there were 66 completed suicides per 100,000 White males above the age of 85 in comparison with a rate of only 5 for White females. In fact, the highest rate of suicide in the United States occurs among older White men.

Characteristic conditions preceding late-life suicides include loneliness, social isolation, diminished economic resources, presence of illness or disability, and, above all, depression (McIntosh et al., 1994). Depression is an important public health problem for older adults and must therefore be taken seriously by clinicians and others who work with older people. Early identification and treatment for depression remain a key measure for suicide prevention.

In considering depression and suicide in old age, it is important to maintain a balanced perspective. Older people, by and large, are not unhappy. In fact, most older people enjoy good mental health and a positive attitude. A 1987 Louis Harris survey found that 72% of those over age 65 reported feeling satisfied with their lives, a finding confirmed by subsequent studies. Even when exposed to stress, older people often show a remarkable capacity for adaptation, for instance, in coping positively with bereavement or chronic illness in later life. Adaptation reflects the capacity of the individual to cope with environmental demands and maintain subjective well-being. But when stress exceeds the capacity for coping, psychotherapy and other mental health interventions may play an important role in maintaining the capacity of those in the last stage of life to make rational decisions about the end of life (Butler & Lewis, 1993).

Exhibit 15   Suicide Rates for People Ages 65 and Over, 2000

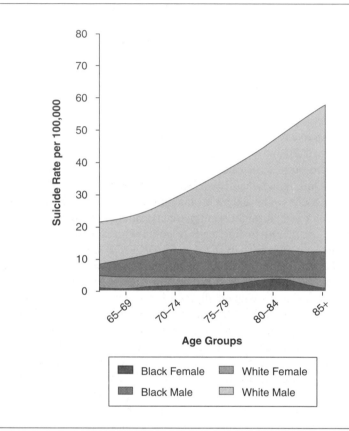

SOURCE: Centers for Disease Control and Prevention, National Center for Health Statistics (2000).

End-of-life choices must also take into consideration what has been learned about the process of death and dying itself. Glaser and Strauss (1965) described the *dying trajectory* by which a person passes from good health to progressively worse health to the point of death. In her influential book *On Death and Dying*, Elisabeth Kübler-Ross (1969) developed a stage theory of dying in which the terminally ill individual moves through stages from denial to acceptance. With respect to end-of-life decisions, it seems clear that older adults who are experiencing a stage of denial or a condition of depression might make different kinds of "rational" decisions about terminating treatment. It would therefore be unwise simply to accept a patient's "spoken choice" at face value. On the other side of the issue, clinicians might well have a less positive view of initiating aggressive medical treatment if they are aware that an older patient is in a period of terminal decline and facing imminent death.

Conclusion

The overall picture of aging and health care today is a mixed one. On the one hand, some optimists hope for a "compression of morbidity," in which disease is postponed and good health continues until late in life. On the other hand, larger numbers of adults with physical or mental frailties are now surviving into old age. The need to make choices about treatment and life prolongation is becoming unavoidable.

The root cause of the problem is that contemporary geriatric medicine is largely symptomatic: Health care responds only after people are sick. Responding to symptoms this way is expensive and frustrating. It proceeds the same way that treatment of polio might have gone if specialists had worked to create evermore complex and refined versions of the iron lung, instead of finding a vaccine to prevent the disease in the first place. In the same way, the "iron lung" approach to geriatric care is bound to be expensive and frustrating.

The American health care system, including geriatric care, spends a great deal of money on acute care conditions such as heart disease and cataracts. In that respect, Medicare simply reflects the same priorities that are favored in health care for the broader population. An expensive procedure, such as a coronary bypass operation, is fully covered by Medicare, but a physical exam to detect hypertension or recommend preventive diet change is not. Such unbalanced emphasis on technology is in some ways perplexing. Contrary to popular belief, it was not medical technology, but largely social interventions—such as sanitation, improved diet, and public health measures—that accounted for the big drop in mortality in infancy and before middle age that occurred at the beginning of the 20th century.

As a universal public insurance program for physical illness, Medicare commands strong public support. By contrast, it has not proved possible to mobilize a consensus behind a universal public program for long-term care, for mental health treatment, or for activities such as early detection that might be beneficial in the long run. Medicare will not pay for regular physical examinations or dental care. Preventive care and health promotion also remain low priorities.

Changing these priorities will be difficult, and solutions to the problems of health care and aging remain elusive. Research on the basic biology of aging will continue, and no one can exclude a dramatic breakthrough that might reshape the conditions of health and sickness in later life. As costs continue to rise, there will be pressure for tough decisions, perhaps even for rationing (Mechanic, 1985). It is unlikely that overt age-based rationing will be adopted in this country, but some form of "backdoor rationing" could come as a result of cost-containment efforts. It seems likely that efforts to liberalize end-of-life decisions will also continue, but we have no way of knowing how many older people or families will decide to deliberately terminate life or where such decisions may lead us as a society. Debates about aging, health care, and society are sure to continue throughout the 21st century.

SHOULD WE RATION
HEALTH CARE FOR OLDER PEOPLE?

To every thing there is a season, and a time to every purpose under the heaven.
A time to be born, and a time to die . . .

—Ecclesiastes 3:1–2

It is no secret that we're spending a lot of money on health care for older adults: Americans over age 65 now account for one third of all national health care expenditures, more than $500 billion on Medicare alone in 2009, and that figure is growing. Health care expenditures for the older population have outpaced general economic growth in recent years, although Medicare covers only half of health care costs. Contrary to what many assume, Medicare is not intended to cover long-term care or many chronic diseases.

As the U.S. population grows older, it seems inevitable that we must spend even more. But what are we getting for all that money? Can we really afford so much health care for an aging population, or are we heading toward a health care crisis in the 21st century (Wolfe, 1993)? These questions would have been unthinkable a few years ago. But today, more and more people are asking such questions. Some have even urged that we cut off expensive health care services for the very old. During the national debate over the Affordable Care Act in 2010, opponents of the new health care law charged that it would lead to "rationing," even though there was nothing in the legislation that would support that claim (Jacobs & Skocpol, 2010). Older people, in particular, were anxious the changes in Medicare would lead to rationing of health care and "killing Grandma."

Rationing health care on the grounds of age is troubling to most Americans. How are we to think about the justice, or the wisdom, of spending vast amounts of money prolonging the lives of the old? Prolonging life seems desirable, but it isn't cheap. With rising costs and new advances in expensive medical technology, decisions about life prolongation are no longer questions just for medical practitioners. The decisions quickly become questions of economics and social justice: Who will get access to expensive health care resources?

Answers to these questions are not easy to find. Some answers that have been given are disturbing and controversial. One of the most controversial is the idea that someday, perhaps soon, we are going to have to "ration" health care to people above a certain age; in effect, we will be telling older people, "You've lived long enough." Philosopher Daniel Callahan has proposed something akin to the sentiment of Ecclesiastes—that there is "a time to be born, and a time to die"—in short, a "natural" human life cycle that people should accept.

Callahan, in his book *Setting Limits* (1987), provoked enormous debate with a serious proposal to ration health care on the grounds of age. But Callahan has not been alone. Others have agreed that age can be a legitimate factor in distributing scarce resources, and some public figures, such as former Colorado Governor Richard Lamm (1993), have called for rationing health care on the grounds of age. Callahan has argued that using age as a way of limiting health care access is unavoidable, and he points to European countries, including England, Switzerland, and some Scandinavian countries, that already engage in rationing.

Callahan's questions are basically these: How much medical progress can Americans afford, and how much money do we want to pay to keep an aging population alive longer and longer? How much should younger generations be prepared to pay for health care of the aged as a group?

PRECEDENTS FOR HEALTH CARE RATIONING

An important question in the health care rationing debate is a practical one: Has it ever been done before? How is rationing based on age likely to be introduced in the United States? Evidence suggests that rationing of health care resources occurs in different countries around the world (Breyer, Kliemt, & Thiele, 2002). A few examples are suggestive here.

Denial of Kidney Dialysis in Britain

In Britain, kidney dialysis has routinely been withheld from people above a certain age, usually 55 (Aaron & Schwartz, 1984). Doctors in Britain's National Health Service simply don't refer such patients to clinics that offer dialysis treatment, and the patients die. CT scans, feeding tubes, hip replacements, and cancer chemotherapy are administered at lower rates than in the United States. In short, British health care authorities use a variety of mechanisms, including deterring people from seeking health care, delaying services, dilution of quality, and outright denial (Harrison & Hunter, 1994).

British primary care physicians have been forced to serve as gatekeepers for the system; they are responsible for denial of lifesaving care or for imposing an age cutoff. Some officials have defended the policy on the grounds that with limited resources, it makes more sense to provide funding to improve quality of life (e.g., offering ample home health care for older adults). But as the covert practice of withholding some treatment for older people became known, public defense of the practice has been abandoned (Halper, 1989).

Waiting Lines in Canada

In Canada, medical care is provided by a national health insurance system, a plan that many believe would be beneficial in the United States. But there have been conflicting views about the Canadian system (Marmor, 1995). Under the Canadian system, virtually no one is deprived of health care because of inability to pay. For some forms of care, however, such as certain surgical procedures that are not needed to save life, it may be necessary to wait long periods. In effect, the waiting list has replaced a market system for allocating some types of medical care (Naylor, 1991).

Life-and-Death Decisions in Seattle

During the 1960s, when kidney dialysis first became available, there were not enough kidney machines in Seattle to take care of all the patients who could benefit from them. For a period, hospitals set up special committees to decide who would have access to dialysis. The committees wrestled with life-and-death decisions and took into account factors such as severity of illness, age, compliance with medical regimen, and social contribution. Decisions by such committees were criticized, and eventually Medicare reimbursement for kidney dialysis made it unnecessary to ration treatment.

A Rationing Plan in Oregon

The state legislature in Oregon passed legislation putting into effect a computer-based ranking of health care problems covered under the state's Medicaid program. According to this ranking system, funding would be made available and services would be rationed not according to individual cases, but according to a consensus reached by democratic means. The state finally obtained federal government approval for the new rationing scheme, but Oregon's proposal received approval over objections that the plan would discriminate against people with disabilities (Oberlander, Marmor, & Jacobs, 2001).

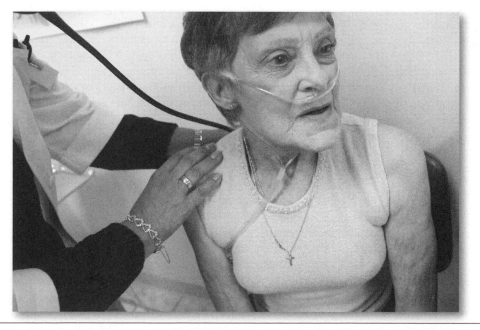

This uninsured woman receives a medical examination at the Open Door Health Center in Homestead, Florida. The Open Door Health Center is a free health clinic that assists the uninsured poor.

These examples show how difficult it is to get public agreement on when or how to ration scarce health care resources. Rationing policies are sometimes put into effect when a clear-cut, unavoidable scarcity exists; organ transplants would be a good example. But if all that is needed is more funding, then rationing health care seems especially open to public criticism. There is evidence that Britain and other European countries have practiced age-based rationing. But virtually none has ever come out publicly and acknowledged this or defended it.

THE JUSTIFICATION FOR AGE-BASED RATIONING

Would age-based rationing be acceptable in the United States? Public opinion surveys tend to show that Americans are concerned about high costs of health care, but are unlikely to attribute these costs to overuse by older adults. A majority of the American public seems willing to withhold life-prolonging medical care for critically ill older persons near the point of death, yet few people would withhold care on the basis of age alone (Zweibel, Cassel, & Karrison, 1993).

Some interesting light is shed on age-based rationing through an opinion survey that asked a British sample the following question: Which of two individuals should be treated if only their ages are different and it is not possible to treat both? Respondents favored treating a 5-year-old over a 70-year-old by a ratio of 84 to 1 and a 35-year-old over a 60-year-old by 14 to 1 (Lewis & Charny, 1989). In short, although rejecting proposals for age-based rationing, people tend to favor choosing younger over older patients for treatment (Kuder & Roeder, 1995).

There are a variety of ways to ration health care besides age. These include ability to pay, anticipated clinical effectiveness, waiting lists or first-come first-served, and productivity to society or social worth. In contrast to these approaches, Callahan believes that chronological age is the best criterion to use because, in his view, each of us has a "natural" life span of 80 to 85 years. When people have completed this natural life span, it is time to "move over" and give others their fair share.

There are some good reasons that can be given in favor of age-based rationing: It would be relatively efficient to administer; from a utilitarian viewpoint, older people are less productive in the economy; from an efficiency standpoint, the likelihood of benefit and years of survival derived from medical care would be less for older than for younger people. Perhaps most important, all people theoretically are members of every age group at some point over a full life course.

But there are powerful reasons against age-based rationing. One major argument against it is the fact that older adults as a group are highly heterogeneous. Chronological age by itself isn't a good predictor of outcome for medical treatments. Once we control for confounding explanations such as disease or functional status, age largely disappears as an explanatory variable.

People on opposing sides of the political spectrum criticize age-based rationing. Those who are more conservative feel that government rationing is morally objectionable and instead favor a market approach, in which each consumer buys insurance coverage appropriate to individually defined need (e.g., medical savings accounts that work somewhat like individual retirement accounts). By contrast, those who are more liberal believe that, instead of the marketplace, we should eliminate the profit motive altogether from health care. They favor access on a more egalitarian basis, perhaps on the pattern of European welfare states.

As we look to the future (see Exhibit 16), U.S. health care spending is likely to rise from 16% in 2012 to 20% by 2016. Medicare currently enrolls just under 15% of the U.S. population, but this figure is expected to rise to 20% by 2025 and even higher after that (see Exhibits 17 and 18). Of all factors, mortality rates have the most powerful influence on Medicare's future because the death rate determines the number of people who survive to become eligible for Medicare and expenditures depend on how long they will live. Would delaying the age of eligibility save Medicare? Apparently not. Even if we raised the age of eligibility from 65 up to 70—a dramatic increase—this change would save less than 15% of total Medicare costs. If Medicare faces financial problems, more far-reaching changes may have to be considered (McKusick, 1999).

Exhibit 16 U.S. National Health Expenditures as a Share of Gross Domestic Product

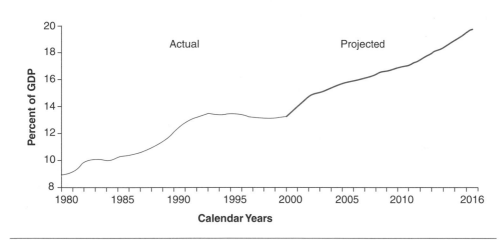

SOURCE: Centers for Medicare and Medicaid Services, Office of the Actuary, National Health Statistics Group.

Others look to strategies such as means testing, using income-related premiums, and seeking alternative sources of revenue beyond the payroll tax for Medicare. Finally, there are those, such as economist Uwe Reinhardt, who believe that some form of rationing is required to improve the efficiency of the system but do not necessarily favor age-based rationing.

Still another strategy is "backdoor rationing," in which implicit or indirect methods limit access to the health care system (Kapp, 2002). For example, when reimbursement rules such as diagnosis-related groups require a patient to leave the hospital for home care, backdoor rationing may be involved. Faced with reimbursement limits, staff members carry out screening procedures that can lead to denial of services. These "gatekeeping" practices have become a familiar part of the practice of hospital discharge planning and case management in geriatric health care. In fact, efficiency and cost control have been motives for adopting case management in many localities (Capitman, 1988).

Exhibit 17 Number of Medicare Beneficiaries, 1970–2030

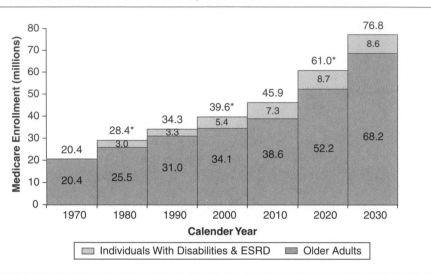

SOURCE: Centers for Medicare and Medicaid Services, Office of the Actuary.

Exhibit 18 Medicare Beneficiaries as a Share of U.S. Population, 1970–2030

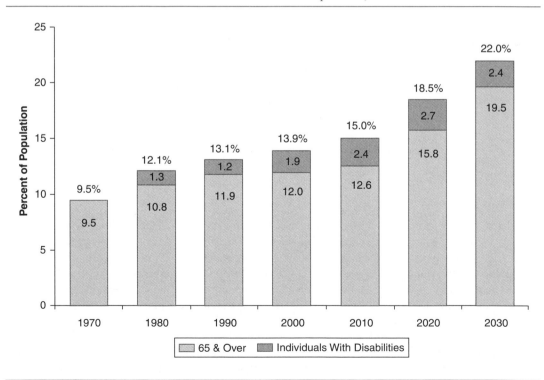

SOURCE: Social Security Administration, Office of the Actuary.

RATIONING AS A COST-SAVING PLAN

One problem with age-based rationing is knowing just how much money it would save. Most of the money spent on health care for older people doesn't go for "high-tech" care in a hospital setting. A substantial share goes for prescription drugs, nursing home care, and home health services. The cost of these last two categories—long-term care for the aged—is increasing rapidly as more and more people survive to advanced ages. Callahan favors spending more on long-term care instead of high-technology medicine.

But the rapid rise in health care costs is not chiefly the result of longevity. Several other forces are also responsible: increases in intensity of services; rate of utilization; introduction of new medical technologies; rise in real wages of health care personnel; general price inflation; and fraud, waste, and abuse, including futile medical treatment. Various strategies have been proposed to contain costs. The most common approach has been **managed care** as a means of combining insurance with health care providers and facilities in a unified network to control costs. During the 1990s, managed care did have a real impact in containing health care costs, although it was unpopular. But will any of these methods enable us to pay for the health care costs of an aging society? Pessimists believe that controlling costs by eliminating unnecessary care, as managed care tries to do, provides only temporary relief because we have already gotten most of the cost savings from managed care. In the long run, population aging and technological innovation may make rationing unavoidable (Schwartz, 1987).

Technology is a major driving force behind the long-term rise of health care spending. Along with medical technology come expensive treatments for life-threatening diseases such as AIDS, heart disease, and cancer. Singling out aging alone seems to miss this larger picture. Even if aggressive, high-cost interventions for older patients likely to die within a year were eliminated, the impact on total U.S. health expenditures would be negligible.

Callahan's proposals, even if fully implemented, would save only $5 billion a year—not a large amount in a $1 trillion annual health care budget (Binstock, 1994).

Others have questioned Callahan's assumption that high-technology care for older adults is inappropriate or wasteful. For instance, coronary artery bypass grafting and angioplasty offer important benefits to older heart patients. Some studies show that older patients can emerge healthy from an intensive hospital stay, proving that age alone is not a good predictor of long-term survival or quality of life among critically ill older patients (Burke, 1993).

Still others have replied to Callahan's proposal by insisting that rationing of health care isn't necessary in the first place (Relman, 1990). They point out that the current health care system is riddled with waste and inefficiency. For example, by comparing statistics with other countries, some analysts have argued that up to half of all the cardiac bypass operations in the United States may not be needed. A 1991 study by the federal government's General Accounting Office found that the present health care system permits unscrupulous providers of services to defraud insurance companies at a staggering rate. It is estimated that 10% ($90 billion) is lost to fraud and abuse every year.

No doubt medical expenses do rise toward the end of life, but costs vary quite dramatically among people over age 65. For example, it turns out that Medicare spends up to 25% of its money on people who are most ill. Are we confronting the principle of diminishing returns? Advancing age brings greater health care expenses, and health care spending is greatest for the oldest-old (85+). Overall personal health care spending for this group is well over $9,000 a year per person, or 2.5 times greater than for persons ages 65 to

69. For nursing home care alone, the ratio is 23 times greater. Yet Callahan is not in favor of cutting off care for people who live in nursing homes, although the annual cost might run up to $50,000 or more.

Behind Callahan's argument is a common stereotypical image of frail older patients subjected to high-technology procedures before being allowed to die. Callahan is concerned that prolonging the lives of these patients is wasteful if the same resources could be used to improve the quality of life of other old people or people of other ages. Overtreatment does occur, of course, partly because of reimbursement incentives and a humanly understandable desire to "do everything possible." However, aggressive treatment of older adults actually seems to decrease as the level of impairment rises. Those with poor quality of life—for example, late-stage dementia patients—are not treated as aggressively as others. A study of heart patients over the age of 75 found that these older patients were more than 12 times *less* likely to receive therapy to dissolve blood clots and 8 times *less* likely to undergo coronary diagnostic procedures compared with patients under 65 years of age (Rosenthal & Fortinsky, 1994). In summary, frail, totally impaired patients seldom receive expensive, high-technology care; instead, they receive supportive care (A. Smith, 1993).

Callahan's proposal for age-based rationing assumes that care for older patients in their last year of life is expensive because of high-technology, life-sustaining medical treatment. But the proportional cost for Medicare beneficiaries in their last month of life remained unchanged between 1976 and 1988, suggesting that expensive high-technology care was not being administered to growing numbers of dying older patients. In 1990, 6.6% of Medicare beneficiaries who died accounted for 22% of Medicare expenditures. These figures seem to support Callahan's argument, but it turns out that patients who are near death are not the biggest cause of large Medicare payments. Instead, the high-cost beneficiaries tended to be survivors. Medicare expenditures for people who died actually *went down* with advancing age. These facts suggest that it may be difficult to develop a policy limiting expenditures for people in the last year of life without also curtailing health care for sick Medicare beneficiaries who have the potential to survive (Garber, MaCurdy, & McClellan, 1998).

URBAN LEGENDS OF AGING

"Health care costs are high because we spend most of the money on old people in the last year of life."

This one can be dubbed "The Last Year of Life Fallacy" because it confuses treatment for severe illness with unreasonable extension of life for people who are dying. Actually, it's not easy to know when "the last year of life" will turn out to be. When we look at the data in retrospect, it turns out that Medicare spends around 25% of its money on people who are the sickest, that is, in "the last year of life." It's just another illustration of the familiar 20-80 rule: 20% of your customers account for 80% of your revenues. Of course there are cases when dying people are unreasonably kept alive, just as there are many cases of undertreatment. But we only know "the last year of life" in retrospect. For comparative purposes, in 2011 the National Institute for Health Care Management Foundation found that 5% of the U.S. population accounted for nearly half of all health care expenditures—another "discovery" that those who are sickest end up costing the most. Contrary to stereotype, after the age of 80, the use of expensive intensive care actually declines.

Perhaps by voluntarily avoiding unneeded care or treatment that prolongs dying, we could avoid rationing health care. The problem here, however, is that it is not so easy to predict how long a given patient will live. Medical costs in the last year of life amount to approximately 18% of total lifetime medical costs and nearly 25% of the entire Medicare budget. That proportion has not changed much in two decades despite advancing technology. The trouble is that we only know that we've spent money on the "last year of life" when that life is over—that is, in retrospect. Clinical studies of medical care at the end of life confirm what doctors have admitted for a long time: Medical science lacks any realistic way of determining who would have died if they hadn't gotten the care they received. In a careful study of Medicare expenditures, among the 1% of beneficiaries who had the highest costs, a majority survived, and of the 5% with the highest costs, nearly two thirds survived. Those in favor of rationing health care on the grounds of age cannot claim any special new power of prediction.

There are many myths about the cost of care in the last year of life (Alliance for Aging Research, 1997). One study of the last year of life found that older adults who received expensive, high-technology care were those patients with good functional status ages 65 to 79. By contrast, frail patients with poor functional status tended to receive mainly supportive care in their final year. In other words, despite the difficulty of predicting when death will occur and despite the lack of explicit rationing criteria, it may be that high-cost medical services are already being provided to the older people with age and functional status being taken into account. Moreover, high technology is not the only factor responsible for high expenses. The frail and debilitated older-old are likely to have high expenses even without high-technology care. In short, the biggest factor in high costs at the end of life may not be inappropriate high technology after all (Scitovsky, 1988).

The real solution, some critics argue, would be a system of national health care combined with careful cost controls to ensure that appropriate care, but not overtreatment, is provided to people at all levels throughout the health care system, not simply in the last year of life. A variety of proposals for providing more health care in a cost-effective manner have been put into practice in recent years. These include new forms of managed care, popular with private industry, and case management, practiced by community-based health service programs. Both are methods for deciding how much care to provide individuals based on some verified assessment of individual need.

The Impetus for Rationing

Will more efficient management of health care distribution solve the problems of access and allocation in an aging society? The answer we give involves some forecast about the health status and needs of the aging population in decades to come (e.g., what is the likely impact of health promotion, such as reductions in smoking, or the probability of a breakthrough in understanding the biology of aging or the causes of specific diseases?).

A prime factor in the rationing debate is *economics*, which can be defined as the "science of scarcity." It is only when scarcity is at hand, when the wolf is at the door, that rationing is seriously considered. In times past, some societies have deprived older people of resources, sometimes even life itself, to make way for the young. One example often cited is the Aleut (Eskimo) tribes who at the point of starvation were sometimes forced to put an older adult out on an ice floe to die in order to have enough food for the remainder of the group. Similarly, in Leningrad during World War II, hundreds of thousands of people, including the very old, died of starvation so that young children might survive.

These life-threatening conditions that prompted rationing have become rare as economic conditions have improved. Today, we face a different kind of scarcity created by the fact that medical technology can save and extend the lives of the very sick and old. Even when the technology is cheap, caring for older people with chronic diseases such as stroke or Alzheimer's disease can be very expensive. As new medical technologies enable us to prolong the lives of the chronically ill, the expense continues to rise.

Exhibit 19 shows a large actual and projected increase in the population ages 85 years and older. This group of people age 85+, sometimes called the "oldest-old," also has the greatest number of health problems and costs the most in terms of health care. If expensive health care resources were rationed on the grounds of age, as Callahan proposes, then this group would probably be the group denied care.

Exhibit 19   Actual and Projected Increase in the 85+ Population, 1900–2100

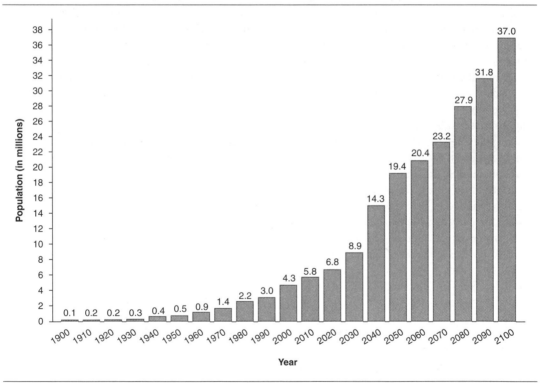

SOURCE: U.S. Census Bureau (2001).

Cost Versus Age

As medical technology advances, Callahan fears that life-extending technology will run up against the law of diminishing returns. We end up spending more and more to achieve limited incremental gains, often with poor quality of life, while other social needs go unmet. But is this assumption about the cost of technology correct? Some life-extending technology, such as a penicillin shot, is actually quite inexpensive. However, keeping a patient alive and cared for in a low-technology environment such as a nursing home can be very expensive: $50,000 or $60,000 a year or more. If we really want to cut down on the expense created by the increase in the older population, why wouldn't we also withhold inexpensive life-extending technologies? As long as we look exclusively at the economic aspect of health and aging, it is difficult to avoid thinking of choices in terms of cost-benefit or cost-effectiveness standards (Avorn, 1984). But once we adopt those standards, don't we tend to downgrade the value of life in old age? Callahan believes that society owes older adults a decent minimum of health care, at least up to a certain age.

The high cost of more health care for the old is part of the problem. But cost alone is not the whole story for Callahan because he would accept paying for certain expensive procedures for younger people. The basic principle that Callahan wants us to consider is chronological age. In the end, Daniel Callahan, along with the verse from the Book of Ecclesiastes cited at the opening of this chapter, believes that human life has a natural rhythm or cycle: a time to live and a time to die. Callahan argues that this "natural" life span comprising the traditional 70 years and then 10 years, or maybe a bit more, should be the basis for thinking about the goals of health care. He believes we should do what we can to enable people to live out a full life span, however defined, but nothing more. After that, we should not expend scarce resources on the very old. Instead, we should let them die.

Pushing this argument further, some have even urged a "duty to die cheaply." If preserving a life of diminished quality in old age is less of a benefit to the aged patient than the resources saved for others, then it would be in the mutual self-interest of all to have death come more quickly (Menzel, 1990). This idea of solidarity and altruism makes sense in some situations, but not in others. For instance, rationing within the British National Health Service is more easily justified than in the United States because Britain operates a closed system, providing universal access to care under a regionally centralized budget. Denial of treatment for one patient means money is available to treat other patients.

Callahan's goal might be seen as fair and even idealistic. He wants to guarantee older people, along with everyone else, access to universal health care and help everyone avoid an early, premature death. He proposes to reform the health care system by achieving a better balance of caring and curing, specifically by improving long-term care and home care. Only after accomplishing these goals, he insists, would it be time to introduce an age-based cutoff of life-extending technologies under Medicare (Callahan, 1994).

However, critics of Callahan argue that age-based rationing actually affects only those who depend on government-run health programs—that is, older people who can't afford private care. Thus, proposals for rationing health care actually amount to a rationale for spending less for the neediest in society—in no way a just solution. Age-based rationing would tend to perpetuate or make worse the problem of access according to ability to pay.

But Callahan and others who favor age-based rationing reply that invisible forms of rationing already take place for the 40 million Americans who lack health insurance. They believe that adopting an explicit rationing policy would force everyone to face up to the allocation decisions already in effect, but kept invisible. The result of this honesty would be greater fairness for all. Indeed, Oregon's rationing plan for Medicaid was put into effect with this idea in mind.

URBAN LEGENDS OF AGING

"The 2010 health care law introduced 'death panels' and rationing of Medicare."

It seems that a third of Americans believe this. But there's nothing remotely like this in the Affordable Care Act that was passed. Earlier in the legislative process, a Georgia Republican did introduce a provision permitting payment for Medicare patients to talk with their doctor about end-of-life choices, but it was stricken after controversy. The health care law did not cut Medicare, but it did slow the growth of future Medicare spending, thereby extending the life of the Medicare Trust Fund from 7 to 19 years.

GLOBAL PERSPECTIVE

Age-Based Rationing of Health Care in Britain

In their landmark book, *The Painful Prescription* (1984), Henry Aaron and William Schwartz describe how age-based rationing operated in Great Britain. Compared to the United States, they found much lower rates for coronary artery surgery, X-ray exams, CT scans, kidney dialysis, hip replacement, and cancer chemotherapy. Has the situation changed in the years since then?

There are many forms of rationing, ranging from outright denial of service to long waiting lines that also limit use of resources. Waiting periods can be particularly critical. The leading British paper *The Guardian* has reported that, with prodding from the National Health Service, emergency rooms in the United Kingdom are trying to meet a new four-hour waiting time goal for emergency room care by the traditional British practice of "queuing" or lining up. Longer waiting time could present particular challenges for very old patients.

Other forms of age-based rationing are more explicit. For example, women over the age of 50 may be regularly encouraged to undergo breast screening, but invitations stop at age 70. A 2006 article in the *British Medical Journal* documented age-based rationing of care for cancer services, coronary care units, prevention of vascular disease, mental health services, and management of minor strokes. Another study by the British advocacy group Age Concern disclosed that half of family doctors said they would worry if an older family member were in a National Health Service facility. In that study many physicians reported that they knew well that age-based rationing of treatment occurs. Sally Greengross, chief of Age Concern, said "the survey provides solid evidence that age-based rationing is the scourge of today's National Health Service." Even official groups can come down in favor of age-based rationing. For example, the British National Institute for Health and Clinical Excellence in 2005 examined issues of rationing related to cost-effectiveness. They concluded that there were no grounds for withholding resources based on gender or sexual orientation, nor for self-inflicted conditions, such as smoking or obesity. However, the recommendations concluded that where age is an indicator of benefit or risk, age discrimination is appropriate. The debate will go on.

SOURCE: Matthews, Eric, and Russell, Elizabeth, *Rationing Medical Care on the Basis of Age: The Moral Dimensions,* Oxford, UK: Trust for Research and Policy Studies in Health Services, 2005.

ALTERNATIVE APPROACHES TO RATIONING

A variety of other approaches to rationing have been put forward by health economists and others concerned with improving the efficiency of the health care system (Jones & Higgs, 1992). One approach is to limit medical procedures based on effectiveness as measured by health outcomes research. For example, angioplasty, an otherwise useful procedure, does not produce any medical benefit when performed on a person already experiencing an acute heart attack. Many procedures of doubtful benefit are employed with the Medicare population. A study by the Commonwealth Fund estimated that more than a third of all procedures reimbursed under Medicare were performed for equivocal or inappropriate reasons. Some procedures are used because doctors are familiar with them and reimbursement is available regardless of whether they are inappropriate.

Other approaches to rationing include **cost-benefit analysis**, in which we ask how much a treatment costs in comparison with the total benefit that will be created if the patient lives. For example, we might measure the patient's future economic productivity. Obviously, a cost-benefit approach would discourage high-cost treatments for older people. Still another approach is known as **cost-effectiveness analysis**. Here we look at which treatment provides the desired outcome for the least cost. But again, depending on the outcome measure used, the lives of the young may be favored over the lives of people who are older (Welch, 1991).

One interesting approach in health care economics is known as **quality-adjusted life years (QALYs)**. The idea behind QALYs is the commonsense view that 10 years of life with disability may not have the same value as 10 years of good health (Nord, 1999); people with more disabilities have a poorer quality of life. If functional assessment determines that some people have poor QALYs, then they should be denied health care. But who decides what counts as "quality of life"? Economists use exercises in which a patient's own priorities and preferences are used to construct an index for comparative purposes. Then different forms of treatment with alternative outcomes can be ranked according to cost. But again, QALYs may result in resources being channeled away from the old or chronically ill—perhaps a form of ageism or discrimination against older people (Tsuchiya, 2000). All these different approaches to the economics of rationing raise a deep and difficult question: What is a human life worth (Dranove, 2003)?

EUTHANASIA AND ASSISTED SUICIDE

Callahan is against deliberately killing people or having doctors collaborate with patients who want help in ending their lives. His rationing proposal calls for holding back treatment, not directly killing people, say, by an injection. But other critics have wondered whether Callahan's argument isn't inherently self-contradictory. Why is it acceptable to hold back treatment when that holding back will predictably result in a patient's death, but it is unacceptable to cooperate with a patient who voluntarily asks for help in ending life?

It seems as if Callahan is calling for involuntary death for people above a fixed age, but, at the same time, he wants to prohibit acts, such as voluntary euthanasia or assisted suicide, that people might adopt as a matter of personal choice. Is it possible that his own proposal could make more headway if he also supported voluntary withdrawal of treatment for people of advanced age (Battin, 1987)? If we moved to a voluntary system, rather than the involuntary one urged by Callahan, what might be the likely consequences for society? For health professionals such as doctors and nurses? For older people?

THE DEBATE OVER AGE-BASED RATIONING

The questions continue, and the debate goes on. Callahan himself has repeatedly maintained his original call for age-based rationing, but he has gone further in calling for wider reform of health care. He makes it clear that, in his view, health care for older adults must be rationed, but we will have to make other difficult choices to have a just system for all. Whether he is right remains the subject of vigorous debate.

In the readings that follow, we see this debate unfold along different lines. Peter Ubel, like Callahan, takes a hard line and insists that it is time to introduce explicit health care rationing, although not necessarily on the grounds of age. In contrast, Nat Hentoff, in his article, vigorously criticizes Callahan and asks what kind of society we would become if Callahan's proposal were adopted. Terrie Wetle and Richard Besdine also reject the proposal, worrying about the danger of a negative view of older people. Finally, Daniel Perry and Robert Butler argue for investment in biomedical research to ensure that old age is not an extended period of expensive frailty, but a time of health and vitality.

Daniel Callahan is serious about his proposal, and this fact has shocked many people. Critics have responded to his proposals by calling them ageist, discriminatory, and dangerous. Callahan denies that he is urging age discrimination. Instead, he wants to guarantee older people, along with everyone else, access to universal health care and thereby help everyone maximize vitality and avoid a premature death. To accomplish this goal, he believes, means we need to achieve a better balance between caring and curing in our health care system. As a practical matter, he wants a trade-off between improving long-term care and cutting off life-extending technologies to be paid for under Medicare. Setting an age limit is tragic, but it is the best we can do, Callahan believes.

Is Daniel Callahan cruel and hard-hearted, or is he instead courageous and farsighted in his willingness to advocate a controversial idea? His own words, and the response of his critics, must be the basis for what the fair-minded reader will conclude.

FOCUS ON PRACTICE

Managed Care

An important feature of health care for older people today is the spread of managed care, such as Medicare health maintenance organizations (HMOs). By 2002, 5 million Medicare beneficiaries—or 12% of the Medicare-eligible population—were enrolled in managed care plans. To qualify for Medicare reimbursement, these managed care plans must provide all the services that Medicare covers; as a result, they have proved attractive to many older people. A growing minority of older Americans are using managed care (Kongstvedt, 2004).

Managed care is already the dominant pattern in health care delivery in the United States. A 1996 survey by the nationally prominent accounting firm KPMG Peat Marwick found that nearly three quarters of all Americans who have health insurance through their employers are enrolled in some type of managed care plan, which is a dramatic rise from 29% in 1988. Today, a majority of HMOs offer some type of Medicare managed care plan.

There are both advantages and disadvantages for an individual to change from conventional Medicare fee-for-service to a managed care plan or Medicare HMO. On the positive side, managed care demands less paperwork. Doctors' visits, hospital bills, and lab tests are covered in full, with low copayments and without high deductibles. Medicare HMOs may also offer extra benefits, such as low-cost prescription drugs or vision care. They also eliminate the need for private Medigap insurance, which is widespread in the Medicare population. Above all, the whole concept of "managed" care is intended to improve coordination of care and services. Finally, managed plans have a clear incentive to offer preventive health care, such as checkups and immunizations. All of these are positive points in favor of managed care.

On the negative side, managed care has the drawback of imposing limits. When people sign up, they must go through the Medicare HMO network to receive their health care. Patients can't choose their own doctors, hospitals, or other service providers. For older people, it may prove a hardship not to be able to continue using a doctor they've known for many years. For those who travel, the plan may limit coverage when outside the service area. Above all, the managed care plan will only pay for preapproved services.

Medicare beneficiaries have long had the option of switching from traditional fee-for-service coverage to HMOs. Since the mid-1990s, managed care providers have been successful in enrolling Medicare beneficiaries in large numbers. Since 1999, the program then known as Medicare Plus Choice allowed beneficiaries to choose from eight standardized health plan categories, including HMOs, fee-for-service plans, and provider-sponsored organizations. Managed care plans remain attractive because they offer no deductibles, extremely low copayments, and almost no paperwork or claim forms. In many instances, they provide eye and ear exams or low-cost prescription drugs.

Enrollment in Medicare HMOs remains voluntary. The young-old who are healthy today may find managed care plans attractive, but their attitudes could change later, especially if they are denied coverage or if they face a limited choice of doctors. In the new environment of managed care, physician–patient relationships are likely to take on importance for older adults because their expectations about medical encounters have developed over a lifetime under the traditional fee-for-service approach (S. Putnam, 1996). Older consumers will have to educate themselves to evaluate health-marketing appeals and become capable of choosing the most appropriate managed care provider for themselves (see Exhibit 20).

There is an ongoing debate about whether encouraging more older Americans to enroll in managed care organizations will actually reduce Medicare costs (Gold, 2003). Some critics fear that managed care is just a means of "backdoor rationing." Yet surveys suggest that older patients enrolled in Medicare HMOs actually may be more satisfied with their coverage than those enrolled in traditional fee-for-service plans (Margolis, 1995).

Yet the issue of hidden or "backdoor" rationing persists because of reimbursement under managed care. Under a complex formula, the federal government pays a Medicare managed care plan a fixed sum of money on a per-person rather than a per-service basis. This reimbursement method is known more broadly as **capitation** (per head). What this means is that, after receiving a fixed amount under capitation, a managed care plan then becomes responsible for each beneficiary's full health care costs. If a person stays healthy, the managed care group gets to keep

(Continued)

(Continued)

the extra money. But if someone gets sick, even if treatment costs $100,000, the plan is responsible for covering the cost, just as with any insurance plan. The profit motive, then, may introduce incentives for backdoor rationing in unexpected ways.

For example, Medicare managed care plans are motivated to avoid signing up those likely to incur large medical costs. This form of backdoor rationing can be accomplished indirectly by marketing techniques. But refusing to enroll frail older people is against federal law. Furthermore, even people who sign up while healthy are likely to get sick later on. Another approach is to cut costs by denying coverage to those who are very sick. Backdoor rationing may take the form of refusing treatments on the grounds of "medical necessity." If a treatment fails that test, it doesn't qualify for coverage.

Exhibit 20   Primary Reason a Beneficiary Joined a Medicare Risk HMO, 2000

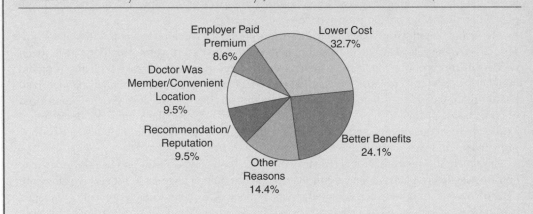

SOURCE: Centers for Medicare and Medicaid Services, Office of Research, Development, and Information: Data from the Medicare Current Beneficiary Survey (2000).

NOTE: Does not include beneficiaries in facility care.

What happens if someone in a Medicare HMO is denied coverage? Studies by the General Accounting Office have found that few people appeal a denial of coverage. An appeal can take up to six months, and that is a significant time factor for someone who is 70 or 80 years old. Another federal study reported that a quarter of Medicare HMO members weren't even aware that they had the right to appeal a denial of coverage. When beneficiaries do appeal, they win about 40% of the time, but many decide not to go through the process. Nor does the appeals process deal with the question of quality assurance. What happens if backdoor rationing takes the form of substandard care for millions of older adults (see Exhibit 21)?

Exhibit 21   Medicare Managed Care Enrollment Growth, 1990–2009

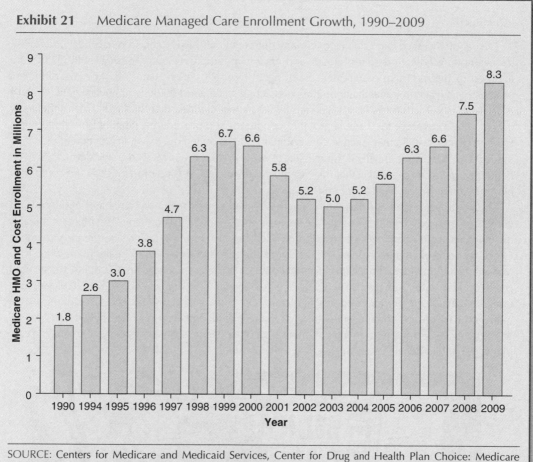

SOURCE: Centers for Medicare and Medicaid Services, Center for Drug and Health Plan Choice: Medicare Advantage data are from the Medicare Managed Care Contract (MMCC) Summary reports for 1990–2009; data development by the Office of Research, Development, and Information.

NOTE: Medicare enrollment numbers are for December of each year, except in 1996 (August data). For all years, the Medicare enrollment includes enrollment in Risk plans—including HMOs, Preferred Provider Organizations (PPOs), and Provider Sponsored Organizations (PSOs)—and in Cost plans other than Health Care Prepayment Plans (HCPPs). For 2004 and 2005, the Medicare enrollment includes enrollment in PPOs that were demonstrations. For all years, the Medicare enrollment excludes enrollment in Private Fee-for-Service (PFFS) plans and demonstrations that were not PPOs. For 2006–2009, the Medicare enrollment excludes enrollment in Regional PPOs.

One problem here is that quality of care is not easy to measure. A study by *Mathematica* researchers found that, compared with a control group under conventional Medicare coverage, stroke patients in a Medicare HMO were discharged from hospitals "quicker and sicker" (R. Brown et al., 1993). The HMO saved money by getting them out of the hospital earlier, but patients faced serious problems because they ended up at home or in facilities that could not provide appropriate rehabilitation.

(Continued)

(Continued)

Denial of coverage can take place in many different kinds of services. In rehabilitative services, for example, Medicare requires "steady and meaningful improvement." Medicare HMOs may be inclined to interpret that requirement in stringent terms. A patient who fails to make sufficient progress gets no more rehabilitation. In a class action suit against the federal government, *Grijalva v. Shalala* (1996), attorneys for a Medicare HMO patient claimed that the health plan repeatedly provided less coverage for a whole group of conditions and procedures, ranging from pneumonia to hip replacement. Some managed care companies have pulled out of Medicare altogether, citing inadequate Medicare payment increases. However, the 2003 Medicare Modernization Act contains provisions suggesting that the "Medicare Advantage" version of managed care could now be more attractive for health care consumers.

The controversy over age-based rationing has shifted to new terrain with debates about how far companies can go in limiting health care services because of the bottom line. The ultimate threat is for managed care companies simply to drop Medicare coverage. In 1999, a General Accounting Office report found that more than 400,000 older adults had to look for new health plans when their managed care providers dropped Medicare contracts. Ethical principles—justice, rights, the greatest good for the greatest number—may not be easy to apply in these circumstances. Backdoor rationing under managed care presents a confusing picture. Or should we even speak about "rationing" when managed care groups control access through indirect requirements that create delays and appeals without any clear resolution? As a larger number of older people sign up for Medicare managed care plans, the public debate about gatekeeping will continue.

An older couple protesting cuts to Medicare at a health care march and rally sponsored by the Keep Patients First, Save Our Health Care Coalition in New York City.

READING 16

Why We Must Set Limits

Daniel Callahan

In October 1986 Dr. Thomas Starzl of Presbyterian University Hospital in Pittsburgh successfully transplanted a liver into a seventy-six-year-old woman, thereby extending to the elderly patient one of the most technologically sophisticated and expensive kinds of medical treatment available (the typical cost of such an operation is more than $200,000). Not long after that, Congress brought organ transplants under Medicare coverage, thus guaranteeing an even greater range of this form of lifesaving care for older age groups.

That is, on its face, the kind of medical progress we usually hail: a triumph of medical technology and a newfound benefit provided by an established health-care program. But at the same time those events were taking place, a government campaign for cost containment was under way, with a special focus on Medicare. It is not hard to understand why.

In 1980 people over age sixty-five—11 percent of the population—accounted for 29 percent of the total American health-care expenditures of $219.4 billion. By 1986 the elderly accounted for 31 percent of the total expenditures of $450 billion. Annual Medicare costs are projected to rise from $75 billion in 1986 to $114 billion by the year 2000, and that is in current, not inflated, dollars.

Is it sensible, in the face of rapidly increasing health-care costs for the elderly, to press forward with new and expensive ways of extending their lives? Is it possible to hope to control costs while simultaneously supporting innovative and costly research? Those are now unavoidable questions. Medicare costs are rising at an extraordinary pace, fueled by an increasing number and proportion of the elderly. The fastest-growing age group in the United States is comprised of those over age eighty-five, increasing at a rate of about 10 percent every two years. By the year 2040, it has been projected, the elderly will represent 21 percent of the population and consume 45 percent of all health-care expenditures. How can costs of that magnitude be borne?

Yet there is another powerful reality to consider that moves in a different direction: Medicare and Medicaid are grossly inadequate in meeting the real and full needs of the elderly. The system fails most notably in providing decent long-term care and home care. Members of minority groups, and single or widowed women, are particularly disadvantaged. How will it be possible, then, to provide the growing number of elderly with even present levels of care, and also rid the system of its inadequacies and inequities, and yet at the same time add expensive new technologies?

The straight answer is that it will be impossible to do all those things and, worse still, it may be harmful even to try. The economic burdens that combination would impose on younger age groups, and the skewing of national social priorities too heavily toward health care, would themselves be good reasons to hesitate.

SOURCE: "Why We Must Set Limits" by Daniel Callahan in *A Good Old Age? The Paradox of Setting Limits* (pp. 23–35), edited by P. Homer and M. Holstein. New York: Simon & Schuster, 1990. Copyright © 1987 by Daniel Callahan. Reprinted by permission.

BEYOND ECONOMICS: WHAT IS GOOD FOR THE ELDERLY?

My concern, however, extends beyond the crisis in health-care costs. "I want to lay the foundation for a more austere thesis: that even with relatively ample resources, there will be better ways in the future to spend our money than on indefinitely extending the life of the elderly. That is neither a wise social goal nor one that the aged themselves should want, however compellingly it will attract them. . . . Our affluence and refusal to accept limits have led and allowed us to evade some deeper truths about the living of a good life and the place of aging and death in that life" (*SL,* 53, 116).[1]

The coming economic crisis provides a much-needed opportunity to ask some fundamental questions. Just what is it that we want medicine to do for us as we age? Other cultures have believed that aging should be accepted and that it should be in part a time of preparation for death. Our culture seems increasingly to dispute that view, preferring instead, it often seems, to think of aging as hardly more than another disease, to be fought and rejected. Why does our culture have such difficulty with this question?

Let me start by saying that "the place of the elderly in a good society is a communal, not only an individual, question. It goes unexplored in a culture that does not easily speak the language of community and mutual responsibility. The demands of our interest-group political life constitute another obstacle. . . . It is most at home using the language of individual rights as part of its campaigns, and can rarely afford the luxury of publicly recognizing the competing needs of other groups. Yet the greatest obstacle may be our almost utter inability to find a meaningful place in public discourse for suffering and decline in life. They are recognized only as enemies to be fought: with science, with social programs, and with a supreme optimism that with sufficient energy and imagination they can

be overcome. We have created a way of life that can only leave serious questions of limits, finitude, the proper ends of human life, of evil and suffering, in the realm of the private self or of religion; they are thus treated as incorrigibly subjective or merely pietistic" (*SL,* 220).

In its long-standing ambition to forestall death, medicine has reached its last frontier in the care of the aged. Of course children and young adults still die of maladies that are open to potential cure, but the highest proportion of the dying (70 percent) are over sixty-five. If death is ever to be humbled, that is where endless work remains to be done. This defiant battle against death and decline is not limited to medicine. Our culture has worked hard to redefine old age as a time of liberation, but not decline, a time of travel, of new ventures in education and self-discovery, of the ever-accessible tennis court or golf course, and of delightfully periodic but thankfully brief visits from well-behaved grandchildren. That is, to be sure, an idealized picture, but it arouses hopes that spur medicine to wage an aggressive war against the infirmities of old age.

As we have seen, the costs of such a war would be prohibitive. No matter how much is spent, the ultimate problem will still remain: People will grow old and die. Worse still, by pretending that old age can be turned into a kind of endless middle age, we rob it of any meaning.

THE MEANING AND SIGNIFICANCE OF OLD AGE

There are various sources of meaning and significance available for the aged, but it is the elderly's particular obligation to the future that I believe is essential. "Not only is it the most neglected perspective on the elderly, but it is the most pertinent as we try to understand the problem of their health care. The young—children and young adults—most justly and appropriately spend their time preparing for future roles and developing a self pertinent to them. The mature adult has the responsibility to procreate and rear the next generation and to manage the present society. What can the elderly most appropriately do? It should be the special role of the elderly to be the moral conservators of that which has been and the most active proponents of that which will be after they are no longer here. Their indispensable role as conservators is what generates what I believe ought to be the *primary* aspiration of the old, which is to serve the young and the future. Just as they were once the heirs of a society built by others, who passed on to them what they needed to know to keep going, so are they likewise obliged to do the same for those who will follow them.

"Only the old—who alone have seen in their long lives first a future on the horizon and then its actual arrival—can know what it means to go from past through present to future. That is valuable and unique knowledge. If the young are to flourish, then the old should step aside in an active way, working until the very end to do what they can to leave behind them a world hopeful for the young and worthy of bequest. The acceptance of their aging and death will be the principal stimulus to doing this. It is this seemingly paradoxical combination of withdrawal to prepare for death and an active, helpful leave-taking oriented toward the young which provides the possibility for meaning and significance in a contemporary context. Meaning is provided because there is a purpose in that kind of aging, combining an identity for the self with the serving of a critical function in the lives of others—that of linking the past, present, and future—something which, even if they are unaware of it, they cannot do without. Significance is provided because society, in recognizing and encouraging the aged in their duties toward the young, gives them a clear and important role, one that both is necessary for the common good and that *only* they can play" (*SL,* 43).

It is important to underscore that while the elderly have an obligation to serve the young, the young and society have a duty to assist the elderly. Before any limits are imposed, policies and programs must be in place to help the elderly live out a "natural life span," and beyond that to provide the means to relieve suffering.

A "NATURAL LIFE SPAN" AND A "TOLERABLE DEATH"

Earlier generations accepted the idea that there was a "natural life span"—the biblical norm of threescore and ten captures that notion. It is an idea well worth reconsidering and would provide us with a meaningful and realizable goal. Modern medicine and biology have insinuated the belief that the average life span is not a natural fact at all, but instead one that is strictly dependent on the state of medical knowledge and skill. And there is much to that belief as a statistical fact: Average life expectancy continues to increase, with no end in sight.

There are, moreover, other strong obstacles to the development of a notion of a "natural life span." This notion "requires a number of conditions we seem reluctant to agree to: (1) that life has relatively fixed stages—a notion rejected on the ground that we are free to make of our different stages of chronological age whatever we want; biology presents no unalterable philosophical and moral constraints or any clear pointers; (2) that death may present an 'absolute limit' to life—an idea repudiated because of the ability of medicine to constantly push back the boundary line between life and death; life is an open-ended possibility, not a closed circle; (3) that old age is of necessity marked by decline and thus requires a unique set of meanings to take account of that fact—a viewpoint that must be rejected as part of the political struggle against ageism, which would make of the old a deviant, marginal, and burdensome group; and (4) that 'our civilization' would be better off if it shared some common view of 'the whole of life'—rejected as a politically hazardous notion, more congenial to authoritarian and collectivist cultures than to those marked by moral and religious pluralism and individualism" (*SL,* 40–41).

I want to argue that we can have and must have a notion of a "natural life span" that is based on some deeper understanding of human needs and possibilities, not on the state of medical technology. I offer a definition of the "natural life span" as "one in which life's possibilities have on the whole been achieved and after which death may be understood as a sad, but nonetheless relatively acceptable event.

"Each part of that definition requires some explanation. What do I mean when I say that 'one's life possibilities have on the whole been accomplished'? I mean something very simple: that most of those opportunities which life affords people will have been achieved by that point. Life affords us a number of opportunities. These include work, love, the procreating and raising of a family, life with others, the pursuit of moral and other ideals, the experience of beauty, travel, and knowledge, among others. By old age—and here I mean even by the age of 65—most of us will have had a chance to experience those goods, and will certainly experience them by our late 70s or early 80s. It is not that life will cease, after those ages, to offer us some new opportunities; we might do something we have never done but always sought to do. Nor is it that life will necessarily cease to offer us opportunities to continue experiencing its earlier benefits. Ordinarily it will not. But what we have accomplished by old age is the having of the opportunities themselves, and to some relatively full degree. Many people, sadly, fail to have all the opportunities they might have: they may never have found love, may not have had the income to travel, may not have gained much knowledge through lack of education, and so on. More old age is not likely to make up for those deficiencies, however; the pattern of such lives, including their deprivations, is not likely to change significantly in old age, much less open up radically new opportunities hitherto missing" (*SL,* 66–67).

A longer life does not guarantee a better life. No matter how long medicine enables people to live, death at any time—at age 90 or 100 or 110—would frustrate some possibility, some as-yet-unrealized goal. The easily preventable death of a young child is an outrage. Death from an incurable disease of someone in the prime of young adulthood is a tragedy. But death at an old age, after a long and full life, is simply sad, a part of life itself, what I would call a "tolerable death."

This notion of a "tolerable death" helps illumine the concept of a "natural life span," and together these two notions set the foundation for an appropriate goal for medicine in its approach to aging. "My definition of a 'tolerable death' is this: the individual event of death at that stage in a life span when (a) one's life possibilities have on the whole been accomplished: (b) one's moral obligations to those for whom one has had responsibility have been discharged; and (c) one's death will not seem to others an offense to sense or sensibility, or tempt others to despair and rage at the finitude of human existence. Note the most obvious feature of this definition: it is a biographical, not a biological, definition" (*SL,* 66).

The Principles and Priorities of a Plan

How might we devise a plan to limit the costs of health care for the aged under public entitlement programs that is fair, humane, and sensitive to their special requirements and dignity? Let me suggest three principles to undergird a quest for limits:

"1. Government has a duty, based on our collective social obligations, to help people live out a natural life span, but not actively to help extend life medically beyond that point. By life-extending treatment, I will mean any medical intervention, technology, procedure, or medication whose ordinary effect is to forestall the moment of death, whether or not the treatment affects the underlying life-threatening disease or biological process.

"2. Government is obliged to develop, employ, and pay for only that kind and degree of life-extending technology necessary for medicine to achieve and serve the end of a natural life span; the question is not whether a technology is available that can save a life, but whether there is an obligation to use the technology.

"3. Beyond the point of a natural life span, government should provide only the means necessary for the relief of suffering, not life-extending technology" (*SL,* 137–38).

What would the actual policy look like? "A full policy plan would include detailed directions, for example, for determining priorities within basic biological research, within health-care delivery, and between research and delivery. That I will not try to provide. I can only sketch a possible trajectory—or, to switch metaphors, a kind of likely general story. But if that at least can be done in a coherent fashion, avoiding the most flagrant contradictions, it might represent some useful movement" (*SL,* 141–42).

Three elements of health policy emerge from my position: "The first is the need for an antidote to the major cause of a mistaken moral emphasis in the care of the elderly and a likely source of growing high costs of their care in the years ahead. That cause is constant innovation in high-technology medicine relentlessly applied to life-extending care of the elderly; it is a blessing that too often turns into a curse. . . . No technology should be developed or applied to the elderly that does not promise great and inexpensive improvement in the quality of their lives, no matter how promising for life extension. Incremental gains, achieved at high cost, should be considered unacceptable. Forthright government declarations that Medicare reimbursement will not be available for technologies that do not achieve a high, very high, standard of efficacy would discourage development of marginally beneficial items" (*SL,* 142, 143).

"The second element is a need to focus on those subgroups of the elderly—particularly women, the poor, and minorities—who have as yet not been well served, for whom a strong claim can be entered for more help from the young and society more generally. . . . The elderly (both poor and middle-class) can have no decent sense of security unless there is a full reform of the system of health care. It may well be that reforms of the sweeping kind implied in these widely voiced criticisms could more than consume in the short run any savings generated by inhibitions of the kind I am proposing in the development and use of medical technology. But they would address a problem that technological development does nothing to meet. They would also reassure the old that there will be a floor of security under their old age and that ill health will not ruin them financially, destroy their freedom, or leave them dependent upon their children (to the detriment of both)" (*SL,* 142, 147).

"The third is a set of high-priority health and welfare needs—nursing and long-term care, prevention—which would have to be met in pursuit of the goals I have proposed. . . . Beyond avoiding a premature death, what do the elderly need from medicine to complete their lives in an acceptable way? They need to be as independent as possible, freed from excess worry about the financial or familial burdens of ill health, and physically and emotionally positioned to seek whatever meaning and significance can be found in old age. Medicine can only try to maintain the health which facilitates that latter quest, not guarantee its success. That facilitation is enhanced by physical mobility, mental alertness, and emotional stability. Chronic illness, pain, and suffering are all major impediments and of course appropriate targets for medical research and improved health-care delivery. Major research priorities should be those chronic illnesses which so burden the later years and which have accompanied the increase in longevity" (*SL,* 142, 149).

EUTHANASIA AND ASSISTED SUICIDE

Some might view my position as an endorsement of euthanasia and assisted suicide. My position "is exactly the opposite: a sanctioning of mercy killing and assisted suicide for the elderly would offer them little practical help and would serve as a threatening symbol of devaluation of old age. . . . Were euthanasia and assisted suicide to be legalized, would there be a large and hitherto restrained group of elderly eager to take advantage of the new opportunity? There is no evidence to suggest that there would be, in either this country or in any other. But even if there might be some, what larger significance might the elderly in general draw from the new situation? It would be perfectly plausible for them to interpret it as the granting of a new freedom. It would be no less plausible for them to interpret it as a societal concession to the view that old age can have no meaning and significance if accompanied by decline, pain, and despair. It would be to come close to saying officially that old age can be empty and pointless and that society must give up on elderly people. For the young it could convey the message that pain is not to be endured, that community cannot be found for many of the old, and that a life not marked by good health, by hope and vitality, is not a life worth living. . . .

"What do we as a society want to say about the elderly and their lives? If one believes that the old should not be rejected, that old age is worthy of respect, that the old have as valid a social place as any other age group, and that the old are as diverse in their temperaments and outlooks as any other age group, an endorsement of a special need for euthanasia for the old seems to belie all those commitments. It would be a way of legitimizing the view that old age is a special time of lost hopes, empty futures, and personal pointlessness. Alternatively, if it is believed that old age can have a special value, that it can—with the right cultural, economic, and political support—be a time of meaning and significance, then one will not embrace euthanasia as a special solution for the problem of old age, either for the aged as individuals or for the aged as a group. It would convey precisely the wrong symbolism. To sanction euthanasia as a special benefit for the aged would signal a direct contradiction to an effort to give meaning and significance to old age" (*SL,* 194, 196, 197). We as a society should instead guarantee elderly persons greater control over their own dying—and particularly an enforceable right to refuse aggressive life-extending treatment.

CONCLUSION

The system I propose would not immediately bring down the cost of care of the elderly; it would add cost. But it would set in place the beginning of a new understanding of old age, one that would admit of eventual stabilization and limits. The elderly will not be served by a belief that only a lack of resources, better financing mechanisms, or political power stands between them and the limitations of their bodies. The good of younger age groups will not be served by inspiring in them a desire to live to an old age that maintains the vitality of youth indefinitely, as if old age were nothing but a sign that medicine has failed its mission. The future of our society will not be served by allowing expenditures on health care for the elderly to escalate endlessly and uncontrollably, fueled by the false altruistic belief that anything less is to deny the elderly their dignity. Nor will it be aided by the pervasive kind of self-serving argument that urges the young to support such a crusade because they will eventually benefit from it.

We require instead an understanding of the process of aging and death that looks to our obligation to the young and to the future, that recognizes the necessity of limits and the acceptance of decline and death, and that values the old for their age and not for their continuing youthful vitality. In the name of accepting the elderly and repudiating discrimination against them, we have succeeded mainly in pretending that with enough will and money the unpleasant part of old age can be abolished. In the name of medical progress we have carried out a relentless war against death and decline, failing to ask in any probing way if that will give us a better society for all.

"There is little danger that the views I advance here will elicit such instant acclaim (or any acclaim, for that matter) that the present generation of the elderly will feel much of their effect. That could take two or three decades if there is any merit in what I say, and what I am looking for is not any quick change but the beginning of a long-term discussion, one that will perhaps lead people to change their thinking, and most important, their expectations, about old age and death" (*SL,* 10).

NOTE

1. Daniel Callahan, *Setting Limits: Medical Goals in an Aging Society* (New York: Simon & Schuster, 1987). References to this volume appear in text here with the notation "*SL.*"

READING 17

--- Pricing Life ---

Why It's Time for Health Care Rationing

Peter Ubel

In the United States, people frequently debate the pros and cons of managed care organizations and whether medicine should be a for-profit business. They almost never debate health care rationing. Instead, they mention it only to accuse managed care organizations or for-profit insurance companies of some egregious crime against humanity. Clearly, there are many important issues to debate about managed care organizations and about the rampant corporatization of American health care. But these debates miss the larger issue of the need to ration health care.

Managed care organizations did not create health care rationing. Instead, the need to ration created managed care. In the United States, managed care organizations proliferated largely because of their presumed ability to contain costs. But governments in Europe, Asia, and other parts of North America are also desperate to control health care costs. Outside the United States, the need to ration health care forced governments to devise other ways besides managed care to contain costs. In Canada, patients wait for months for heart bypass surgery, only to be bumped to the back of the line when another patient becomes urgently ill (Naylor, 1991). Indeed, even in the United States, managed care organizations are not the only groups engaged in rationing. Traditional fee-for-service insurance companies hire hordes of utilization reviewers to have patients discharged from the hospital earlier. State governments change eligibility criteria for Medicaid. Hospitals close down trauma centers to avoid uninsured patients. Rationing is ubiquitous. Managed care is not.

Of course, it is easier for people to argue about greedy managed care organizations, evil insurance companies, and incompetent government bureaucrats than to discuss the need to ration health care. After all, everyone agrees that managed care organizations exist. No one agrees whether rationing exists or has to exist. Instead, to many people concerned about health care cost containment, rationing is an unjustifiable evil. It is wrong that it exists. It is immoral that physicians are allowing it to occur. It is even more evil that physicians are often the ones *doing* the rationing.

Given the relatively recent push to contain health care costs, it should be no surprise that the "R" word is controversial, or that it is often used to discredit political opponents or industry competitors. "Rationing" has become a code word for immoral, inappropriate, or greedy.

Because it is so unpopular, most debates focus more on *whether* we should ration health care than on *how* we should do it. In many cases the debaters do not even agree on what it means.

I want to convince people that rationing is necessary. I expect this view will be unpopular. But I am not running for public office, so the only people I have to be popular with are my wife (who is blindly in love with me) and the members of my tenure review board (who won't read this book). I can afford to be unpopular.

We cannot have it all. We cannot afford to give every health service to every person who could possibly benefit. Most people's health would improve if they had dietitians review what they ate for dinner and physical therapists work the kinks out of their lower backs. Most hospitalized patients and most nursing home residents would benefit from a higher nurse-to-patient ratio. If we could really afford to have it all, standard contrast dyes would no longer be standard, and newer, more expensive dyes would be offered to everyone. Instead, we would only concern ourselves with *effectiveness* analyses—showing us what works best so we could make sure everyone gets it. But there are limits to what we can offer everyone, and we must start figuring out how to set those limits.

In traveling farther down this road of gloom and doom, I am not only going to insist that we have to ration health care, but also that some of this rationing ought to be done by physicians at the bedside, and that our most useful rationing tool (at the bedside and at policy levels) is cost-effectiveness analysis. Although there are ethical problems with physicians rationing... bedside rationing, based on cost-effectiveness, ought to play a larger role.

The moral questions raised by cost-effectiveness analysis deserve to be debated by a broad audience.

REFERENCE

Naylor, C. D., "A Different View of Queues in Ontario," *Health Affairs* (1991), 10(3): 111–128.

READING 18

The Pied Piper Returns for the Old Folks

Nat Hentoff

I expect that the sardonic Dean of Dublin's Saint Patrick's Cathedral, Jonathan Swift, would appreciate Daniel Callahan's *Setting Limits*—though not in the way he would be supposed to. Swift, you will recall, at a time of terrible poverty and hunger in Ireland, wrote *A Modest Proposal*. Rather than having the children of the poor continue to be such a burden to their parents and their nation, why not persuade the poor to raise their children to be slaughtered at the right, succulent time and sold to the rich as delicacies for dining?

What could be more humane? The children would be spared a life of poverty, their parents

SOURCE: "The Pied Piper Returns for the Old Folks" by Nat Hentoff in *The Village Voice,* April 26, 1988. Reprinted by permission.

would be saved from starvation, and the overall economy of Ireland would be in better shape.

So, I thought, Callahan, wanting to dramatize the parlous and poignant state of America's elderly, has created his modern version of *A Modest Proposal.*

I was wrong. He's not jiving. . . .

Callahan sees "a natural life span" as being ready to say goodbye in one's late seventies or early eighties. He hasn't fixed on an exact age yet. Don't lose your birth certificate.

If people persist in living beyond the time that Callahan, if not God, has allotted them, the government will move in. Congress will require that anybody past that age must be denied Medicare payments for such procedures as certain forms of open heart surgery, certain extended stays in an intensive care unit, and who knows what else.

Moreover, as an index of how human the spirit of *Setting Limits* is, if an old person is diagnosed as being in a chronic vegetative state (some physicians screw up this diagnosis), the Callahan plan mandates that the feeding tube be denied or removed. (No one is certain whether someone actually in a persistent vegetative state can *feel* what's going on while being starved to death. If there is a sensation, there is no more horrible way to die.)

What about the elderly who don't have to depend on Medicare? Millions of the poor and middle class have no other choice than to go to the government, but there are some old folks with money. They, of course, do not have to pay any attention to Daniel Callahan at all. Like the well-to-do from time immemorial, they will get any degree of medical care they want.

So, *Setting Limits* is class-biased in the most fundamental way. People without resources in need of certain kinds of care will die sooner than old folks who do not have to depend on the government and Daniel Callahan. . . .

Callahan reveals that once we start going down the slippery slope of utilitarianism, we slide by—faster and faster—a lot of old-timey ethical norms. Like the declaration of the

Catholic bishops of America that medical care is "indispensable to the protection of human dignity." The bishops didn't say that dignity is only for people who can afford it. They know that if you're 84, and only Medicare can pay your bills but says it won't pay for treatment that will extend your life, then your "human dignity" is shot to hell. . . .

It must be pointed out that Daniel Callahan does not expect or intend his design for natural dying to be implemented soon. First of all, the public will have to be brought around. But that shouldn't be too difficult in the long run. I am aware of few organized protests against the court decisions in a number of states that feeding tubes can be removed from patients— many of them elderly—who are not terminally ill and are not in intractable pain. And some of these people may not be in a persistently vegetative state. (For instance, Nancy Ellen Jobes in New Jersey.)

So, the way the Zeitgeist is going, I think public opinion could eventually be won over to Callahan's modest proposal. But he has another reason to want to wait. He doesn't want his vision of "setting limits" to go into effect until society has assured the elderly access to decent long-term home care or nursing home care as well as better coverage for drugs, eyeglasses, and the like.

Even if all that were to happen, there still would be profound ethical and constitutional problems. What kind of society will we have become if we tuck in the elderly in nursing homes and then refuse them medical treatment that would prolong their lives?

And what of the physicians who will find it abhorrent to limit the care they give solely on the basis of age? As a presumably penitent former Nazi doctor said, "Either one is a doctor or one is not."

On the other hand, if the Callahan plan is not to begin for a while, new kinds of doctors can be trained who will take a utilitarian rather than a Hippocratic oath. ("I will never forget that my dedication is to the society as a whole rather to any individual patient.") Already, I have been told by a physician who heads a large teaching institution that a growing number of doctors are spending less time and attention on the elderly. There are similar reports from other such places.

Meanwhile, nobody I've read or heard on the Callahan proposal has mentioned the Fourteenth Amendment and its insistence that all of us must have "equal protection of the laws." What Callahan aims to do is take an entire class of people—on the basis only of their age—and deny them medical care that might prolong their lives. This is not quite *Dred Scott,* but even though the elderly are not yet at the level of close constitutional scrutiny given by the Supreme Court to Blacks, other minorities, and women, the old can't be pushed into the grave just like that, can they?

Or can they? Some of the more influential luminaries in the nation—Joe Califano, George Will, and a fleet of bioethicists, among them— have heralded *Setting Limits* as the way to go.

Will you be ready?

READING 19

—————————— **Letting Individuals Decide** ——————————

Terrie Wetle and Richard W. Besdine

Setting Limits is disturbing in several ways. First, there is the premise that we are justified in setting public policy that determines a "natural life span" for an entire cohort of the population. Referring to the Nazi concept of the *Untermensch,* Callahan notes the evils that result from the political determination that a life is dispensable, but he sets aside the concern far too easily that the elderly—or any other age group, for that matter— would interpret his "natural life span" policy as devaluation of life in old age.

A second concern is whether the program could be applied consistently and fairly. Noting that a policy to limit public payment for life-sustaining care on the basis of age would lead to a two-tiered system in which wealthy older people could still buy such care, Callahan still does not believe that "a society would be made morally intolerable by that kind of imbalance." It was just such an imbalance between those who could pay for care and those who could not that led to the enactment of Medicare and Medicaid 25 years ago.

Many distinctions on which the proposed program would depend are not made clearly or reliably. For example, the distinction between interventions that prolong life and those that relieve suffering is perhaps easy to make conceptually and in situations, but not at the bedside or in that vast middle ground where the majority of cases are found. An 80-year-old man with excruciating abdominal pain and fecal vomiting due to adhesions obstructing his small bowel will have his suffering relieved quickly and best by surgery to release the obstruction. In the process, his life may also be saved. We wonder whether Callahan would urge morphine rather than surgery for such a patient.

Callahan uses the treatment of diabetes to define the rules of his game further. Considering

---

insulin a life-prolonging rather than a symptom-relieving treatment, he states that a diabetic using insulin before the end of his policy-defined natural life span would be "grandfathered" into a continuation of that medication, whereas the person who acquires diabetes after the cutoff age would not be provided such treatment. Similarly, dialysis would be continued indefinitely if it was initiated before the cutoff date, but it would not be provided for late-onset renal disease. Thus, the patient whose diabetes or renal failure develops before the cutoff age and who begins treatment promptly is given preference over the person healthy at that age but in whom illness develops later. This is a peculiar logic.

Much of the book, it seems, is based on the premise that such a policy would save the taxpayer money and allow a reallocation of resources. It is not clear, nor is evidence provided, that the policy actually would accomplish these goals. In fact, it is possible that certain "life-prolonging" interventions also improve function, resulting in the decreased use of other expensive forms of care.

Certainly, the book is worth reading, but with a critical eye. Care must be taken to avoid facile applications of its arguments in support of negative views of older people. Although Callahan has warned against the tyranny of individualism throughout his career, perhaps aging and health care are one arena in which an acute focus on the individual is most appropriate. The decision to provide or withhold life-prolonging interventions may still be best left to the individual patient, family, and care provider.

READING 20

Aim Not Just for Longer Life, but Expanded "Health Span"

Daniel Perry and Robert Butler

Most Americans instinctively recoil at the thought that their government would try to save money by pulling the plug on life-sustaining care when it is needed by older people. In this case, their instincts are correct.

To determine a person's access to medical care solely on the basis of that person's age is clearly unfair, unworkable, and unnecessary. It is wrong to blame the elderly for rising hospital expenses and physicians' fees that are driven principally by other factors or to require older Americans to pay for the failure of government and industry to find a more humane and workable policy to curb health care costs.

President Reagan signed into law the most sweeping Medicare expansion in that program's 22-year history, indicating the nation's strong commitment to providing health care to the elderly. The new catastrophic care program will cost about $31 billion over five years. Even that amount will seem small when compared to proposals for insuring Americans against the costs of long-term care, the next major health care issue to face Congress and the Bush presidency.

SOURCE: "Aim Not Just for Longer Life, but Expanded 'Health Span'" by Daniel Perry and Robert Butler in *The Washington Post*, December 20, 1988. Reprinted by permission of the authors.

As the curtain rose on Congressional debate over long-term care, some came forward to argue that the United States could save billions by simply denying lifesaving medical interventions to people over a certain age—say 65 to 75. But there is a better way to control costs of providing health care to the elderly: work to eliminate the very afflictions of old age, which are costing billions in health care, long-term care, and lost productivity.

By attacking diseases associated with aging—such as Alzheimer's disease, stroke, osteoporosis, arthritis, and others—the need for many costly medical procedures, lengthy hospital stays, and financially draining long-term care could be ended or reduced.

Why not start with a real commitment to scientific research that could extend the healthful middle years of life and compress the decline of aging into a very short time?

Why not redirect federal research efforts to aim for scientific and medical discoveries to reduce frailty, improve health status, and increase independence in older people? It's a far better goal—and more realistic—than rationing medical treatment.

At present, however, aging research is not where the U.S. government is placing its biggest bets. Most people don't believe much can be done to change aging. Therefore, research funds generally go elsewhere.

There is every reason to fear spiraling health costs if effective ways to lengthen healthy years and delay the onset of debilitating age are not found before the baby boomers become the biggest Medicare generation in history.

Americans already are paying billions because medical science lacks the ability to cure, prevent, or postpone many chronic maladies associated with aging. And national investment in research to avoid these costs is minuscule when compared to the billions spent for treatment.

Of the $167 billion a year spent on health care for people over age 65, far less than one half of 1 percent of that amount is reinvested in research that could lead to lower health care costs for chronic diseases and disabilities. That is a poor investment strategy for a nation soon to experience the largest senior boom in history.

Tinkering with changes in the health care delivery system can save some money, but these savings will not equal the long-term benefits of dramatic medical and scientific changes that alter the way people experience old age.

If scientists do not find a way to treat Alzheimer's, for instance, by the middle of the next century, there will be five times as many victims of this disease as there are now simply because of the demographic shift that is occurring. Incontinence, memory loss, and immobility are the main factors driving long-term care and high health costs to the elderly. If no advances occur in these and other conditions of aging, up to 6 million older Americans will be living in nursing homes, instead of the 1 million who are there today.

Unfortunately, there may be no way to prevent aging per se. However, there are conditions that occur only as a person ages. Many of these can be prevented. The risk of suffering a chronic disease such as arthritis or osteoporosis is very slight at middle age. But from the forties onward, that risk doubles exponentially about every five years until someone in the mid-eighties has about a one-in-three chance of having dementia, immobility, incontinence or other age-related disabilities.

If medicine could delay the beginning of decline by as few as five years, many conditions and the costs they incur could be cut in half. The ability to re-set biological clocks to forestall some of the decline of aging may be closer than anyone realizes, thanks to new knowledge in immunology and in the molecular genetics of aging.

Answers may be near. Help for immobility, osteoporosis, and incontinence can be achieved with only a modest extension of present technologies. If the U.S. doubles its present meager $30 million for osteoporosis research, by the year 2010 this condition could be eliminated as a major public health problem, which now affects 90 percent of all women over 75.

Learning how to postpone aging could help lower health care costs and improve the health of older Americans at the same time. The goal here is not just longer life span but extended "health span," with fewer problems caused by chronic disease.

FOCUS ON THE FUTURE

Scenarios for Rationing: Fiction or Forecast?

Sometimes our most provocative images of the future come from science fiction, which has explored some of the issues involved in rationing health care resources in a society coping with burdens of an aging population.

Holy Fire is a science fiction novel by Bruce Sterling (1996), a story set in the year 2096. The book depicts a world after years of plague and natural disasters. Its heroine is a 94-year-old medical economist named Mia Zemann, who is a member of the gerontocracy, or ruling elite, composed of older adults. She has been able to afford the best that technology has to offer for life extension.

As the story unfolds, Mia has undergone a technique known as Neo-Telomeric Dissipative Cellular Detoxification (NTDCD), which is a revolutionary technique for reversing aging and promising virtual immortality. But one day, Mia wakes up convinced she is another person, and she explores a world in ways that call her basic identity into question. *Holy Fire* contains other messages worth pondering. The gerontocracy of the late 21st century is a society that has allocated extraordinary resources to keeping the older people alive. But at what cost to other age groups? At what cost even to the rejuvenated old themselves? These are questions that Daniel Callahan also posed in his proposal for age-based rationing of health care.

A feature film, *Logan's Run*, takes up the prospect of age-based extinction as a means of promoting the greater social good. This 1976 movie, starring Michael York and Peter Ustinov, portrays a futuristic world set in the year 2274. It is a society of unlimited pleasure, followed by extinction at the age of 30: a macabre fulfillment of the 1960s slogan "Don't trust anyone over 30!" In this youth-worshiping society, young people don't know what awaits them on their 30th birthdays. They believe that, after participating in a ritual ceremony, they will be "taken up" and transported into another state of being. In fact, they are to be killed to make room in a society threatened by overpopulation.

As the film unfolds, we discover that the entire society is located underground, a remnant left over after a nuclear holocaust has destroyed life on the surface of the earth. Forced to ration resources, this future world arranges for a limited lifespan, at a much lower age than Daniel Callahan ever anticipated in his rationing proposal.

Another film on the theme of overpopulation and planned death is *Soylent Green* (1973), starring Charlton Heston and Edward G. Robinson. The film is an adaptation of Harry Harrison's (1966) science fiction novel *Make Room! Make Room!* set in the year 2022. Manhattan has become a crowded nightmare world (just as in a later film, *Escape from New York*, 1981). Charlton Heston plays a police officer who investigates the murder of a prominent figure and stumbles onto a vast conspiracy. The government, it turns out, has plans for people to die and then arranges for their bodies to be taken away, "recycled," and used as part of the food supply for a world gripped by shortages of everything.

(Continued)

(Continued)

Some of these science fiction images of the future could have been scripted by Thomas Malthus, an 18th-century clergyman and writer who warned of a race between overpopulation and dwindling natural resources. Malthus advocated population control, but never urged killing as a solution. However, Jonathan Swift, writing a few decades earlier than Malthus, actually wrote down an extinction plan in his *A Modest Proposal*, suggesting one way to solve the population problem in Ireland. In an elaborate joke, Swift asked his readers to consider the possibility of cannibalism carried out on infants killed to keep down the Irish population. But genocide in the 20th and early 21st centuries was no joke, and we are forced to take seriously the prospect that science fiction scenarios could one day become fact.

Questions for Writing, Reflection, and Debate

1. Is Callahan right in his suggestion that our modern U.S. culture thinks of aging as "hardly more than another disease"? Does it make sense to talk about aging as a "disease," the cause of which might be identified and then perhaps even "cured"? How do we decide whether something is a disease? Following Callahan's own argument, would it be a good idea to promote anti-aging research if this might reduce the expenses of geriatric care?

2. Antibiotic therapies such as erythromycin today are inexpensive. According to Callahan's own argument, would they have to be withheld from the very old just because they are "life-extending"? Or is it only expensive therapies that should be withheld? What happens if a cheap therapy helps people to survive an illness, but then it turns out to be expensive to take care of them?

3. How do we know when "rationing" starts taking place? During World War II, everyone knew that butter, gasoline, and other commodities were being rationed. But some critics argue that rationing of health care is already going on in the United States. Is it possible for resources to be rationed without public knowledge of it? As a hypothetical exercise, assume that you are a journalist who has just discovered that a local hospital routinely makes decisions about health care based on the age of the patients. Write a short newspaper article bringing public attention to the practice.

4. Hentoff argues that Callahan's proposal is class biased—that is, it discriminates against the poor—because people with money can purchase any amount of medical care they want. Is this argument convincing? Is there any alternative to this arrangement? Does Hentoff's point, if valid, destroy Callahan's argument?

5. At the end of his article, Hentoff argues that the Callahan proposal deprives an entire class of people of the "equal protection of the laws," and he cites the Dred Scott case (1857), in which the U.S. Supreme Court, in essence, decided in favor of slavery. Is age-based rationing, like slavery, a kind of discrimination? In what ways is age discrimination like or unlike race discrimination? Assume that you are a lawyer arguing this issue of age-based rationing before the U.S. Supreme Court. Write a "brief" based on Hentoff's general idea and offer your strongest possible arguments to convince the justices.

6. Wetle and Besdine, like Hentoff, cite the case of the Nazis and their program of killing off certain groups of people who were judged unworthy to live. Is it fair to judge Callahan's proposal by comparing it to what the Nazis did? Assume for a moment that you are Daniel Callahan and write a "letter to the editor" defending yourself against this charge of being like the Nazis.

7. Some critics argue that figures show that Callahan's proposal, if adopted, won't really save

much money as long as rationing is limited to people over age 75. If their figures are correct, should Callahan be willing to lower his age limit to 70 or 65? Why or why not?

8. Managed care depends on some form of "gatekeeping" to decide who will get services. What are the ethical dilemmas involved in gatekeeping? What can be learned from considering the activities of gatekeepers in other domains—for example, college admissions officers or caseworkers in the welfare system? Imagine that you are a gatekeeper and you're faced with a situation in which it might not be possible to provide a questionable service needed by an older person. Write a memorandum to your boss giving arguments on why the service should be provided.

Suggested Readings

Aaron, Henry J., and Schwartz, William B., with Melissa Cox, *Can We Say No? The Challenge of Rationing Health Care,* The Brookings Institution, 2005.

Binstock, Robert H., and Post, Stephen G. (eds.), *Too Old for Health Care? Controversies in Medicine, Law, Economics and Ethics*, Baltimore: Johns Hopkins University Press, 1991.

Hackler, Chris (ed.), *Health Care for an Aging Population*, Albany: SUNY Press, 1994.

Homer, Paul, and Holstein, M. (eds.), *A Good Old Age? The Paradox of Setting Limits*, New York: Simon & Schuster, 1990.

Student Study Site

Visit the Student Study Site at **www.sagepub.com/moody7e** for these additional learning tools:

- Flashcards
- Web quizzes
- Chapter outlines
- SAGE journal articles
- Web resources
- Video and audio resources

SHOULD FAMILIES
PROVIDE FOR THEIR OWN?

When problems arise in old age, most people turn to their families and friends for help (Feder, Komisar, & Niefeld, 2000). According to a recent report, "in 2009, about 42.1 million family caregivers in the U.S. provided care to an adult with limitations in daily activities at any given point in time, and about 61.6 million provided care at some time during the year. The estimated economic value of their unpaid contributions was approximately $450 billion in 2009, up from an estimated $375 billion in 2007" (Feinberg et al., 2011).

The vast bulk of care for frail older adults, perhaps 80%, is furnished by families and other private individuals (Shanas, 1979). But the American family is changing at the same time that American society is witnessing changes in the proportion and character of the aging population (Burton, 1993; Cantor, 1992). Families are facing new challenges to give care and help, as well as bearing the cost of long-term care for older members (Brubaker, 1990).

AGING AND THE AMERICAN FAMILY

Older Americans have a rich and extended family life. For example, in the United States in 2009, 72% of men and 42% of women over age 65 were married, and most of them have adult children (Administration on Aging, 2010). An equivalent proportion have at least one brother or sister, and three quarters are grandparents. According to the U.S. Department of Health and Human Services Administration on Aging, more than two thirds of older noninstitutionalized people live in a family setting. These facts show that the popular image of old people as lonely and abandoned is inaccurate.

But advanced age frequently brings a need for caregiving. Indeed, caring for older adults has long been a major and predictable part of the life cycle of Americans (Glick, 1977). Among married couples, the primary caregiver tends to be the healthy spouse (Stephens & Christianson, 1986). A big problem for old-old couples is that with advancing age, older spouses are increasingly likely to be impaired. In that event, older people typically turn for help to adult children, who are also the chief caregivers for older men or women who are no longer married or whose spouse can no longer care for them (Brubaker, 1985).

Some patterns of caregiving over the life span are illuminated by the **exchange theory of aging,** which is the idea that interaction in social groups is based on reciprocal balance

(Dowd, 1975). Thus, parents care for children and spouses care for one another because they are motivated by both moral obligation and the knowledge that they can count on reciprocal help in times of difficulty (Raschick & Ingersoll-Dayton, 2004).

Many different kinds of family members can be involved in caregiving (Hays, 1984), but responsibilities still tend to be divided according to gender, as Exhibit 22 demonstrates. The overwhelming majority of care for aged relatives is still provided by women, typically wives, daughters, or daughters-in-law, who must balance the burden of care for the aged with the demands of employment and their own families. The term **sandwich generation** describes the impact of such caregiving responsibilities on middle-aged women (E. Brody, 1985, 2004; Neal & Hammer, 2007).

Watkins, Menken, and Bongaarts (1987) estimated that the average American woman will spend more years caring for older parents than she will spend caring for children under age 18. Of course, this generalization overlooks the fact that *caregiving* may entail different levels of responsibility: For some, it is a weekly telephone call; for others, it is round-the-clock support for someone with Alzheimer's disease or for someone recovering from a stroke). Still, it remains true that for many women during the middle-age years, that time will be spent with caregiving obligations. Among those 45 to 54 years of age, for example, 17% have some responsibility for an older parent with a disability (Stone & Kemper, 1989).

In cases of extreme frailty or dependency, the burden on family members may prove exhausting, leading to burnout and perhaps even elder abuse or neglect (Stone, Cafferata, & Sangl, 1987). The burdens created by Alzheimer's disease and other varieties of dementia are a case in point. As the disease progresses and the patient's behavior becomes more extreme, caregiving stress can become almost unbearable in a home setting (Corbin & Strauss, 1988; Springer & Brubaker, 1984). These conditions have led gerontologists to speak of family

Exhibit 22   Caregivers and Their Relationship to Older Adult Care Recipients

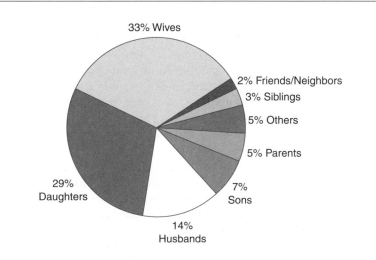

SOURCE: California's Caregiving Resource Centers Assessment database, 1997–1998.

NOTE: Caregiver population includes primary and secondary caregivers who are caring for a brain-impaired adult.

caregivers as the hidden victims of the disease (Zarit, Orr, & Zarit, 1985). There are important family and social justice questions to be asked about the life course impact for women of being expected to assume elder caregiving responsibilities in addition to other family caregiving and professional responsibilities, not to mention personal development (Wisensale, 2005).

The gloomy portrait of caregiver burden, however, should not be exaggerated. Many caregivers remain in their roles for a long time and never "burn out." There is a normalcy to family caregiving, especially between spouses, that makes it seem nonextraordinary to those who render care.

Moreover, significant help for caregiver burden exists. Social supports, especially the informal support of family or friends, can prove helpful for caregivers under stress. In addition, caregivers may benefit from **respite care**: temporary care for dependent older people to allow the caregiver some time off (Klein, 1986). Such programs can relieve some of the strain involved in efforts to delay placing an impaired or severely ill relative in a nursing home (Montgomery, 1989). Mutual-aid groups, such as those sponsored by the Alzheimer's Association, have also proved effective for caregivers (Mace & Rabins, 1981). In all of these cases, formal support services complement informal care, serving not to replace it, but to support it (Litwak, 1985). Last, there are national and state organizations that provide current information and resources to family caregivers, such as the Family Caregiver Alliance.

ABANDONMENT OR INDEPENDENCE?

There is a persistent stereotype that older people are abandoned by their children, but this stereotype is largely inaccurate. In an important, early study on family support in later life, nearly half of older adults reported that they lived or expected to live in proximity to their children, two thirds within 30 minutes of a child (Shanas, 1980). Currently, more than 40% are in daily contact with their children, and three fourths talk on the phone at least weekly with their children. Older people whose families are spread out geographically do not necessarily consider their families broken up or believe that the young have abandoned the old.

Nonetheless, a clear trend toward independence in living arrangements among older adults has been apparent for a long time. For instance, in 1960, only one fifth lived alone; by 1984, the proportion had increased to one third, and this trend has continued into the first part of the 21st century. Sharing a household in an **extended family** has also dropped significantly in recent years. In 1960, 40% of older people were residing with their adult children, but by 1984, the proportion had dropped to 22% (U.S. Congressional Budget Office, 1988), and the trend continued unchanged until relatively recently. During the first part of the 21st century, largely because of economic recession, there has been an increase in multigenerational families (Fleck, 2010), and a new report on multigenerational family households suggests that, since the 1980s, there has been a revival of multigenerational households and that this trend may continue (Pew Research Center, 2010). See Exhibit 23.

Nonetheless, the sentimental image of the multigenerational family in the "good old days" is mistaken in several ways. In Europe and the United States, multigenerational living arrangements were never common, even in agrarian societies centuries ago (Laslett, 1972). Idealizing the extended family—that is, several generations—living under one roof is part of the **"world-we-have-lost" myth** (Laslett, 1965/1971), in which we idealize the golden age of preindustrial society. Yet Western societies have tended toward a separate residence

Exhibit 23 Share of Adults 65+ Living in a Multigenerational Family Household, by Gender, 1900–2008

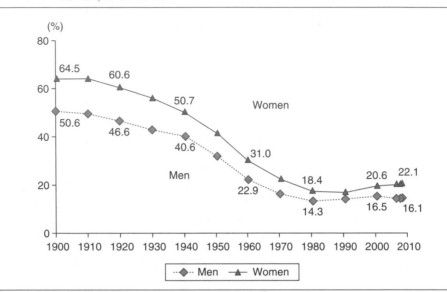

SOURCE: Pew Research Center, The Return of the Multigenerational Family Household, 2010. http://www .pewsocialtrends.org/2010/03/18/the-return-of-the-multi-generational-family-household/. Pew Research Center tabulations of the U.S. Decennial Census data, 1900–2000, and 2006, 2007, 2008 American Community Survey (IPUMS).

for the **nuclear family**—that is, only parents and children—for a long time. There is also a common stereotype of older people as isolated from others. Yet a majority of older people live with others: around half with their spouses, another 14% with other relatives, and smaller numbers in other living situations. Even those alone are usually within a close distance of relatives or only a phone call away. Fewer than 1 out of 20 are socially isolated, and usually they are so because they have lived that way most of their lives.

URBAN LEGENDS OF AGING

"Only 4% of older people live in nursing homes."

This half-truth was propagated by well-intentioned advocates. What is famously called the "4% fallacy" number applies only to a single point in time. But on a life span basis, 40% or more of those who reach age 65 will spend time in a long-term care facility before they die—one more example of the fallacy of cross-sectional versus longitudinal comparison.

Families today typically remain in close and frequent contact, increasingly through the Internet and social media (Hogeboom & Bell-Ellison, 2010). This pattern has been called **intimacy at a distance**, and it reflects a common desire by older people to live independently yet still remain close enough to have regular contact with grown children. When illness or need arises, a spouse, an adult child, or another relative is typically the first to help.

Most older adults who live alone have connections to family and friends.

We do need to recognize that the living arrangements of older people today are different from those of a century ago. One reason for change is simply demographic. Today, unlike in the past, vastly larger numbers of older people survive into advanced age and thus require sustained help with activities of daily living. In cases of debilitating chronic illness, such as stroke and Alzheimer's disease, older people may live many years in conditions of dependence that exceed the capacity of family caregivers. Other older people may never have been married or may simply outlive available family members. The result is that we must increasingly rely on government to provide what families are not in a position to give.

FAMILY RESPONSIBILITY

The development of social welfare programs for older people, such as Social Security and Medicare, has meant that health care and income support for the aged have become a societal responsibility, rather than a family obligation. But in the United States, unlike other advanced industrialized countries, long-term care has remained largely a family responsibility (Buchanan, 1984). Government has been reluctant to provide coverage for long-term care, so families remain an important source of both hands-on care and financial support. When it comes to long-term care needs, older people first rely on spouses; **spousal responsibility** is deeply embedded in our culture as a matter of both ethics and law. If a spouse is not available to provide care, then other family members such as children and siblings take responsibility.

In some cultures, such as the Chinese, Confucian teachings inculcate filial piety or strong reverence for parents, including the duty to support parents over one's own children (Cowgill, 1986). In the United States, **filial responsibility**—that is, responsibility for care of the aged by adult children—is treated ambiguously as a matter of law, custom, and ethics (Callahan, 1985; Pillemer & Luscher, 2004; Post, 1989). In fact, half the states have laws on the books that could compel children to give financial support to aged parents, but these laws have rarely been enforced (Garrett, 1980; Lammers & Klingman, 1986) partly because of deeply conflicting public attitudes toward filial responsibility (Seltzer & Troll, 1982). In contrast, Singapore has begun enforcing stringent laws of compulsory filial responsibility that allow older people to sue their adult children for support (Liu & Kendig, 2000).

In the United States, however, filial responsibility continues to be practiced not as a matter of law, but as a matter of ethics or custom, and gerontologists have documented rich intergenerational ties in American families (Pfeifer & Sussman, 1991). The unresolved question is how government programs should interact with spousal and filial caregiving duties and financial responsibilities, and this question is faced by societies around the world (Bengtson & Lowenstein, 2003).

GLOBAL PERSPECTIVE

Singapore's Law Requiring Support of Aged Parents

The Bible's Fifth Commandment says "Honor Thy Father and Thy Mother." At least one country, Singapore, has passed laws making this duty legally enforceable. Singapore, a small city-state in Southeast Asia, holds adult children legally responsible for support of their aging parents. Other nearby countries, such as Malaysia, subsidize adult day care or other support services helping children care for older parents. In traditional Asian societies, the old would live in extended, multigenerational households and depend largely on their adult children for support and care. But today that traditional family support system is less viable. Singapore is unique in its Maintenance of Parents Act, passed by its legislature in 1995.

In its first years of operation, more than 400 parents petitioned a tribunal to compel their children to support them. The court has some discretion in case children were neglected or abused, and orders for maintenance for parents take into account an adult child's duty to maintain spouses and children of their own. But four out of five elders have obtained court orders compelling their children to support them. Court administrators note that often, the adult children who were negligent had become unemployed or were too busy to pay in a timely fashion.

Singapore's law reflects several assumptions: first, that the family, rather than government or society, should provide care for older people; second, that children have the financial means to support their aged parents; and, third, that older people actually want more direct care from their families. Although Singapore is distinctive in its approach to filial piety, its law is not entirely unique. Countries including India, Israel, and Taiwan have had laws enforcing support for aged parents, and Britain had such a law itself until 1967. A number of states in the United States, including California and Illinois, have similar laws, but they are rarely enforced. Those who supported Singapore's law argue that it promotes traditional family and religious values and offers government sanction as a last resort.

SOURCE: Tiong, T. N., & Bentelspacher, C. E., "Elder Care in Singapore: Not by Law Alone," *Ageing International* (December 1995), 22: 4.

MEDICAID AND LONG-TERM CARE

Under Medicare, the U.S. health care system provides near-universal coverage for acute diseases among the old. A majority of Medicare beneficiaries do not realize that, in fact, Medicare does not cover long-term care to any great extent. Financing of acute and long-term care remains separate. About half the money spent on long-term care in nursing homes comes from some branch of government, chiefly Medicaid, and Medicaid is the primary payer for two thirds of nursing home residents (Hagen, 2004).

Medicaid, a joint government program supported by federal and state funds, was created in 1965 to provide health care for the poor. But over the years, it has become the primary government mechanism to pay for long-term care for older adults and people with disabilities. Medicare pays only 2% of those nursing home costs; Medicaid pays 36%. Medicaid is a large and expensive program, and its cost is increasing rapidly. As the number of oldest-old in the 85+ population increases, long-term-care expenses are likely to become even higher.

Although created as a health care program for poor people, Medicaid has, in fact, become a key factor in nursing home coverage for middle-class older people. Three fourths of Medicaid recipients are low-income parents with children, but these families receive only about a quarter of total Medicaid dollars. About two thirds of all that Medicaid spends goes to institutional care for people who have physical or mental disabilities or are older.

FINANCING LONG-TERM CARE

Presumably, middle-class families rely on Medicaid for long-term care because they do not have the financial capacity to bear the cost of care (Cohen et al., 1987). Long-term care already consumes a larger portion of the private health care dollar for older adults than any other type of expenditure. The cost of a year in a nursing home today can range up to $75,000 or more. Few individuals or families can afford that cost on an extended basis. Of those who enter a nursing home as "private-pay" patients, after only 3 months, nearly 70% have reached the poverty level, and within a year, 90% are impoverished.

In the likely event that long-term-care costs exceed savings, those who face such costs have few options. One option is to qualify for Medicaid. But Medicaid is a means-tested entitlement program. That is, it makes use of eligibility rules based on income and assets to determine whether people qualify for Medicaid coverage. A nonmarried applicant for Medicaid can keep nonexempt assets of only $2,000 or less, excluding the value of a home. Married couples, taking advantage of recent changes in the law, could keep up to $95,000 under the community spouse asset allowance, with all but this limited portion of assets assumed to be available to pay for the partner's long-term care (O'Brien, 2005).

Many of those who do not qualify for Medicaid still do not have enough assets to pay for long-term care themselves. They face a cruel choice: struggle to provide home-based care or do what is necessary to obtain Medicaid. To qualify for Medicaid, it is necessary to *spend down* lifetime accumulated assets to become impoverished and thereby eligible for assistance (Liu & Manton, 1991). Under regulations of the Medicaid law, spouses of those thus impoverished may obtain some protection, but children and grandchildren lose their share of accumulated life savings. One major problem with Medicaid financing of long-term care is that it introduces inequities across families, age groups, and social classes (Arling et al., 1991). For example, should people who become poor in old age be treated the same as those with a lifetime of poverty? Should families that contribute their own labor for caregiving have that contribution taken into account?

According to public opinion surveys, 82% of the general public recognize that they cannot afford to pay the cost of long-term care either at home or in a nursing home. They also know that they cannot rely on the family alone: 86% want the government to help pay for long-term care instead of leaving it entirely up to the family. Significantly, in an era of strong sentiment against taxes, a majority of adults say they would be willing to pay for a long-term-care program with increased taxes.

But despite such clear public sentiment, a universal public insurance program for long-term care is still not available in the United States. On the contrary, Medicaid has become the public program of last resort to pay nursing home costs. In fact, it is the fastest-growing component of state budgets, and it is increasingly becoming an old-age program. Nearly 40% of all Medicaid benefits go to the older adult, chiefly for nursing home care (see Exhibit 24).

The growing burden of Medicaid on the government has prompted a search for more affordable alternatives. For years, aging advocates have sold the idea of home care to legislators and to the public with the argument that home care is more humane, in keeping with people's preferences to stay at home, and also more cost-effective. Unfortunately, the facts are not so clear. National demonstration projects and other studies have shown that home care may be more desirable, but it doesn't necessarily save money. One reason may be the so-called **woodwork effect** (Fama & Kennell, 1990): Government policy makers are afraid of people coming "out of the woodwork" to demand services that families would have provided otherwise or that weren't provided before (Arling & McAuley, 1983). Once the

Exhibit 24   Distribution of Funding for Freestanding Nursing Home Expenditures for All Payers, 2000

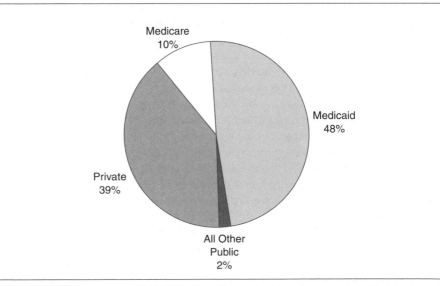

SOURCE: Centers for Medicare and Medicaid Services, Office of the Actuary: National Health Statistics Group (2005).

NOTE: Hospital-based skilled nursing facilities are excluded.

government is willing to pay, people may ask "someone else" to pick up the tab (Doty, 2000). Many of the same issues arise when we consider the question of whether government should pay family members to give care that they might be giving anyway (Simon-Rusinowitz, Mahoney, & Benjamin, 1998).

URBAN LEGENDS OF AGING

"Home care is cheaper than nursing homes."

Would that it were so because most people prefer home care. But for decades economic studies have shown that it's not true, certainly for full-time care, whenever all relevant costs, such as rent and unpaid caregiver's labor, are taken into account. Home care may be preferable, but it's not necessarily cheaper.

It is often said that older people should go into nursing homes only as a matter of their own choice, not for the convenience of the family. But when family members give the bulk of hands-on care, it is not so simple to say that the legitimate interests of the family are to be disregarded (Dill et al., 1987). The reality is that virtually no one enters a nursing home as a matter of choice. Government financing of nursing home care under Medicaid does introduce a certain financial incentive, but few would be inclined to take advantage of this incentive without some compulsion. People go into nursing homes, by and large, as an act of desperation, when the needs of the older care recipient exceed the abilities of the family to provide care and the family members have reached a point where they have no other alternative.

MEDICAID PLANNING

As older people and their families have become more aware of the cost of long-term care, middle-class families have found ways of qualifying for Medicaid. In doing so, families have tried to avoid the harsh requirements of **Medicaid spend down**—that is, impoverishing themselves by spending all income and assets to qualify for Medicaid coverage. Both attorneys and financial planners urge a variety of strategies to enable middle-income families to qualify for Medicaid coverage of nursing home costs. The heart of these strategies comes down to *divestment planning*, that is, appearing to be poor by taking advantage of legal loopholes to "avoid the Medicaid trap" (Budish, 1989).

The following are some of the key strategies the specialists recommend:

- Transfer assets at least 30 months ahead of applying for Medicaid
- Transfer assets between husband and wife
- Seek protection through a court order
- Keep assets in a form exempt from Medicaid
- Set up a trust account

It is not known exactly how many middle-class older people presently take advantage of loopholes in the Medicaid law to appear impoverished and thus protect their family wealth.

But the numbers are large enough to have sustained a rapidly growing body of **elder law** attorneys, who have now established their own association, the National Academy of Elder Law Attorneys. As Medicaid divestment planning has become more widespread, the practice has attracted criticism. Some equate the practice with the deception used by some poor people to qualify for welfare payments and with schemes by rich people who dodge taxes through loopholes. In 1997, Congress enacted legislation making it a criminal offense to use professional consultation as a way of spending down to qualify for Medicaid coverage.

Regardless of whether middle-class spend down is technically legal, the critics argue that this form of Medicaid planning is socially irresponsible. This moral criticism is another way of insisting that taxpayers in general should not pay to protect the inheritance wealth of affluent families (Freedman, Lomasky, & May, 1983). When Medicaid spend down becomes a form of estate planning, critics say, it is a means of cheating the government and using public money intended for those in genuine poverty. However, it is not possible to understand what is at stake in this about family responsibility without seeing the importance of inheritance and the intergenerational transfer of assets (Lee, 2003).

It is easy enough to recognize that family abandonment of older members is a myth and to acknowledge that families are already taking care of their own by giving time and effort. But it is not easy to agree on whether families have an obligation to go further and make use of a portion of their assets to pay for long-term-care costs even if doing so eliminates inheritance.

Similarly, we know that informal supports enable frail elders to remain at home when the only alternative might be entering long-term care. But people disagree about whether families should be paid in cash by the government for giving the kind of hands-on care that they customarily give. Knowledge of the facts is essential for debating the issue. However, because at the heart of the debate is the relationship between family members and thus feelings and values are involved, the debate will not be settled by facts alone.

In the readings that follow, which cover the major contours of the debate over the past couple of decades, strong opinions are expressed about whether families should take care of their own and what role government should have in providing long-term care. The first issue is whether people should take advantage of Medicaid eligibility laws. On one side of the debate are Peter Strauss and Nancy Lederman, both elder law attorneys who believe that families should realistically plan ahead for nursing home costs and that this planning may involve Medicaid. Planning ahead to qualify for Medicaid, they believe, is not immoral or illegal. Don't make the mistake of thinking that Medicaid is only for the poor, they tell us.

On the other side of the debate, Jane Bryant Quinn finds transfer of assets to be a troubling practice at the borderline of legality and social morality. Quinn voices a widely shared opinion. It seems wrong for some families to preserve an inheritance while others are forced to impoverish themselves to pay for long-term care. In his article, Stephen Moses focuses directly on this phenomenon of impoverishment due to nursing home expenses. Moses urges us to look carefully at the facts about paying for long-term care.

What about directly paying family members to take care of older people at home? C. J. Blaser rejects any proposal to have taxpayers support payments to family caregivers. The author argues that experience has shown that there are dangers of exploitation, fraud, and abuse in such arrangements. For example, when a family member is paid for providing home care services, who verifies that the services were actually given? The vast majority of care is already provided by family members, so if the government were to pay them to do what they do anyway, it would just drive up the total cost of care without any new services being added. Blaser concludes that, instead of direct payments, it is better for government to support family members in their normal caregiving activities.

In contrast, Suzanne Kunkel and coauthors argue that recent research suggests that concerns about neglect, safety, and fraud have been exaggerated. New evidence suggests that older people at home are quite satisfied if they are able to pay other family members instead of strangers to give them the care they need. For some, this kind of arrangement seems desirable, and most older people in need of care do prefer to have the right to decide who will provide them with what they need. Giving them the right to pay members is the best kind of consumer-directed health care and should be supported by government policy.

FOCUS ON PRACTICE

Long-Term-Care Insurance

Many middle-income older people and their relatives have been following the debate over Medicaid coverage of long-term care with great interest. In recent years, there has been dramatic growth in privately paid long-term-care services, especially assisted living and home- and community-based long-term care (Bonifazi, 1998). Those who don't have enough financial resources to pay for an extended stay in a nursing home or who have a strong desire to pass along an inheritance have had to face the dismaying prospect of bankrupting themselves to qualify for Medicaid. Today, however, families that want to plan ahead for long-term-care costs have another option: purchasing private long-term-care insurance. More than 100 insurance companies now offer these policies, and more than 7 million policies are now in force. Still, present private long-term-care insurance provides only 7% of total funding for long-term care in the United States while out-of-pocket expenditures cover nearly 25%.

Private long-term-care insurance typically covers nursing home care and sometimes other community-based services. The best of policies pay for medically necessary services for a period from one year up to a lifetime, but usually with a maximum period of coverage. Good policies are guaranteed to be renewable and need not require a prior hospital stay, as Medicare does (see http://www.longtermcare.gov).

Long-term-care insurance is bought mainly by people over age 55; half the current policyholders are in their 60s. The age when a policy is first purchased is important because the premium paid, although it remains level once the policy is purchased, rises sharply with age at purchase. For example, the same policy that goes for $250 a year at age 50 would cost up to $2,000 a year at age 70. But as more employers include such insurance in benefit programs, it is likely that more younger people will participate, and group rates may bring costs down. Less than 4% of the older population is covered by such policies (U.S. General Accounting Office, 1991).

Because paying for long-term care is an expensive and widespread problem, one might wonder why more people do not buy private long-term-care insurance. One answer is that older people and their families mistakenly believe that Medicare will cover such expenses. Another reason is that many policies are simply not affordable when purchased at an advanced age. The best estimates are that only a tiny proportion of the older population can afford to buy private long-term-care insurance (Wiener, 1998). Furthermore, many consumers, as well as state governments, lack confidence in the products on the market.

(Continued)

(Continued)

Is long-term-care insurance a good buy? Consumer advocacy organizations have raised doubts. Long-term-care policies have many exclusions and limitations that complicate comparisons of competing products. For example, insurance companies generally will not write insurance for preexisting conditions—an exclusion that covers many chronic diseases. Still another problem is that long-term-care policies generally pay a fixed dollar amount for each day of care covered, in contrast to Medicare, which pays a percentage of customary or reasonable fees regardless of inflation. Without inflation protection, a person buying long-term-care coverage at age 60 may find the policy inadequate when it is needed, say at age 80.

Private long-term-care insurance is now regulated entirely by state governments, but states differ in the regulations they apply. Without some kind of federal regulation, many states appear unlikely to implement standards for long-term-care insurance to protect vulnerable consumers, such as protection from forfeiture of the policy should a single payment be missed.

As long as government funding for long-term care is explicitly income based or means tested, as under Medicaid, middle-class older people will naturally look to private insurance to protect themselves. Government policies have begun to encourage private long-term-care insurance. The Health Insurance Portability and Accountability Act of 1996 gives more favorable tax treatment to long-term-care insurance (Guttchen & Pettigrew, 1998). Still another option might be to organize public long-term-care coverage funded through some combination of tax revenues and individual contributions/premiums. But until such insurance is available, older people and their families will need to be familiar with all practical options for covering long-term-care costs. Private long-term-care insurance is certainly one of those options.

READING 21

Medicaid and Long-Term Care

Peter J. Strauss and Nancy M. Lederman

Medicaid is a public assistance grant program, the twin Great Society benefit signed into law along with Medicare by President Lyndon B. Johnson in 1965. Financed jointly by federal and state moneys, it provides health benefits to 35 million low-income people who are aged, blind, or disabled as well as those who are poor.

Although designed to serve low-income people of all ages, Medicaid has become a lifeline for the older adult, providing essential services that Medicare doesn't—home health care and long-term nursing home care.

Medicaid is the major payer of long-term care for those who can't afford the average

SOURCE: "Medicaid and Long-Term Care" is reprinted from *The Elder Law Handbook: A Legal and Financial Survival Guide for Caregivers and Seniors* by Peter J. Strauss and Nancy M. Lederman (pp. 99–102, 112). Copyright © 1996 by Peter J. Strauss and Nancy M. Lederman.

yearly cost of $40,000 or more (more than double that amount in some areas). Medicaid covers the "nonskilled" but unbelievably expensive custodial care services that Medicare doesn't. Thousands of middle-income Americans have found themselves divesting themselves of their assets or "spending down" to eligibility levels in order to qualify for benefits under Medicaid. Medicaid pays for 42 percent of all nursing home costs nationwide. It covers over 60 percent of all nursing home patients.

Administered by the states, Medicaid represents the fastest-growing component of many state budgets. States pay approximately 45 percent and the federal government 55 percent of Medicaid costs. The federal share amounts to 6 percent of federal outlays. Nearly one-third of Medicaid spending goes to home health services and long-term nursing home care. . . .

Medicaid rules must be understood in context. One target for budget cuts has been the benefits furnished to middle-income people who have made themselves eligible for Medicaid to avoid the astronomical costs of long-term care. Stricter eligibility rules have made qualifying for benefits harder than ever before, and service and program cuts have affected both home care and nursing home care as well as programs such as adult day care. Additional changes and restrictions are being contemplated by Congress and several states. These events make it more important than ever that you understand the rules and how Medicaid works.

Before learning how to qualify for Medicaid, you should understand what benefits you're trying to obtain. In general, Medicaid pays for doctors and hospital stays, like Medicare. It also provides coverage for long-term nursing home care not covered by Medicare. Nursing homes operate according to a Medicaid plan which requires doctor certification of the need to enter the facility and periodic review of the need for continued care.

Medicaid also covers home health care services, medical supplies, and equipment. It commonly pays for at-home services supplied under state plans for people who would otherwise be institutionalized, covering part-time skilled nursing, home-health, and homemaker services provided by certified home health agencies.

"Spending Down" for Medicaid

For millions of older Americans, Medicaid is the only means by which long-term custodial care can be supported. Each year half a million people "spend down" their assets in order to qualify for long-term care assistance available under Medicaid. Some actually pay for their care until their assets are used up. Others purchase "exempt" items or transfer their assets, all legitimate Medicaid-planning strategies to allow them to keep their independence and autonomy without sacrificing their life savings.

These strategies all follow one basic rule. *All your income and assets above specified levels must be spent to pay for care before you will qualify for Medicaid.* Various planning strategies using statutory exemptions, spousal protections, and asset transfers are described below.

Plan ahead! This is one area in which advance planning is critical. Depending on your circumstances, it may take time to spend down or otherwise divest yourself of your assets in order to meet eligibility levels—and the law imposes penalty periods after transfers before you can qualify for benefits. Last-minute action may not work.

Although financing long-term care in this manner is entirely legal, Medicaid planning has become a major political issue. So many people have been forced to try to qualify for Medicaid benefits that the various methods for achieving that goal are under continuous attack.

Applying for Medicaid

Applying for Medicaid entails verification of your financial resources as well as citizenship and residency requirements.

Financial requirements are strict. You will be asked for your bank statements, tax returns, and other financial records reflecting your income, assets, and expenses.

More and more older people seeking Medicaid for nursing home and home care services are enlisting professional help in obtaining Medicaid, using lawyers or social workers familiar with the process and its requirements. We recommend this approach highly. Regardless of whether you need to avail yourself of other Medicaid-planning strategies to protect assets, you want to ensure that you get the health care coverage you need.

As we've described, Medicaid is a *means*-tested program. In order to qualify, you must establish financial eligibility by meeting [an] income and assets test set by the states. Depending on where you live, you can qualify for Medicaid under one or more of these programs: Eligibility rules in the states are confusing even for experts. Many states distinguish between nursing home and other home- and community-based care in determining Medicaid eligibility for those services. In some states there is a "medically needy" option for nursing home care only but not for general services.

Don't make the mistake of thinking that Medicaid benefits are limited to the very poor. Medicaid is a complex, confusing, but important government program which has become the life-line for many middle-income families who need help paying for long-term care in order to avoid impoverishment.

Originally intended for the poor, it has become the payer of last resort for persons of modest means. Spousal protections for resources and income clearly indicate Congressional intent that Medicaid continue as a program for middle-income Americans. Yet the use of Medicaid to finance long-term care, particularly nursing home costs, has had a profound impact on state budgets and resulted in attempts to restrict access to the program and limit benefits through calls for block grants.

Despite intended overhaul, with the federal government unlikely to increase Medicare benefits in the foreseeable future, Medicaid is likely to continue as the only government-funded program that deals with long-term care.

There may be any number of reasons not to apply for Medicaid, including personal, family, and psychological reasons; quality-of-care issues; and tax consequences. Nevertheless, it is an option that deserves serious consideration. While access to Medicaid is complicated and may not be available to or right for everyone, it is a possible source of financing for long-term care—even for those who may not now imagine they can take advantage of it.

READING 22

——— Shame of the Rich ———

Making Themselves Poor

Jane Bryant Quinn

As the population ages, Medicaid spending on nursing homes could easily lurch out of control. That is, unless it's limited to the people who really need it.

Medicaid is supposedly for the poor. But increasingly, it's being exploited by the well-to-do. Instead of buying nursing-home insurance or using their personal savings, they're getting the government to cover their bills.

Medicaid is a state and federal welfare program, providing various kinds of medical assistance to low-income people. Its charter includes nursing-home coverage.

If you need nursing-home care and are too poor to pay, Medicaid picks up the cost. But if you have personal savings, you're supposed to cover your own expenses.

When your savings drop below a certain level, Medicaid steps in. From that point on, the taxpayers support you for the rest of your life.

Growing numbers of middle- and upper-middle-class people don't like these rules.

They're willing to take care of themselves as long as they maintain their health. If a nursing home looms, however, they decide to quit being responsible. They look for ways of leaving their own money to their children, while forcing the taxpayer to provide their care.

I suspect that some of the well-off people who weasel their way onto Medicaid are vigorous supporters of big income-tax cuts. But where do they think the money for Medicaid comes from? Chocolate bars?

In general, their gambits are legal. The state laws on who's eligible for Medicaid conceal many weak points that let moneyed people onto the rolls.

To me, exploiting these weaknesses is unethical. The question for families is whether money will always trump morals.

In determining whether you're eligible for Medicaid, the states look mainly at your assets. If your assets are too high, you can't go on the welfare program. What's "too high" varies by state.

Married couples might not qualify if their savings exceed $87,000, plus house, car, personal property, pension income and other items. (That assumes one person in a nursing home and the other at home.)

Singles might not qualify if they have more than $2,000 to $4,000 in savings. They, too, might be able to keep a paid-up home and other assets, even though they'll be in the nursing home for the rest of their lives.

If you give away money to get yourself under the savings limit, there's a waiting period before you can collect benefits. Anyway, that's the way the law is supposed to work.

But there are loopholes. So-called "Medicaid planners" use the loopholes to make you instantly "poor." You can qualify for taxpayer help without a waiting period.

That turns Medicaid into an "inheritance insurance plan for the middle-class," says Stephen Moses of the Center for Long-Term Care Financing in Bellevue, Wash. Parents go on welfare so they can leave their money to their kids.

Here are some of the Medicaid-planning ideas promoted at a recent Elder Law Symposium in Vancouver, sponsored by the National Academy of Elder Law Attorneys:

- *Cut your spouse loose.* When one spouse enters a nursing home, assets can be moved into the name of the healthy spouse, says attorney Daniel Fish of Freedman and Fish in New York City. The healthy spouse signs a statement, refusing to support the nursing-home spouse. That spouse then goes on welfare (Medicaid).

The state can sue the healthy spouse to recover the money, but Fish says that's "not frequent." He thinks there are around 100 such lawsuits a year in New York City.

- *"Buy" lifetime personal care from your child.* You use your savings to pay for that care in advance, and solidify the deal with a written contract. You're now out of money and can go on Medicaid immediately, says attorney Scott Solkoff of Solkoff & Zelien in Boynton Beach, Fla.
- *Put your assets into a small business or farm.* Medicaid generally doesn't require you to use business assets to help pay the nursing-home bill, as long as you or your spouse are active in the business in some way. You could even hire someone to start a business for you, says attorney Lee Holmes of Oklahoma City, although he says he's never done that.

- *Put your assets into an annuity.* The income would go toward the nursing-home cost. But you could arrange for an heir to get payments if you died early.

Bottom line—Medicaid is in serious trouble. The government isn't spending enough for quality care. The more people with money exploit the system, by not paying for themselves, the worse the care is going to be for everyone.

READING 23

The Fallacy of Impoverishment

Stephen Moses

The debate over how to solve the long-term care financing crisis has reached a new and perilous stage: stalemate. Big government solutions are out of favor, but private sector initiatives appear inadequate. When a problem seems intractable, wise counsel is to check your premises. Is there a missing piece in the long-term care financing puzzle? Does a false assumption underlie the deadlock on this issue?

Conventional wisdom holds that eligibility for Medicaid, the nation's single largest financier of long-term care, requires the spenddown of assets and income to impoverishment. Federal and state laws, regulations, and policies seem to say that a person must spend down to the poverty level or below before qualifying. If this is true, elderly people and their heirs should seek private risk-sharing protection aggressively, but they do not. If it is false, private sector options such as long-term care insurance are severely handicapped—people only insure against real risks.

This [essay] describes research that casts doubt on the belief that Medicaid requires impoverishment. It also explores the broader social and policy ramifications of such a finding.

BACKGROUND

Although originally intended to ensure access to mainstream health care for the poor, Medicaid has become the major payor of nursing home care for the middle class (Rymer, Burwell, Adler, & Madigan, 1984, p. 122). The program funds 44% of America's nursing home costs (Letsch, Levit, & Waldo, 1988) and is the principal payor for over 63% of patient days (Dean, personal communication, Oct. 20, 1989). Nevertheless, from the enactment of the Medicaid statute in 1965 until 1981, no federal rules existed against transferring assets to qualify for assistance. People who needed nursing home care could give away their property in order to qualify for Medicaid.

In 1981, Congress took the first step to limit this practice with the Boren-Long amendment. This statute allowed states to restrict asset transfers

SOURCE: "The Fallacy of Impoverishment" by Stephen Moses, *The Gerontologist*, vol. 30, no. 1, pp. 21–25, 1990. Copyright © The Gerontological Society of America. This essay is based on research conducted by the Office of Inspector General of the Department of Health and Human Services and published in Medicaid Estate Recoveries (OAI-0986-00078), June 1988, and Transfer of Assets in the Medicaid Program: A Case Study in Washington State (OAI-09-8801340), May 1989. The views expressed herein are those of the author. Neither the Office of the Inspector General nor the Department of Health and Human Services reviewed the paper in draft and no endorsement by them should be inferred. The author was the project director for the studies cited above.

made for the purpose of qualifying for Medicaid. But Boren-Long did not apply to exempt property. Inasmuch as Medicaid exempts the home, and 70% of the net worth of the median elderly person is in a home (U.S. Bureau of the Census, 1986), Boren-Long excluded large amounts of property from the transfer restrictions.

Congress corrected this shortcoming and developed a more comprehensive approach in 1982 with the Tax Equity and Fiscal Responsibility Act (TEFRA). The TEFRA authorized states to (1) restrict asset transfers within 2 years of Medicaid nursing home eligibility, (2) place liens on the property of living recipients, and (3) recover from the estates of deceased recipients. Each of these procedures was optional for state Medicaid programs. Nevertheless, the expressed intent of Congress was "to assure that all of the resources available to an institutionalized individual, including equity in a home, which are not needed for the support of a spouse or dependent children will be used to defray the cost of supporting the individual in the institution" (U.S. Code, 1982, p. 814).

In 1985, a draft report of the Health Care Financing Administration revealed lax state enforcement of the TEFRA asset control authorities (Moses & Duncan, 1985). For example, transfer of assets rules were fraught with loopholes, only one state (Alabama) fully used the lien power, and although 18 states recovered from estates, most did so with very little success. Based only on telephone inquiries to state Medicaid programs and a valid random sample of cases in Idaho, this report speculated that estate recoveries could leap from $36 million nationally per year to $535 million if all states followed the asset control methodologies of Oregon's exemplary program. Although it remained unpublished, the report was released to the Office of Inspector General (OIG) of the Department of Health and Human Services and to the General Accounting Office (GAO). Both the OIG and the GAO began national studies of Medicaid estate recoveries in 1986.

OIG AND GAO MEDICAID ESTATE RECOVERY STUDIES

Office of Inspector General (1988)

The OIG sent a 17-page questionnaire to all 50 state Medicaid programs probing their policies and practices on transfer of assets, liens, and estate recoveries. The objective of the study was

> to find out exactly what States have done since 1982 to implement TEFRA's asset control authorities . . . to determine the extent and effectiveness of Medicaid estate recovery programs throughout the country . . . to report on "best practices" . . . [and] to examine State Medicaid eligibility policy with regard to transfer of assets and liens, because estate recoveries are obviously moot if no property is retained in recipients' possession that can be recovered after their deaths. (OIG, 1988, p. 2)

Both the Health Care Financing Administration and the Office of Management and Budget approved the OIG's survey instrument in advance. All 50 states and the District of Columbia responded. The study team did extensive telephone follow-up with Medicaid eligibility and estate recovery staff in three-fourths of the responding programs.

The OIG found very weak enforcement of asset transfer restrictions: "The States report that Medicaid eligibility rules permit knowledgeable individuals to transfer or shelter property from Medicaid resource limitations in a manner reminiscent of income tax avoidance" (p. ii). Three pages of quotations from Medicaid eligibility staff across the country supported this conclusion. For example: "People are starting to use a lot of fancy footwork to avoid losing the 'family fortune'" (Maryland). "Many, many, many attorneys call on a daily basis looking for 'loopholes.' There are lots of welfare specialists who help people avoid welfare resource limits" (Minnesota). "We recover from people who are not clever enough to transfer their property, and everyone else goes scot-free" (California).

Only two states had implemented TEFRA's lien provisions to secure property for estate recovery. Most states found the lien authority too restrictive to administer cost effectively. For example: "Liens are too difficult to administer because of Federal restrictions. Other property retention techniques, such as aggressive identification of assets, reversing illegal transfers, and challenging every possible resource shelter, are more effective under the circumstances" (Oregon, p. 20). Commenting on the ineffectuality of transfer of assets and lien rules, the OIG observed, "States cannot recover what is not there" (p. 23).

Twenty-three states and the District of Columbia recovered $42 million from the estates of Medicaid recipients in 1985, according to the OIG. But most states were very inefficient at recoveries. Even under the existing restrictive laws, regulations, and policies, the OIG concluded that "if all States recovered at the same rate as the most effective State (Oregon), national recoveries would be $589 million annually" (p. 46).

General Accounting Office (1989)

The GAO study sought to "assess the extent and effectiveness of state efforts to reduce program costs by using the estates of Medicaid nursing home recipients or their surviving spouses to recover all or part of the costs of care paid for by Medicaid" (GAO, 1989, p. 14). The agency reviewed 200 randomly selected nursing home cases in Oregon and seven other states. Oregon was chosen "to identify the key elements of a successful estate recovery program because it reported annual recoveries per nursing home recipient more than twice those reported by any other state" (p. 14). It recovered "about $10 for every $1 spent administering the program . . . " (p. 3). Projections of potential estate recoveries in the other states were based on their use of Oregon's policies and procedures.

The GAO found that " . . . two-thirds of the amount spent for nursing home care for Medicaid

recipients who owned a home could be recovered from their estates or the estates of their spouses. If implemented carefully, estate recovery programs can achieve savings, while treating the elderly equitably and humanely" (p. 3). The six states GAO studied that lacked recovery programs "could recover $85 million from recipients admitted to nursing homes in fiscal year 1985" (p. 4). Only "about 14 percent of the Medicaid nursing home residents in the eight states GAO reviewed owned a home . . . " (p. 4). The GAO did not account for the discrepancy between this percentage and the well-known statistic that three-quarters of elderly people own their homes (Rivlin & Wiener, 1988, p. 123). One presumes that people are either selling their homes and "spending down" before going on Medicaid or they are effectively transferring or sheltering the home's value.

Thus, both the OIG and the GAO studies confirmed that large amounts of private resources ($589 million nationally and $85 million in six states, respectively) pass to heirs each year instead of being used for long-term care costs or to reimburse Medicaid. Additionally, for reasons discussed in the next section, the OIG and GAO projections may be vastly underestimated.

MEDICAID ASSET SHELTERS

Assets that do not remain in an estate until the death of a Medicaid recipient are obviously not recoverable and would not show up in studies like the OIG's and GAO's. Therefore, the extensive anecdotal evidence of asset transfers and shelters discovered by the OIG in its estate recovery study raised another serious question. If people are jettisoning property before they apply for Medicaid nursing home care, how could we possibly know how much money is diverted from private to public long-term care costs? The OIG conducted further research that bears on this question (OIG, 1989). It found that people initially denied but subsequently approved for Medicaid nursing home benefits in Washington state for 1 year possessed $27.5 million in assets at the time of their denial. These assets had to be disposed of before they could qualify for assistance. Over 80% of the assets had been sheltered: 59% were transferred to a spouse, 11% were transferred to adult children, and 11% were retained as exempt. Only 8% were consumed for long-term care. The remainder was of uncertain disposition.

To account for the magnitude of these figures, the OIG interviewed 32 professional advisers on Medicaid eligibility. These people described a network of private "elder law" attorneys, publicly funded legal services attorneys, social workers, and even Medicaid staff who counsel families on how to qualify an infirm elder for Medicaid while preserving income and assets. The sheltering techniques recommended by such advisers included: interspousal and other legal transfers, trusts, purchase of exempt assets, "intent to return" to the home, life estates, joint tenancy with right of survivorship, gift and estate planning, durable power of attorney, guardianships, divorce, relocation, care contracts, and nonsupport suits. One attorney, whose bag of tricks is highlighted in the OIG report, guaranteed Medicaid eligibility within 30 days for a $950 fee. . . .

Finally, the OIG observed that "financial abuse of the elderly, according to study respondents, is 'commonplace,' 'bigger than anyone thinks,' 'rife.' We heard many stories about people forced onto Medicaid when their income or resources were taken" (OIG, 1989, p. 11).

MEDICAID ASSET SPENDDOWN

A common understanding, often referenced in the literature, is that half or more of all nursing home patients on Medicaid were private pay until they spent down to poverty (Branch et al., 1988, p. 649; Burwell, Adams, & Meiners, 1989, p. 2; Davis, 1984, p. 3; DHHS, 1987,

p. 19; Dobris, 1989, p. 10; NAIC, 1987, p. 2). Tragically, many people actually do sell their homes and spend their life savings on nursing home care before they qualify for Medicaid. Evidence is mounting, however, that such draconian measures are both unnecessary and less common than previously supposed. Impoverishment is not the only path to Medicaid nursing home eligibility, according to the OIG work. Financially sophisticated people who are accustomed to dealing with attorneys, accountants, and financial planners can find ways to protect their assets and still qualify for Medicaid. Others, with less financial savvy, often lose what little they have before they learn how the system works (OIG, 1988, p. ii).

Very little is known about the magnitude of asset spenddown. For example, Branch reported how fast people would become impoverished if they spent down in nursing homes (Branch et al., 1988). He did not tell us how often or to what degree they actually do spend down. Several recent studies have found that spenddown is actually much smaller than previously believed (Burwell, Adams, & Meiners, 1989; Liu & Manton, 1989; Liu, Doty, & Manton, 1989; Spence & Wiener, 1989). Like the Branch study, however, these studies assume that people who had significant assets at one time, but ended up on Medicaid, must have had catastrophic care costs. None of them developed the possibility that assets were transferred or sheltered in order to qualify for nursing home assistance. Nevertheless, the techniques to transfer and shelter assets, and the counseling to learn them, are readily available, according to the OIG.

SPOUSAL IMPOVERISHMENT

Impoverishment of the spouse at home caused by institutionalization of a disabled husband or wife used to be a serious problem. Medicaid rules allowed only a few hundred dollars of income per month to be shifted from an institutionalized to a community spouse who had little or no separate income (Neuschler, 1987, pp. 48–49). The new community spouse "minimum monthly maintenance needs allowance" was designed to solve that problem at considerable public expense. We should keep in mind, however, that the same people whose income and resources will now be protected may live in homes they own free and clear. Their problem is not poverty per se, but rather cash flow.

MEDICARE CATASTROPHIC COVERAGE ACT

The Medicare Catastrophic Coverage Act of 1988 changed Medicaid long-term care eligibility in several ways that affect asset shelters and estate recovery potential. Some of the changes, such as more generous treatment of community spouse income and resources, will make Medicaid benefits easier to obtain. This could mean that more assets will remain to be recovered from estates or, alternatively, that people will have longer to find ways to protect the assets. Other changes, such as mandatory and lengthened transfer of assets restrictions, make eligibility somewhat more difficult. This could lead to liquidation of assets and greater spenddown or, alternatively, to wider use of better planning and qualifying techniques. Most of the methods used to shelter or transfer assets legally in the past are still intact. The basic condition remains unchanged: Families can preserve significant assets while qualifying elders for Medicaid nursing home care. Without estate recovery programs, these assets pass unencumbered to noncontributing heirs and Medicaid shoulders the full brunt (minus mandatory contributions to cost of care) of the long-term care costs.

IMPLICATIONS

Medicaid requires impoverishment. Few scholarly papers or popular articles on long-term care financing say otherwise or explain further.

Yet, impoverishment is neither a sufficient nor a necessary cause of Medicaid eligibility. Two-thirds of the elderly poor in America are not covered by Medicaid (Holahan & Cohen, 1986, p. 99). On the other hand, people with median and even higher income and resources often qualify for the program's most expensive benefit (nursing home care) while preserving the bulk of their assets for heirs (Neuschler, 1987, p. 20; OIG, 1988; OIG, 1989).

Looming in the background of last year's "catastrophic" debate was the question: What shall we do about *long-term care* costs? That predicament is front and center now. Most of the work done by the federal government on this issue has encouraged the development of private risk-sharing solutions. Private sector answers, however, have been much slower to develop than anticipated. This is a puzzle, because most of the obstacles to market-based solutions do not seem insurmountable. Experts on long-term care financing have assumed that Medicaid is not a major impediment.

In light of the findings discussed here, however, consider that elderly people are often unclear about catastrophic long-term care risks. They deny their personal jeopardy and do not plan ahead to protect privately against financial catastrophe. They do not plan to rely on public assistance either, but once they get sick, welfare is their only option. Under today's system, they can avoid paying insurance premiums (often in excess of $100 per month) or risk-sharing membership fees, wait to see if they are stricken by a long-term debilitating illness, and still receive nursing home care paid for by Medicaid while preserving their assets for heirs. Because of the shame felt by families forced to qualify their elders for welfare, the negative aspects of Medicaid nursing home care—dependency, loss of income, access and quality problems, institutional bias, and stigma—often go uncommunicated to others.

Therefore, the elderly population perceives no urgent need to purchase insurance, join a Social/Health Maintenance Organization or Continuing Care Community, convert the equity in their home, or save toward long-term care costs. Without a compelling need among consumers to buy (low demand), sellers of such services lack sufficient reason to invest in the necessary research, development, and marketing of private protection (low supply). This in itself could explain why the impact of private sector long-term care financing options has been disappointing.

CONCLUSION

A plan to eliminate this impasse between public and private long-term care financing options is quite simple conceptually: Give middle-class elderly people a clear choice between access to public funding of long-term care or preservation of their estates—not both. The rudiments of such a plan are evident in the OIG report's (1988) recommendations:

- Change Medicaid rules to permit families to retain and manage property while their elders receive long-term care.
- Strengthen the transfer of assets rules so that people cannot give away property to qualify for Medicaid.
- Require a legal instrument as a condition of Medicaid eligibility to secure property owned by applicants and recipients for later recovery.
- Increase estate recoveries as a nontax revenue source for the Medicaid program while steadfastly protecting the personal and property rights of recipients and their families. (p. ii)

Underlying these recommendations is the belief that we should eliminate the indignities and inequities associated with qualifying for nursing home assistance. We should not pressure people to divorce, impoverish their spouses, liquidate their property, or hire estate planners in order to qualify. Elderly people, financially independent all their lives, but stricken by catastrophic illness in their most vulnerable years, should not be compelled to rely on welfare because of temporary cash flow problems. To correct such deficiencies in the existing program, however, we would have to pay for the solution. We can do this by closing the loopholes in transfer of assets restrictions, requiring legal encumbrances on property as a condition of eligibility, and mandating cost-effective estate recoveries as a prerequisite for federal financial participation. Alternatively, we could offer middle-class seniors a line of credit secured by their estates with which to purchase home or nursing care and get them off welfare entirely.

If implemented, these recommendations would increase Medicaid estate recoveries substantially. But this new nontax revenue is not the most important aspect of the recommendations. We would also be sending a message to America's senior citizens and their families: If they do not or cannot protect themselves privately against the risk of catastrophic long-term care costs, their government will provide the necessary care. But, if they own property, they must understand that it will be recovered—when it is no longer needed for the livelihood of their immediate dependents—to pay for publicly funded care and ensure that the same benefits will be available for others. Only the remainder after reimbursement of costs will pass to their heir and beneficiaries. So if they do not want to encumber their estate, then they or their heir should purchase protection in the private marketplace.

If we send this message, we can expect the demand for private risk-sharing products to increase. Greater demand means more suppliers, increased competition, better products, leaner pricing, thriving new industries, and, therefore, increased employment and tax revenues.

The last piece in the puzzle is to explain how seniors will pay for private risk sharing. The experts say older people lack the cash flow to purchase private long-term care protection (Rivlin & Wiener, 1988). Yet people over 65 possess more than $800 billion in home equity (Rivlin & Wiener, 1988, p. 131). Seniors are "house rich" and "cash poor." Home equity conversion experiments intended to solve this problem have failed. These experiments have failed, however, because Medicaid pays for nursing home care and exempts the home. Why encumber the house to buy insurance you may not need when the government will pay for your care if you need it and save the house anyway? When people know they can save the house or get Medicaid, but not both, they will be more likely to seek home equity conversion to provide the cash flow to purchase private protection. This change could make home equity conversion economically viable as private enterprise.

Finally, faced with the potential loss of their inheritances, adult children of elderly people will contribute voluntarily toward long-term care insurance premiums or other forms of financial protection for their parents. They have the cash flow and their aging parents have the assets. Both parties have an intense interest in preserving the estate, including the family home. Under the current system, the adult children of elderly people reap a windfall from Medicaid for ignoring the risk of catastrophic costs.

The proper role of government in this arena is to help those who cannot help themselves. It is not to transfer wealth from tax payers to indemnify heirs. If we make long-term care assistance more readily available than now, but require a payback from estates, middle-class elderly people will have better access to care and stronger reasons to seek nonwelfare protection. In time, they will be freed entirely from the indignity of legal maneuvering to qualify for public assistance.

REFERENCES

Branch, L. G., Friedman, D., Cohen, M., Smith, N., & Socholitzky, E. (1988). Impoverishing the elderly: A case study of the financial risk of spenddown among Massachusetts elderly people. *The Gerontologist, 28,* 648–652.

Burwell, B., Adams, E., & Meiners, M. (1989). *Spenddown of assets prior to Medicaid eligibility among nursing home recipients in Michigan* (contract 500–86–0016). Washington, DC: SystemMetrics/McGraw-Hill for the Health Care Financing Administration.

Davis, C. (1984). Long-term care financing and delivery systems: Exploring some alternatives (conference proceedings). Washington, DC: Health Care Financing Administration.

Department of Health and Human Services (DHHS). (1987). *Report of the task force on long-term health care policies.* Washington, DC: U.S. Government Printing Office.

Dobris, J. C. (1989). Medicaid asset planning by the elderly: A policy view of expectations, entitlement, and inheritance. *Real Property, Probate and Trust Journal, 24,* 1–32.

General Accounting Office. (1989). *Medicaid, Recoveries from nursing home residents' estates could offset program costs* (GAO/HRD-89–56). Washington, DC: U.S. Government Printing Office.

Holahan, J. F., & Cohen, J. W. (1986). *Medicaid: The trade-off between cost containment and access to care.* Washington, DC: The Urban Institute.

Letsch, S. W., Levit, K. R., & Waldo, D. R. (1988). National health expenditures, 1987. *Health Care Financing Review, 10,* 109–122.

Liu, K. Doty, R., & Manton, K. (1989). *Medicaid spenddown of disabled elderly persons: In nursing homes or in the community?* (unpublished paper prepared under Cooperative Agreement no. 18-C-98641/4–02). Washington, DC: Health Care Financing Administration.

Liu, K., & Manton, K. (1989). The effect of nursing home use on Medicaid eligibility. *The Gerontologist, 29,* 59–66.

Moses, S. A., & Duncan, J. (1985). *The Medicaid estate recoveries study* (unpublished report). Seattle, WA: Health Care Financing Administration.

National Association of Insurance Commissioners (NAIC). (1987). Long-term care insurance: An industry perspective on market development and consumer protection.

Neuschler, E. (1987). *Medicaid eligibility for the elderly in need of long-term care.* Washington, DC: National Governors' Association.

Office of Inspector General. (1988). *Medicaid estate recoveries* (OAI-09–86–00078).

Office of Inspector General. (1989). Transfer of assets in the Medicaid program: A case study in Washington State (OAI09–88–01340).

Rivlin, A. M., & Wiener, J. M. (1988). *Caring for the disabled elderly: Who will pay?* Washington, DC: The Brookings Institution.

Rymer, M., Burwell, B., Adler, G., & Madigan, D. (1984). *Grants and contracts report, short-term evaluation of Medicaid. Selected issues* (contract no. HHS-100–82–0038). Washington, DC: Health Care Financing Administration.

Spence, D. A., & Wiener, J. M. (1989). *Medicaid spenddown in nursing homes: Estimates from the 1985 national home survey* (unpublished draft). Washington, DC: The Brookings Institution.

U.S. Bureau of the Census. (1986). Household wealth and asset ownership; 1984 (Current Population Reports Series P-70, No. 7). Washington, DC: Author.

U.S. Code. (1982). *Congressional and Administrative News,* 97th Congress—Second Session, Legislative History (Public Laws 97–146 to 97–248, Vol. 2). St. Paul, MN: West.

READING 24

The Case Against Paying Family Caregivers

Ethical and Practical Issues

C. Jean Blaser

Paid family caregiving can be the best of care and, unfortunately, the worst of care. This article will detail how paying family can produce the worst of care, and why taxpayers should not support such payments.

This position is based on experience derived from managing the Community Care Program in Illinois, which provides home- and community-based care to over 35,000 older people a month. Eligibility is based on a need for care, as measured by a standardized instrument termed the Determination of Need. The instrument assesses functioning with fifteen activities of daily living and instrumental activities of daily living, and for each activity with which the older person has difficulty, the availability of family and informal supports is addressed. Need for care is determined by a look at those activities with which the applicant has difficulty and lacks necessary assistance. In this way, the program is designed to complement and supplement family support, but not replace it.

In the first years of the program, as a result of a policy decision by another state agency, a significant portion of the caseload was served by family members who were paid as personal care attendants. When the program was transferred to the Illinois Department on Aging, detected abuses led the department to close that subprogram, allowing no more clients to have personal care attendants, and to allow payments only to contracted agencies. Since that time, however, a number of agency providers have elected to hire family members as "preferred" workers, assigning them to care for an elderly family member. As a result, the department has a considerable history to draw upon regarding the problems that can occur when a family member is a paid caregiver.

EXPLOITATION

Advocates may argue that a policy of paying the family caregiver supports and strengthens basic family values. On the other hand, it can be argued that such a policy exploits family values by paying the family member less than the going "market" rate for provided services.

Under the banner of "consumer-directed care," states can reduce the costs of home- and community-based care by providing vouchers or direct payments to clients who, in turn, hire their own workers, termed personal care attendants. By avoiding the administrative costs of recruiting, hiring, training, and supervising workers, the cost per unit of service is substantially reduced. The cost is further reduced by not having to pay mandated fringe benefits such as unemployment and workman's compensation, although most states may pay Social Security taxes on behalf of the client. And, of course, no health insurance, retirement benefits, sick leave, or vacation are offered. Indeed, the states are careful not to pay for these benefits lest they be open to a charge that these workers are state employees and subject to all the benefits state workers enjoy.

It is a well-established fact that reimbursements to homecare workers are inadequate in most areas of the country. In these times of full or nearly full employment, workers can demand and receive higher wages. Because fewer are willing to accept the low salary and lack of benefits paid to personal care attendants, there is a severe shortage of homecare workers.

However, family workers, who can be considered to be a subset of the larger class of personal care attendants, can be an exception to this general finding. Family members are more likely to be trapped into accepting such employment because they are unable to recruit and hire a nonfamily worker. Faced with the prospect of placing their family member in a nursing home, these family workers will sacrifice higher wages to care for their family member at home.

One such family member detailed this problem in a public hearing on providing a "living wage" for homecare workers. She reported a long and fruitless search for a competent and reliable worker. After many experiences of workers not showing up, not performing the requested tasks, or even stealing from the client, she reluctantly decided to quit her higher paying job with benefits to stay with the client as a paid family worker. She was paid minimum wage and received no benefits for this sacrifice. In addition, she again faced the difficult task of finding a replacement whenever she was ill and unable to work, her car broke down, or she needed respite. She felt trapped by a system that did not value caregiving and did not provide sufficient reimbursement to attract a qualified and quality workforce.

In a society that already exploits the in-home worker, the policy of paying family to provide the care simply continues the exploitation and, in fact, may remove any incentive to change. If family members agree to provide the care at a less-than-adequate wage, and if the policy that allows them to do so can be cloaked in the "feel good" language of consumer choice, the pressure to increase wages and benefits for all in-home workers is reduced. And, with other potential workers able to obtain jobs with higher wages and benefits, the client and family are likely to have very little choice but a family caregiver.

POTENTIAL FOR FRAUD AND ABUSE

The above discussion focused on the better side of paid family caregiving, where the family member is more reliable, competent, and caring than a nonfamily worker. On the other side of the picture are instances in which the family member defrauds or abuses the client and program.

While the potential for fraud and abuse exists in any social service program, a program in which family members are paid to provide care creates an environment that is particularly ripe for fraud. The most common type of abuse is financial fraud, where the client and the family member collude to report services that were not delivered, in order to collect payments. In some instances, the benefits of the fraudulent payments are shared. Other times, the older client allows the family member to receive the payment, perhaps through a distorted sense of intergenerational transfer.

A recent example of collusion was detected when a case manager conducted an annual re-determination of eligibility for an elderly woman who had been served by the Community Care Program for five years. The assessment was conducted in the home of the grand-daughter, who had been hired by a contracted service agency to care for her grandmother. The client was lying on the sofa and reported she was in great pain and able to do very little for herself. The case manager, who did not speak the language of the client, used the grand-daughter as an interpreter and, when the assessment was completed, the client was found to have scored 79 points, which on a scale of 0 to 100 is very impaired and represents less than 4 percent of the service population. As a consequence, the case manager authorized fifty hours of service a week, to be provided by the grand-daughter.

An alert homecare supervisor, unable to contact the worker or the client at times when the worker was supposed to be serving the bed-bound client, made an unannounced in-home visit and learned from a building manager that the client did not reside in the apartment but, rather, lived in a senior highrise. The supervisor alerted the case manager, who visited the senior housing site and observed the same client participating vigorously in an activity. Upon inquiry, the case manager was advised that the client had lived in the highrise for five years, and was able to function independently. In fact, the supposedly very confused bed-bound client who did not speak English had taken English classes.

In this example, the client and the grand-daughter colluded to defraud the state of more than $48,000 in service payments. In other cases, however, the department has found the family caregiver defrauding the state without client involvement. Through a match of service records with state death records, the department has found cases in which the client has died but the family member continues to report services, forging the client's name to the service verification records. In another case, the client moved to another state, but the family caregiver continued to bill the state as if services were still being provided. Unfortunately, these examples are not all that uncommon.

A more troubling problem arises when the older person is coerced through intimidation into signing the service receipt. Most often, the older person is fearful of losing support and is threatened with nursing home placement and so signs for receipt of services. But, in some instances, the older person has been subjected to physical abuse or neglect or financial exploitation. Neglect is the most common type of abuse. Department staff have seen numerous examples of care provided in early morning or late evenings because the family member is holding down another full-time job, or the grandchild is the supposed worker and is using the funds in order to pay for college.

In other instances, the abuse takes the form of financial exploitation. The family member may be dependent on the pension or Social Security check of the client as well as the payment for services to the client. Case managers have reported instances in which the older person is very impaired and in need of more intensive or skilled care than can be provided by the family member, but is denied this needed care because the family member would then lose control of the client's financial resources. Staff who had talked with one such client reported that she begged for someone to get her into a nursing home and away from her daughter, who was the paid caregiver.

INCREASED ADMINISTRATIVE COSTS

With such potential for fraud and abuse, homecare providers report having to take extra measures to assure quality service from "preferred" or family workers. First, the agencies report more difficulty in assuring that the workers are trained before they start service and that they participate in required quarterly in-service training sessions. Second, the agencies have had to increase their monitoring efforts, making more calls to the home or making unannounced visits to the home when the worker is supposed to be on duty. Indeed, it is this sort of monitoring that brings to light many of the cases of fraud, as was seen in the case described earlier.

There are limits, however, to how successful such training and monitoring measures can be. In cases in which the family worker fails to attend the required training session or is not providing the care as directed, the agency will often follow its personnel policies for employee discipline and may terminate the worker. When this happens, the worker will simply go to another homecare agency and secure employment. The client will then request to transfer to the second agency and request services from the family member. This "employer hopping" can continue until the worker finds an agency that is willing to hire family members as workers and that is less than diligent in monitoring the delivery of services.

Advocates of consumer choice will argue that such behavior is an example of the client exercising the right to choose a family member as worker rather than a stranger. A less sanguine interpretation is that the family member is exploiting the client and the service system. Otherwise, why is the family worker not content to make the same salary serving a different client, while an unrelated worker serves the family member?

Thus, agencies not only incur increased administrative costs in monitoring workers but may lose clients as a result of either refusing to assign workers to care for family members or detecting and acting upon fraud. And, the agencies that do not diligently monitor the delivery of services may be subject to loss of contracts or even payments for damages as a result of poor or nonexistent care.

INCREASED PROGRAM COSTS

In addition to the potential for fraud and abuse of the system and the client, there is the potential program cost of a policy to pay family caregivers. The financial impact of such a policy could be significant. If we are to believe the literature, about 80 percent of the care provided to older people is informal and is provided most often by family members. A systematic program to pay these family and informal caregivers, then, could increase program costs as much as five times, with no increase in actual care provided.

In states where the home- and community-based services are entitlements, which is the case in Illinois, a new entitlement, for families with older family members in the area, would be created if a formal policy of paying family workers were to be instituted. It is not too difficult to imagine not only a significant number of families applying for the benefit once they learn of it but also family intrigues about who gets to "claim" Granny.

On the other hand, the more usual case is that the state caps the amount of funds available for home- and community-based services. In such states, the limited resources could no longer be targeted only to those who had needs beyond those that the family could meet or who had no family nearby to provide assistance.

AN ALTERNATIVE APPROACH

The issue of family responsibility has plagued policy makers for decades. Several years ago, as a response to an advocacy effort to establish payments for family caregivers, the Illinois Department on Aging commissioned an opinion survey of the provider network. The results were interesting, with an almost equal number of respondents agreeing with each of the following statements: "strongly support," "somewhat support," "somewhat oppose," and "strongly oppose" paying families to care for their older members. With such a clear lack of consensus, the department sought the middle ground.

Current department policy does not allow direct payment to family members for care but offers services to complement and support the family members in their efforts. Eligibility for services is based upon both impairment and informal support, so that individuals with moderate impairment but no informal supports are eligible, as are those with strong family support but high impairment. In this way, the program acknowledges the need for support and respite for the family. And, if the family is absent or not able or willing to provide assistance, the state will provide for the needed services. With these policies, the family is supported but the negative consequences of direct payments to the family are avoided.

READING 25

For Love and Money

Paying Family Caregivers

Suzanne R. Kunkel, Robert A. Applebaum, and Ian M. Nelson

The recent emergence of options for compensating family caregivers has raised a host of new issues. Paying family members for providing care has brought to the forefront policy questions about the intrusion of public systems into family life; ethical and ideological issues about obligation and accountability; and pragmatic concerns about health, safety, and quality of services. In addition, compensating family members who provide long-term care has added to the growing dialogue about economics, family values, and the nature of "care work." Understanding the tensions about paying family members to provide care requires an examination of some fundamental assumptions about care and work. What is the difference between the work people do for love and the work people do for money? What does society expect and require families to do for love, without expectation of money? What are the reasonable limits to those expectations? Proponents of compensating family caregivers argue that it is a way to strengthen, expand, and sustain the natural support system. Critics of paid family care worry that compensation for some of the work will erode family obligation, create a strain on the public system, and put older people at greater risk of abuse and poor care.

While direct payment to caregivers is the model of compensation that brings the debate into sharpest focus, it is only one of several types of compensation. In the broadest sense, financial supports for caregivers can include

SOURCE: "For Love or Money: Paying Family Caregivers" by Suzanne R. Kunkel, Robert A. Applebaum, and Ian M. Nelson. Reprinted with permission from *Generations*, Winter, 2003–2004, 27:4, pp. 74–80.

direct payment for services provided, tax credits, unpaid leave, and cash allowances to cover expenses related to caregiving.

Here we focus only on direct payments to family members for caregiving work. The prevalence of this phenomenon has increased significantly over the past decade, as a direct result of the consumer-direction movement in home and community-based long-term care. In a recent inventory, Doty and Flanagan (2002) identified 139 home and community-based support programs with some consumer-directed option; half of these programs offered the option to older clients. Consumer direction is both a philosophy and a practice that emphasizes the right and ability of consumers to assess their own care needs, decide how best to have those needs met, and evaluate the quality of the services provided. One of the meaningful ways that consumer direction has been put into practice is in care-provision programs in which payers (either government or private plans) allow individuals to hire and manage their own workers. A significant majority of older people who have services that they themselves direct choose to hire a family member (Doty et al., 1999; Dale et al., 2003). Consumer-directed long-term care, then, provides a focus for a review of the issues and evidence related to paying family caregivers. For purposes of discussion, we have categorized the issues as ideological, ethical, professional, and personal-interpersonal. To provide evidence on each of the issues raised under these categories, we rely heavily on two consumer direction programs in Ohio and on the National Cash and Counseling Demonstration and Evaluation project.

IDEOLOGICAL CONCERNS AND EMPIRICAL REALITIES

U.S. culture places great importance on the primacy of the family. However, our public policies related to caregiving reflect a reluctance to legislate supports for family care. For example, the United States was very late among industrialized nations to adopt employment

policies in support of family care. The number of programs and the amount of public expenditure in support of family caregiving (for children, older people, or others who need assistance) are very low. For these reasons, paying family caregivers represents a significant shift for U.S. policy: use of public dollars to support what had been considered a private and obligatory activity, and the involvement of government in family life.

Critics of paying for family care have voiced a range of specific concerns about how this practice might undermine social values. If informal caregivers are paid, critics hypothesize, there would be a major shift away from caring as part of normal family responsibility. They suggest that in this and other ways, payment would decrease the quality of the caregiving experience for care recipient and caregiver, with paid services substituting for unpaid care now provided. Because family care is the dominant mode of provision in long-term care, such a shift would place tremendous burdens on public expenditures. Critics also anticipate that the cost increase could be compounded by a likely increase in the number of homecare recipients choosing this more flexible benefit.

To address these concerns, we rely on evidence from recent evaluation studies of consumer-directed programs, in which a high proportion of consumers have chosen to hire family members.

In both the Cash and Counseling demonstration and the Ohio projects, consumers received the same dollar allocation that they would have under the traditional service system. This allocation is based on health, functional, and cognitive status. Consumers are then able to decide on a payment rate for workers, but the total cost is fixed. In some instances, consumers who directed their own care used a higher rate than that paid to agencies; in other cases, the rate paid in consumer-directed care was lower.

In the Ohio programs and in Cash and Counseling there were differing policies on who could be a paid worker. Some programs did not allow spouses to be paid, while others did. All programs paid worker compensation, unemployment insurance, and Social Security taxes. Training needs were determined by the consumer.

Data from the National Cash and Counseling Demonstration and Evaluation found significant increases in the satisfaction levels of both consumers and caregivers (Foster et al., 2003a; Foster et al., 2003b). Consumers in the demonstration, about 80 percent of whom hired family members, reported large and consistently higher rates of satisfaction compared to a randomized control group (Dale et al., 2003; Foster et al., 2003a). For example, more than 90 percent of the demonstration's consumers older than 65 were very satisfied with their relationship with their paid caregiver, compared to close to 80 percent for the control group (Foster et al., 2003a). Just over one-quarter of control group members felt neglected by their paid caregiver, compared to 11 percent of demonstration consumers. When comparing consumers who hired family members to those who hired nonfamily workers, findings showed significantly higher satisfaction rates (99 percent versus 91 percent) for those with family workers (Simon-Rusinowitz et al., 1998). Interviews with caregivers also showed large and significant differences in favor of the option of hiring family members. Demonstration program caregivers' reports of satisfaction showed them to be significantly more satisfied with overall care arrangements and significantly less worried about whether the care recipients had enough help in their absence (Foster et al., 2003b).

There did not seem to be any negative effects on the overall relationship between the paid family caregiver and the consumer. In response to questions such as whether the caregiver and care receiver get along very well and whether the current relationship is better than at enrollment, there were no differences between the two groups (Foster et al., 2003b). Caregivers participating in the demonstration program were significantly more likely to talk with consumers about personal care needs, and the program consumers were significantly more cooperative. Program caregivers also reported significantly lower emotional strain and significantly higher satisfaction with life. In combination, these data indicate that there is no evidence in the demonstration that family relationships are negatively affected by the payment option.

In a telephone survey of Medicaid personal care clients in New Jersey, researchers (Mahoney et al., 2002) found that about 40 percent were potentially interested in a cash option, but older people were 2.7 times less likely than the younger clients to be interested in this option. In addition, in all of the consumer-directed demonstrations, even consumers who do choose to hire their own workers do not always hire a family member. Taken together, the findings suggest that these programs do not bring consumers "out of the woodwork" to use services they would not otherwise seek.

ETHICAL CONCERNS AND EMPIRICAL REALITIES

Some of the concerns about paid family caregiving are related to the values of beneficence and avoidance of maleficence, the desire to do good and to do no harm. These values translate into a heavy emphasis on protection and minimizing risk for those receiving publicly funded services—which critics call well-intentioned but paternalistic. In the early days of homecare, some suggested that in-home service recipients would be at greater risk of receiving poor quality care. Worker fraud, abuse, and neglect were expected to be a much greater problem in the home when compared to the nursing home setting, because there was only limited agency supervision in that venue. Ironically, this same logic has been expanded to suggest that family and other non-agency-based workers present a higher degree of risk than agency-based workers. Anecdotal reports from providers and homecare program administrators have identified concerns about fraud and poor quality provided in consumer-directed programs (Blaser, 1998). In a survey of state-contracted homecare agency administrators, Linsk and colleagues (1992) found that fraud and abuse were the most frequent concerns about paying family members. The concerns ranged from potential exploitation of the system and of the consumer to failure to provide the services that were paid for.

Findings from studies of the previously mentioned demonstration projects and from an evaluation of the California In-Home Supportive Services Program show no significant differences in safety risks between clients receiving agency-based services and those using consumer-directed services. On many variables, consumers under the self-directed model have better health and safety outcomes. For example, in the Cash and Counseling demonstration program there were no differences in accident rates or falls, but consumers with self-directed care were significantly less likely to have bedsores or to have seen a doctor because of a cut or burn (Foster et al., 2003a). The project also reported large and significant reductions in the proportion of consumers with self-directed care reporting helpers arriving late or failing to arrive at all and in rates of theft. A study of the California program, the largest consumer-directed option in the country, reported no differences on a series of health and safety measures that examined such areas as abuse, harmful behaviors, theft, injury, and neglect (Doty et al., 1999).

Data on quality of service also indicate that consumers hiring family members report better care. For example, among consumers hiring family members, the Cash and Counseling demonstration reported large and significant reductions in the proportion of consumers feeling neglected or being rudely treated by workers. These sizable differences also were evident in satisfaction rates in the delivery of care (Foster et al., 2003a). Findings from the California study found self-directed consumers of care to be more satisfied with the quality of their workers and the services provided (Doty et al., 1999). Preliminary results from the Ohio demonstration suggest that consumers who directed their own care rated the quality of services highly, at a level equivalent to those receiving agency-based services (Kunkel and Nelson, 2003).

A final area of concern involved fraud on the part of the consumer or their family. The three sites participating in the Cash and Counseling demonstration invested considerable resources in monitoring the development of the service plan and in reviewing expenditures. Using a social service professional in a support and monitoring role with consumers and a systematic book-keeping system to assist and review expenditures, the effort found minimal auditing concerns.

These demonstration programs clearly support the notion that consumers can make good decisions about their own care, even when family members are providing that care. Paid family workers did not abuse the system, exploit the consumer, or fail to provide good services. Consumers hiring primarily family members were healthier, safer, and are more satisfied with services. Program funds appeared to be spent according to plan.

Based on evidence from consumer direction, we can argue that paying family caregivers provides an "acid test" for the notion that beneficence and lack of harm can only be achieved in a formal, public system. The success of the consumer-employed family caregiver arrangement suggests that the values of health and safety might be reframed in ways that engage, and give primary voice to, the consumer of services, moving us from paternalism to participation, with no loss of good care and no increased harm.

PERSONAL/INTERPERSONAL ISSUES: CARE AS A COMMODITY

The caregiver–care recipient relationship can be emotional, intense, and challenging, whether the individual providing services is a family member or not. "Care work" is an inherently problematic concept in U.S. culture. How can something so clearly emotional in content and motivation as "care" be considered "work"? "Paid caregiving" is similarly incongruous. If care is something we give, from the heart, doesn't the introduction of payment demean that dimension of the relationship? These questions are magnified when the person being paid for care work is a family member.

These difficult philosophical concerns can be better tackled when put into cultural context and translated into more specific questions. When considering the cultural values that shape the debate, we find that the deep concern over maintaining lines between the work we do for love and the work we do for money, between "care" and "work," is unique to the United States. Linsk and colleagues (1992) document the "remarkable worldwide expansion of provisions in support of caregivers," and the prevalence of policies of government compensation for family and other informal care providers that has been widespread for a number of years, whereas, as noted earlier, in the United States we have placed greater value on the separation of government and family as a reflection of the value we place on the primacy of the family. That the introduction of public dollars into the private family domain of caregiving would cause concern in the United States is, therefore, predictable.

We can address that concern more directly in the form of two specific and interrelated issues: how and whether payment changes the relationship between caregiver and care receiver, and the difficulties old and frail consumers may have in taking on the role of employer of their own family members. Does payment change the caregiver–care recipient relationship? Probably so. However, the assumption that these changes must be negative has not been borne out by the demonstration projects. In focus groups and phone interviews, consumers consistently talk about the sense of empowerment that they get from being in charge of their own workers and their own services. They also consistently report that hiring their own worker—very often family members—makes them feel more secure and more in charge of their lives. They were more confident that their workers would show up. They were hiring people they knew, people who knew them and their preferences, people they trusted.

In exchange for receiving services from a trusted worker, these consumers are able to give them something tangible in return: money. The demonstration programs typically arrange for the workers' paychecks to be sent to the consumers, who, as employers, can hand them to their employees. This practice helps to make roles and responsibilities clear and helps to even out the balance in relationships between the caregiver and the care receiver. Empowered consumers seem to be successful at managing their workers, giving feedback, and making sure that their services are being provided in the best way possible. Earlier discussion in this paper pointed out that fraud and abuse were not significant problems in consumer-employed provider models. This finding, and the overall sense of empowerment and responsibility voiced by consumers, supports the notion that consumers, even when they are old and frail, can manage their workers successfully, even with the overlay of family dynamics.

CONCLUSION

Families have long been the bedrock of long-term care, and all indicators suggest that they will continue to be so in the future. Despite this strong foundation, societal changes in such areas as longevity patterns, workforce participation, and family composition suggest that caregiving will grow in both importance and difficulty. Social policy in support of family care must continue to evolve.

Empirical evidence from well-designed research demonstrates that recipients of paid family care are more satisfied, as are the caregivers. Anecdotal concerns about neglect, safety, and negative effects on family relationships have been dispelled in the studies now available. Although the policy debates about paying family members will continue, this work reinforces earlier studies in concluding that caring for love *and* money is possible, and, for some, desirable. Compensating family workers, and having consumers hire and manage their own workers, can be good for consumers, family members, and the long-term-care system overall.

REFERENCES

Applebaum, R, et al. 2002. "Quality and Consumer Directed Care: Lessons Learned from the New Jersey Preference Program." Oxford, Ohio: Scripps Gerontology Center, Miami University.

Blaser, C. J. 1998. "Case Against Paid Family Caregivers: Ethical and Practical Issues." *Generations* 22(3): 65–9.

Dale, S., et al. 2003. "The Experiences of Workers Hired Under Consumer Direction in Arkansas." New Jersey: Mathematica Policy Research Inc.

Doty, P., et al. 1999. "In-Home Supportive Services for the Elderly and Disabled: A Comparison of Client-Directed and Professional Management Models of Service Delivery." U.S. Department of Health and Human Services. Non-Technical Summary Report or [http://www.aspe.hhs.gov/daltxp/reports/ihss.htm].

Doty, P., and Flanagan, S. 2002. "Highlights: Inventory of Consumer-Directed Support Programs." U.S. Department of Health and Human Services [online: http://aspe.hhs.gov/daltxp/reports/highlight.html].

Foster, L., et al. 2003a. "Improving the Quality of Medicaid Personal Assistance Through Consumer Direction. Project Hope—the People to People." *Health Foundation Corporation.* [www.healthaffairs.org/freecontent/vzzn3/53.pdf].

Foster, L., et al. 2003b. "Easing the Burden of Caregiving: The Impact of Consumer Direction on Primary Informal Caregivers in Arkansas." New Jersey: Mathematica Policy Research, Inc.

Kunkel, S., and Nelson, I. 2003. "Consumer Direction in Ohio's Passport and Elderly Services Program: Preliminary Report." Internal Report. Oxford, Ohio: Scripps Gerontology Center, Miami University.

Linsk, N. L., et al. 1992. *Wages for Caring: Compensating Family Care of the Elderly.* New York: Praeger.

Mahoney, K. J., et al. 2002. "Consumer Preferences for a Cash Option Versus Traditional Services: Telephone Survey Results from New Jersey Elders and Adults." *Journal of Disability Policy Studies* 13(2): 74–86.

Nelson, I., et al. 2002. "Implementing Consumer Direction in Home Care for Older People in the Elderly Services Program in Southwest Ohio." Draft Report of Focus Groups with Case/Care Managers. Internal Report. Oxford, Ohio: Scripps Gerontology Center, Miami University.

FOCUS ON THE FUTURE

Genetic Screening for Alzheimer's Disease?

The year is 2022, and the time is 7 a.m. A voice on your home computer wakes you up. A blinking light on the screen indicates you have overnight e-mail: two messages. One is from your Great-Aunt Mabel. She's gotten back her genetic screening testing results for Alzheimer's disease, and the news is not good. They've told her the odds of her getting Alzheimer's are 90%, and she's pretty depressed by the news. She's decided that she doesn't want to live with that prospect ahead of her. Now she's having trouble finding a doctor who'll do assisted suicide for her. There are plenty of doctors who do it, of course, but she needs one who'll accept Medicare assignment for the procedure. Now she wants advice from you.

The other message is from your HMO: It's time to come in and have a blood test to determine your genetic susceptibility to hypertension and colorectal cancer. Your HMO now requires these new tests for everyone in their plan. They assure you that the genetic test is not for purposes of discrimination: You're already enrolled in their plan, so you'll be covered. But by getting genetic information about you, they insist, they'll be able to tell you how to engage in preventive health practices to limit your risk of a heart attack or cancer. The e-mail message promises that the HMO has your best interests at heart. But if you don't comply, they warn, you'll be subject to penalties. Seems like they know everything about you. It's a good thing they have your best interests at heart, you tell yourself. Time to get on the phone to Aunt Mabel.

Questions to Ponder

Researchers have recently identified a link between Alzheimer's disease and a specific genetic pattern known as apolipoprotein (ApoE-4). This form of the gene, ApoE-4, appears in around half of those with Alzheimer's. The other half of those with Alzheimer's do not carry ApoE-4, so another factor besides the gene must also be at work. Nonetheless, 90% of people who have double copies of the ApoE-4 gene will develop Alzheimer's disease by the time they reach age 80. People with double copies of ApoE-4 are only 1% of the total population, but their chances of developing Alzheimer's are about 10 times what they are for people with a different distribution of ApoE genes.

The presence of the ApoE gene in any of its forms doesn't give absolute prediction of Alzheimer's. An official statement on genetic testing was issued by a working group of the American College of Medical Genetics warning that DNA tests for Alzheimer's should not be used for routine clinical diagnosis or predictive testing (Wagner, 1996).

At present, then, the ApoE blood test doesn't give a definitive prediction of Alzheimer's, but it is possible, even likely, that a better genetic screening test will be developed in the future. The development of such a genetic test for Alzheimer's disease raises troubling questions, however:

- There is presently no cure for Alzheimer's. Are we justified in testing for diseases before any treatment is available?
- Who should be tested, and who will pay for testing? Who will have access to the results?
- How will insurance and health care systems be affected? Will legislative antidiscrimination safeguards be necessary?

The discovery of a new genetic screening test for Alzheimer's disease raises questions about an individual's right to know, and the right *not* to know, about a diagnosis. If there were a reliable test for predicting Alzheimer's, why shouldn't people have a right to know the results? But what if that genetic test weren't as reliable as people believe? Consumers may not understand that the current test yields only a probability estimate. People may mistakenly believe that the test is like a pregnancy test or like screening for Huntington's disease—tests that yield definitive knowledge. There has already been controversy about using genetic tests for Alzheimer's disease.

The case of Alzheimer's testing seems different from, say, cholesterol or hypertension screening, for which knowledge about genetic markers or other predictors can motivate patients to change behavior and reduce the likelihood of illness. In the case of genetic markers for colon cancer, genetic screening can lead to actions that might actually reduce the risk of disease, which is different from Alzheimer's disease, for which preventive measures are limited. In the scenario presented previously, Aunt Mabel believes that the genetic test has value for her. She has lived a full life and now prefers to end her life, rather than face some likelihood of developing dementia. We might recall that the first patient who died at the hands of Dr. Jack Kevorkian was Janet Adkins, a woman age only 54 who feared that she had Alzheimer's disease because of ambiguous symptoms.

(Continued)

(Continued)

The availability of a genetic predictor for Alzheimer's raises important questions for private long-term-care insurance. In the prior scenario, the HMO wants people to have genetic tests for hypertension and colorectal cancer. Private insurers someday might want applicants to undergo Alzheimer's genetic testing. We could pass laws prohibiting discrimination based on genetic tests. But is it fair to prohibit insurers from using genetic-risk data on Alzheimer's while the test results remain available to individuals? In that case, it seems likely that individuals who find they have a higher-than-average probability of developing Alzheimer's are likely to purchase long-term-care insurance, which could overload insurers with big claims—a classic instance of a pattern called *adverse selection*.

Furthermore, once a genetic screening test for Alzheimer's becomes widely used, would there be pressure for using it with older people under consideration for prominent positions? For example, John McCain was 72 when he ran for president of the United States in 2008. Should he have been urged to have an Alzheimer's screening test? In asking that question, we cannot forget that Ronald Reagan was diagnosed with Alzheimer's soon after leaving office and may already have had it during his second term.

We are only at the beginning of debate over the implications of genetic testing for society, and Alzheimer's is one of many diseases that will be at the center of this debate. The Human Genome Project has given us detailed knowledge of the entire genetic code, but genetics is not the whole story about human health and well-being. As we have seen, a genetic screening test for Alzheimer's yields a prediction of probability alone—a message of chance, not of fate. We misunderstand the test if we see too much of our destiny in it. It has been said that it is a blessing that prevents us from seeing our future, especially our future suffering, because each of us can bear more suffering than we can presently imagine. But new knowledge of genetics is likely to enlarge our ability to see into the future in ways we've barely begun to consider.

Questions for Writing, Reflection, and Debate

1. Elder law attorneys often argue that transfer of assets is perfectly right because it is permitted by law. Is this argument a convincing one? Imagine that you are an elder law attorney who has been suddenly questioned about your practice by a reporter from a local newspaper. Write a detailed statement defending your practice to be distributed to the newspaper.

2. Elder law attorneys sometimes defend transfer of assets by arguing that Medicare treats physical illnesses differently from Alzheimer's disease or similar impairments. Is this argument a persuasive one? If Medicare were amended to provide full coverage for Alzheimer's and related disorders, would transfer of assets no longer be justified?

3. Critics like Jane Bryant Quinn have charged that for older people to deliberately transfer assets to qualify for Medicaid is a form of "middle-class welfare." Is this charge a fair one? List each of the arguments in favor of and against this charge. Then look over what you've written and produce a rebuttal for each argument.

4. Some who favor the idea of transfer of assets from aged parents to adult children to qualify for Medicaid argue that older people have a "right to leave an inheritance." Is this a "right" that should be encouraged or discouraged by either Medicaid or the tax system? Who would benefit and who would be harmed if we were to expand that right? Who would benefit and who would be harmed if we were to limit it?

5. Many believe that frail older people should be able to select anyone, including a family member, to provide the services to which they are entitled and have the government pay for that care. Are there any valid reasons for prohibiting the hiring of family members to perform home care services? Draft a letter to your congressional representative suggesting why you think this practice should be permitted or why you believe such a practice is mistaken.

6. Assume you are an assistant to a U.S. senator responsible for drafting an expanded version of a national health care law to cover the whole range of long-term care, from community care to the nursing home. Write a "bill" describing the kinds of services that might be provided to the public under the new law, including the types of conditions covered. Then write an accompanying memorandum for the senator suggesting ways the new services could be paid for. What combination of taxes and fees would cover the full package of long-term-care services?

7. Rhonda Montgomery has written about what most families immediately recognize—namely, that women end up handling most caregiving for frail older adults. Is this fact about gender differences something that the government should be concerned about, or is it an issue best left for families to work out for themselves? If we wanted to correct this apparent unfairness in the burden of caregiving, how could the government make things fairer? What are the risks of setting up a new government program to correct the problem of fairness?

8. Assume that you have a close family member who may need long-term care. Visit the following websites and identify the factors that seem most attractive about the services described on those sites: LeadingAge (Formerly American Association of Homes and Services for the Aging; http://www.leadingage.org/) and Assisted Living Federation of America (http://www.alfa.org). What questions *aren't* well addressed by the information you found at those websites?

Suggested Readings

Biegel, David E., and Blum, A., *Aging and Caregiving: Theory, Research, and Policy,* Newbury Park, CA: Sage, 1990.

Kane, Rosalie, and Kane, Robert, *Long-Term Care: Principles, Programs, and Policies,* New York: Springer, 1987.

Linsk, Nathan, and Keigher, Sharon, *Wages for Caring: Compensating Family Care of the Elderly,* New York: Praeger, 1991.

Neal, M. B., & Hammer, L. B. 2007. *Working Couples Caring for Children and Aging Parents: Effects on Work and Well-Being.* Mahwah, NJ: Lawrence Erlbaum.

Wiener, Joshua M., *Sharing the Burden: Strategies for Public and Private Long-Term Care Insurance,* Washington, DC: Brookings Institution, 1994.

Student Study Site

Visit the Student Study Site at **www.sagepub.com/moody7e** for these additional learning tools:

- Flashcards
- Web quizzes
- Chapter outlines

- SAGE journal articles
- Web resources
- Video and audio resources

SHOULD OLDER PEOPLE BE PROTECTED FROM BAD CHOICES?

Uncle Bert's bad judgment. The trouble started when Bert, 79 years old and a widower for 5 years, arranged for a housekeeper to come in a few days each week to take care of the house. Lilly at first seemed dedicated to her job, and gradually Bert began paying her extra money to do other jobs for him—like reading to him when his eyes began to fail. Soon it seemed natural for her to help him with his checkbook. Before long, Lilly was staying overnight at the house, and it seemed like Bert was becoming her only job. But Bert brightened up whenever Lilly was in the house.

Several times, when Bert's nephews or niece dropped by early in the morning, they found Lilly in bed with Uncle Bert. Something was going on with the two of them, but no one knew exactly what it was. One of the nephews told another he was afraid that Uncle Bert was becoming "a dirty old man."

Another factor in their concern was Lilly's brother, Shawn, who got involved in the household. Shawn ran a multilevel marketing program selling vitamins and health products door to door. Uncle Bert began taking megadoses of vitamins in the belief that they could reverse his diabetes. He also invested his substantial life savings in Shawn's business, which was soon being operated out of Uncle Bert's garage.

When Lilly wasn't around, Uncle Bert didn't stay at home. He had a lifelong habit of taking long walks through the city, and sometimes he still walked after dark through dangerous neighborhoods. He frequently walked all the way across town to the dog track, where he would lose a lot of money gambling. Uncle Bert's nephews and niece were worried about all this and told him so.

But Uncle Bert dismissed their concerns. He said he had a right to spend his money the way he wanted and asked simply to be left alone. "I'm old enough to live as I please," he told everyone, "and the quality of my life has never been better."

Some questions about the case of Uncle Bert. Bert's nephews and niece didn't like the way he was living. They thought he was acting in ways that were misguided, even dangerous. Specifically, Uncle Bert was acting in ways that put his health and maybe his life at risk. But what do *risk* and *danger* actually mean in this case? True, Bert was following unorthodox medical treatments and walking after dark in dangerous neighborhoods. But don't younger people act in risky ways, too? Does making bad choices or running a risk constitute some kind of self-neglect? Are we ever justified in interfering when a competent adult acts that way?

Does age make any difference in a case like this? The ill-defined relationship with Lilly makes one nephew ridicule his uncle as a "dirty old man." What can he and his siblings do about any of this? When all is said and done, does Uncle Bert have a right to be left alone? Does he have a right to pursue a romantic relationship even if others think it is ill advised? Bert stoutly defends his own definition of *quality of life.*

If these questions are troublesome, the problems only became worse as this case progressed: Shawn's vitamin and health product business was apparently going well, but Bert's nephews and niece could not see any evidence that the vitamins were helping their uncle. On the contrary, one of them happened to see on the table a copy of Uncle Bert's income tax return—filled out by Lilly—and it indicated that all his savings had been depleted. The nephews and niece were dismayed in part because, ever since they were little, their father told them that his brother Bert, who was childless, planned to leave his fortune to his favorite nephews and niece, who were his only family.

Still worse, Bert had some small strokes and began using a walker to get around the house. He couldn't go out much anymore, but still kept taking his vitamins, although not his blood pressure medication. Lilly started staying at the house all the time. Whenever members of the family visited, they tried to ask how things were going, but Uncle Bert seemed confused and also fearful about answering questions. He would act uneasy whenever Lilly or Shawn was in the room. The nephews and niece noticed that his back and neck were bruised. When asked about the bruises, Lilly said Bert had fallen but hadn't seriously injured himself. It was nothing to worry about, she added.

Uncle Bert's nephews and niece decided to have Bert's doctor look at his bruises. In the course of the examination, Bert said things that made the doctor believe Shawn and Lilly might have abused him, so the doctor reported the case to the local adult protective services agency. Within a few days, a social worker came to the door and asked to interview Bert. But Uncle Bert, standing in the doorway with Shawn by his side, quietly told the social worker to go away—he didn't want any interference in his life.

THE VULNERABILITIES OF OLDER PEOPLE

The case of Uncle Bert presents a dilemma. He asserts his right as an adult to make his own choices, but the signs that he is being manipulated or abused are growing stronger and stronger. Bert might be subject to financial exploitation by people he trusted, perhaps foolishly, in an investment scheme. His money is rapidly being depleted, and there are bruises that may indicate physical abuse. Family members are concerned about his welfare, but they also have an interest in protecting their inheritance. They seem to have reasons for suspecting "undue influence" on Uncle Bert's decisions (M. Quinn, 2002).

How has Bert's quality of life changed as events have unfolded? From Bert's point of view, his quality of life seems to be strengthened by being able to do things he has always enjoyed, whether wandering through city neighborhoods or gambling at the dog track. His relationship with Lilly, from Bert's point of view, was also a favorable factor at first. But now his quality of life has been diminished because he is limited to staying at home. The relationship with Lilly and her brother seems to have introduced an element of fear into Bert's own home.

Other questions also need to be answered. Was the doctor right to have reported Bert's case to the local agency that responds to suspected elder abuse? Was Bert acting freely when he rejected outside interference in his life? What should we make of the fact that the family members are concerned about preserving the money for their inheritance? What can be done in a complex case like this?

The decision is never easy. Aging does tend to make people more vulnerable physically, socially, and sometimes emotionally. At the same time, however, we know that individuals have different competencies even as they age. We are finding that some of our stereotypes about the vulnerabilities of older people are simply not accurate. Why should the old be protected from risky choices that may be important for their quality of life?

INTERFERING WHEN PEOPLE MAKE BAD CHOICES

The dilemmas in this case boil down to a single question: When is it right to interfere with other people's actions if those actions constitute a danger to themselves? The law tells us that we are not permitted to curtail a person's liberty just because that person lacks the ability to carry out decisions without outside help. Thus, a quadriplegic, for example, would not be legally incompetent—just a person unable to carry out decisions made. But it could be proper to intervene if the individual is unable to make decisions at all, say, because of an incapacity to understand what's going on. That may happen, for instance, in a delusional state or in dementia, where a person cannot understand or evaluate information.

As the case of Uncle Bert began, Bert seemed able to understand the risks he was taking by walking in dangerous neighborhoods or by taking vitamins to manage his diabetes. He simply evaluated matters differently than his family. There was no evidence at all that Uncle Bert was mentally incompetent. Sometimes competent people make bad choices, as we all do from time to time. As the case unfolded, Bert continued to engage in risky behavior—for example, not taking his blood pressure medication. But there was evidence of other threats: first, the suggestion that his life savings was being depleted and, second, signs that Bert might be subject to physical abuse. Even more disturbing, he seemed "fearful" in the presence of Lilly and Shawn. But in the eyes of the law, Uncle Bert still remained a free and competent adult. If a person wants to put money at risk or remain in a dangerous living situation, he is free to do so and is also free to reject help offered from outside.

One reason for intervening might be that he seemed "confused." If indeed Uncle Bert is unable to make decisions or to understand the implications of his decisions, that might be a sign of diminished mental capacity and could be grounds for intervening, regardless of whether Bert agrees. Note here that there is a difference between a case of potential elder abuse or self-neglect and a case of potential child abuse. An adult, unlike a child, is always presumed mentally competent until proven otherwise. Uncle Bert may be fearful, and he may be in what seems like an intolerable condition of exploitation, but unless we have grounds for doubting his mental competence, we cannot override his liberty.

The case of Uncle Bert's bad judgment raises many questions and suggests the need for some basic concepts to understand the issues presented by this case (Bonnie & Wallace, 2003):

- *Elder Abuse and Neglect.* Older people, like other adults, have a basic right to live in ways that others judge risky or ill advised. At what point does such behavior become categorized as *self-neglect* and justify intervention? What happens when others are involved who may be the cause of physical, psychological, or financial abuse? When harm is threatened, how do we assess a person's mental capacity for decision making? Is there a difference between neglecting one's self and being neglected or abused by another?

- *Quality of Life.* What is it that constitutes *quality of life*—in short, what makes life worth living? Could an older person make a judgment that quality of life is more important than quantity— even if it means running risks to health and safety or having relationships that others might regard as unorthodox?
- *Sexuality.* What patterns of sexual behavior are most prevalent in old age? What is the relationship between sexuality and mental health in later life? What are the signs that an older person is being exploited sexually?
- *Crime and Safety of Older People.* Are older people more likely than others to be crime victims? Do older people fear crime more than other age groups? What is the impact of crime and the fear of crime on the quality of life of older Americans?

ELDER ABUSE AND NEGLECT

Just how common is elder abuse? Reliable statistics are difficult to obtain, but the evidence suggests that physical violence, chronic verbal aggression, and neglect do fester among a small but significant segment of the older population (Jogerst et al., 2003). A comprehensive community study found that elder abuse was prevalent at a rate of 1.6% in the population of people over age 65 living outside institutions (Lachs et al., 1997). Nearly half of the perpetrators of mistreatment were adult children, and one quarter of them were spouses. An earlier study found that two thirds of elder abuse cases were physical and the remainder verbal (Pillemer & Finkelhor, 1988).

URBAN LEGENDS OF AGING

"We need more regulation of nursing homes to prevent elder abuse."

Actually, on paper, the nursing home industry is already more highly regulated than any industry in the United States, except for nuclear power. As for elder abuse, it's more likely to happen—and remain undetected—in a home care setting. Citizens' groups could certainly contribute by closer consumer scrutiny and advocacy of nursing homes. But more regulations may not help since we're not enforcing the ones we already have.

Research on elder abuse suggests that this phenomenon is complex (National Center on Elder Abuse, 1998; Quinn & Tomita, 1987; R. S. Wolf & Pillemer, 1989). There are difficulties in measuring just how much elder abuse occurs because of the way statistics are collected (Heller, 2000). Moreover, we must look carefully at victims and perpetrators and at the different types of abuse. There is a series of known risk factors for elder abuse, including the presence of psychopathology, especially alcohol and substance abuse; family history of violence; a family member dependent on others for financial support; caregiving burdens; social isolation; and the recent occurrence of stressful life events (Bloom, Ansell, & Bloom, 1989). Lifelong patterns of domestic violence are often reversed when the parent or the formerly dominating spouse becomes less powerful and is now the victim instead of the abuser (Anetzberger, 1987).

Physicians and home care workers are often in a position to identify preliminary signs of potential elder abuse, and lawmakers have addressed the problem of reporting elder abuse. All states now have formal reporting arrangements for suspected cases of elder abuse. Today, laws in 42 states require **mandatory reporting** whenever there is evidence of abuse (Daly et al., 2003).

But some advocates have questioned the effectiveness of mandatory reporting. Reports do not necessarily lead to any follow-up action because in-home services may not be readily available (Silva, 1992). Furthermore, as in Bert's case, not all instances of abuse are readily apparent. When we broaden the issue to include self-neglect, it becomes even more difficult to draw clear lines, especially when an older person simply insists on living his or her own life, independent of interference by outsiders.

When professionals come across cases of suspected abuse or neglect, they face significant challenges and complex questions (Gergeron & Gray, 2003). If they report the matter to authorities, are they violating privacy or confidentiality? How serious do the signs have to be to prompt reporting people against their will? What happens if accusations turn out to be false? Could reporting a case of potential abuse make matters worse? These questions bring us to the heart of the controversy about whether older people should be protected from bad choices.

PERCEPTIONS OF QUALITY OF LIFE

In the case study of Uncle Bert's family, his doctor and a social worker questioned whether Bert was taking unreasonable risks—making bad investments, exposing himself to the danger of crime and abuse—to live independently. Bert believed that he alone had the right to define the quality of his life. At age 79, he may have felt he had lived long enough to make his own decisions about risks and benefits. With regard to money, there is an important relationship between subjective well-being and financial control of one's life. Quality of life in old age may be tied to an individual continuing to exercise control over money (Cutler, Gregg, & Lawton, 1992), but exercising control and making choices mean running certain risks.

Debates about risky behavior and mental health in later life often come back to an ambiguous phrase: *quality of life.* Defining quality of life and measuring well-being in old age are serious problems for gerontology and present many challenges for measurement and theoretical interpretation (R. L. Kane & R. A. Kane, 2000). Broadly speaking, **life satisfaction** can be defined as a person's attitude toward past and present life as a whole. In contrast, morale is a specific feeling, whether optimistic or pessimistic, about the future. These psychological concepts, along with happiness and mood, are obviously important. But notice that they address the issue of subjective well-being in later life, which is ultimately an individual's own perception.

To measure subjective well-being, gerontologists make extensive use of an instrument called the Life Satisfaction Index, which considers items such as zest and apathy, self-blame, attainment of life goals, and mood. The Philadelphia Geriatric Center Morale Scale has also been used to measure subjective well-being. Chronological age by itself is only weakly correlated with subjective well-being; in other words, old age in and of itself is not necessarily a cause of poor morale or unhappiness. Physical health, however, especially self-rated health, is a strong predictor of subjective well-being. Health problems, of course, are more common in later life, but subjective rating of one's own health reflects individual coping style and not simply objective physiological function. In fact, self-report measures often are more positive than objective health status measures for the same individual. We cannot ignore individual capacity for adaptation.

Relationships are another factor in subjective well-being. Is Uncle Bert's living situation—cuddling in bed with his housekeeper—a pattern of adaptation that we could call *successful aging?* The mere existence of relationships does not prove that people are aging well; more important is the quality of those relationships (Adams & Blieszner, 1995). We can't simply analyze the numbers of social networks or kinship patterns to measure well-being. Older adults need to develop relationships with people who help them in ways they want to be helped, whether those people are family, friends, or others. No single pattern of personal relationships is optimal for all people as they age, and human beings require intimacy and love throughout the life course, including later life.

SEXUALITY IN LATER LIFE

Uncle Bert's nephews and niece were uneasy about the possibility of their uncle having a sexual relationship with Lilly. To them, it seemed inappropriate for a 79-year-old man to have sex with a much younger woman without being married. In all likelihood, Uncle Bert's nephews and niece had some misconceptions and stereotypes about late-life sexuality (Butler & Lewis, 1993). Sex among older adults has long been a topic for humor, even in Roman comedy and the poetry of Chaucer. But of 106 traditional societies studied around the world, in only 3 did sexual behavior among the aged violate social mores. In Western culture, however, the sexuality of older people, although usually appropriate and a normal part of life, is still considered unusual or a subject for humor or judgment by younger people (Kaye, 1993).

The normal aging process is not the determining factor of adult sexuality as much as marital status, general physical health, or the feelings of an older person about sexuality are (Hillman, 2000). Complete cessation of sexual activity is most often a result of a decline in the physical health of one or both partners. But sexuality is much broader than sexual intercourse. We need to remember that older people, like younger adults, need intimacy of all sorts to be happy and healthy, and that there is individual variability in terms of what people need, want, and enjoy.

In fact, studies by sex researchers Masters and Johnson (2010) have shown that female sexuality has no time limit and that male sexual capacity may extend into the 80s. However, sexual activity in late life is most clearly accounted for by the continuity theory of aging. The best predictor of sexual behavior in later life is earlier sexual behavior. The biggest limiting factor, especially for women, is the availability of a partner.

A key feature of late-life sexuality is its multidimensional character: sensuality, intimacy, and touching are all at least as important as intercourse. Nevertheless, one Swedish population health study found that, among 70-year-olds, 46% of the men and 16% of the women still found enjoyment in sexual intercourse.

The contrast between the figures for men and women mostly reflects the changing sex ratio in later life. Exhibit 25 shows the discrepancy in the numbers of older men and women now, and projected into the future. The reason for the imbalance is that women tend to outlive men by an average of 6 years, and they tend to marry men about 2 years older. With advancing age, therefore, widowhood takes its toll: In 2009, 41% of all older women were widows, compared to 13% of all older men, four times as many widows as widowers (Administration on Aging, 2010). Older widowers are more likely than older widows to remarry quickly; 7 in 1,000 men remarry each year, whereas only 1 in 1,000 women remarries each year (U.S. Census Bureau, 1996).

Older people, like younger adults, need intimacy of all sorts to be happy and healthy.

Assuming that the opportunity exists, most older men and women can enjoy sexuality in later life. Men can maintain fertility and generate sperm into their 80s, although men in their 70s are more likely to be worried about sexual function and sexual performance than men in middle age (Panser et al., 1995). Like women, men experience hormonal changes as they age, but men do not experience a distinct climacteric phase during which they completely lose reproductive capacity. Older men do experience a slowing down of the speed of sexual responsiveness, and they and their partners need to adapt to these changes. As continuity theory suggests, men's pattern of earlier sexuality is the best predictor of their sexuality in old age (Whitebourne, 1990).

A major difference between the sexes is that, by the end of middle age, women experience menopause and loss of fertility. The decline in hormones that causes menopause is usually gradual, and thus menopause need not be a traumatic event. With longer life expectancy, menopause can actually be an opportunity for a second adulthood. Menopause does bring distinct physical changes, however. For example, decreased production of the hormone estrogen can increase the risk of osteoporosis (Shapiro, 2003). Traditional images of women's late-life sexuality have sometimes been quite negative (Covey, 1989), but literature abounds with examples of vigorous postmenopausal female figures, such as Penelope in Homer's *Odyssey,* Chaucer's Wife of Bath, and several heroines in Shakespeare. Modern writers such as Toni Morrison and Alice Walker have picked up on these positive themes of late-life

Exhibit 25   Percent Female for the Older Population by Age for the United States: 2010, 2030, and 2050

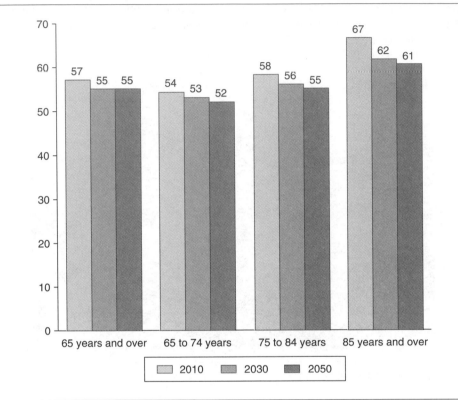

SOURCE: U.S. Census Bureau, 2008.

development (Banner, 1993), and Gabriel García-Márquez's *Love in the Time of Cholera* offers a vivid portrayal of passion in later life embodied in the character of Femina.

CRIME AND OLDER ADULTS

In the case study that opened this chapter, Bert seemed unaware that he might be taken advantage of, whereas his nephews and niece thought his attitude proved his incompetence. Who is most qualified to determine whether an older person is likely to be a crime victim? Actually, well-established statistics show that Uncle Bert's experience and judgment might better match reality here. Perhaps his nephews and niece are fearful because the media give heightened attention to crime against the old. But the facts suggest a different picture. Annual figures from the National Crime Victimization Survey show that people over age 65 have *lower* rates than other age groups for serious crimes such as robbery, personal theft, assault, and rape (McCabe & Gregory, 1998). When all crimes are taken together, the victimization rate for older adults is lower than the rate for the rest of the population (Klaus, 2000). Is Uncle Bert safe when he walks in the dangerous parts of the city at night? Not necessarily. Crime against persons of any age may be more likely in certain neighborhoods. If an older person does become a crime victim, then the aftereffects—for example, from serious injury—can be more long lasting than for a younger person.

Whatever the actual incidence of crime, fear of crime is widely believed to be more common among older people than among younger people. Nonetheless, serious studies have failed to find a clear relationship between age and fear (Ferraro & LaGrange, 1992). The popular stereotype of older people becoming "prisoners in their own homes" because of fear of crime is exaggerated.

URBAN LEGENDS OF AGING

"Older people are more likely to be victims of crime."

According to Justice Department data and the FBI, older people are *less* likely to be crime victims than any other age group in the population. However, they're more afraid of crime, perhaps because they watch more television than other age groups. Older people should probably be more afraid of falling, since falls injure more seniors than crime does.

The biggest threats to older people may actually come from financial exploitation (Jordan, 2002). Older adults are vulnerable to frauds and scams, as well as pickpockets, purse snatching, and stealing checks from the mail. Then, too, there are borderline abuses by con artists and unscrupulous advisers. A study of randomly selected older people in Canada found a much smaller proportion—only one half of 1%—suffering physical abuse, but also a rate of 2.5% for financial exploitation, which may be the most common form of abuse. Another study found that individuals over age 65 make up approximately 30% of all scam victims, although they comprise only 12% of the total population (T. Davis, 1993).

Because of the assumption that older adults have nest eggs set aside for retirement, they are tempting targets for swindlers (Sharpe, 2004). Women, especially the newly widowed, are among the most vulnerable because of the assumption that they may never have entered the workforce, balanced a checkbook, or paid a bill. Lonely and frightened, the new widow sometimes trusts too quickly someone who offers to handle her finances. But clearly older women are not the only victims of crime in later life—remember Bert's involvement with Shawn's vitamin business. Scams may involve real estate, stocks, mutual funds, or business investments. Is Shawn's vitamin business like this, or is it just an investment with some risk?

Strangers may perpetrate these scams, but opportunities for exploitation are also found closer to home. Defining exploitation is not always easy. For example, in many cultures, older people share their financial assets with younger people in the family. But loved ones can exert emotional pressure that may be just as detrimental to the older person's interests as the deliberate manipulation of strangers and acquaintances. Because abusers are often dependent on their older victims for financial support, exploitation can be linked to verbal or physical abuse. We saw the slippery slope toward exploitation in Bert's case. But if a presumed victim doesn't report financial exploitation, then how can one prove a crime was committed? Are we justified in interfering for a person's own good? How do we go about determining what a person's "own good" actually might be?

GLOBAL PERSPECTIVE

Ponzi Schemes Around the World

A *Ponzi scheme* is a financial scam named for Charles Ponzi, an immigrant to Boston who in the 1920s swindled thousands of people by promising a 40% return on their money. Ponzi simply used money from previous investors to pay new victims, until the chain was eventually broken. Ponzi schemes remain one of the most prevalent consumer frauds. Today's turbulent stock markets and low interest rates make older people a tempting target for Ponzi schemes and other financial frauds. Older people have even been villains as well as victims: 81-year-old John Heath was sentenced to 28 years in prison for promoting a $190 million Ponzi scheme.

Ponzi schemes are found all over the world. Ponzi himself was an Italian immigrant who began his fraud career with financial elder abuse in Montreal, Canada. International financial fraud today is a huge business, familiar to anyone who has ever received e-mail messages from Nigeria promising huge investment rewards. One of the biggest frauds occurred in Albania in the 1990s, when government officials endorsed investment plans that eventually attracted two thirds of the Albanian population. The plans turned out to be Ponzi schemes that bilked investors of $1.2 billion until the government stepped in to stop protests and a civil upheaval that claimed 2,000 lives.

According to U.S. Postal Inspectors, older people are among the biggest victims of financial fraud, including Ponzi schemes as well as chain letters, illegal contests, phony billing scams, and foreign lotteries. Homebound elders constantly receive solicitations by mail as well as from telemarketers. This sad story raises again the challenge of how to protect people from making bad choices.

SOURCE: Zuckoff, Mitchell, *Ponzi's Scheme: The True Story of a Financial Legend,* Random House, 2005.

INTERVENTION IN THE LIVES OF VULNERABLE OLDER ADULTS

How can society and loved ones intervene to protect an older person who seems incapable of functioning safely or who has been targeted as a victim by exploiters? One answer is that it is possible to restrict people's liberty for their own good through legal procedures intended for that purpose (Quinn & Heisler, 2002).

Civil commitment is a legal procedure whereby people can be placed in psychiatric hospitals against their will. The rationale for civil commitment is to diagnose or treat a mental disorder if there is reason to believe a person may be in danger of causing harm to self or others. Civil commitment is an extreme measure causing loss of personal liberty. In previous decades, disoriented older people were often sent against their will to mental hospitals, but today, more guarantees are in place to protect against such actions. In earlier times, "old age" alone might be a sufficient reason for labeling someone incompetent to make decisions. But today, instead of relying on vague labels like "senility," we must undertake a formal court proceeding to declare someone mentally incompetent.

When a person is capable of some independence but is unable to manage money or make decisions, another approach is to appoint a guardian or conservator for the incapacitated person's affairs (Zimny & Grossberg, 1998). **Guardianship** can take two forms: guardianship of the person, in which the guardian has the power to determine where the older person will live and what treatment or services he or she may receive, and guardianship of the estate, in which the guardian has power to manage property and take over financial affairs (Ritter, 1995). Appointing a guardian is less restrictive than committing someone to a mental hospital, but it still represents a restriction of liberty.

Guardianship is overwhelmingly a procedure applied to help older people as compared with members of other age groups. One empirical study from Florida revealed an average age of individuals referred for guardianship of 73 years; nearly three quarters were over age 70 (Peters, Schmidt, & Miller, 1985). Two thirds of those who received guardianship were women, and a comparable number lived in private homes. Wards, or those subject to guardianship, were moderately wealthy; nearly half had estates greater than $50,000. Guardianship was very much a family affair: Relatives were available and willing to serve as guardians in 70% of these cases. All petitions to declare a person incompetent were affirmed by the probate court, but in 90% of the cases, reports did not provide the specific behavioral information for the court that is required by law.

Despite the legal safeguards, both civil commitment and guardianship proceedings are viewed by many informed critics as a kind of *new paternalism* that can be oppressive to older people (Regan, 1981). Moreover, some studies have shown that depicting people in age-related terms is a common, even routine feature of civil commitment proceedings, and age depictions are rarely challenged. Specifically, behaviors construed as age inappropriate—for example, certain kinds of sexual activity—are frequently cited as grounds for viewing a range of behaviors as symptoms of mental illness (J. Holstein, 1990). Laws intended to protect older people have vague standards and weak accountability for those who are appointed as guardians or conservators. Civil commitment proceedings may fail to grant older adults an opportunity for a fair hearing, and statutes may not require legal notice, the presence of the older individual at the hearing, or the right to counsel (Rein, 1992). The result is that programs intended to protect the old can end up disregarding their rights.

Some of the worst failings of the guardianship system are its disregard for the rights of older adults. The courtroom hearings for incompetency are too often one sided and superficial, and vague laws permit a ruling of incompetency based on flimsy evidence. Moreover, follow-up supervision of guardians who are appointed by the courts remains minimal. Financial incentives make matters worse because guardians and conservators are paid out of the ward's assets. Appointments to conservatorship jobs may be awarded to politically well-connected lawyers. Once a guardian is appointed, guardianship is rarely revoked before the death of the ward. Efforts to reform the guardianship system have been frustrated, so advocates today tend to recommend guardianship only as a last resort.

In the classic readings that follow, we see some of the dilemmas and controversies that result from efforts to protect older people from vulnerability. The first reading, by Robert Brown, emphasizes the basic constitutional right to freedom from restraints. Writing from a civil liberties perspective, Brown raises a basic question: When is government justified in intervening in any of our lives? The excerpt from Terrie Wetle and Terry Fulmer looks at the ethical dilemmas of how to balance the best interest of people with their right to autonomy. Instead of looking at this question in abstract or societal terms, Wetle and Fulmer stress the social context of families and health professionals who may be obliged to report suspected cases of elder abuse.

On one side, we have the human service professions, but on the other side, we have the legal system. Candace Heisler and Mary Joy Quinn describe the presuppositions of the legal system, in particular its adversarial nature. By defining ideas of *autonomy, least-restrictive autonomy,* and *mental competency,* the legal system sets the boundaries around the debate about when it is possible to intervene to protect people from their own bad choices. Finally, Dorothy Fabian and Eloise Rathbone-McCuan examine the concept of *elder self-neglect*. In this instance, there may be no outside party engaged in abuse or exploitation. Instead, an older adult may simply insist on living independently under conditions of danger that frustrate our ability to provide protection to vulnerable people.

FOCUS ON PRACTICE

Adult Protective Services

We have seen in Uncle Bert's case that an older person's questionable judgment raises a difficult problem: How do we respect an older person's right to make personal decisions while safeguarding that person from situations of abuse, exploitation, and neglect? There are no easy answers to that question, but practitioners who work with older adults have come up with some practical strategies that deserve attention. One of these is **adult protective services**, a service program that attempts to balance individual rights with concern for safety and welfare (Byers & Hendricks, 1993).

The vast majority of states now have special legislation and social service programs aimed at protecting people like Uncle Bert—that is, people who seem vulnerable to physical or financial abuse and who may be impaired and unable to protect themselves. Laws that protect these vulnerable adults are administered by local departments of aging, departments of social welfare, and even police departments.

Adult protective services workers begin a case by trying to define exactly what they are facing:

- *Is it abuse?* Is someone else in a position to harm a vulnerable adult either physically or psychologically?
- *Is it exploitation?* Is financial abuse, which involves misappropriating another person's money, a possibility?
- *Is it neglect?* Has a caregiver or someone responsible for a frail elder's welfare failed to provide minimal support?

Another possibility is self-neglect or endangerment: An individual, entirely independently, may have fallen into a life-threatening situation—for example, wandering around outside in cold weather without adequate clothing or failing to take essential medication. Bert's case shows how difficult it is to classify a situation as abuse, exploitation, neglect, or endangerment.

What usually happens is that someone in a position to know—a physician, a home care worker, a neighbor, or a friend—reports to authorities that an older person appears to be abused or neglected. Adult protective services then begins a prompt investigation, usually by visiting the home and talking with the people who are involved.

Following this preliminary investigation, adult protective services would prepare a report and decide what to do next. Further fact-finding may be required. For example, it may be necessary to look at financial records or test an individual's mental status or competency to make decisions. If the facts warrant doing so, the agency will intervene in the case.

Intervention can take a variety of forms, some more aggressive than others. For example, a social worker might try to arrange for home health care or work with the caregivers to make sure that services are delivered smoothly. In Bert's case, adult protective services would want to make sure that his bruises aren't the result of harm inflicted by the people who are supposed to care for him. Adult protective services would also look at Bert's own actions that seem to endanger his welfare, such as wandering in dangerous neighborhoods and engaging in questionable health practices.

This review would try to strike a balance between ensuring the client's safety and welfare and taking serious account of individual autonomy and quality of life. Striking such a balance is not easy, and adult protective services professionals face a challenging professional task. The agency has to weigh competing interests of family members, older persons, and the values of society as a whole.

In more extreme situations, adult protective services can bring in the police or go to court to obtain a formal order of conservatorship or guardianship. But such serious measures involve risks, which are usually justified only if clear harm is threatened and if a person seems mentally incapable of making decisions about harm and risk. Many cases are borderline, and some individuals simply lead eccentric lifestyles. In such cases, adult protective services workers need great skill and judgment to find a good solution. But the availability of adult protective services gives hope that cases like Uncle Bert's can be resolved.

READING 26

The Right to Freedom From Restraints

Robert N. Brown

We know that people, even those without mental or physical limitation, possess differing abilities and have differing needs. Sometimes physical or mental impairments leave people totally unable to manage their personal or financial affairs, as for example a person in a coma. Sometimes impairments affect only part of a person's ability to care for himself. A person with quadriplegia may be unable to care for himself physically but can direct others in providing care. Some impairments are temporary, such as the mental confusion that may occur as a side effect of medication, while some impairments seem to be permanent, such as the progressive physical and mental deterioration associated with Alzheimer's disease. Most impairments pose harm only to their victim, who will suffer the effects of not being able to provide for his needs. Some impairments, however, are alleged to pose a risk of harm to others, such as a patient with schizophrenia whom psychiatrists may believe to be nearly certain to physically assault another. In short, there are a variety of physical and mental conditions which may affect, to varying degrees, for a limited or lasting amount of time, a person's ability to prevent harm to himself or to restrain himself from harming others.

When such conditions exist, and a person no longer can manage to prevent harm to herself or to others, the question is whether, and at what point, the legal system ought to intervene. The means by which legal intervention may be sought, broadly referred to as "protective proceedings," include such processes as guardianship, conservatorship, and civil commitment. These procedures seek to prevent harm from occurring to an impaired person by legally removing, to one extent or another, the impaired person's right to make decisions on her own behalf and appointing a surrogate to make decisions in her place. To appreciate the impact of this, some understanding of what "decision making" entails is needed.

Life presents each of us with a constant succession of choices requiring decisions. Most are minor, such as what to wear in the morning, what newspaper to read, or whether to order fish or fowl in a restaurant. Some are major, such as whether to marry, or whether, near the end of one's life, to move from one's home into a nursing home, or to undergo painful chemotherapy to fight a terminal cancer. All these decisions, for the purpose of understanding protective proceedings, can be placed into two categories: those that affect financial interests and all others. These latter decisions are considered personal decisions. Financial matters that may be acted on can range from buying a candy bar in a vending machine, to giving a gift to a favorite niece, to selling one's home to pay for medical care or a college education. Personal matters that may be acted on range from deciding what book to read or television show to watch to deciding where to live, with whom to associate, or whether to have an organ transplant.

When a person loses part or all of the ability to make decisions, it may become necessary for someone else to have the legal authority to act on his or her behalf. If the impaired person had, at an earlier time, entered into an arrangement (such as a power of attorney or a multiple-party bank account) that would allow another person to act on his behalf, further legal intervention may not be required. If not, however, a protective proceeding may need to be sought. The consequence of such an action is that the impaired person's right to make independent decisions is diminished or eliminated. The right to make decisions is turned over to the guardian or conservator.

Whether losing one's right to make decisions and having to abide by the decisions of another is unfair or unjust depends on the circumstances. It may not be a great loss to a person in a coma, who is unaware of any loss of control, to have his choices made by a guardian. It is a good deal less clear what the impact is of making a mentally ill person who makes "not very good" financial choices subject to a style of life dictated by the financial decisions of a court-appointed conservator, on the presumption that the conservator's decisions will better protect the person's assets. A person who is not mentally ill but who makes the same "not very good" financial decisions could not be subjected involuntarily to such control. Furthermore, the actual outcome of a decision is never certain. All decisions entail a certain amount of risk; well-thought-out plans often fail, and ill-advised schemes often succeed. There is, in addition, the question of values. What is best for an individual may not be the safest and most long-lasting life, but may include the chance to do what that individual believes is best and most important.

Guardians and conservators are often ignorant of, or fail to heed, the concerns and wishes of their wards. Protective proceedings often give little weight to these concerns. These proceedings seem unable to accommodate the differing natures and degrees of impairments and may not recognize that an impairment which affects one area of a person's ability to care for himself may not affect other areas. The result is that the law, in seeking to "help," sometimes takes away rights from those who are merely "different" and not disabled and often fails to respond appropriately to those who need some, but not complete, help.

WHAT IS THE JUSTIFICATION FOR GOVERNMENT INTERVENTION INTO OUR LIVES?

A general rule of our legal system is that people are allowed to exercise self-determination; people have both the right and responsibility to make their own decisions about how they will live, on what they will spend their money, and how they will spend their days. For the most part, people are free to think, speak, and do as they please without interference from our government. Yet there are limits. Laws limit our freedom by prohibiting us from injuring others or from harming their property. The legal authority for such laws is the "police power" of the state. This power authorizes the state to proscribe activities which are dangerous to others in order to protect society. A person convicted of violating such laws may, for the protection of society, be deprived of his or her freedom. The loss of liberty suffered by persons "civilly committed," or confined involuntarily in state mental hospitals, is justified in part by the police power, inasmuch as such persons are thought to be dangerous to others. Another limitation on individual freedom is the power of the state to protect individuals incapacitated by disease or other causes who consequently cannot care for themselves, their dependents, or their property. This power is called *parens*

patriae, or "parentage of the state." Unlike police power, which is aimed at protecting others, this power focuses on the incapacitated individual and declares that the state has the responsibility of protecting those who cannot protect themselves. Under the *parens patriae* power, the state has the authority to determine when a person cannot care for him- or herself adequately and has the authority to appoint another person, usually called a guardian or conservator, to provide such care.

Because the exercise of police power entails a great loss of liberty (for example, confinement to a prison), procedural protections have long been established to ensure that a person involved in a criminal proceeding is not mistakenly imprisoned or penalized. Examples of these protections are rights to a lawyer, against self-incrimination, to trial by jury, and to cross-examination of witnesses. Unfortunately, not all of these procedural safeguards are provided to persons threatened with such protective proceedings as civil commitment, guardianship, or conservatorship. The reason is that protective proceedings are, for the most part, theoretically founded on the idea that the intervention will benefit the person, in effect "protect" him from himself. Such benevolent proceedings may be felt to pose no real loss of liberty or harm to a person unable to exercise such rights independently. However, in practice, these proceedings are often brought against persons at least partially if not fully able to exercise their rights and for reasons far from benevolent. The resulting genuine deprivations of liberty and property that occur, such as loss of control of one's finances or confinement in an institution such as a psychiatric facility or nursing home, suggest there is no justification for failing to offer procedural and substantive safeguards in protective proceedings. Lawmakers should ensure that the most appropriate and least restrictive processes available are used in seeking intervention in an individual's life.

Ethical Dilemmas in Elder Abuse

Terrie T. Wetle and Terry T. Fulmer

The application of values and ethical concepts to specific cases of elder mistreatment often engenders ethical dilemmas involving conflict among two or more values or concepts. Several such dilemmas are outlined below including the balancing of individual autonomy with beneficence and paternalism, concerns of confidentiality and legal reporting requirements, the impact of reporting on patient and professional relationships, and the issue of respect for autonomy for patients with diminishing cognitive capacity.

BALANCING PATIENT AUTONOMY AND THE BEST INTERESTS OF PATIENTS

Perhaps the most difficult ethical dilemma for health professionals is the effort to respect the expressed wishes of the patient (autonomy) while protecting the patient from harm (beneficence) (Wetle et al., 1991). It is not uncommon for individuals to behave in ways that place themselves at risk or are personally injurious, or to choose to remain in risky or abusive circumstances. Respecting the autonomous wishes of such an individual may be in direct conflict with the professional's judgment of what is in the patient's best interest. On the one hand, the principle of respect for the autonomy of an individual would prevent the health professional from intervening with a service or action that is in opposition to the expressed wishes of a competent patient. On the

other hand, health professionals are required, by law, to report cases of suspected elder mistreatment. Moreover, there is also an obligation to determine that the patient's refusal of assistance is an autonomous decision, free of coercion or undue duress. These cases are often complicated by questionable decisional capacity or cognitive impairment of the elder patient. Disagreement with a health professional's judgment or advice is not sufficient evidence to make a determination of decisional incapacity. Certainly, we have moral obligations to protect incompetent elders from incapacitated decision making, but we are also obligated to determine if they are truly incapacitated before intervening against their wishes. Such a determination requires a formal evaluation of decisional capacity. If the patient is determined to be competent, even if suspected mistreatment is confirmed, the competent patient has the right to refuse interventions.

Confidentiality and reporting requirements

Another aspect of patient autonomy is respect for confidentiality. Once again, the elder abuse reporting laws override the obligation for confidentiality under very specific circumstances. Nonetheless, this is not a blanket override, in that such reported information must be handled with extreme care, and information identified in subsequent investigations should be shared only on a "need to know" basis.

SOURCE: "A Medical Perspective" by Terrie T. Wetle and Terry T. Fulmer, *Journal of Elder Abuse and Neglect,* 7(2/3), pp. 36–38. Copyright © 1995 by The Haworth Press. Reprinted by permission of the Taylor & Francis Group, http://informaworld.com.

Impact of reporting on patient/ professional relationship

Many professionals express justified concern that an elder abuse report and the subsequent investigation will have deleterious effects on their relationships with patients (and patients' families). Much progress has been made in improving the response system and investigating approaches used to determine whether or not mistreatment has occurred and in planning appropriate interventions. Nonetheless, even the most skilled response to such a report may damage the relationship between professional and patient. Several steps can be taken to reduce negative impacts, including informing the patient that such a report is to be made, describing the process, identifying potential positive outcomes of the process and suggested interventions, working with the family as a unit if possible, and recognizing the needs and concerns of all involved.

ETHICAL ISSUES RELATED TO FAMILIES AND HEALTH PROFESSIONALS

Family members are often intimately involved in the care of frail elders. Several ethical issues and conflicts relevant to elder mistreatment are faced as health professionals care for the individual patient in the context of family caregiving. These issues include questions regarding just who is the patient, caregiver burden and inadequate family supports, changing dependency relationships and longstanding patterns of family interaction, and ethical approaches to families. Many of these issues are discussed elsewhere in this volume in substantial detail, but are briefly discussed here from the perspective of the health professional.

Who is the patient?

It is not uncommon for health professionals to become involved in the treatment of the family as a unit, as well as in care of an individual patient. The burden of caregiving can be substantial, and some cases of elder mistreatment occur when a well-intentioned primary caregiver becomes overwhelmed by caregiving tasks. The health professional faces a particularly thorny dilemma when treating two members of a family as patients and encounters evidence of mistreatment of one by the other (Fulmer, 1991). This raises the question of just who is the primary patient, and how are the professional's obligations to be balanced among various members of the family. Certainly, there is a first-level responsibility to the patient who appears to have been mistreated but successful approaches include offering information and services to all involved parties.

Dependency relationships in families change as one member becomes increasingly dependent on the family for care

The health professional may be unaware of longstanding patterns of interaction among family members. The stresses of family caregiving may exacerbate a relationship that has always been characterized by verbal and physical abuse. In some cases, changing dependency relationships may turn the tables, and the previously abused may become the abuser. Being aware of these patterns and the specific relationship history of caregiver and care receiver may help explain observed dynamics and provide crucial information for developing effective intervention strategies.

Caregiver burden and spousal and intergenerational responsibilities involve differing personal and societal expectations

Society remains unclear as to the reasonable level of caregiving to be expected from a spouse or adult child. It is not unusual for family caregivers to drain all physical and emotional resources before accepting help or assistance. One responsibility of health professionals is to assist caregivers in recognizing the limits of their own health and to promote their well-being by identifying supportive services and encouraging their use and by assisting in balancing caregiving among all available family members. This may in fact have a double protective effect. First, it may prevent the elder abuse or neglect that might have been triggered by caregiver overburden and stress. Second, such interventions may protect the caregiver from the overburdening that also could be considered a form of "caregiver mistreatment" or abuse.

REFERENCES

Fulmer, T. (1991). "Elder Mistreatment: Progress in Community Detection and Intervention." *Family and Community Health,* 2, 26–34.

Wetle, T., Crabtree, B., Clemens, E., Dubitzky, D., Eslami, M., and Kerr, M. (1991). "Balancing Safety and Autonomy: Defining and Living With Acceptable Risk." *The Gerontologist,* 31(11), 237.

READING 28

A Legal Perspective on Elder Abuse

Candace J. Heisler and Mary Joy Quinn

Most practitioners working with older adults have had limited experience with the American legal system. As a result, it is underutilized in the prevention and resolution of elder mistreatment. . . .

GOALS OF THE LEGAL SYSTEM

The civil and criminal legal systems approach the prevention and resolution of elder mistreatment matters with common *goals* and certain *rules* that govern their handling. The *goals* of the legal response to elder mistreatment are to: (1) Stop the unlawful, improper, or exploitive conduct that is being inflicted on the victim; (2) Protect the victim and society from the perpetrator and further inappropriate or illegal acts; (3) Hold the perpetrator accountable for the conduct and communicate a message that the behavior is unacceptable and exceeds societal norms; (4) Rehabilitate the offender, if possible; and (5) Make the victim whole by ordering restitution and/or the return of property as well as the payment of expenses incurred by the victim as a result of the perpetrator's conduct.[1] The two parts of the legal system also seek to act in ways that create the least disruption or invasion into the victim's life; that take into account that person's individual situation, competency, wishes, and desires; and that keep the situation from becoming worse. . . .

The legal system is adversarial in nature. In the clash between the parties in a search for the truth, the legal system mandates roles for the participants. On the civil side, the person who files the lawsuit is termed the plaintiff and that person must meet one of the two legal burdens previously described in order to prevail. Because the lawsuit generally concerns the interaction of

SOURCE: "A Legal Perspective" by Candace J. Heisler and Mary Joy Quinn, *Journal of Elder Abuse and Neglect,* 7(2/3), pp. 131–140. Copyright © 1995 by The Haworth Press. Reprinted by permission of the Taylor & Francis Group, http://informaworld.com.

two or more parties, the parties themselves retain control over the case. The plaintiff and the defendant can settle the case, come to agreements, and otherwise direct what takes place. Where guardianships are concerned, the person filing the case is typically termed the petitioner because traditionally, guardianship hearings have not been adversarial. Rather, there has been an assumption that the petitioner is acting on behalf of an older adult who needs help. Traditional views are giving way to many system reforms as it is realized that sometimes persons seeking guardianship are acting in their own interests rather than to assist an elder in need. The result is that guardianship proceedings are becoming more complex and, at times, adversarial in nature. Legal rights of proposed wards are protected to a greater degree than occurred previously. . . .

ETHICAL ISSUES

The ethical issues most apparent in legal proceedings with elder mistreatment are (1) promotion of autonomy and (2) least restrictive alternative. Ethical dilemmas arise when evaluating these issues in light of an older adult's mental competency.

Autonomy

American legal and ethical values place a high priority on autonomy. The issue is self-governance: "being one's own person, without constraint either by another's action or by psychological or physical limitations."[7] The high value placed on autonomy is reflected in the Constitution and the professional ethical protocols of the medical, legal, and nursing professions.[8] Autonomy may be an idealized notion given the very real interdependence most elders have within their families and their communities and given the reality of high levels of impairment in old age.[9]

The issue of autonomy is always present whether the older adult is in the civil or the criminal justice system. In the criminal justice system, the issue of autonomy comes into play when the elder is informed and consulted about the various options available to keep him or her safe and when the sentencing of the offender is under consideration. It is not a victim's responsibility to determine if criminal charges will be brought against an offender and in many jurisdictions, victims are not asked to "press charges." The prosecuting attorney, as the state's representative, makes that decision in order to communicate the message that the conduct is criminal, not simply a "private matter," and to ensure that the victim is protected, rather than manipulated, exploited, or threatened.

In guardianship matters, autonomy is the main issue when an adult is thought to be incapable of managing his or her affairs, whether due to abuse or neglect by self or others. In guardianship, decision making is placed in the hands of surrogates. With changes in state laws and the dissemination of current gerontological thought, courts now carefully consider autonomy and the wishes of the elder as expressed in the past, the present, and for the future. Forward-looking courts seriously consider these expressions in their deliberations, regardless of the mental status of the older adult.

Ethical practice requires that practitioners are careful not to err on the side of failing to take action to protect an elder. Blind adherence to the concept of autonomy can lead to resolutions that fail to ask even the most basic questions and can result in the abandonment or death of a client who declines the first offer of help or who has an unpleasant personality or is "difficult."

Least Restrictive Alternative

The concept of the least restrictive alternative, a legal doctrine first articulated in the field of mental health, has gained wide acceptance among courts and service professionals. It creates an ethical duty for practitioners to fashion individualized solutions that are least intrusive upon their client's personal freedom.

The concept applies to the personal and the environmental care of the elder and the handling of material resources. It recognizes that elders may have capacities in some areas and lack capacity in others. Ideally, the more restrictive the option, the greater the due process protections and the opportunities for the individual to object and state preferences.[10] The doctrine is primarily civil in nature, though on occasion it is applied in criminal matters, particularly in the placement of persons found to be criminally insane.[11]

In civil law, legal options begin with the client handling [his or her] own affairs and then move up the ladder to more restrictive options in the following manner: client signs name to checks but someone else fills out the checks, direct deposit to bank accounts, representative payee arrangements for certain checks, joint tenancy on bank accounts and/or real property, trusts, the various powers of attorney, protective orders (for placement or medical treatment), guardianship of estate, and, lastly, involuntary placement in a locked mental health facility.[12]

Legal alternatives to guardianship are used more frequently, especially since 1987 when the Associated Press conducted a nationwide investigation into the guardianship system and found it woefully lacking in protections for frail elders. In most cases, a neutral person never spoke to elders or advised them of their legal rights despite the fact that a guardianship often deprives them of many rights such as the rights to vote, to make a will, to select a physician, or to control finances. Following that investigation, most states amended their guardianship laws.[13] Also as a result of the investigation, practitioners began relying on less restrictive legal options, sometimes without fully considering their benefits and burdens. For instance, there has been heavy reliance on powers of attorney with little understanding that there is no bonding, no notice to other relatives, no monitoring by a third party, and no way to regain misappropriated or mishandled assets short of a civil lawsuit. There has also been reliance on trusts that may be beneficial financially in some situations but make no provisions for personal care or abusive trustees. Additionally, individuals who become trustees or attorneys-in-fact seldom understand what is required and as a result, they assume responsibilities that they are not prepared to handle. Ethical practice requires that practitioners in the position of recommending legal options have basic knowledge of the benefits and burdens of each legal option and mandates that practitioners fully discuss all available options with the client.

The concept of least restrictive alternative is not easily transported to the criminal justice system. It is not a guiding principle for the criminal system though most courts decide sentences by balancing what a perpetrator did with available sentencing alternatives while attempting to protect the public, hold the offender accountable, and protect the victim and society at large. The concept of least restrictive alternative can be seen in criminal matters that are relatively minor and may have elements which are both civil and criminal in nature. Interventions may begin in the civil arena and move to the criminal side if the offender does not comply. For instance, the offender may first be warned and urged to make changes, then officially admonished, then cited, and eventually arrested. This process is sometimes used in situations involving public nuisances or neighborhood disputes. Finally, the least restrictive alternative approach is applied in the criminal justice context with those found criminally insane. Persons are placed in facilities according to the amount of treatment and control they need and the degree of protection the community requires.

Competency

The issue of competence is critical in the legal context. It determines whether a witness will be allowed to testify in a criminal case, if a guardianship will be imposed, and if a lawsuit is "winnable" based on the allegations and evidence a victim produces. And yet, the term is as poorly understood in legal circles as it is in mental health and clinical gerontological settings. Marin et al. note . . . that there is growing recognition of the difference between *legal competence* and *clinical competence*. All adults are thought to be *legally competent* until a court of law rules otherwise. This means they can execute legal documents, make medical decisions, decide where to live, and in general, have full control of their lives. *Clinical competence* relies on observations of health and social services practitioners. It is task and time specific. It is interesting to note that recent guardianship reform laws have tended to incorporate concepts of clinical competence into the law.

Over the years, competency has been variously measured by old age, the quality of decision making, medical or psychiatric diagnosis, risk of impoverishment through heedless spending, and physical endangerment.[14] In truth, the search for a commonly accepted definition of competence can be likened to the search for the Holy Grail. The journey is ongoing but as yet there is no mutually agreed upon conclusion. Each discipline functions with its own definition. For instance, the legal profession focuses on what an elder is *incapable* of doing while psychology looks at what the elder is *capable* of doing.[15]

Many state laws provide that a guardianship can be imposed on an individual who is "subject to undue influence," but that concept does not appear in the medical or psychiatric reports upon which criminal prosecutors and civil attorneys must rely. It does not appear to be a concept that is familiar to practitioners outside the legal field. Undue influence situations can occur even when an elder is alert, oriented, and capable of carrying out activities of daily living. Undue influence has been defined as the substitution of one person's will for the true desires of another. It can be accompanied by fraud, duress, threats, or the application of various types of pressure on susceptible persons including frail elders.[16]

Current trends conceptualize "competency" in ways other than simply evaluating the quality of the elder's decision making. There is growing reliance on a constellation of factors to determine competency. There is more focus on what the older adult *actually does* to take care of the needs of daily living including the management of material assets. There is consideration of the elder's past decision making. For example, was the elder in the habit of giving large sums of money to perfect strangers or is this new behavior? Medical and psychiatric diagnoses offer valuable information as to conditions that impinge on mental functioning and are able, with some measure of accuracy, to predict the course of a given condition. There is recognition that mental and physical functioning is subject to a variety of factors such as nutritional status, the presence of mental illness such as clinical depression, the time of day, isolation, grief states, substance or medication intake, relationship status, and self-esteem. There is also recognition that "competence" is dependent on the types of personal and environmental support an older adult may be receiving.[17]

Given all these complexities, it is understandable that there is no single tool to determine competency. Nevertheless, practitioners working with frail older adults who are the victims of elder abuse and/or neglect must try to determine if their client is "competent," often in order to take decision making away for what are usually good and benevolent reasons.[18] In general, practitioners look for two elements when assessing competency: (1) Does the individual have the capacity to assimilate the relevant facts? and (2) Can the person appreciate or rationally understand his or her own situation as it relates to the facts at hand? Relevant questions include: Can the person make and express choices regarding his or her life? Are the outcomes of these choices "reasonable"? Are the choices based on "rational" reasons? Does the person understand the personal implication of the choices made?[19]

The reality is that dealing with competency is less about creating the perfect definition and more about working with the characteristics of the individual older adult.[20] Practitioners must deal with the definitions that are set forth in their respective laws and disciplines while understanding the everyday realities of "competence." For instance, the criminal prosecutor or civil attorney may realize that the elder victim is less cognitively clear in the afternoon and therefore will attempt to have him or her testify in the morning. Those working within civil and criminal courts may attempt to ensure that judges who hear elder mistreatment cases are familiar with their special dynamics. In the absence of valid tools to accurately measure the various features of competence, we must rely on a variety of pieces of knowledge about elders and the conditions that affect them. Ethical practice demands no less.

NOTES

[Only the notes that are included in the excerpted material appear here.]

1. Modified from Carter, J., Heisler, C., Lemon, N. K., *Domestic Violence: The Crucial Role of the Judge in Criminal Court Cases.* Family Violence Prevention Fund, San Francisco (1991); Heisler, C., "The Role of the Criminal Justice System in Elder Abuse Cases," *Journal of Elder Abuse and Neglect* 3(1), pp. 5–33 (1991).

7. Beauchamp, T. L., Childress, J. F., *Principles of Biomedical Ethics,* 2nd ed., Oxford University Press, New York (1983).

8. American Nurses Association, *American Nurses Association Code for Nurses* (1950); Wood, E. F., "Statement of Recommended Judicial Practice," adopted by the National Conference of the Judiciary on Guardianship Proceedings for the Elderly, Commission on Legal Problems of the Elderly, American Bar Association and the National Judicial College (1986); American Medical Association, *American Medical Association Principles of Medical Ethics* (1980).

9. Caplan, A. L., "Let Wisdom Find a Way," *Generations* 10(2), 10–14 (1985); Moody, H. R., "Ethics and Aging," *Generations* 10(2), 5–9 (1985).

10. Quinn, M. J., "Elder Abuse and Neglect," *Generations* 10(2), 22–25 (1985); Quinn, M. J., Tomita, S. K., *Elder Abuse and Neglect: Causes, Diagnosis, and Intervention Strategies.* Springer Publishing, New York (1986).

11. See, e.g., Calif. Penal Code Sections 1026 et seq.

12. Quinn, M. J., Tomita, S. K., *Elder Abuse and Neglect: Causes, Diagnosis, and Intervention Strategies, supra.*

13. Keith, P. S., Wacker, R. R., "Guardianship Reform: Does Revised Legislation Make a Difference in Outcomes for Proposed Wards?" *Journal of Aging and Social Policy* 4(3/4), 139–155 (1992).

14. Quinn, M. J., "Everyday Competencies and Guardianship: Refinements and Realities," in Snyder, M. A., Kapp, M. B., Schaie, K. W. (Eds.), *Older Adults' Decision-Making and the Law.* Springer Publishing, New York (1996).

15. Willis, S. L., "Assessing Everyday Competence in the Cognitively Challenged Elderly," in *Older Adults' Decision-Making and the Law.* Ibid.

16. Grant, L. H., Quinn, M. J., "Guardianship and Abuse of Dependent Adults," in Zimny, G. H., Grossberg, G. T. (Eds.), *Guardianship of the Elderly: Psychiatric and Judicial Aspects.* Springer Publishing, New York (1998).

17. Willis, S. L., *supra.*

18. Caplan, A. L., *supra.*

19. Kapp, M. B., *Geriatrics and the Law: Patient Rights and Professional Responsibilities,* 2nd ed., Springer Publishing, New York (1992); Gutheil, T. G., Applebaum, P. S., *Clinical Handbook of Psychiatry and the Law,* 2nd ed., Springer Publishing, New York (1991); Roth, L. H., Meisel, A., Lidz, C., "Tests of Competency to Consent to Treatment," *American Journal of Psychiatry,* 134, 279–283 (1977).

20. Moody, H. R., "Ethics and Aging," *Generations,* 10(2), 5–9 (1985).

READING 29

Elder Self-Neglect

A Blurred Concept

Dorothy R. Fabian and Eloise Rathbone-McCuan

Self-neglect has been emerging poignantly as one of the many troublesome problems that beset the elderly. It is a source of anxiety and frustration for practitioners, family members, and the community because self-neglecting elders may present themselves in antisocial and life-threatening situations. That is because self-neglect is frequently manifest by disregard of the needs of both the self and the environment. Usually, self-neglect results from physical and/or mental impairments that reduce the elder's ability to perform essential life tasks. There may be no shame about the situation, and outside help may be refused or passively sabotaged. . . .

SELF-NEGLECT IN THE LITERATURE

Self-neglect is probably not a new phenomenon. Isolated, frequently unkempt, and often eccentric individuals such as hermits, witches, tramps, and recluses have long been depicted in the social history, literature, folklore, and opera of Western society. These were the early sources of cultural stereotypes depicting extreme examples of self-neglecters. Often arousing fear, discomfort, and disgust, some of these men and women were seen as mentally ill; some were expected to be able to call on fearful supernatural powers; some were thought to hoard treasure. All were considered strange and lived on the social, if not the physical, periphery of their communities. The local populace tended to subject such individuals to jeers, taunts, beatings, and ostracism. Imprisonment, banishment, or violent death at times followed attention by authorities.

The stereotyping of elderly self-neglecters continues through mass media efforts to keep the U.S. homeless population viable as a newsworthy phenomenon. An increasing amount of contemporary medical and psychiatric literature has addressed gross neglect as a manifestation of individual dysfunction even though it is not clear that the rate and severity of self-neglect in the aging population have increased. Clinical case studies and reports of small samples attempt to verify that self-neglecters become embroiled with the police, wander into emergency rooms, and plague social service agencies. Lurid published accounts lead the public to conclude that high-cost care to improve the condition of these persons provides no long-term benefits and that many of those who do receive community resources will eventually die in very deteriorated conditions.

In attempts to understand self-neglect, certain observers noted an age-related dimension. Macmillan and Shaw (1966) used the senile breakdown syndrome to describe persons failing to maintain levels of cleanliness that the community found acceptable:

> The usual picture is that of an old woman living alone, though men and married couples suffering from the condition are also found. She, her

SOURCE: Fabian, D. R., & Rathbone-McCuan, E. (1992). Elder self-neglect: A blurred concept. In E. Rathbone-McCuan & D. Fabian (Eds.), *Self-neglecting Elders: A Clinical Dilemma,* pp. 3–12, New York: Auburn House.

garments, her possessions, and her house are filthy. She may be verminous and there may be feces and pools of urine on the floor. These people are often tolerated for years by the neighbors, who may suddenly decide that they cannot stand this state of affairs any longer and report the case to various organizations, such as the police or the health department. (p. 1032)

Clark, Mankikar, and Gray (1975), referring to the fourth-century Greek philosopher Diogenes, who reportedly admired lack of shame, outspokenness, and contempt for social organization, suggested the phrase "Diogenes Syndrome" to characterize elderly patients who appeared filthy and unkempt, whose homes were dirty and untidy and usually full of hoarded rubbish, but who showed no shame for these circumstances. Although not necessarily poor or in substandard housing, these self-neglecting individuals were usually known to social service agencies, whose efforts were frequently resisted.

The term "Diogenes Syndrome" became a stereotype for self-neglecting elders perceived as grossly neglectful of their person and the environment, who displayed not only lack of shame but also contempt for, or at least lack of interest in, the recommendations of neighbors, family members, health providers, or the community at large. The Diogenes Syndrome is used to describe patients in at least one nursing care study (Cornwall, 1981) as well as in a paper on psychotic disturbances (Klosterkotter and Peters, 1985). It is also referred to in a study of the social breakdown syndrome in community-dwelling elderly (Radebaugh, Hooper, and Gruenberg, 1987). . . .

CONCEPTUAL AMBIGUITIES

Insights into how to treat or resolve self-neglect among the elderly are subject to as much ambiguity as are the diverse explanations of causality. Cybulska and Rucinski (1986) comment:

Regrettably, when one is faced with a clinical decision whether to intervene or not, the scanty research, medical textbooks, and professional training offer little help. If a crisis occurs in the community, it is often difficult to determine whether the neglect was a result of a consciously determined free choice, some deeply rooted unconscious factors, helplessness, or mental or physical illness. (p. 25)

This ambiguity and the contradictions that surround the problem of self-neglect among elderly persons result in decisions regarding intervention that may become mired in a morass of ethical dilemmas. The desire to guarantee a client's personal safety is often pitted against that client's right to self-determination. The struggle to provide some responsible intervention may be experienced at several levels. The first is between the practitioner and the would-be client who is not amenable to assistance but is in rapid decline or dangerous circumstances that can be attributed to self-neglect. Also, the practitioner's effort to obtain access to what the client needs may be frustrated because agencies and organizations controlling resources have not responded in accordance with client need. A third issue may involve matters of jurisdiction over a client, the resources needed by the client, or a combination of the two.

Important ethical questions seem to surface at almost every turn when practitioners are attempting to work with self-neglect problems. Debates about lifestyle and judgments made by others do not entitle society to develop a general policy of pitchforking people into institutional tidiness (Roe, 1987). Even the mentally ill have an increasing amount of protection from those who would help out of concern and from others who would act out of a blatant or subtle desire to control behavior that is considered unacceptable by some sector of the community.

The field of aging needs to consider what guidelines are appropriate to direct intervention around matters of elder self-neglect. Part of that process will involve helping clinicians to gain the expertise to engage the self-neglecting elderly in a process that respects client autonomy to make a choice even if that choice is counter to clinical opinion, that engages clients in a process of decision making rather than mere debate over the decision, that helps clients accomplish steps toward health and well-being, and that facilitates the best outcomes of those decisions, once made and implemented. . . .

IS THERE A "TYPICAL" CASE OF SELF-NEGLECT?

In the image that comes to mind when health professionals think of self-neglect, both person and environment are consistently and persistently neglected, and help from the community is either refused or passively sabotaged by noncompliant behaviors. The individual shows no shame regarding this situation and does not, therefore, understand the concerns of the community. Usually, no clear-cut psychopathology is apparent that explains the behavior. Finally, the behavior often places the elder in severe and life-threatening risk, if not immediate, then relatively imminent. However, the pure case rarely exists.

Although a situation is sometimes discovered that exemplifies all these characteristics, most cases involve many causes with many possible directions for solutions. Sometimes organic brain impairment is involved. Strong intervention may be required although the individual's rights need to be protected. Additionally, mental illness may be diverting the elder's attention from the lacks of cleanliness and adequate nutrition. Lifestyle plays a part in the situation. Alcohol and drug abuse may also contribute to, or form the central core of, the problem in some cases. Thus, the stereotypic or "typical" case is actually

exceptional. In order to understand the nature of the entire phenomenon of elder self-neglect, we must examine all of its possible causal factors and the socio-medico-legal conditions that currently affect its clinical deposition. We need to resist the impulse to accept stereotypic and simplistic solutions to complex and multicausal problems if we are to make an impact on the issue of self-neglect, which has remained so impervious to current intervention approaches.

REFERENCES

Clark, A. N. G., Mankikar, G. D., and Gray, I. (1975). Diogenes Syndrome: A clinical study of gross neglect in old age. *The Lancet 1* (790): 366–68.

Cornwall, J. A. (1981). Filth, squalor, and lice. *Nursing Mirror 153*(10): 48–49.

Cybulska, E., and Rucinski, J. (1986). Gross self-neglect in old age. *British Journal of Hospital Medicine 36*(12): 21–25.

Klosterkotter, J., and Peters, U. H. (1985). Das diogenes-syndrome. *Fortschr. Neurol. Psychiat. 53* (1): 427–34.

Macmillan, D., and Shaw, P. (1966). Senile breakdown in standards of personal and environmental cleanliness. *British Medical Journal 2* (5521): 1032–37.

Radebaugh, T. S., Hooper, F. J., and Gruenberg, E. M. (1987). The social breakdown syndrome in the elderly population living in the community: The helping study. *British Journal of Psychiatry 151:* 341–46.

Roe, P. F. (1987). A letter. *British Journal of Hospital Medicine 37*(1): 83–84.

FOCUS ON PRACTICE

Inheritance in an Aging Society

How important to American families is the transfer of wealth from one generation to another? The question is not just one for individual families; it is part of a broader issue about the role of inheritance in society. In previous times, inheritance was of interest mainly to the wealthy, but inheritance now commands attention from more and more middle-class people. A rise in real estate values of homes

(Continued)

(Continued)

bought in the 1950s or 1960s, combined with high stock market values in the 1990s, make many older parents more wealthy than they, or their children, ever imagined they would be.

Inheritance can permit adult children of today's retirees to put their own children through college, start a business, or prepare for their own retirement. Laws curbing Medicaid spend down and the growth of private long-term-care insurance are two signs of concern around inheritance issues. Much of the practice of elder law attorneys is concerned with inheritance in one way or another.

Inheritance plays a large role in late-life financial planning, and it can have a major impact on family wealth, which is partly why guardianship law came into being. It is estimated that transfers of assets account for at least 20% of total family net worth (Gale & Scholz, 1994). There has been much speculation about whether today's baby boomers will inherit substantial wealth from their elders, but this remains uncertain, and, in any case, inheritance is unevenly distributed across the population (Gokhale & Kotlikoff, 2000).

Although a majority of Americans do not have wills, those who do are most likely to be older: About 70% of Americans over the age of 70 do, and the proportion that have wills rises with household income and assets as well as educational background (O'Conner, 1996). Eighty-nine percent of these wills provide for family members apart from the spouse, mostly leaving property to children.

Gerontologists have recently identified how important it is to older people to leave some kind of legacy to their children, but they note that the meaning of legacy has been changing in recent years (R. A. Kane, 1996). In fact, some analysts point to what they call an *inheritance revolution.* Parents transfer wealth to their children by paying for their education, rather than transferring property like the family farm. As a result, their children may not expect to receive an inheritance upon their parents' death. In addition, with people living longer, more of their assets may be consumed—for example, by long-term-care expenditures. Other elders may take the view that surviving generations do not have a valid claim on assets in the first place: A popular bumper sticker reads, "I'm spending my children's inheritance." Still others take the view of wealthy financier Warren Buffett, who has limited what his own children will receive because he feels that inheritance can have a negative influence.

Questions about inheritance are always related to the distribution of wealth and income across the broader population. In general, wealth is unequally distributed in the United States, and this inequality has increased in recent years. More than half of all wealth is held by those in the top tenth of the income distribution, and most of this wealth is held by the richest 1%. Estate taxes, which would tend to even the score, amount to less than 1% of federal tax receipts. More than half of the states have some form of estate or inheritance tax, but many exempt transfers to surviving spouses. The federal estate tax has a $1 million exemption ($2 million for couples), and, as a result, only around 2% of estates are susceptible to federal taxation (Mieskiel, 1996). Recent efforts to raise this exemption threshold even higher make it likely that inheritance and intergenerational transfers will become an even more important part of family life in years to come.

Questions for Writing, Reflection, and Debate

1. List the things that define *quality of life*—what makes life worth living for you personally. Then rank the terms on your list with the most important first. Now try to imagine yourself as an older person in your 80s living alone. List the things that would constitute *quality of life* for you at that age. Is your second list the same as the first one, or is it different? What conclusions do you draw from this fact?

2. When it comes to deciding on matters of personal liberty, the law is supposed to treat adults alike regardless of age: There are no exceptions based on age or personal circumstance. The only basis for civil commitment is the combination of danger plus mental incapacity. Do you think the law should be rewritten to make an exception for frail older people who are vulnerable to mistreatment? Write a newspaper editorial to make your case one way or the other.

3. Suppose that one of Uncle Bert's nephews or nieces was spending many hours each week doing errands for and taking care of him. Should such caregiving be a basis for expecting some share of an inheritance? Defend your opinion.

4. Imagine that you are the social worker from the welfare department who knocked on Uncle Bert's door and was turned away. Write a memorandum to your supervisor stating what you believe should be done next, using all the facts you know about issues of elder abuse. What is the biggest risk you can think of if your boss agrees and you do what you've recommended?

5. Imagine you are a local police precinct captain in a city that has recently had some much-publicized crimes against older people. You are getting ready to give a speech to the local Kiwanis Club about the police department's plans for responding to the situation. What facts about crime and older adults would you cite in preparing your speech?

6. Make a visit to the website for the National Center on Elder Abuse: http://www.aoa.gov/AoARoot/AoA_Programs/Elder_Rights/NCEA/index.aspx. How would you assess the reliability of the statistics cited on this website? Give some reasons that the number of victims of elder abuse in the United States could be *much smaller* than the statistics cited here. Can you think of reasons that the number of victims of elder abuse could be *much larger* than the statistics cited here?

7. Financial exploitation of elders is becoming a much bigger problem today. Make a visit to the website for the National Consumers League Fraud Center: www.fraud.org. What frauds and scams are most likely to target elders in the future as the older population becomes more familiar with computers?

Suggested Readings

Butler, Robert N., and Lewis, Myrna I., *Love and Sex After 60*, New York: Ballantine Books, 1993.

Dejowski, Edmund (ed.), *Protecting Judgment-Impaired Adults*, New York: Haworth, 1990.

Iris, Madelyn, Ridings, John W., and Kendon, Conrad J. The Development of a Conceptual Model for Understanding Elder Self-Neglect. *Gerontologist* (2010), 50(3), 303–315.

Kane, Rosalie (guest ed.), Special Issue on "Legacy," *Generations* (Fall 1996), 20(3).

Payne, Brian K., *Crime and Elder Abuse: An Integrated Perspective*, Springfield, IL: Charles C Thomas, 2000.

Rothman, Max B., Dunlop, Burton David, and Entzel, Pamela (eds.), *Elders, Crime, and the Criminal Justice System: Myth, Perceptions, and Reality in the 21st Century*, New York: Springer, 2000.

Smyer, Michael, Schaier, K. Warner, and Kapp, Marshall B. (eds.), *Older Adult Decision-Making and the Law*, New York: Springer, 1996.

Student Study Site

Visit the Student Study Site at **www.sagepub.com/moody7e** for these additional learning tools:

- Flashcards
- Web quizzes
- Chapter outlines

- SAGE journal articles
- Web resources
- Video and audio resources

SHOULD PEOPLE HAVE THE CHOICE TO END THEIR LIVES?

It is just as neurotic in old age not to focus upon the goal of death as it is in youth to repress fantasies which have to do with the future.

—Carl Jung, "The Soul and Death"

The play *Whose Life Is It Anyway?* tells the story of a patient suffering from paralysis and confined to bed (B. Clark, 1978). In the play, the patient, played by Mary Tyler Moore, engages in a spirited debate with her doctor, asking for help in ending her life. She is no longer able to live as the kind of person she has always known herself to be, so the drama of the play centers on the question: What to do?

Increasingly, this kind of question is asked not about people like the character in the play, but about their grandparents. More than two thirds of all deaths in the United States occur among people over age 65. More and more, the timing of death is not an event that happens according to nature, but is a decision made by human beings.

End-of-life decisions are rapidly becoming our choice to make, regardless of whether we want to make them. Medical advances force us to make decisions unforeseen just a few decades ago. For example, in times past, a person who was unable to breathe without help would die within minutes. Today, mechanically assisted respiration or artificial nutrition and hydration (tube feeding) can sometimes sustain life for years. But medical technology that is a benefit to some can become a burden to others. The decision, in any event, is not easily avoided.

How we understand the decisions to be made will help shape the kind of decisions we make. Consider the moral problem of *euthanasia*, a term that originally came from ancient Greek, meaning simply a "good death." The question put forth by the play *Whose Life Is It Anyway?* is whether a doctor should help the patient end her life. Will the doctor, in other words, engage in *active euthanasia*—sometimes called *mercy killing*? The term *active euthanasia* is used here to denote some deliberate intervention to end the patient's life, such as giving a fatal dose of painkilling medication. *Passive euthanasia*, by contrast, means not doing something, such as withdrawing life-supporting therapy, with the result that the patient dies (Rachels, 1986). Finally, there is the option of *assisted suicide*, in which a doctor or family member actively provides the means or carries out the instructions required for an individual to end his or her life (Moreno, 1995).

People interpret the morality of these acts or omissions in different ways. Some answers depend on how we ask the question and the terms that we use. Most people sharply condemn involuntary euthanasia—that is, killing someone without his or her consent because one believes that person would be better off dead. But there is much more controversy about voluntary euthanasia. Moreover, some critics wonder whether there really is a valid ethical distinction between active and passive euthanasia. Others ask whether there is any difference between direct killing and assisting someone who takes his or her own life.

These questions are not abstract or hypothetical at the moment when it is time to terminate life-sustaining treatment (Hastings Center, 1988). Does it make any difference whether we withhold a treatment from the beginning or withdraw that treatment once it has already started—for example, "pulling the plug" in the case of a mechanical respirator? Then again, what really counts as treatment anyway? For instance, would food and water be considered a treatment in the same way as administering antibiotics is a treatment?

These are some of the ethical issues involved in end-of-life decisions. As the timing of death has been displaced more and more in later life, older people obviously have a vital interest in this debate. On one side are those who argue that the right to self-determination means patients should have the means to end their lives at a time of their own choosing. On the other side are those who warn that the right to suicide or euthanasia runs grave moral risks. For example, should we encourage depressed older people to end their lives instead of changing the conditions that gave rise to the problem?

DEPRESSION AND SUICIDE

Depression is certainly an issue that needs consideration for people who express a wish to die (Steinberg & Youngner, 1998). Psychologists have identified a number of common predictors of suicide: intolerable psychological pain and frustration, a feeling of hopelessness or helplessness, and communicating the intent to kill oneself (Osgood, 1992). Old age is not a time of happiness for everyone. According to the National Institute of Mental Health (2010), "Estimates of major depression in older people living in the community range from less than 1 percent to about 5 percent, but [depression] rises to 13.5 percent in those who require home healthcare and to 11.5 percent in elderly hospital patients." A much larger percentage of older adults may experience periods of depression that aren't severe enough to receive a medical diagnosis. But what exactly does *depression* mean? **Clinical depression** is different from the "down" state that is a common response to setbacks, but is usually temporary. Depression following bereavement and depression among residents of nursing homes may be a reaction to the fact that it is difficult to "start over" in later life.

Clinical depression remains the most important cause of suicide among the old (Blazer, 1993). But depression is rarely the result of social isolation or withdrawal alone. In fact, most older suicide victims either live with family or are in contact with family and friends. Nearly three quarters of older people who commit suicide have had a recent visit to their primary care provider, but rarely has any older person who committed suicide received mental health services, specifically treatment for depression (Ellison & Verma, 2003).

Of all age groups, older people are most at risk for suicide (National Institute of Mental Health, 2010). Suicide rates do not rise with increasing age for women, but they do go up with advancing age for men. Older people make up around 13% of the U.S. population, but account for 20% of suicides. Epidemiological studies estimate the rate of major depression at under 1% in the population over age 65. But at least a quarter of older people living in the community show significant depressive symptoms that have a functional impact on their lives. For older people who are isolated, however, lack of social support can make it difficult for them to cope with depression and overcome it.

Clinical depression remains the most important cause of suicide among the old.

Depression is difficult to diagnose because it can manifest itself in a variety of symptoms. Older patients may report insomnia, fatigue, inability to concentrate, anxiety, and other physical or emotional discomforts. Depression is also difficult to diagnose among older adults because its expressions are transient or appear along with other problems, such as dementia. In addition, some symptoms of depression can be confused with changes associated with normal aging, such as withdrawal from activities. Finally, late-onset depression may differ from depressive disorders that begin earlier in adulthood (Roose & Sackeim, 2004).

Psychiatrists use the term *clinical depression* to describe the presence of five or more symptoms, such as loss of appetite, sleep disturbance, and so on. But late-life depression might better be viewed along a continuum, with an individual's place determined by the number of symptoms experienced during a defined period of time (George, 1993). Older adults report higher levels of depressive symptoms than younger adults, but they are less likely to meet all the criteria for diagnosis of clinical depression. Age by itself is not necessarily a risk factor for depression, and depression is not necessarily more common among the aged than among younger adults.

The majority of older people do not feel sad or unhappy most of the time. Contrary to stereotype, a third of respondents said old age was the happiest period of their lives. In short, although there are differences between older individuals and a wide variety of experiences, the stereotype of old-age misery is wrong.

The psychological problem is that multiple losses and the expectation of further losses can be damaging to self-esteem and can weaken healthy psychological defense mechanisms (Vaillant, 1995). Denial may no longer be possible when an older person faces deterioration and dependency in the course of illness. In such cases, it is not easy to say whether an older patient's rejection of lifesaving treatment, for instance, represents an informed choice to be respected or instead is a sign that the patient needs help for a depressive disorder.

Depressed older adults do not differ from nondepressed elders in the number and types of medical interventions they want when their overall outlook is poor. Counterintuitively, depressed patients are more likely to refuse procedures in situations where the medical prognosis is actually good. The factor that explains most of the difference may be not clinical depression but broader quality of life, which a clinical depression diagnosis may not capture (M. Lee & Ganzini, 1992).

A serious ethical dilemma arises for older people and their families. Many doctors are intensely committed to keeping patients alive and may therefore doubt that anyone who rejects a life-sustaining treatment can be fully rational. The mere fact of deciding not to continue living becomes "proof" of irrationality. But this attitude of "treatment at all costs" fails to take seriously the possibility that some people in the last stage of life may decide that they simply have lived long enough. Should we therefore fail to respect their decisions?

The dilemma was clear in the case of Theresa Leguerrier, a resident in the Good Samaritan Nursing Home in upstate New York. She was in her 80s, but was not suffering from any serious medical problems. One day, she began to refuse food and water, expressing a wish to die of starvation. Staff in the nursing home were divided in their opinion about what to do. The nursing home administration claimed that, by refusing food, a resident might make the home vulnerable to legal penalty for assisting suicide. Against this view, a physician and social worker involved in the case maintained that Theresa Leguerrier was rational and competent to make her own decision in the matter. The nursing home petitioned a court to institute artificial feeding, but the court eventually agreed with the patient's right to refuse treatment in this case.

Some patients who reject a treatment or behave in another way that puts life at risk, such as refusing food, may be suffering from a treatable depressive disorder. They simply lack enough "tender loving care." Surveys of nursing home residents, for example, suggest that depression is widespread among residents. To adopt a laissez-faire attitude—"Well, it's their choice to make"—fails to take seriously the way depressive disorders can impair judgment. Failing to diagnose and treat late-life depression could consign untold numbers of older people to self-imposed death by neglect under the label of *self-determination,* but regarding anyone who refuses treatment as suffering from mental illness is disrespectful of the patient's autonomy. A difficult dilemma, indeed.

The "Right to Die"

Widespread public discussion of the ethics of death and dying began during the late 1960s (Filene, 1998). The first stimulus to this debate was the problem of brain death, a condition in which a critical part of the brain loses its ability to function. When mechanical respirators were developed that could be used to keep patients in this condition alive, states passed laws defining when death could be said to occur under these conditions. But defining the moment of death was not as vexing a problem as resolving the ethical dilemmas around end-of-life choices.

The first major "right-to-die" case was that of Karen Ann Quinlan (*In re Quinlan,* 1976). The patient, who was 21 years old at the time, was in a coma. Her family asked the court's permission to discontinue the use of a mechanical respirator, which they termed an "extraordinary means" of sustaining life. Upon appeal, the New Jersey Supreme Court ruled that there was a constitutional "right to privacy" to permit withholding or withdrawing life-sustaining treatment. Karen Ann Quinlan was then removed from the ventilator and brought to a nursing home, where she remained for 9 years, sustained by feeding tubes and antibiotics until her death in 1985.

The Quinlan case was not an isolated incident, but stimulated new legislation. The first important "right-to-die" law passed was the California Natural Death Act (1976). Since then, other states have been the scene of both legislation and court decisions that pushed far beyond the California law.

Under widely recognized principles of common law in the United States, people have a basic right to accept or reject medical treatment and therefore a right to refuse treatment. This principle of autonomy remains the foundation of legal and ethical thinking about end-of-life decisions. Many state courts, along with the U.S. Supreme Court, have found that a constitutional right to refuse treatment can be exercised by another person on behalf of someone who has been deemed legally incompetent. Where no family member is present, courts have relied on a *guardian ad litem*: a designated spokesperson who represents the interests of the incapacitated person and reports to the court.

Courts have relied on two kinds of standards to determine when it is proper to withhold or withdraw life-sustaining treatment from patients deemed legally incompetent: the standard of **substituted judgment** (What would this patient have wanted under these conditions?) and the **best-interest** standard (What is the balance of benefits and burdens that a "reasonable person" might want under these conditions?). Some evidence suggests that both of these approaches have appeal to different groups of older persons (Moore et al., 2003).

The American Medical Association (AMA, 2011) issued a statement approving, in appropriate cases, the removal or withholding of life-prolonging medical treatment. The AMA stated that discontinuing all means of life-prolonging treatment was "not unethical" even if the patient's condition was not terminal, but the patient was instead in an irreversible coma. This view echoed the finding of some courts.

Typical of a whole range of end-of-life decisions in later life was the 1985 case of Claire Conroy, age 84, a nursing home resident (*In re Conroy,* 1985). Her nephew, appointed as her guardian, asked a New Jersey court for permission to remove a nasogastric tube. The trial court first granted permission on the grounds that life for the patient had become too great a burden. But the New Jersey Supreme Court, reviewing the *Conroy* case, rejected that reasoning and instead defined a range of procedures incorporating tests of substituted judgment and the patient's best interest. Another case of this kind was that of Earle Spring, age 77 and suffering from both dementia and kidney failure. The Supreme Judicial Court in Massachusetts ruled that court approval is not necessarily required before withholding treatment from a patient deemed incompetent to make his or her own decisions (*In re Earle Spring,* 1980).

But decisions in one state court may not be consistent with those in other states. In the O'Connor case (*In re Westchester County Medical Center,* 1988), New York's Appellate Court adopted a strict interpretation of the idea of substituted judgment. In this case, the court insisted that there is a duty to preserve life in all cases unless there is "clear and convincing evidence" that the patient intends to refuse treatment under a particular circumstance. The New York court was fearful that, without a strict standard of proof, the vulnerable older person could be abused by family members interested in inheriting property or by caregivers exhausted by the burden of care. Their concern was *geronticide*, or the killing of the old, a practice prevalent in some premodern societies facing conditions of extreme scarcity (Brogden, 2001).

In an attempt to guarantee that their wishes regarding life-sustaining treatment will be carried out, some people write living wills, but a living will does not answer all questions that can arise in end-of-life decisions. Consider the case of Estelle Browning of Florida (*In re Guardianship of Estelle M. Browning*, 1990), who suffered a massive stroke at age 86. In a previously written living will, Mrs. Browning had stated that she wanted medical treatment, including artificial nutrition, withheld or withdrawn in the event that her condition was terminal and death was imminent. But did these provisions of her living will apply to the present condition? Mrs. Browning was not in a coma, but damage from the stroke was extensive and irreversible. But is *irreversible* the same thing as *terminal?* The court eventually agreed to permit withdrawal of Mrs. Browning's feeding tube on the grounds of a right to privacy and in recognition of the substituted judgment rendered by a proxy decision maker.

One of the most important cases was that of Nancy Beth Cruzan (*Cruzan v. Director of Missouri Department of Health*, 1990), a young woman who suffered brain damage following an accident and was kept alive with artificial nutrition and hydration. A Missouri court denied her parents' request to discontinue treatment, maintaining that clear and convincing evidence of Nancy's wishes was not available. In 1990, this Missouri decision was upheld, by a 5–4 vote, by the U.S. Supreme Court. But by an 8–1 vote, the Supreme Court endorsed "the principle that a competent person has a constitutionally protected liberty interest in refusing unwanted medical treatment." At the same time, the court ruled that there were legitimate state interests in preserving life and preventing potential abuse in terminating treatment. For this reason, the court judged it proper to permit states to impose a high standard of evidence in determining whether an action by a surrogate decision maker (e.g., Nancy Cruzan's parents) actually reflects the wishes of the patient.

The *Cruzan* case was the first time the U.S. Supreme Court had rendered a verdict on right-to-die cases. In its decision, it upheld the right to refuse life-sustaining care, including artificial nutrition as a medical treatment. A majority of the court found that an appointed proxy or surrogate decision maker, just like a competent patient, would be legally entitled to refuse treatment. Significantly, the court finally endorsed the use of **advance directives** (King, 1996), such as the living will and the durable power of attorney for health care. But the court also left procedural requirements to the states, ensuring that both legislation and litigation will continue for years to come.

In many ways, the recent evolution of right-to-die laws and court cases is a continuation of long-held cultural ideals: above all, the idea of self-determination and protection of rights by due process of law. For a competent adult, the right to refuse medical treatment, even at risk of death, has been widely recognized in U.S. common law tradition. The *Natanson v. Kline* case (1960) expressed the ideal of self-determination in these words: "Each man is considered to be master of his own body, and he may, if he be of sound mind, expressly prohibit the performance of life-saving surgery, or other medical treatment." But these developments do not confer on anyone a right to have active euthanasia performed, nor do they confer a right to involve other people in assisting with suicide. The right to die in all cases under law has involved some variety of passive euthanasia.

Should the right to die be extended further? Where to draw the line between a passive right to die and more active forms of assisted suicide or active euthanasia remains an unresolved issue. An early case, *Perlmutter v. Florida Medical Center* (1978), illustrates the point. A 73-year-old man was terminally ill, but deemed competent, and sought to have his respirator disconnected. The Florida state attorney argued that anyone helping to disconnect the respirator could be criminally charged with assisting a suicide, an argument that was firmly rejected by the Florida Supreme Court.

Today, few people would regard withdrawing treatment from the terminally ill as equivalent to suicide. Courts have repeatedly concluded that termination of life-sustaining treatment is not homicide, suicide, or assisted suicide. At the same time, assisting a suicide does remain a crime in most jurisdictions of the United States. This fact is important in weighing the actions of Dr. Jack Kevorkian, who in the 1990s developed a "suicide machine" that delivers a lethal dose of drugs to patients requesting it. The self-styled suicide manual, titled *Final Exit* (Humphry, 2002), remains popular among some older adults and those with terminal illnesses. Although a jury finally convicted Dr. Kevorkian, public opinion polls showed substantial support for him.

In 1997, in the *Washington v. Glucksberg* case, the Supreme Court ruled that terminally ill individuals do not have a constitutional right to assisted suicide. The result of this decision is that states may make their own regulations in this area, as Oregon did in the same year, becoming the first state to permit physician-assisted suicide; Washington state and Montana followed later. In the first years after the law was passed, few people took advantage of the Death With Dignity Act (1997), which permits doctors to prescribe lethal medications for patients with a terminal illness who want to end their lives. The average age of those making use of the law to date has been 71, so the Oregon experience does have implications for an aging population (Sullivan, Hedberg, & Fleming, 2000). One interesting outcome of the new law has been vastly greater interest in Oregon in the use of pain medications, as well as hospice. The Oregon assisted-suicide law remains controversial.

Whatever the outcome of public debate over assisted suicide, many older people are taking matters into their own hands. Suicide rates increase with advancing age among Americans age 65 and reach a peak for those over age 75 (McIntosh et al., 1994). Older men are more likely to commit suicide than any other group in the United States, most often using guns. But old-age suicide is a complex phenomenon (Leenaars et al., 1992). For instance, there are dramatic differences in the suicide rate in later life among different ethnic groups. White men are the group most likely to kill themselves in old age, and Black women are the least likely to die by their own hands (Baker, 1994).

Despite wide public discussion, Americans lack a clear consensus about exactly how "dying well" might be defined (Kearl, 1996). At a minimum, dying well would mean having the right to know one's medical condition and the choice to accept or reject life-prolonging treatment. At the same time, there is growing public approval for the right of the terminally ill to have the options of euthanasia or assisted suicide. Some surveys suggest that two thirds of Americans favor legalizing physician-assisted suicide. One factor pushing public opinion in this direction is the loss of control that terminally ill patients experience in hospitals: Even completing an advance directive does not guarantee that it will be honored in practice.

What is distinctive about the current debate is that some critics are coming to view suicide in old age not as a problem, but perhaps as a rational choice in a desperate situation. Compassion & Choices argues that suicide and assisted suicide in the face of a terminal illness that causes unbearable suffering should be ethically and legally acceptable. The Netherlands has gone furthest in legalizing euthanasia and physician-assisted suicide, and advocates of assisted suicide look favorably on the Dutch case (Pierson, 1998). But some observers see the experience of the Netherlands as proof of how difficult it is to effectively regulate euthanasia and suicide (Gomez, 1991). For instance, one study of the practice showed that for more than half the patients in the Netherlands who had legal euthanasia, the procedure was performed without their full consent (Butler, 1996). Critics have argued that, instead, the real focus should be on *palliative care*, such as hospice and better pain medication (R. Morrison & Meier, 2003). Some opinion polls showing that a majority of Americans favor legalizing assisted suicide give different results if respondents are given a choice of better palliative care at the end of life (Shelanski, 1998).

GLOBAL PERSPECTIVE

Assisted Dying in Europe

Ms. Bettina Schardt, age 79, was a retired X-ray technician living in Würzburg, Bavaria, in Germany. She was old, but she was neither sick nor dying. Still, she wanted help in committing suicide. The reason? She did not want to move into a nursing home. In response to her request, an advocate of assisted suicide helped her to end her life and thus set off a furor of public debate in Germany. It turns out that many Germans do not want their country to follow the example of Switzerland, where liberalized laws on euthanasia have brought about a booming trade in assisted suicide. In recent years, almost 500 Germans have crossed the border into Switzerland in order to end their lives: a striking example of "medical tourism," where people flock to foreign countries for medical services. Medical tourism is usually driven by lower prices, but in the case of assisted suicide, the motive is different. Advocates of assisted suicide believe that people are seeking a dignified way to end their lives. Critics point to Bettina Schardt as a case where unreasonable fears about nursing homes led to an irrevocable decision.

European countries present different options when it comes to assisted suicide and euthanasia. Assisted dying is legal in Belgium and the Netherlands, and it is tolerated and openly practiced in Switzerland. In Great Britain, laws have been proposed to liberalize assisted dying in the case of the terminally ill. It is in the Netherlands where legalized euthanasia has been longest established, but that practice has been the subject of vigorous debate among bioethicists around the world. In the Netherlands, a request for euthanasia is not unusual, yet the vast majority of patients who initiate the discussion actually die from natural causes. The law permits a competent adult suffering from a terminal illness to request and receive assistance in dying from a physician. The reasons for requests for euthanasia have gradually been broadened beyond terminal illness to include loss of autonomy, control, independence, and the ability to pursue pleasurable activities of life. It appears that there has been a growing European acceptance of assisted dying, but it remains a topic of significant debate.

SOURCE: Mark Landler, "Assisted Suicide of Healthy 79-Year-Old Renews German Debate on Right to Die," *The New York Times* (July 3, 2008). Retrieved September 16, 2011 (http://www.nytimes.com/2008/07/03/world/europe/03germany.html).

OUTLOOK FOR THE FUTURE

By the time Karen Ann Quinlan died in 1985, public opinion in the United States had undergone a substantial change that paralleled dramatic developments in the legal sphere. Public opinion has continued to evolve in favor of greater choice about end-of-life decisions (Yankelovich & Vance, 2001). Withdrawal or withholding of heroic measures such as artificial respiration and cardiopulmonary resuscitation has become more acceptable to a majority of Americans, at least in cases of terminal illness. But most health care professionals and vocal elements of the public still disapprove of active euthanasia.

The issues remain controversial. In 2003, there was enormous publicity around the case of Terri Schiavo, a Florida woman who had for years been in a persistent vegetative state and kept alive by a feeding tube. Family members disagreed about whether to remove life support, and the Schiavo case was the subject of extended litigation and vast national publicity. Public opinion was polarized. Terri Schiavo was eventually removed from life support and died in 2005.

The Schiavo case revealed that some issues around the right to die provoke vigorous dispute. For example, is artificial nutrition or hydration to be considered in the same category as other medical treatments? The AMA has claimed that there is no ethically significant difference between withdrawing food and water and using other life-supporting measures. Most medical ethicists and courts have agreed. But many laypeople and professionals remain unconvinced (Solomon et al., 1993).

In a similar way, ethicists have argued that withholding life-supporting treatment is no different morally from withdrawing treatment once it has already begun. But, again, most families and health care practitioners remain persuaded that there is an important psychological difference: It is easier not to start a treatment than it is to withdraw it once begun. Clearly, the social and interpersonal context of a decision continues to make a difference to those who are involved in end-of-life decisions.

A major step in the process of resolving these issues came with the enactment of the Patient Self-Determination Act (PSDA), which went into effect in 1991. This law requires hospitals, nursing homes, and other health care facilities to advise all patients at the point of admission about their right to accept or refuse medical treatment. The PSDA in effect creates new requirements for hospitals, but it does not create new rights for patients. Under the law, patients are specifically to be told about their right to determine in advance whether they wish to have life-sustaining treatment if they become ill without hope of recovery. The staff members of health facilities are required to document and implement policies that respect the wishes of patients (LaPuma, Orentlicher, & Moss, 1991).

Despite this law, however, relatively few patients actually complete an advance directive of any kind. Public opinion polls have revealed that close to 90% of American adults would not want to be maintained on life support systems without prospect of recovery. Yet a survey by the AMA revealed that not even 15% of the general public had actually completed a living will, and a low proportion holds true for persons over age 65.

Why is the proportion so low? It appears that physicians in general favor the idea of advance directives, but they remain reluctant to engage in an open discussion with patients on the topic. Recent research confirms the assumption that many Americans today prefer to be told the truth about their medical diagnoses. Yet substantial proportions of Asian American and Hispanic elders are less likely than others to favor an autonomy model of decision making, and practitioners need to take account of these ethnic differences in communicating with older people (Braun, Pietsch, & Blanchette, 2000). The PSDA is unlikely to be the final answer to helping patients make end-of-life decisions. In 1997, the U.S. Supreme Court (in *Vaco, Attorney General of New York, v. Quill*) issued a landmark decision that assisted suicide is not a constitutional right, thus leaving the matter up to individual state legislatures. Neither court decisions nor legislative action will put an end to the ethical debate about right-to-die issues, which is certain to continue for many years to come.

In the readings that follow, we hear impassioned voices in the debate over the right to die. Dr. Charles McKhann speaks in favor of physician-assisted suicide under certain circumstances. Robert Pearlman and Helene Starks offer some powerful case histories along with data from surveys that help answer the question of why some people seek physician-assisted death. The answer, it turns out, has less to do with physical pain and more to do with threats to our sense of personal integrity. A mixed message comes from Sue Woodman in her article, "Last Rights: Aunty's Story." She explores how the same person's feelings toward suicide and euthanasia can change over time. Woodman's article shows how complicated these decisions can be in practice.

In contrast, Leon Kass decisively rejects the idea of assisted suicide. Kass offers strong arguments for why doctors must never kill and why physician-assisted suicide is morally wrong. Kass's arguments force us to think carefully about a basic question arising in end-of-life decisions. To what extent is the decision to end one's life a purely individual matter, and to what extent does it involve other people? For example, does the ending of one's life become a different matter because physicians, family members, or others participate or assist in the act of suicide? As individuals and society debate the issue in years to come, we will be preoccupied with the question of how individual choices are connected to wider values, such as the cost of health care and the condition of old age in our society.

FOCUS ON PRACTICE

Advance Directives

The ethical dilemmas of end-of-life decisions have increasingly led to demands for a greater measure of control in how those decisions are made. One practical response has been the spread of so-called advance directives, or written statements prepared well before a serious illness arrives, in which an individual can state the choice to be made when a decision is necessary.

Under the law, everyone already has the right to refuse all or any part of medical treatment if it isn't desired. But sometimes, because of diminished medical capacity, people are unable to let others know whether they would want a specific treatment. An advance directive solves this problem by giving specific directions or designating someone else in case an individual is unable to express a wish about medical treatment.

There are two general types of advance directive: Both are intended to give guidance and retain validity even after the person who executed the document has lost mental capacity. As mentioned, these are the living will and the durable power of attorney for health care. A **living will** is a written statement expressing an individual's wish for what should be done in a life-threatening situation. It permits physicians to stop undesired medical treatment even if stopping treatment might result in death. If a clear living will has been drafted beforehand, everyone can know that the patient would want life-sustaining treatment withdrawn.

A second legal instrument naming a specific person to make health care decisions is the **durable power of attorney for health care**. This instrument permits an incapacitated person to designate, in advance, another person who is trusted to make health care decisions for the patient.

A living will permits an adult of sound mind to stipulate the kind of life-prolonging treatment to be provided in the event of an emergency when the patient cannot indicate a choice—for example, "I desire food and fluids but no cardiopulmonary resuscitation." It is a legal document and should be prepared on a standard form, although a simple written declaration of preference may also suffice. Different states have different legal requirements about witnesses and other rules for writing a living will.

A durable power of attorney is also a witnessed legal document. But an individual need not specify all the details about treatment to be given or withheld. Instead, the individual designates someone else—called a *proxy* or a *health care agent*—to make those decisions in the event the patient is incapacitated. This proxy could be a family member, a close friend, or any trusted person. As with living wills, different states have varying rules and regulations applying to power of attorney documents and their acceptability for health care decision making.

Federal law now requires hospitals, nursing homes, and other health providers covered by either Medicare or Medicaid to give people information about advance directives. No one is required to fill out an advance directive, but organizations are required by law to let patients know that they have the right to do so.

If you have an advance directive, this legal tool goes into effect only if you are unable to speak on your own behalf. An advance directive doesn't prevent you from changing your mind at a later date if you are conscious and able to act on your own behalf. An advance directive can be changed or canceled by the person who wrote it at any time.

People sometimes ask whether a living will or a durable power of attorney is better. Living wills were the first type of advance directive to become popular, but health care powers of attorney have turned out to be more flexible. The exact kind of document required depends on state laws, which have become a critical factor in light of the *Cruzan* decision. Such variations in law from state to state can, of course, be a problem because it is unclear whether one state will always honor a directive from another state.

When end-of-life decisions must be made in the absence of a written directive, the decision about discontinuing treatment is typically made by the family and health care providers in consultation. It is rare for such decisions to end up being made by a court, which is a cumbersome way to proceed. Written advance directives can be helpful in keeping such cases out of the legal system and in the hands of those closest to patients.

At this point, while awareness of the importance of preparing an advance directive has increased, only a small proportion of Americans—fewer than one in five—have actually prepared any kind of written advance directive. Why don't more older people use advance directives? Studies of educational interventions promoting advance directives have turned up some answers. One careful demonstration project achieved success by combining a moderate level of information along with practical help completing the documents (High, 1993). Over the 4-month period, the completion rate for the living will increased from 25% to 50%, and the completion rate for the health care proxy rose from 14% to 30%. These results suggest that older people don't simply lack information or need encouragement. They need better help in achieving communication with family and health professionals.

(Continued)

(Continued)

Advance directives in any case may not be the total solution for end-of-life decisions. For example, in a study of patients on kidney dialysis, 61% said they wanted the doctor or proxy decision maker to have "leeway" to disregard the patient's own previously expressed preferences. Instead of simply following a living will, decision makers were supposed to take into account circumstances and the patient's best interest.

Some critics have argued that the move toward advance directives may end up making decisions to refuse treatment easier only for those who have filled out the necessary "paperwork." Those who have not completed the proper legal documents—for instance, persons without the necessary information or education—may find themselves lacking rights theoretically granted to them by law.

Still other critics argue against advance directives because people often change their minds about end-of-life decisions. It is common for older people to say, "I would never want to live in an old-age home." Yet after living for a time in a retirement residence or nursing home, the same people often discover that life is quite satisfactory there. That same point could hold true for living with chronic illness. People fear they would be unable to go on living under an extreme disability, yet adaptation to loss and living with disability does take place.

Recent critics have gone so far as to speak of the "failure" of the whole idea of living wills, favoring instead a stronger use of the durable power of attorney, which, of course, has its own problems (Fagerlin & Schneider, 2004). These problems, however, do not make advance directives a bad idea. Instead, they suggest a need to improve the way advance directives are used in practice. For example, we should strengthen communication between health care professionals and patients and not simply treat written directives as another form of paperwork. The need for better communication also suggests that advance directives can be an occasion for family members to share with one another their own values and expectations. Used in this way, advance directives can help older people and their families better approach end-of-life decisions.

URBAN LEGENDS OF AGING

"Advance directives would have prevented the tragic case of Terri Schiavo."

Not at all. Her husband Michael would almost surely have been appointed proxy decision maker, since the spouse is already presumed to be as such by the courts and common law. As for living wills, they have been shown by empirical studies to be mostly underutilized and therefore ineffective. In any case, Terri Schiavo was not terminally ill, and an advance directive almost certainly would not have prevented her tragic case.

READING 30

———— Why Do People Seek Physician-Assisted Death? ————

Robert A. Pearlman and Helene Starks

Despite the illegality of physician-assisted death in the United States (except in the state of Oregon), many primary care providers and oncologists have been asked to provide aid in dying.[1] Popular explanations for physician-assisted death include inadequate treatment for pain or other symptoms, depression, hopelessness,[2] and socioeconomic stressors, such as concerns about the burden of increasing dependency on other members of the family and the economic hardship associated with the costs of health care.[3]

What motivates people to pursue physician-assisted death has proved to be both a controversial and a foundational issue in debates regarding the legality of and appropriate clinical response to requests for aid in dying. To date, insights regarding the motivations for assisted death have come largely from three sources: provider impressions, patients who transiently consider physician-assisted death or other means to end their lives under certain conditions, and forced-choice reporting (that is, responding to a checklist of reasons) from patients in Oregon who have pursued physician-assisted death. However, there has been limited direct reporting from patients about the motivation and process that lead them to pursue assisted death.

To address the current gaps in the understanding of physician-assisted death, we conducted a longitudinal, qualitative study of patients who seriously pursued assisted death and their family members.[4] The patients in our study lived an average of ten months between their first request for aid in dying and the time of death. Patients used this time to acquire the means to end their lives. For many, however, simply having medications did not result in their immediate use. These patients engaged in an ongoing evaluation of the value of living versus dying and repeatedly assessed the benefits and burdens of their current experience. Moreover, among the individuals in our study, the pursuit of physician-assisted death was not motivated by any single factor, nor was depression reported as a reason. Rather, the decision to hasten death culminated from an interaction of illness-related experiences, threats to the person's sense of self, and fears about the future.

The reports from these patients and their family members identified opportunities for improving palliative care that might have reduced patients' perceived need to choose aid in dying. However, these accounts also illustrate why a small number of patients will continue to view assisted death as a desirable choice. Some will decide that hospice and palliative care are not for them or that some of the choices offered by palliative care, such as the trade-off between pain management and cognitive function, are not acceptable ones. In addition, some issues, such as the desire to control the dying process and the suffering associated with a loss of sense of self, are not easily addressed even by the most capable health care providers.

"I Can't Do This Anymore"

When Anna was sixty-two years old she was diagnosed with metastatic ovarian cancer.

Throughout her life, Anna was organized, energetic, and athletic. She was actively involved in community activities; she was a professional and an involved grandparent. About her illness she said, "I'm trying not to change my life and let cancer steal any more of it than it has to." She also expressed long-standing beliefs about having control over her life and death: "It should be up to me to decide ... when I've had enough suffering. . . . If I'm at the point where all I can do is lie on a bed all day long, then to me that's probably not living anymore."

Over four years, Anna underwent multiple surgeries and rounds of chemotherapy and radiation. Many of her treatments were quite uncomfortable. She reported that she has been "deathly ill after every [chemotherapy] treatment, just not even able to read or barely even watch television. I would wake up in the morning with dry heaves and being incontinent and rolling out of bed so I wouldn't get the bed messed up. It was really wretched."

Anna talked to her family and sought medications for assisted death in the event she decided to hasten her death. Once she got them, she reported, "I felt I had more energy to fight the cancer and just to live in the present time. It just took a big weight off my shoulders somehow, knowing at least that that was one thing that maybe I didn't have to worry about." It was another three years before Anna used the medication to end her life.

During her illness Anna also experienced painful complications, including bowel obstruction and spinal cord compression. Despite enduring significant amounts of pain, Anna never cited this as the motivation for a hastened death. Her primary concern was that she would die in a hospital, "away from my home with familiar people and familiar surroundings and some privacy and some control."

After exhausting her anticancer treatment options, Anna became very weak, sleeping much of the day, and was unable to perform many of her routine functions. She started bleeding uncontrollably from her bowel as a late side effect of total abdominal radiation and was told that the only treatment was constant transfusions. She told her husband, "Honey, this is it. I can't do this anymore." Over the next thirty-six hours, Anna gathered her family together to say good-bye. She ingested the medication to hasten her death, with twelve loved ones in attendance, and died within two hours.

FACTORS MOTIVATING PHYSICIAN-ASSISTED DEATH

In our interviews with thirty-five families, we asked questions about the history of the patient's illness, the patient's stated reasons for seeking aid in dying, and other factors influencing the pursuit of physician-assisted death, as well as the manner of death. Our analysis identified nine common factors. No single factor on its own ever accounted for a serious interest in a hastened death. Rather, interest usually arose out of an interactive process involving multiple factors in three broad categories: illness-related experiences (symptoms, functional losses, effects of pain medication, and the like), threats to the person's sense of self (as revealed by his or her desire for control over the circumstances of dying and long-standing beliefs in favor of hastened death), and fears about the future. Figure 1 presents our assessment of the frequency and importance of these factors in the patients' deliberations.

Illness-Related Experiences

Weakness, tiredness, and discomfort made up one set of motivating concerns for patients who sought aid in dying. Excellent end-of-life care can often ameliorate these concerns. However, the functional losses caused by advanced illness are often less amenable to successful intervention, especially when the patient is dying. Approximately two-thirds of the patients in our study described the relation between symptoms brought on by illness and treatment and the resultant loss of function as considerations in their pursuit of assisted death.

Figure 1   Motivations for Seeking a Hastened Death

Motivating Factor	Total Patients[a] (*N* = 35)
Illness-related experiences	
Feeling weak, tired, and uncomfortable	24 (69%)
Loss of function	23 (66%)
Pain or unacceptable side effects of pain medication	14 (40%)
Threats of sense of self	
Loss of sense of self	22 (63%)
Desire for control	21 (60%)
Long-standing beliefs in favor of hastened death	5 (14%)
Fears about the future	
Fears about future quality of life and dying	21 (60%)
Negative past experiences with dying	17 (49%)
Fear of being a burden on others	3 (9%)

SOURCE: Data from R. A. Pearlman, C. Hsu, H. Starks, A. L. Back, J. R. Gordon, A. J. Bharucha, B. A. Koenig, and M. P. Battin, "Motivating Factors for the Pursuit of Physician-Assisted Suicide: Patient and Family Voices," unpublished manuscript.

NOTE: Motivating factors were rated independently by two investigators as to their role in influencing the pursuit of a hastened death. Four response categories were recorded: "not mentioned," "present but not judged to be influential," "influential," and "very influential." The table records response ratings inferred to be "influential" or "very influential" based on the reading of the transcripts.

[a] Total patients included prospective and retrospective cases. Prospective cases included patients and family members recruited while the patient was alive. Retrospective cases included only surviving family members recruited after the patient's death.

For example, a woman with ovarian cancer spoke of the effects of chemotherapy: "Of course, with me, with the chemo and things, [there's] just the terrible weakness and the nausea and just not feeling like you can do anything. . . . If it'd been like two weeks after and I was going out to do things—[but] you were still shaky, and you couldn't quite predict how you were going to feel, or you were afraid to make commitments because you weren't sure you were going to be able to carry them out." Physical symptoms often led to a lack of function. For example, one woman described her response to steroids for her chronic lung disease:

My thighs are so weak I can't get up from the floor, and I don't have energy. . . . My arms are withering away. It's ridiculous. And also I take stuff with [prednisone] that gives me stomach problems, and I have bowel problems—there is no part of me that is functioning routinely. I'm very, very tired, and I cough a lot. My cough is worse when I talk a lot, and early in the morning it's horrible. I cough all this yucky stuff up and can't talk a lot. And I'm so weak, for example, that although I can drive a car, I find it just too much to lift my own oxygen tank up and in and out of the car. . . . So I'm not living; I'm existing.

Pain and the side effects of treatment for pain were another set of motivating issues for more than a third of the patients in our study. As others have argued, pain can be better managed with excellent end-of-life care.[5] However, the side effects of treatment, especially effects on cognition, will continue to be challenging. For some patients, these side effects are totally unacceptable, and feeling that they must choose between being pain free and being mentally competent creates the desire for another alternative—a hastened death. One woman described her mother's situation as follows:

> She was in a lot of pain. . . . What she feared more than the pain was the effect that the pain drugs would have on her. . . . She didn't want to lose her rational mind. She didn't want to lose any of her personality or capacity. . . . Certainly, [she had a] fear of the pain increasing and causing her to have to take more medication . . . enough that she didn't really know what she was doing all the time or that she might start drooling or saying stupid things or coming across as drunk.

In another case, a man had metastases to his spine causing severe pain that was difficult to control, despite large doses of morphine. Hospice nurses had attempted to control his pain but ended up sedating him completely for twenty-four hours, which he found absolutely unacceptable. His wife told us,

> Pain was an extraordinary factor. He would go out and split kindling . . . and I kept saying, "I'm going to have more kindling than Carter has pills." And he says, "It gives me something to do, and if I have something to do, I don't have to think about how badly it hurts." Pain was an extreme factor . . . and he was not a sissy, not at all. . . . [But] the morphine pump didn't do it. He was still taking Roxicet. . . . This had been going on for months, and so by that time his body had built up a tremendous tolerance. . . . They could knock him out; he could be a vegetable; but that was not what he wanted. . . . If he couldn't function and at least think somewhat clearly, life wasn't worth it. And he did not want to leave me. He did everything possible to set me up, because

he knew he was going to die. . . . There was no way you could stop that. And he did everything imaginable to make my life as easy as it could be. But it got to a point where the pain was just intolerable.

Threats to Sense of Self

The literature suggests, indirectly, that depression and hopelessness may motivate some patients to seek a hastened death.[6] However, acute depression was not judged to be an influential factor for any of our participants. The more common motivating issues of a psychological nature pertained to loss of control and sense of self (both experienced by nearly two-thirds of our patients). Most experts in end-of-life care acknowledge that redirecting the loss of sense of self in a dying patient is a daunting challenge and one that is rarely achieved. This form of suffering is profound and is related to losses in relational, social, and community involvement.[7] These losses affect not only what one can do but how one perceives oneself. One woman described her mother's response to progressive losses owing to an autoimmune disorder:

> My mother was at age seventy-two before she got sick. She was still very energetic and vibrant, youthful and active and looked ten years younger than she was. And then when she got sick, she started losing her hair, and she became disfigured with this skin condition. . . . She would, literally, have spells when her muscles would be so bad that she would feel paralyzed, and she was in incredible fear of becoming paralyzed and becoming completely incapacitated and immobile. . . . [She wanted to have] control over her own dignity and her own independent life. The things that were meaningful to her in her life were her art, her ability to do her art, and her friends and spending time with her friends and cooking and eating. . . . She was . . . convinced that when she couldn't do any of those things anymore, her life would be meaningless, and she wouldn't want to live anymore.

In another example, a man described his mother's final sense of loss when she could no longer visit with her extensive network of family and friends:

> She lost her appetite, her pleasure for food. She lost her strength to go outside and even having . . . people come over. She'd try. She hated to say no, they couldn't visit. . . . She wanted to, but a visit was a painful process. She just didn't have the strength or the concentration . . . to even visit with people anymore. . . . So basically, even having company is a strain, a pain to you, almost more than it's a pleasure to enjoy the people. And that was about the last thing she had left, was just sitting around and enjoying people, and . . . even that went away. Like she said, you get so sick of being sick and tired.

For most patients, sense of self is inextricably connected to ability to function. Moreover, the loss of sense of self creates appreciable anxiety. For example, one woman had metastases to the bone and was at risk for a spinal compression that would lead to paraplegia and incontinence. This functional loss represented more to her than just another accommodation to her illness. Her daughter explained the patient's view about her future:

> She saw this declining physical curve, and that at some point along that she was going to lose significant ability to be the person that she was. And she had already lost a lot—she couldn't go for the long hike anymore, . . . she couldn't go for the long walks, and now she wasn't supposed to work full time, and she was completely exhausted all the time, because she was working and taking care of her big house and all. So she had already seen some things go, and that was acceptable, but she knew at some point she'd be somewhere down on the curve [where] she had given up so many things that it wasn't okay. . . . Rather than wait to get there and have to figure out where that point was, she wanted to just die before it got any worse. And then when her spine started to go, of course, the big threat was losing control of her bowels, [which] was clearly not acceptable.

Fears About the Future

Fears about the future, including fears about the experience of dying, also motivate patients who seek to hasten death. Many of the patients projected that the course of their illness would result in a fate worse than death and preferred to end their lives before they reached this condition. Their judgments about this poor quality of life were often based on their own prognosis, accumulated losses, and fears of becoming a burden on others. For about half of our patients, their fears were also influenced by having witnessed loved ones go through terrible deaths involving pain, what they viewed as pointless use of medical technologies, and images of tubes coming out of every orifice. With better palliative and end-of-life care, fears like these should abate as people see others having better dying experiences. Nonetheless, even the best palliative care cannot prevent all functional losses and loss of sense of self, and thus for some, these fears will remain and motivate them to seek aid in dying.

IMPLICATIONS FOR END-OF-LIFE CARE

Several important implications for clinicians emerge from these case reports. First, that multiple interacting factors prompt patients to seek assisted death challenges health care providers, including those in palliative care, to understand patients' illness and dying experiences holistically. Our patients reported intricate and subtle interactions between physical and functional decline and existential concerns that could not be easily separated or compartmentalized. Our data confirm the familiar recommendation that providers repeatedly assess the patient's concerns about losses and dying in order to understand patients' concerns and tailor end-of-life care of the patient's evolving personal experience.

Second, these narrative accounts also demonstrate the importance to patients of their sense of self and of control over the manner of death. Clinicians need to be sensitive to these deeply personal psychological and existential issues and differentiate them from clinical depression. Although it is important to determine whether depression is driving a request for assisted death, it is equally important to examine other psychological processes.

Finally, the factors that motivate an interest in assisted death are similar to those that prompt people to complete advance directives and forgo life-sustaining treatment: the desire to control the timing and circumstance of one's death.[8] Many people view assisted death as another option that should be available at the end of life. The topics identified in Figure 1 can serve as a guide in talking to patients about the far-reaching effects of illness, including the quality of the dying experience. Clinicians should explore patients' fears and how they see themselves in light of current and future physical decline and functional losses. Clinicians frequently shut down discussion about aid in dying and, in so doing, thwart opportunities for understanding patients' responses to their dying experience. Instead, clinicians must explore the motivations for these patients' interest in this option and identify ways to ameliorate their suffering.[9] For some patients, however, the amelioration of suffering will come only with death, and these patients may be unwilling to endure a prolonged dying process.

Our study suggests that some dying patients will continue to desire a hastened death in spite of excellent palliative end-of-life care. Some suffering cannot be relieved and will continue or worsen. The question that must be asked is, Are we as a society sufficiently compassionate to allow the choice of a hastened death to terminally ill, competent patients who are receiving state-of-the-art end-of-life care but are still suffering?

NOTES

1. Doukas, D. Waterhouse, D.W. Gorenflo, and J. Seid, "Attitudes and Behaviors on Physician-Assisted Death: A Study of Michigan Oncologists," Journal of Clinical Oncology 13 (1995): 1055–61; D. E. Meier, C. A. Emmons, S. Wallenstein, T. E. Quill, R. S. Morrison, and C. K. Cassel, "A National Survey of Physician-Assisted Suicide and Euthanasia in the United States," New England Journal of Medicine 338 (1998): 1193–1201; E. J. Emanuel, D. L. Fairclough, B. R. Clarridge, D. Blum, E. Breera, W. C. Penley, L. E. Schnipper, and R. J. Mayer, "Attitudes and Practices of U.S. Oncologists Regarding Euthanasia and Physician-Assisted Suicide," Annals of Internal Medicine 133 (2000): 527–32.

2. Emanuel, D. L. Fairclough, E. R. Daniels, and B. R. Clarridge, "Euthanasia and Physician-Assisted Suicide: Attitudes and Experiences of Oncology Patients, Oncologists, and the Public," Lancet 347 (1996): 1805–10; K. M. Foley, "Competent Care for the Dying Instead of Physician-Assisted Suicide," New England Journal of Medicine 336 (1997): 54–58; T. E. Quill, D. E. Meier, S. D. Block, and J. A. Billings, "The Debate over Physician-Assisted Suicide: Empirical Data and Convergent Views," Annals of Internal Medicine 128 (1998): 552–58; H. M. Chochinov, K. G. Wilson, M. Enns, and S. Lander, "Depression, Hopelessness, and Suicidal Ideation in the Terminally Ill," Psychosomatics 39 (1998): 366–70; L. Ganzini, W. S. Johnston, B. H. McFarland, S. W. Tolle, and M. A. Lee, "Attitudes of Patients with Amyotrophic Lateral Sclerosis and Their Care Givers toward Assisted Suicide," New England Journal of Medicine 339 (1998): 967–73; W. Breitbart, B. Rosenfeld, H. Pessin, M. Kaim, J. Funesti-Esch, M. Galietta, C. J. Nelson, and R. Brescia, "Depression, Hopelessness, and Desire for Hastened Death in Terminally Ill Patients with Cancer," Journal of the American Medical Association 284 (2000): 2907–11.

3. Emanuel, Fairclough, Clarridge, et al., "Attitudes and Practices of U.S. Oncologists Regarding Euthanasia and Physician-Assisted Suicide"; M. E. Suarez-Almazor, M. Belzile, and E. Bruera, "Euthanasia and Physician-Assisted Suicide: A Comparative Survey of Physicians, Terminally Ill

Cancer Patients, and the General Population," Journal of Clinical Oncology 15 (1997): 418–27.

4. Pearlman, C. Hsu, H. Starks, A. L. Back, J. R. Gordon, A. J. Bharucha, B. A. Koenig, and M. P. Battin, "Motivating Factors for the Pursuit of Physician-Assisted Suicide: Patient and Family Voices," unpublished manuscript.

5. Field and C. K. Cassel, eds., Approaching Death: Improving Care at the End of Life (Washington, D.C.: National Academy Press, 1997); R. S. Morrison, A. L. Siu, R. M. Leipzig, C. K. Cassel, and D. E. Meier, "The Hard Task of Improving the Quality of Care at the End of Life," Archives of Internal Medicine 160 (2000): 743–47; D. E. Meier, "United States: Overview of Cancer Pain and Palliative Care," Journal of Pain and Symptom Management 24 (2002): 265–69.

6. Emanuel, Fairclough, Daniels, et al., "Euthanasia and Physician-Assisted Suicide"; Quill et al., The Debate over Physician-Assisted Suicide: Empirical Data and Convergent Views"; Chochinov et al., "Depression, Hopelessness, and Suicidal Ideation in the Terminally Ill"; Ganzini et al., "Attitudes of Patients with Amyotrophic Lateral Sclerosis and Their Care Givers toward Assisted Suicide";

Breitbart et al., "Depression, Hopelessness, and Desire for Hastened Death in Terminally Ill Patients with Cancer"; S. D. Block and J. A. Billings, "Patient Requests to Hasten Death: Evaluation and Management in Terminal Care," Archives of Internal Medicine 154 (1994): 2039–47; E. J. Emanuel, D. L. Fairclough, and L. L. Emanuel, "Attitudes and Desires Related to Euthanasia and Physician-Assisted Suicide among Terminally Ill Patients and Their Caregivers," Journal of the American Medical Association 284 (2000): 2460–68.

7. Lavery, J. Boyle, B. M. Dickens, H. Maclean, and P. A. Singer, "Origins of the Desire for Euthanasia and Assisted Suicide in People with HIV1 or AIDS: A Qualitative Study," Lancet 358 (2001): 362–67.

8. Pearlman, K. C. Cain, D. L. Patrick, H. E. Starks, M. Applebaum-Maezel, N. S. Jecker, and R. F. Uhlmann, "Insights Pertaining to Patient Assessments of States Worse Than Death," Journal of Clinical Ethics 4 (1993): 33–41.

9. Back, H. Starks, C. Hsu, J. R. Gordon, A. Bharucha, and R. A. Pearlman, "Clinician-Patient Interactions about Requests for Physician-Assisted Suicide: A Patient and Family View," Archives of Internal Medicine 162 (2002): 1257–65.

READING 31

A Time to Die

The Place for Physician Assistance

Charles F. McKhann

My interest in physician-assisted dying and the earliest impetus to write this book grew out of my father's death in one of the best hospitals in the country. A physician himself, he died in 1988, at the age of eighty-nine, with widespread intra-abdominal cancer. In the process, he was kept alive for more than a month when his outlook was clearly hopeless. At one point he asked whether I thought he was on his deathbed. When I said that he probably was, he replied, "That's what I think, too, and I wish they would just let me go." He complained to us on several occasions that too much was being done, that he just wanted to be left alone. In truth, we don't know what he said directly to his doctors, but we passed his concerns on to them, along with our own. We were assured that they were "doing everything possible" for our father. Everything possible included palliative surgery, blood transfusions, and intravenous feedings. Even his two physician sons, one of whom is on the staff of the same hospital, were unable, or too timid and conflicted, to influence the decision making so that he might be *allowed* to die sooner.

It seems unfair that people who manage their own affairs successfully in life should be required to turn over so much of their death and dying to others. We direct the disposition of our belongings through wills and trusts, but except for the limited protection provided by "living wills," we have no such control over the conditions of our death, and the physicians who have this responsibility may be deaf to our entreaties and those of our families. Behind them stand tradition, a conservative profession, and the law. Regardless of a physician's personal compassion, helping people to die is risky in today's atmosphere. This atmosphere is changing, though, and one can already distinguish between the conservative views of medical societies—representing "the profession"—and those of practicing physicians who are concerned about the suffering of their patients. Many physicians have helped people to die, and many more would be willing to do so if it were legal.

Given a choice, most of us would like to live to old age, satisfied that we have accomplished what we could, and then to die peacefully, perhaps with enough warning to say our good-byes, but without undue suffering from prolonged illness. Old age, however, can be marred by severe disability, and death may not be kind or peaceful. Acute infections that killed swiftly and relatively painlessly fifty years ago have been replaced by organ failures and chronic diseases that take their victims slowly and sometimes very painfully. Through the passing of friends and relatives we have learned that there are good deaths and bad deaths, even horrible deaths.

What term then is best applied to the person who wishes to die and asks for help? The death is intended and expected, both by the person dying and by the person providing the help. Both also anticipate that the death will be accomplished in a gentle and humane fashion, not a violent one. The assistance is a compassionate response to an individual's considered wishes. The term that I will use is physician-assisted dying, rather than assisted suicide. *Dying* is a much more appropriate term than *suicide* for people who are already terminally ill. More accurately, assisted dying includes both assisted suicide, in which the patient must bring about his own death with materials provided by a physician, and euthanasia, in which the physician directly causes the death. Morally, they are quite similar activities, and eventually I think that both should be made legal. But there are also important differences, and in the immediate future the public, the medical profession, and our legal processes will be looking only at assisted suicide. Any consideration of assisted dying, however, requires acceptance of the concept of rational suicide—namely, that there are circumstances when death is clearly preferable to continued suffering. The question of whether suicide is ever morally acceptable is really the heart of the controversy.

The current movement for assisted dying began in direct response to the requests of individual patients and has expanded to the level of public demand. Society's concern about unnecessary suffering at the end of life is reflected in recent polls, which show that about 65 percent of people in the United States favor legislation that would permit physician-assisted dying. Many feel that their own needs should be placed above those of their physicians or the medical profession. They feel that even the promise of help would increase their confidence in their physicians and would allow them to better enjoy their remaining days.

Public concerns about assisted dying include the roles of various financial interests in prolonging or shortening life, the possibility of medical error, the potential for abuse, and the chance of a misstep onto a "slippery slope" leading to irreversible moral decay of our society. Abuse could come at the hands of family, custodians, or even physicians. The slippery slope could be a devaluation of life by individual physicians, the medical profession, even society as a whole. As a result, euthanasia could be legally extended beyond the limits of voluntary subjects to incompetents, and then still further to a spectrum of disadvantaged people: the poor, racial minorities, and the handicapped. These important concerns, which have become the cornerstones of most of the opposition to assisted dying, must be analyzed, understood, and addressed.

The differing views on assisted dying must eventually be assimilated into constructive laws that meet the needs of those immediately affected while protecting all others.

Oregon was the first state to pass a law permitting physician-assisted dying. For such laws to succeed they must include appropriate safeguards, but not so many restrictions as to make them unworkable. Several models are evolving through legislative attempts in different states. Some will succeed.

A rational decision or action is one that is well thought out by a competent individual, for reasons that are understood and can be explained to others. To wish to die in order to be spared unendurable pain from illness is seen by many as perfectly rational.

Quality of life to the last and control over the circumstances of dying are issues that touch everyone, and assisted dying will become legal and accepted when the public wants it to be. It is essential that the issues be understood, recognizing that today's needs are not necessarily met by yesterday's laws, that reason can counteract dogma, and that hypothetical fears of tomorrow's abuse and the slippery slope can be tested against today's reality in our own society. Assisted dying is destined gradually to be accepted as an end-of-life option. The trend will begin with assisted suicide in a few states, then spread to other states as voters and legislators see that it is desirable and socially safe. After it becomes acceptable in many states, the courts may step back in to provide similar protection for people residing in more conservative parts of the country. Eventually, assisted dying should be extended to include euthanasia for some people, and the range of underlying disorders should be extended to include neurological diseases that entail severe suffering but are not necessarily fatal, dementia, and even severe debility from old age. Assisted dying must be an option that can be requested by those who have lived with dignity and are determined to die the same way.

READING 32

Last Rights

Aunty's Story

Sue Woodman

INTRODUCTION

I received a call summoning me back to London, where I grew up, to the hospital bedside of my 90-year-old aunt. Aunty, as she is known to family and friends alike, had fractured her leg during a fall, and the physicians decided they would have to operate to reset the bone. When I

SOURCE: From *Last Rights: The Struggle Over the Right to Die* by Sue Woodman. Copyright © 1998 by Sue Woodman. Reprinted by permission of Basic Books, a member of the Perseus Books Group.

arrived at the hospital the following morning, the physician told me that Aunty had survived the operation itself, but there was a problem. Perhaps related to her fracture, perhaps not, she had developed a blood clot in the broken leg, and no blood was reaching her toes or lower calf. The leg would have to be amputated some way above the knee. If not, he said, Aunty would die of gangrene, probably within a week.

And now, at age 90, she had come on another horror. At one point, she wondered aloud what evil she had ever done to deserve such a cruel fate.

I asked her whether she was afraid to die. "No, no, I'm not afraid," she answered emphatically. "I would love it—just to go to sleep, to be finished. . . ." I tried to explain that this would be more or less possible: She could refuse the amputation, and allow her poor, battered, worn-out old body to die. The nursing staff promised they would make the process as easy and painless as possible; the morphine was already on order. Out of my love for her, I secretly prayed that she would choose this course of action.

It took Aunty a year and a half to die after her amputation. It was a heartbreaking time during which she lost not only her physical independence and the ability to do the things she loved—traveling, going to parks, and cooking meals for her family—but ultimately also her will to live.

From the time she was admitted to hospital after her fall, she was never able to return to her home again; she was moved from the hospital to a long-term rehabilitation unit, and after 7 months there, to an old-age home. Although she worked hard at the rehab hospital to learn to use a prosthetic leg, she didn't have enough upper-body strength to manage it. Confined to a wheelchair, she could no longer look after herself and felt demeaned and appalled every time she had to summon a nurse to take her to the bathroom.

So Aunty would sit gazing out of her window for days on end, her mood alternating between blunt depression and a frantic, lashing rage about how wretched her life had become. By Christmastime, she was telling everyone who would listen that she wanted to die. She begged her nurses and her physician to help her. Of course, they said they couldn't.

She told me that had she known what her life would be like after the amputation, she would never have gone ahead with it. But I wonder: When she was faced with that decision, she was not ready to die; she couldn't imagine how diminished her life would become—what it would mean when stripped of all of the activities she prized. How difficult it must be for anyone to choose death when, up until that point, they have only experienced life under normal circumstances. After all, the survival instinct is powerful indeed: I saw for myself how tenaciously Aunty clung to hope, even at age 90; the hope of walking again, and when that failed, the hope that she could use an electric wheelchair to gain some kind of freedom. She didn't give up hope for a very long time.

When she did, she was ready: ready to die, prepared and unafraid. And by being with her during those last days, I saw how sometimes, dying is not tragic and terrible, but simply an inevitable end, a relief and a release. I had never experienced a death like that before: a quiet, natural culmination of a long life fully lived.

Neither for Love nor Money

Why Doctors Must Not Kill

Leon Kass

CONTEMPORARY ETHICAL APPROACHES

The question about physicians killing is a special case of—but not thereby identical to—this general question: May or ought one kill people who ask to be killed? Among those who answer this general question in the affirmative, two reasons are usually given. Because these reasons also reflect the two leading approaches to medical ethics today, they are especially worth noting. First is the reason of *freedom* or *autonomy*. Each person has a right to control his or her body and his or her life, including the end of it; some go so far as to assert a right to die, a strange claim in a liberal society, founded on the need to secure and defend the unalienable right to life. But strange or not, for patients with waning powers too weak to oppose potent life-prolonging technologies wielded by aggressive physicians, the claim based on choice, autonomy, and self-determination is certainly understandable. On this view, physicians (or others) are bound to acquiesce in demands not only for termination of treatment but also for intentional killing through poison, because the right to choose—freedom—must be respected, even more than life itself, and even when the physician would never recommend or concur in the choices made. When persons exercise their right to choose against their continuance as embodied beings, doctors must not only cease their ministrations to the body; as keepers of the vials of life and death, they are also morally bound actively to dispatch the embodied person, out of deference to the autonomous personal choice that is, in this view, most emphatically the patient to be served.

The second reason for killing the patient who asks for death has little to do with choice. Instead, death is to be directly and swiftly given because the patient's life is deemed no longer worth living, according to some substantive or "objective" measure. Unusually great pain or a terminal condition or an irreversible coma or advanced senility or extreme degradation is the disqualifying quality of life that pleads—choice or no choice—for merciful termination. Choice may enter indirectly to confirm the judgment: If the patient does not speak up, the doctor (or the relatives or some other proxy) may be asked to affirm that he would not himself choose—or that his patient, were he *able* to choose, *would* not choose—to remain alive with one or more of these stigmata. It is not his autonomy but rather the miserable and pitiable condition of his body or mind that justifies doing the patient in. Absent such substantial degradations, requests for assisted death would not be honored. Here the body itself offends and must be plucked out, from compassion or mercy, to be sure. Not the autonomous will of the patient, but the doctor's benevolent and compassionate love for suffering humanity justifies the humane act of mercy killing.

SOURCE: "Neither for Love nor Money: Why Doctors Must Not Kill" by Leon Kass in *The Public Interest* (Number 94, Winter 1989), pp. 26–37, 42–45. Copyright © 1989 by The Public Interest. Reprinted by permission of The Public Interest and the author.

As I have indicated, these two reasons advanced to justify the killing of patients correspond to the two approaches to medical ethics most prominent in the literature today: the school of autonomy and the school of general benevolence and compassion (or love). Despite their differences, they are united in their opposition to the belief that medicine is intrinsically a moral profession, with its own immanent principles and standards of conduct that set limits on what physicians may properly do. Each seeks to remedy the ethical defect of a profession seen to be in itself amoral, technically competent but morally neutral.

For the first ethical school, morally neutral technique is morally used only when it is used according to the wishes of the patient as client or consumer. The implicit (and sometimes explicit) model of the doctor-patient relationship is one of *contract:* The physician—a highly competent hired syringe, as it were—sells his services on demand, restrained only by the law (though he is free to refuse his services if the patient is unwilling or unable to meet his fee). Here's the deal: for the patient, autonomy and service; for the doctor, money, graced by the pleasure of giving the patient what he wants. If a patient wants to fix her nose or change his gender, determine the sex of unborn children, or take euphoriant drugs just for kicks, the physician can and will go to work—provided that the price is right and that the contract is explicit about what happens if the customer isn't satisfied.[2]

For the second ethical school, morally neutral technique is morally used only when it is used under the guidance of general benevolence or loving charity. Not the will of the patient, but the humane and compassionate motive of the physician—not as physician but as *human being*—makes the doctor's actions ethical. Here, too, there can be strange requests and stranger deeds, but if they are done from love, nothing can be wrong—again, providing the law is silent. All acts—including killing the patient—done lovingly are licit, even praiseworthy. Good and humane intentions can sanctify any deed.

In my opinion, each of these approaches should be rejected as a basis for medical ethics. For one thing, neither can make sense of some specific duties and restraints long thought absolutely inviolate under the traditional medical ethic—e.g., the proscription against having sex with patients. Must we now say that sex with patients is permissible if the patient wants it and the price is right, or, alternatively, if the doctor is gentle and loving and has a good bedside manner? Or do we glimpse in this absolute prohibition a deeper understanding of the medical vocation, which the prohibition both embodies and protects? Indeed, as I will now try to show, using the taboo against doctors killing patients, the medical profession has its own intrinsic ethic, which a physician true to his calling will not violate, either for love or for money. . . .

ASSESSING THE CONSEQUENCES

Although the bulk of my argument will turn on my understanding of the special meaning of professing the art of healing, I begin with a more familiar mode of ethical analysis: assessing needs and benefits versus dangers and harms. To do this properly is a massive task. Here, I can do little more than raise a few of the relevant considerations. Still the best discussion of this topic is a now-classic essay by Yale Kamisar, written thirty years ago.[4] Kamisar makes vivid the difficulties in assuring that the choice for death will be *freely* made and adequately *informed*, the problems of physician error and abuse, the troubles for human relationships within families and between doctors and patients, the difficulty of preserving the boundary between voluntary and involuntary euthanasia, and the risks to the whole social order from weakening the absolute prohibition against taking innocent life. These considerations are, in my view, alone sufficient to rebut any attempt to weaken the taboo against medical killing; their relative importance for determining public policy far exceeds their relative importance in this essay. But here they serve also to point us to more profound reasons why doctors must not kill.

There is no question that fortune deals many people a very bad hand, not least at the end of life. All of us, I am sure, know or have known individuals whose last weeks, months, or even years were racked with pain and discomfort, degraded by dependency or loss of self-control, isolation or insensibility, or who lived in such reduced humanity that it cast a deep shadow over their entire lives, especially as remembered by the survivors. All who love them would wish to spare them such an end, and there is no doubt that an earlier death could do it. Against such a clear benefit, attested to by many a poignant and heartrending true story, it is difficult to argue, especially when the arguments are necessarily general and seemingly abstract. Still, in the aggregate, the adverse consequences—including real suffering—of being governed solely by mercy and compassion may far outweigh the aggregate benefits of relieving agonal or terminal distress.

The "Need" for Mercy Killing

The first difficulty emerges when we try to gauge the so-called "need" or demand for medically assisted killing. This question, to be sure, is in part empirical. But evidence can be gathered only if the relevant categories of "euthanizable" people are clearly defined. Such definition is notoriously hard to accomplish— and it is not always honestly attempted. On careful inspection, we discover that if the category is precisely defined, the need for mercy killing seems greatly exaggerated, and if the category is loosely defined, the poisoners will be working overtime.

The category always mentioned first to justify mercy killing is the group of persons suffering from incurable and fatal illnesses, with intractable pain and with little time left to live but still fully aware, who freely request a release from their distress—e.g., people rapidly dying

from disseminated cancer with bony metastases, unresponsive to chemotherapy. But as experts in pain control tell us, the number of such people with truly intractable and untreatable pain is in fact rather low. Adequate analgesia is apparently possible in the vast majority of cases, provided that the physician and patient are willing to use strong enough medicines in adequate doses and with proper timing.[5]

But, it will be pointed out, full analgesia induces drowsiness and blunts or distorts awareness. How can that be a desired outcome of treatment? Fair enough. But then the rationale for requesting death begins to shift from relieving experienced suffering to ending a life no longer valued by its bearer or, let us be frank, by the onlookers. If this becomes a sufficient basis to warrant mercy killing, now the category of euthanizable people cannot be limited to individuals with incurable or fatal painful illnesses with little time to live. Now persons in all sorts of greatly reduced and degraded conditions—from persistent vegetative state to quadriplegia, from severe depression to the condition that now most horrifies, Alzheimer's disease—might have equal claim to have their suffering mercifully halted. The trouble, of course, is that most of these people can no longer request for themselves the dose of poison. Moreover, it will be difficult—if not impossible—to develop the requisite calculus of degradation or to define the threshold necessary for ending life.

From Voluntary to Involuntary

Since it is so hard to describe precisely and "objectively" what kind and degree of pain, suffering, or bodily or mental impairment, and what degree of incurability or length of anticipated remaining life, could justify mercy killing, advocates repair (at least for the time being) to the principle of volition: The request for assistance in death is to be honored because it is freely made by the one whose life it is, and who, for one reason or another, cannot commit suicide alone. But this too is fraught with difficulty: How free or informed is a choice made under debilitated conditions? Can consent long in advance be sufficiently informed about all the particular circumstances that it is meant prospectively to cover? And, in any case, are not such choices easily and subtly manipulated, especially in the vulnerable? Kamisar is very perceptive on this subject:

> Is this the kind of choice, assuming that it can be made in a fixed and rational manner, that we want to offer a gravely ill person? Will we not sweep up, in the process, some who are not really tired of life, but think others are tired of them; some who do not really want to die, but who feel they should not live on, because to do so when there looms the legal alternative of euthanasia is to do a selfish or a cowardly act? Will not some feel an obligation to have themselves "eliminated" in order that funds allocated for their terminal care might be better used by their families or, financial worries aside, in order to relieve their families of the emotional strain involved?

Even were these problems soluble, the insistence on voluntariness as the justifying principle cannot be sustained. The enactment of a law legalizing mercy killing on voluntary request will certainly be challenged in the courts under the equal-protection clause of the Fourteenth Amendment. The law, after all, will not legalize assistance to suicides in general, but only mercy killing. The change will almost certainly occur not as an exception to the criminal law proscribing homicide but as a new "treatment option," as part of a right to "A Humane and Dignified Death."[6] Why, it will be argued, should the comatose or the demented be denied such a right or such a "treatment," just because they cannot claim it for themselves? This line of reasoning has already led courts to allow substituted judgement and proxy consent in termination-of-treatment cases since *Quinlan*, the case that, Kamisar rightly says, first "badly smudged, if it did not erase, the distinction between the right to choose one's own death and

the right to choose someone else's." When proxies give their consent, they will do so on the basis not of autonomy but of a substantive judgment—namely, that for these or those reasons, the life in question is not worth living. Precisely because most of the cases that are candidates for mercy killing are of this sort, the line between voluntary and involuntary euthanasia cannot hold, and will be effaced by the intermediate case of the mentally impaired or comatose who are declared no longer willing to live because someone else wills that result for them. In fact, the more honest advocates of euthanasia openly admit that it is these nonvoluntary cases that they especially hope to dispatch, and that their plea for *voluntary* euthanasia is just a first step. It is easy to see the trains of abuses that are likely to follow the most innocent cases, especially because the innocent cases cannot be precisely and neatly separated from the rest.

DAMAGING THE DOCTOR-PATIENT RELATIONSHIP

Abuses and conflicts aside, legalized mercy killing by doctors will almost certainly damage the doctor-patient relationship. The patient's trust in the doctor's wholehearted devotion to the patient's best interests will be hard to sustain once doctors are licensed to kill. Imagine the scene: You are old, poor, in failing health, and alone in the world; you are brought to the city hospital with fractured ribs and pneumonia. The nurse or intern enters late at night with a syringe full of yellow stuff for your intravenous drip. How soundly will you sleep? It will not matter that your doctor has never yet put anyone to death; that he is legally entitled to do so—even if only in some well-circumscribed areas—will make a world of difference.

And it will make a world of psychic difference too for conscientious physicians. How easily will they be able to care wholeheartedly for patients when it is always possible to think of killing them as a "therapeutic option"? Shall it be penicillin and a respirator one more time, or perhaps just an overdose of morphine this time? Physicians get tired of treating patients who are hard to cure, who resist their best efforts, who are on their way down—"gorks," "gomers," and "vegetables" are only some of the less than affectionate names they receive from the house officers. Won't it be tempting to think that death is the best treatment for the little old lady "dumped" again on the emergency room by the nearby nursing home?

Even the most humane and conscientious physician psychologically needs protection against himself and his weaknesses, if he is to care fully for those who entrust themselves to him. A physician friend who worked many years in a hospice caring for dying patients explained it to me most convincingly: "Only because I knew that I could not and would not kill my patients was I able to enter most fully and intimately into caring for them as they lay dying." The psychological burden of the license to kill (not to speak of the brutalization of the physician-killers) could very well be an intolerably high price to pay for physician-assisted euthanasia, especially if it also leads to greater remoteness, aloofness, and indifference as defenses against the guilt associated with harming those we care for.

The point, however, is not merely psychological and consequentialist: It is also moral and essential. My friend's horror at the thought that he might be tempted to kill his patients, were he not enjoined from doing so, embodies a deep understanding of the medical ethic and its intrinsic limits. We move from assessing the consequences to looking at medicine itself. . . .

THE ESSENCE OF MEDICINE

Healing is the central core of medicine: to heal, to make whole, is the doctor's primary business. The sick, the ill, the unwell present themselves to the physician in the hope that he can help them become well—or, rather, as well as they can become, some degree of wellness being possible always, this side of death. The physician shares that goal; his training has been devoted to making it possible for him to serve it. Despite enormous changes in medical technique and institutional practice, despite enormous changes in nosology and therapeutics, the center of medicine has not changed: it is true today as it was in the days of Hippocrates that the ill desire to be whole; that wholeness means a certain well-working of the enlivened body and its unimpaired powers to sense, think, feel, desire, move, and maintain itself; and that the relationship between the healer and the ill is constituted, essentially even if only tacitly, around the desire of both to promote the wholeness of the one who is ailing.

Can wholeness and healing ever be compatible with intentionally killing the patient? Can one benefit the patient as a whole by making him dead? There is, of course, a logical difficulty; how can any good exist for a being that is not? "Better off dead" is logical nonsense—unless, of course, death is not death at all but instead a gateway to a new and better life beyond. But the error is more than logical: to intend and to act for someone's good requires his continued existence to receive the benefit.

Certain attempts to benefit may in fact turn out, unintentionally, to be lethal. Giving adequate morphine to relieve the pain of the living presupposes that the living still live to be relieved. This must be the starting point in discussing all medical benefits: no benefit without a beneficiary.

To say it plainly, to bring nothingness is incompatible with serving wholeness: one cannot heal—or comfort—by making nil. The healer cannot annihilate if he is truly to heal. The boundary condition, "No deadly drugs," flows directly from the center, "Make whole."

But there is a difficulty. The central goal of medicine—health—is, in each case, a perishable good: Inevitably, patients get irreversibly sick, patients degenerate, patients die. Unlike—at least on first glance—teaching or rearing the young, healing the sick is *in principle* a project that must at some point fail. And here is where all the trouble begins: How does one deal with "medical failure"? What does one seek when restoration of wholeness—or "much" wholeness—is by and large out of the question? . . .

Although I am mindful of the dangers and aware of the impossibility of writing explicit rules for ceasing treatment—hence the need for prudence—considerations of the individual's health, activity, and state of mind must enter into decisions of *whether* and *how vigorously* to treat if the decision is indeed to be for the patient's good. Ceasing treatment and allowing death to occur when (and if) it will seem to be quite compatible with the respect that life commands for itself.

Ceasing medical intervention, allowing nature to take its course, differs fundamentally from mercy killing. For one thing, death does not necessarily follow the discontinuance of treatment; Karen Ann Quinlan lived [nearly] ten years after the court allowed the "life-sustaining" respirator to be removed. Not the physician, but the underlying fatal illness becomes the true cause of death. More important morally, in ceasing treatment the physician need not *intend* the death of the patient, even when the death follows as a result of his omission. His intention should be to avoid useless and degrading medical *additions* to the already sad end of a life. In contrast, in active, direct mercy killing the physician must, necessarily and indubitably, intend *primarily* that the patient be made dead. And he must knowingly and indubitably cast himself in the role of the agent of death. . . .

The enormous successes of medicine these past fifty years have made both doctors and laymen less prepared than ever to accept the fact of finitude. Doctors behave, not without some reason, as if they have godlike powers to revive the moribund; laymen expect an endless string of medical miracles. It is against this background that terminal illness or incurable disease appears as medical failure, an affront to medical pride. Physicians today are not likely

to be agents of encouragement once their technique begins to fail.

It is, of course, partly for these reasons that doctors will be pressed to kill—and many of them will, alas, be willing. Having adopted a largely technical approach to healing, having medicalized so much of the end of life, doctors are being asked—often with thinly veiled anger—to provide a final technical solution for the evil of human finitude and for their own technical failure: If you cannot cure me, kill me. The last gasp of autonomy or cry for dignity is asserted against a medicalization and institutionalization of the end of life that robs the old and the incurable of most of their autonomy and dignity: Intubated and electrified, with bizarre mechanical companions, helpless and regimented, once proud and independent people find themselves cast in the roles of passive, obedient, highly disciplined children. People who care for autonomy and dignity should try to reverse this dehumanization of the last stages of life, instead of giving dehumanization its final triumph by welcoming the desperate goodbye-to-all-that contained in one final plea for poison.

NOTES

[Only the notes that are included in the excerpted material appear here.]

2. Of course, any physician with personal scruples against one or another of these practices may "write" the relevant exclusions into the service contract he offers his customers.

4. Yale Kamisar, "Some Non-Religious Views Against Proposed 'Mercy-Killing' Legislation," Minnesota Law Review 42: 969–1042 (May, 1958). Reprinted, with a new preface by Professor Kamisar, in "The Slide Toward Mercy Killing," Child and Family Reprint Booklet Series, 1987.

5. The inexplicable failure of many physicians to provide the proper—and available—relief of pain is surely part of the reason why some people now insist that physicians (instead) should give them death.

6. This was the title of the recently proposed California voter initiative that barely failed to gather enough signatures to appear on the November 1988 ballot. It will almost certainly be back.

FOCUS ON THE FUTURE

Neighborhood Suicide Clinics?

Imagine that it's the year 2020. Assume that you're having a conversation with your cousin Michael, who has spent the past three years in a remote rural village in Central Africa, working on village agriculture projects for the Agency for International Development. He's just gotten back to the United States, and on the way from the airport, he asks you about something new he's noticed—signs for local "Suicide Clinics."

"What are these places, anyway?" asks Michael, in shock.

"I guess you've been a little out of touch," you reply. "These clinics offer assisted suicide or euthanasia on demand. They provide a public service."

Michael is astonished. "How is that possible? When did they get started?"

"You remember that a long time ago, the courts ruled that everyone has a constitutional right to assisted suicide? Later rulings expanded on that idea and opened up federal funding for life termination. Soon entrepreneurs moved in to fill the need. There are over 50,000 suicides each year now. It's a growing market, and funded by Medicaid."

"I thought that sort of thing happened in hospitals, in intensive care units?"

"Oh, it still does, lots of times, and in nursing homes, too. Many of those people are better off dead. But what about people who aren't institutionalized? The idea is that everyone should have a right to death with dignity and have access to professional suicide services. People who are unhappy nowadays find these clinics a godsend. You know, if you don't have good alternatives, ending your life can be the best thing. Anyway, everybody has a right to decide for themselves."

"Why doesn't the regular health care system handle this new service?"

"Well, some doctors do it, and insurance companies naturally pay for it because it keeps their costs down. But the medical establishment has never been keen on euthanasia: It looks bad for business. So a specialist group has taken up end-of-life practice. Remember back in the 1990s when Dr. Jack Kevorkian was hooking people up to his 'suicide machine'? He's the one who had the original idea: specialists in end-of-life practice."

"There are enough people who want to kill themselves to justify a new medical specialty like that?"

"Oh, you'd be surprised, especially when you include executions. Thousands of people every year are being executed by lethal injection since the new Omnibus Capital Punishment Law went into effect. Most physicians didn't want to have anything to do with executions or killing patients, so a specialty group started up. Those doctors got in early as investors with the Thanatos Corporation, which franchises most of the suicide clinics. Those early investors made a killing, so to speak."

"But what about people who really shouldn't be ending their lives?"

"Well, it's 'buyer beware' in the marketplace, you know. Nobody forces anyone to go. Anyway, people would rather make these decisions instead of letting the government decide for them."

"What about people who are mentally ill?"

"Oh, the clinics don't discriminate against them. Discrimination against the mentally ill is against the law. There was a big court case on that issue two years ago. If you're mentally ill, you have the same right to use suicide services as anyone else. Well, welcome back to America, Michael."

Questions for Writing, Reflection, and Debate

1. In the early 1980s, when Sidney Hook was close to death, he asked the doctor to discontinue life support, but the doctor refused. Hook recovered and went on to publish his autobiography as well as other writings before he finally died in 1989. Are these examples of Hook's later productivity enough to justify the doctor's refusal to honor Hook's request? If not, is there any other reason to justify the doctor's refusal?

2. A main point in Sidney Hook's argument for voluntary euthanasia is that he dreaded imposing a burden on his family. Is this a convincing reason for encouraging infirm older people to end their lives? Imagine that you are Sidney Hook's son or daughter, and then proceed to write a detailed letter to your "father" explaining why you agree or disagree with Hook's fear about becoming a burden on the family.

3. Are Dr. McKhann's reasons for a physician to help a terminally ill patient die convincing? Would exactly the same reasons apply if the patient were not terminally ill, but instead were suffering from a chronic condition—for example, the aftereffects of a stroke—that diminished the quality of life?

4. Advocates for physician-assisted suicide generally believe that if a person is depressed and not rational, others should not help to end that person's life. Is it possible for someone to be deeply gloomy about life yet still be rational and therefore rationally decide to commit suicide? Imagine that you have just received a letter from an older friend expressing such gloomy thoughts in favor of suicide. Write a detailed response giving your reasons for agreeing or disagreeing with the conclusions reached.

5. Is there really a difference between a doctor going along with a request to terminate treatment that will result in a patient's death and a doctor intentionally giving a deadly drug? What about the case of withdrawing artificial nutrition or hydration from a patient?

6. Leon Kass offers a "slippery slope" argument against allowing doctors to engage in mercy killing. That is, Kass believes that, once we set a precedent and get used to the idea of deliberate killing, we will have no way to stop the practice from expanding. Is Kass's fear of the danger a realistic one, or is it exaggerated? What steps could be taken to avoid the dangers?

7. Look carefully at the different decisions that Woodman's older "Aunty" makes at various times over the course of her last illness. She changes her mind and sometimes says contradictory things. In your view, does this fact suggest that advance directives aren't useful in cases like hers? How big a problem would this be for people in general?

8. Assume that your employer, the chief administrator of a nursing home, has asked you to draft a statement of policy expressing what the nursing home should do when a resident says he or she no longer wants to go on living. In developing your policy statement, be sure to give guidance to doctors, nurses, and social workers on how they should act when they come in contact with such a situation.

9. Visit the website for the Euthanasia Research & Guidance Organization at http://www.assistedsuicide.org/. If you are worried about the suicide rate among older people, can you find anything to criticize in the material available on this site?

Suggested Readings

Annas, George J., *The Rights of Patients: The Authoritative ACLU Guide to the Rights of Patients,* Totowa: New York University Press, 2004.

Byock, Ira, *Dying Well*, New York: Riverhead, 1998.

Lynn, Joanne, *Handbook for Mortals: Guidance for People Facing Serious Illness,* New York: Oxford University Press, 2011.

Meier, Diane E., Isaacs, Stephen L., and Hughes, Robert (Eds.), *Palliative Care: Transforming the Care of Serious Illness,* Jossey-Bass, 2010.

Moody, Harry R., *Ethics in an Aging Society*, Baltimore, MD: Johns Hopkins University Press, 1992.

Nuland, Sherwin B., *How We Die*: *Reflections on Life's Final Chapter*, New York: Vintage, 1995.

Student Study Site

Visit the Student Study Site at **www.sagepub.com/moody7e** for these additional learning tools:

- Flashcards
- Web quizzes
- Chapter outlines

- SAGE journal articles
- Web resources
- Video and audio resources

Social and Economic Outlook for an Aging Society

In centuries past, when few people survived to old age, it wasn't essential for society to give much thought to planning for later life. Traditional societies throughout history had only a small proportion of people who were old. But today, larger numbers of people are living longer, and population aging raises new questions about roles and responsibilities in an aging society. Because government has assumed a role in providing older people with income, health care, and social services, more resources are now going to the older generation than in the past. That fact prompts unavoidable questions. How has government expenditure affected the condition of older people? What about resources needed by other age groups? What has been the effect of age-based political advocacy in shaping government response to aging in the United States?

Social Security and Medicare, along with private pensions, have decisively improved the economic well-being of older Americans. Nonetheless, we cannot make generalizations that all older people today are comfortably retired, traveling, and playing golf while the rest of society supports them. Generalizations overlook dramatic differences within the older population. Those called "old" are divided by social class, ethnicity, and gender. How should we take into account those differences? How can government best target resources to the most vulnerable or least advantaged groups among the older adult population?

With increased life expectancy, more people spend a larger part of their lives in old age, and later-life income therefore becomes a major consideration. How will individuals plan and provide for later life? Within the next few decades, as baby boomers enter later life, we will see a rapid aging of the American population. To cope with this huge demographic shift, we have already begun to rethink some major institutions of our society. For instance, the age of eligibility for retirement under Social Security will gradually rise over the next two decades.

As a rapidly aging society, we face many new questions. There are still disputes about whether Social Security fairly treats subgroups among the older population. But there is no dispute that Social Security, along with private pensions, has given older people more freedom to leave the workforce at earlier ages. Will this positive outlook for the nation's elders endure into the 21st century and prove sustainable in years to come (Generations Policy Initiative, 2004)?

With larger numbers of older people eligible for Social Security and Medicare, these programs already claim a major share of the federal budget. Questions have therefore been raised about whether age or need should be the basis for these entitlement programs. It is not surprising that these questions provoke intense controversy.

Other issues of an aging society are also claiming our attention, such as the meaning of work and leisure in the later years. Why do we as a society encourage older people to withdraw from productive roles after a certain age? Though retirement is still a social and individual expectation in the United States in recent decades, there has been an increase in midlife and older adults participating in the labor force, either because they continue working past "retirement age" or because they have entered new "encore careers." In 2009, 17.2% of adults 65+, or 6.5 million, were participating in the labor force, a 20% increase since 2002 (Administration on Aging, 2010). In terms of encore careers, according to a survey conducted by MetLife Foundation and Civic Ventures (2008), anywhere from 5.3 million to 8.4 million adults aged 40 to 70 are working in an encore career. It remains to be seen if this trend will continue as the baby boomers reach their later years, but it certainly raises some interesting questions about older adulthood in the 21st century.

These issues about the diversity of the aging population, the role of government, and the future of work and retirement all provoke vigorous debate. An understanding of basic concepts of aging in society can help us ask the right questions and be familiar with the key facts involved in the debates. Understanding the facts and clarifying our values can be vital in shaping what the aging society of the future will be like for all of us.

THE VARIETIES OF AGING EXPERIENCE

People often speak about aging as if it were a universal human experience, but describing broad trends and discussing older people as a group may obscure marked differences between subgroups and individuals. Stereotypes suggest that age is a great leveler; as a result, older adults are commonly thought to be much alike. In fact, just the opposite is true. As people grow older, they tend to become less and less similar. Heterogeneity increases with age, and this tendency is in keeping with accumulated advantage or disadvantage from earlier in life.

We are all, as members of North American society, experiencing population aging. As we move through the adult life course, we also experience small, telltale signs of aging, such as gray hairs and reduced visual and auditory capacities. Those who have reached age 65 will likely begin to experience what it means to grow old in a more pronounced way. With advancing age, there may be gains, such as increased leisure in retirement, but also losses, such as the increasing threat of chronic illness or the death of a spouse or friend. The balance of gains and losses is difficult to judge from the outside. For instance, two individuals with the same objective indicators for health status may interpret and even experience their own health in completely different ways, as we know from surveys that ask for self-reporting on subjective health status. Aging remains very much an individual affair, manifesting itself in subtle ways, so that we may fail to recognize systematic ways in which society gives structure to the variations we see.

In a sense, old age in contemporary society is well described as a *roleless role,* a status with no clearly defined purpose or rules of behavior (Blau, 1981; Rosow, 1967). Some individuals find that freedom exhilarating, whereas others interpret it as a matter of role loss, as social scientists would label it. In fact, role losses are real. Leaving a valued position in the workforce, losing parental authority as children leave home, and experiencing bereavement with the death of family or friends—these losses can create problems for those who are unable to establish new sources of meaning and satisfaction.

A big problem in reestablishing identity in the face of change and loss comes from attitudes about aging, attitudes held by other people and often shared by older people. Telltale signs of aging after middle age need not signify decline, but older people are often the victims of *ageism*, or stereotyping and discrimination based on age (Butler, 1969; T. Nelson, 2002). The typical ageist stereotypes suggest that people over age 65 are all sick and frail, or worse: impoverished, impotent, senile, and unhappy (Palmore, 1999).

One important step toward overcoming stereotyping is to recognize the varieties of aging experience and to see how these variations are tied to structures of social class, gender, and ethnicity. Another step is to see today's elders in the context of historical forces. When looking at an old face, for example, try to imagine the life experience that led to conditions encountered by that person today. Last, all generalizations, including understanding conveyed by gerontology, should be balanced by acknowledgment of individual differences. By combining understanding of social structure with history and individuality, we can better appreciate the experiences of aging in a diverse society.

Objectively, it may not make much sense to speak of the older adults as a homogeneous group. It may make more sense to distinguish between at least two broad groups in the 65+ population: the "ill-derly" and the "well-derly." The first group of older people tends to be poor and subject to chronic illness. The second group tends to be well off, both physically and financially (Cook & Kramek, 1986). These two groups have been described as the "two worlds of aging" (Crystal, 1982, 1986). But what are the implications of categorizing older adults in such a way?

Issues of social class, race and ethnicity, and gender create further distinctions. Old men and old women in general are treated differently in our society, as are older persons of various races and ethnic backgrounds (Ruiz, 1995). Some older individuals are more socioeconomically disadvantaged than others, and some may be both privileged and disadvantaged at the same time. A good example is the old widow played by Jessica Tandy in the film *Driving Miss Daisy,* who continued to enjoy advantages, in comparison with her African American chauffeur, by virtue of her race and wealth. At the same time, she faced a certain measure of discrimination as a Jew and as a woman. Finally, as Miss Daisy became very old and frail, she maintained the advantages of social class, but encountered other problems based on stereotypical images of old age. What we see in any concrete case, such as the story of Miss Daisy, is that age, gender, race, and social class interact in complex ways to define individual experiences (Dressel, 1988).

Moreover, socioeconomic disadvantage alone is not a good way to determine whether an older person is socially well off. Many disadvantaged older people are still able to cope, find meaning in life, and be productive. For example, aged African Americans often draw self-esteem and a sense of well-being from their involvement in the Black church, which historically has provided a critical informal support system (Barer & Johnson, 2003). Many women today experience old age as a time of independence and self-affirmation, a time when they can come into their own at last (Martz, 1987; Thone, 1992). A full account of the aging experience must acknowledge the adaptive strength of many disadvantaged older people.

In years to come, the aged population will reflect the increasing diversity of U.S. society, and the disadvantages related to social class and gender will be important issues. Another will be our ability to equitably accommodate the millions of immigrants who came to this country in the 1990s, most of whom are Asian or Hispanic. As these immigrants age, they will make the older population far more diverse than it is today. Increasing ethnic diversity and a growing population of older adults will present a challenge for decades to come and will demand thoughtful attention to the aging experience for all members of our society (Rogers & Raymer, 1999).

Social Class

Social class is a key factor influencing the experience of old age in all societies. The concept of social class involves unequal shares of wealth, status, and power in a society. In adulthood, social class tends to be related to employment, but in old age, the impact of social class is largely a matter of accumulated advantage or disadvantage built up over a lifetime.

Four elements influence the class position that a person inhabits in old age: occupation, income, property, and education (Streib, 1985). It is easy to see how economic factors such as savings, a home, and a private pension can determine an individual's experience of later life. But less tangible factors can also be linked to social class; educational attainment and family support networks, for instance, can influence the resources available to older people. A good example is occupational status. A retired governor or judge may still command prestige even if his or her available income is reduced, and such a person might get better access to health care or other special treatment not available to less favored people who lack connections.

Yet old age does not simply reproduce the pattern of social class that holds in earlier years. Consider what happens when a person of high social class, such as a physician or a prominent writer, becomes ill or impoverished to the point where his or her living conditions are seriously degraded. Such downward mobility is not unusual among older people who outlive their economic resources, as when widowhood or serious illness causes a drastic depletion of assets. As a general rule, after the point of retirement, income tends to decline. Thus, the oldest-old (85+) are also likely to be the poorest-old.

It is a mistake, however, to equate old age with poverty and economic vulnerability. Until quite recently, the aged in most societies have formed a large segment of the poor. At the same time, however, the richest people have also been more likely to be old: The median age of the richest people in the United States is 65. Significantly, about half the millionaires in the United States are women, perhaps reflecting the role of inheritance and the fact that women typically outlive men. One clear conclusion is that social class has a complex influence on the experience of aging.

Race and Ethnicity

Today, the aged population in the United States is largely White, but by 2030 minority elders will comprise a significantly larger proportion, projected to be 23% of the population (Administration on Aging, 2010). According to 2009 census data, non-Hispanic Whites comprised approximately 80% of the U.S. population over age 65. The largest aged minorities were Blacks (8.3%), Hispanics (7%), and Asians/Pacific Islanders (3.4%). But the census estimates that by the year 2050, the older population in the United States will look different. Within 10 to 20 years, ethic minorities will be the new majority. Hispanics will be the largest minority, and Hispanics and Asians/Pacific Islanders will have the highest rates of growth numerically. In essence, the ethnic minority aged population is expected to more than double in proportion during the next half-century. These population changes will greatly affect services for the aging, such as those provided under the Older Americans Act (OAA), which were developed with a White, middle-class population in mind (Jacobson, 1982). Increasingly, service providers need to take into account differences in language and ethnic customs (Gelfand, 2003).

Any discussion of race and ethnicity as it applies to the aging population presents some definitional problems. For one thing, not every distinctive ethnic group is thought of as a minority group requiring special consideration; consider the Irish Americans of today, for example. Nor is a minority group necessarily limited to one definable ethnic or racial group; for example, among Hispanics, there are many nationality subgroups, all of which have distinctive customs and dialects. The term *minority group,* which emphasizes social disadvantage or discrimination, may therefore be more useful than the term *ethnic group* in a discussion of how different subgroups experience aging.

The *Columbia Documentary History of Race and Ethnicity in America* lists numerous ethnic groups that have been significant in U.S. history (Bayor, 2004). In this discussion, however, we focus on four groups that have been studied extensively because of their history of disadvantage in the United States.

African Americans

African Americans constitute the largest minority group among the aged, comprising 8.3% of all Americans over age 65. By 2025, it is estimated that 14% of the older population will be African American. In comparison with Whites, African Americans generally face a lower life expectancy at birth and in most decades of life. Older African Americans also tend to experience more functional impairment from chronic illness, yet they are far less likely than Whites to be admitted to nursing homes. Is this disparity attributable to the **crossover phenomenon**, to discrimination in long-term care facilities, or to some other factor, such as family caregiving patterns (Stoller & Gibson, 2000)? This last possibility points to a source of strength among many Black families, namely, extensive informal support networks (Dilworth-Anderson, 1992).

Although African Americans face many disadvantages (Jackson, 1988), all are not disadvantaged to the same degree. For instance, some are quite well off at retirement, especially those who have had college educations and professional careers. Out of every 10 older African Americans, 9 now receive Social Security, a proportion comparable to the rest of the older population. But 25% receive Supplemental Security Income, reflecting a higher poverty rate.

Hispanics

The Spanish-speaking older persons currently represent 7% of the older population of the United States, and their numbers are growing rapidly (Administration on Aging, 2010). According to the U.S. Census Bureau, the number of Latinos in the U.S. population aged 65+ is expected to reach 16% by the year 2050. In terms of religion, Hispanic or Latino older adults are overwhelmingly Catholic, but they show great diversity on other dimensions, such as immigrant status, poverty, and educational level. Generalizing can therefore be dangerous, although some tendencies seem clear. For instance, traditional Latino cultures tend to encourage respect for older persons, and women fulfill key caregiving roles (Coles, 1974).

The older Hispanic population is quite definitely a group at risk (Martin & Soldo, 1997). For example, findings from the "Latinos and Social Security" project sponsored by the UCLA Center for Policy Research on Aging indicate Latinos have a higher life expectancy at age 65 compared to all other racial/ethnic groups, but they have lower lifetime earnings and fewer years of contributing to Social Security, as well as below-average education levels (Torres-Gil, 2006). These factors, which reflect a complex combination of the impact of systems of social and economic inequality, culturally specific experiences, and, for some members of the Latino community, issues related to immigration, result in higher rates of poverty, lower health status, and higher incidence rates of cognitive impairment in later life compared to other racial/ethnic groups. Last, like African Americans, Hispanics are more likely to rely on informal supports than on formally organized services. This tendency may give rise to an assumption by service providers that Hispanics and other minorities simply "take care of their own" (Gratton, 1987).

Asian Americans

Like Hispanics, Asian Americans come from many different countries of origin and thus have many differences in languages and customs. The largest subgroups are those of Chinese (24%), Japanese (24%), Filipino (8%), and Korean (5%) origins. Older Asian Americans, like other minority group members in the United States, have faced discrimination over the course of their lives. In addition, older Asian American immigrants face difficulties in reconciling their cultural heritage with the American values adopted by their children and grandchildren (Cheung, 1989). In East Asian societies influenced by Confucian religion, the obligation to honor one's parents in old age is still taken seriously, despite modernization and industrialization (Kim, Kim, & Hurh, 1991; Palmore, 1975). American attitudes toward filial piety are likely to be different. Today's older Asian immigrants must therefore cope with the erosion of the status and traditional roles of older adults (Yee, 1992). As a result, they commonly experience family strains (Kao & Lam, 1997; Koh & Bell, 1987).

Native Americans

Native Americans, or American Indians, make up a relatively small proportion of the total aged population in the United States—in 2009, less than 1% (Administration on Aging, 2010). Members of this group have their origins in more than 500 distinct tribes in North America (Kunitz & Levy, 1991); therefore, it is difficult to generalize about them. It is common, however, for older Native Americans, especially those who remain on isolated reservations, to suffer from major economic disadvantages. These problems are sometimes partly offset by family support and a strong degree of social integration (John, 1995). In addition, tribal elders may take the role of cultural conservators, gaining respect for their memories of old ways (S. Johnson, 1994). This role not only enhances their prestige but also helps the next generation to achieve a better future (Weibel-Orlando, 1990).

Although today's minority older adults are better off than they once would have been, they still carry the accumulated disadvantages of a lifetime spent in a harsher social environment. For example, the civil rights movement began to produce school desegregation and equal employment opportunity only after the 1950s and 1960s. Thus, today's older Blacks carry with them the effects of prejudice, including poverty and poor health status, that limited their life chances many decades ago. In addition, immigrants of all ethnic groups were once encouraged to assimilate into the dominant culture and discard their old customs, even their first language.

Today, we have come to celebrate ethnic differences more readily, but people who are members of minority groups still carry with them the impacts of earlier discriminatory attitudes and policies. For instance, in the past, Native American children were often sent away from the reservations to boarding schools, where they were discouraged from maintaining traditional languages or tribal ways (Stoller & Gibson, 2000). The results included disruptive changes in the relationships between the generations.

Gender and Aging

The role of gender in aging deserves special consideration not least because so many more women survive into old age than do men (Coyle, 1997). Women who reach age 65 can expect to live nearly 20 more years, whereas men at age 65 have only about 17 years ahead of them. In all parts of the world, women comprise the majority of the older population (Gist & Velkoff, 1997). Indeed, the sex ratio, or proportion of men to women in the population, shifts dramatically each decade after age 65 (see Exhibit 26). In the United States today, there are 22.7 million older women and 16.8 million older men—a ratio of 135 women per 100 men 65+. By age 85, the ratio increases to 216 women per 100 men (Administration on Aging, 2010). The typical fate is for men to die earlier and for women to survive with chronic diseases.

The experience of growing old is different for men and women in our society in some obvious ways and other ways that are less obvious. For instance, physical signs of aging bring more severe social consequences for women than for men (Bell, 1989). Media images celebrate youth and sexuality in younger women, so older women become virtually invisible in general society. In the family division of labor, older women typically play a vital role of "kinkeeping" through social networks and caregiving (Rosenthal, 1985). However, the family caregiving role taken on by women often has the consequence of removing them from the paid labor force so that they accumulate lower pension benefits than men do.

In gender roles, we can see a pattern similar to that noted among minority groups, namely, that cumulative disadvantage means diminished economic security in retirement. Retirement income for older women is on average only about 59% of what it is for comparable men, and 10% of older women fall below the poverty line, compared to 6.6% of older men (Administration on Aging, 2010). The feminization of poverty in old age in fact has many causes, including sex discrimination, patterns of economic dependency, and widowhood (Older Women's League, 1998; Smolensky, Danziger, & Gottschalk, 1988).

Longevity and living arrangements have significant impacts on older women's quality of life. Statistically, women tend to marry men who are older than they are, and women's longevity is greater. Thus, women are much more likely than men to be widowed and to live alone in old age. Among Americans ages 65 and over, in 2009, 72% of men were married and living with their spouses; for women in this group, the figure was only 41%; the number of older women married decreases with increasing age. The number of women living alone increases from 38% for those between ages 65 and 74 to 49% for those ages 75 and older. Of 11.4 million noninstitutionalized older adults living alone, around two thirds are women (see Exhibits 27 and 28).

Exhibit 26 Age and Sex Structure of the Population for the United States: 2010, 2030, and 2050

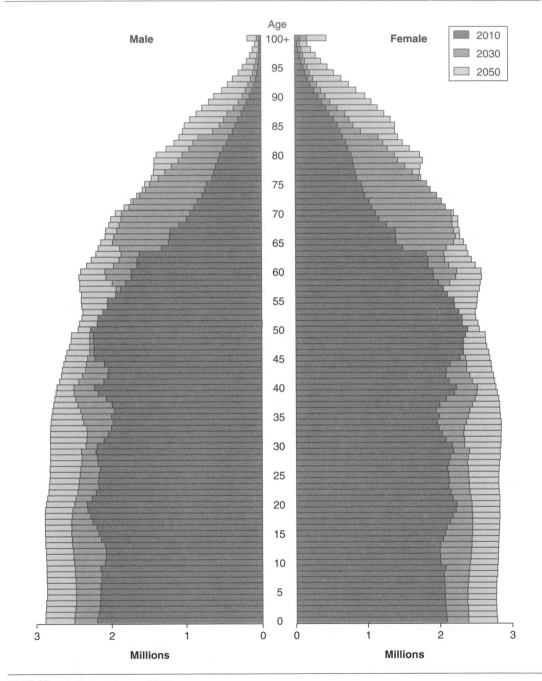

Exhibit 27   Marital Status of Persons 65+, 2009

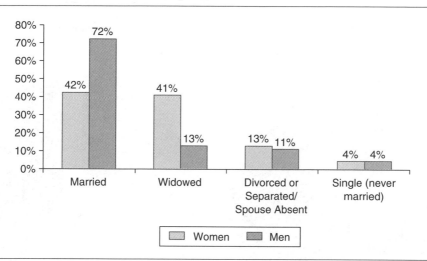

SOURCE: Based on Internet releases of data from the 2009 Current Population Survey, Annual Social and Economic Supplement of the U.S. Census Bureau.

Divorce is also becoming an increasingly prevalent influence on older women's living arrangements. Divorced or separated older persons constituted 11% of those over age 65 in 2009, but since the 1980s, the proportion of older people who are divorced has been increasing four times as fast as the growth of the older population overall. Divorced women usually experience a sudden reduction in their financial circumstances, and they are less likely than older men to remarry. Thus, the dramatic increase in the number of older divorced women today could mean serious socioeconomic problems in the decades ahead (Uhlenberg, Cooney, & Boyd, 1990).

Older women's problems remain unresolved, but some positive steps have been taken. For example, mutual self-help groups for widows have shown great effectiveness in helping isolated older women. Widow to Widow is one outstanding example of this kind of group (Silverman, 1986). Another step toward improving the lives of older women takes the form of organized advocacy, such as that demonstrated by the Older Women's League (OWL), founded in 1980 to address issues faced by middle-aged and older women. Along with advocacy by feminists, there has also been growing recognition that the impact of gender, along with race, ethnicity, and social class, must be looked at over the entire life course, not just old age alone (Hatch, 2000).

Multiple Jeopardy

Patterns of inequality involving social class, race or ethnicity, and gender reinforce one another. An older person who is simultaneously a member of two (or more) disadvantaged groups faces what has been called **double jeopardy** (Dowd & Bengtson, 1978; Minkler & Stone, 1985) or even multiple jeopardy. The consequences of multiple jeopardy are understandable in terms of cumulative disadvantage. If women earn less than men and if

Exhibit 28 Living Arrangements of Persons 65+, 2009

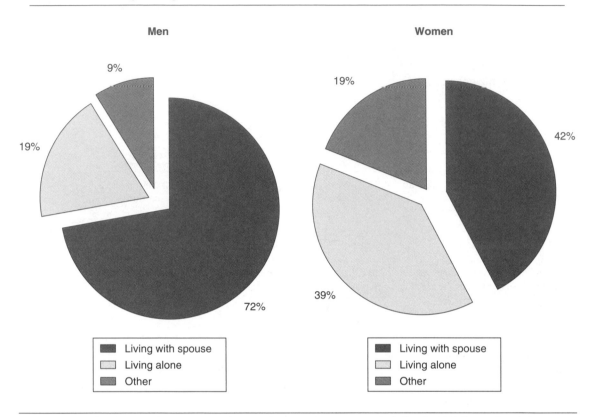

SOURCE: Administration on Aging (2010); based on data from U.S. Census Bureau including the 2009 Current Population Survey, Annual Social and Economic Supplement and the 2009 American Community Survey. See: March 2009 Current Population Survey Internet releases, detailed tables and unpublished data from the 2007 Medicare Current Beneficiary Survey.

minority group members are subject to prejudice over their lifetimes, it is not surprising that older minority women suffer multiple problems in the areas of health status, income, housing, and so on.

Empirical support for the assertion that multiple jeopardy exists has not been convincing to all. Some point out that old age, instead of widening disadvantage, may perhaps serve as a leveling influence; in other words, older people become more similar in their economic and social circumstances than younger people are. Moreover, not all minority groups are alike. African American and Native American older persons tend to have serious health problems, but Asian/Pacific Islander older persons are usually better off in this area. Perhaps the older Asians who immigrated to this country in their earlier years had to be relatively healthy in order to undertake such traumatic change. Clearly, more research is needed to clarify just how the multiple disadvantages of aging, gender, and minority status shape later life. Just as clearly, differential disadvantages among subgroups of older adults remain a potent challenge for public policy in an aging society (Markides & Black, 1996).

Economic Well-Being

The poverty rate is one important yardstick for inequality. Exhibit 29 compares the poverty rates for different age groups over the past several decades. Note that in 2009, the overall poverty rate for all ages was 14.3%, the highest poverty rate since 1994. For older adults 65+, the poverty rate has been decreasing in the first part of the 21st century, to 8.9% in 2009, while the poverty rate for children has increased to 20.7%.

Exhibit 29 Poverty Rates by Age: 1959–2009

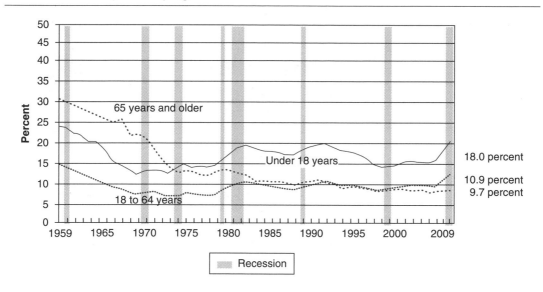

Recession

SOURCE: U.S. Census Bureau, Current Population Survey, 1960 to 2009 Annual Social and Economic Supplements.

NOTE: Data for people aged 18 to 64 and 65 and older are not available from 1960 to 1965.

As Exhibit 30 reveals, the rate of poverty among the aged tends to vary dramatically according to different characteristics. For example, the poverty rate among older women living alone is much greater than for older people in general who live alone. The poverty rate for the old-old and oldest-old (75+) and for women living alone is much higher than average. Regardless of age, the poverty rate among minorities is more than twice as high as that among Whites (Wu Ke Bin, 1998). It is not surprising, then, that higher poverty rates are found among older women, minorities, and those with chronic illnesses—all groups that could be candidates for those who are least advantaged.

The differences in economic well-being among subgroups of older adults are important to keep in mind when we come to consider debates about age versus need in distributing benefits and entitlements. For example, the great strength of social insurance programs such as Social Security and Medicare has been universality and egalitarianism. Growing old is seen as part of a universal human experience, like birth and death, a great leveler. The aged are thought of as being "our future selves," and it is hoped that Americans are not reluctant to pay into a Social Security system from which we all hope to benefit one day.

Exhibit 30 Percentage of People Aged 65 and Older Below Poverty Level, by Selected Characteristics: 2005

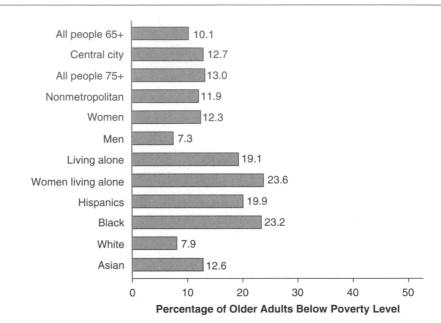

Percentage of Older Adults Below Poverty Level

Characteristic	Percentage
All people 65+	10.1
Central city	12.7
All people 75+	13.0
Nonmetropolitan	11.9
Women	12.3
Men	7.3
Living alone	19.1
Women living alone	23.6
Hispanics	19.9
Black	23.2
White	7.9
Asian	12.6

SOURCE: Based on data from the Current Population Survey, Annual Social and Economic Supplement, "Income, Poverty, and Health Insurance Coverage in the United States: 2005" issued by the U.S. Census Bureau, 2006.

Nearly 3 out of 4 older Americans who fall below the poverty line are women.

When we focus on heterogeneity and diversity, a different reality comes into view. There are sharp differences between men and women, Whites and ethnic minorities. How then should universal programs, such as Social Security or the Older Americans Act, take account of such differences? For example, if average life expectancy for Blacks is lower than it is for Whites, does that mean the age of eligibility for Social Security should be reduced for Blacks? If we did that, would we then have to do the same for men, who live, on average, fewer years than women? But then, what about the fact that women on average have lower earnings and income in retirement or a higher poverty rate? Should that mean we revise Social Security to give more to women in contrast to men?

The Social Security system does redistribute income, in modest ways, toward lower-income people regardless of gender or ethnicity. But the redistributive function in Social Security is balanced by other elements in the system. There is persisting controversy about how to take account of women's experience in family roles and how those characteristic gender differences should influence Social Security (Estes, 2004).

An important lesson here is that some degree of redistribution and targeting of benefits to the most needy can be acceptable as long it remains within certain bounds. But as the long controversy over race and affirmative action shows, once we begin targeting benefits too explicitly or giving preference to one group over another, then principles of need versus universalism do come into conflict.

In years to come, the older population in the United States will reflect the diversity of American society. An important issue in the 21st century is our ability to equitably accommodate an influx of immigrants, most of whom are Asian or Hispanic. The 8 million immigrants who came to America during the 1980s represent a figure comparable to the number who came to these shores early in the 20th century. Most of those in this recent immigrant group are young; but as they age in decades to come, they will make the older population far more diverse than it is today.

Finally, there are unresolved questions about the meaning of ethnicity in relation to citizenship and national identity. In the last decade of the 20th century, we saw the breakup of large multiethnic nation states, most notably the Soviet Union and Yugoslavia. The question is whether different racial and ethnic groups can live together in the United States without disintegrating tensions. Can we do better than other societies? The verdict is not yet in. But the history of the Untied States has been one of a society produced by successive waves of immigrants and guided by a political ideal of equal treatment under law. Increasing ethnic diversity and a growing population of older adults will present a challenge for the future and will demand thoughtful attention to the varieties of aging experiences.

The Economic Status of Older Americans

Economic circumstances, we know, vary sharply among different subgroups of older Americans, but along with variation, we now need to address a basic question: How has the economic condition of older people as a group fared in recent decades? Since the early 1980s, the income of older people in the United States has grown faster than the income of younger people (Radner, 1987). This fact has led some analysts to speak of the "declining significance of age" as a factor in poverty (Smolensky et al., 1988). The rate of poverty among the aged has been drastically cut, and the soaring value of some assets, such as homes purchased 20 or 30 years ago, has brought dramatic rises in net worth.

But here we should distinguish between two different economic concepts: income and wealth. *Income* denotes available money or its equivalent in purchasing power, whereas *wealth* denotes all economic assets of value, regardless of whether they produce cash. A person's wealth and income are not necessarily related. For example, an older person might continue to live in a home that has substantially increased in market value, but not be able to maintain it on a modest fixed income. Older people do tend to have more wealth than younger people, and they receive more in government benefits than other age groups. On average, however, older people have lower incomes than other adults in the United States (see Exhibit 31).

Exhibit 31   Poverty Status, by Age, 2008

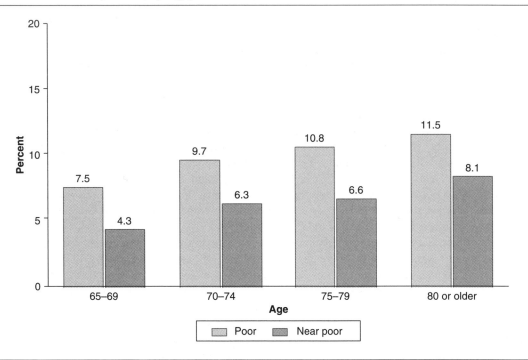

SOURCE: *Income of the Aged Chartbook, 2008,* Social Security Online (2010).

NOTE: Poverty status was determined for everyone except those in institutions, military group quarters, or college dormitories, and unrelated individuals under 15 years old.

The issue is further complicated by enormous diversity in the financial status of older people. Both their assets and their incomes have a much wider range than is seen in other age groups; that is, more extremes of wealth and poverty exist among the old (Quinn, 1987). Most studies look only at average or mean measures, and means are skewed by extreme values (very rich or very poor). To get a truer picture of the status of a typical individual, we should look at median figures (half the cases are above the median, and half are below). Median cash income is highest for middle-aged families and lowest for the very young and for those ages 85 and over. When noncash income (such as Medicare benefits) or wealth is considered, the median financial status of the aged as a group looks better (Radner, 1992).

Sources of Retirement Income

Retirement income policy in the United States has often been described as a "three-legged stool." The three components are Social Security, private pensions, and individual savings and other assets that yield income (see Exhibit 32). Some older people also have earnings from employment, but in the last three decades, the contribution of earnings to the income of older people has clearly declined. Thus, this discussion focuses on Social Security, pensions, and assets.

Note the trends in sources of income. As Social Security has grown to become a larger portion of income for older people, earnings from employment have sharply diminished. For instance, in the late 1960s, earnings were still the leading income source for married couples

Exhibit 32   Percentage Receiving Income From Specified Source, 2008

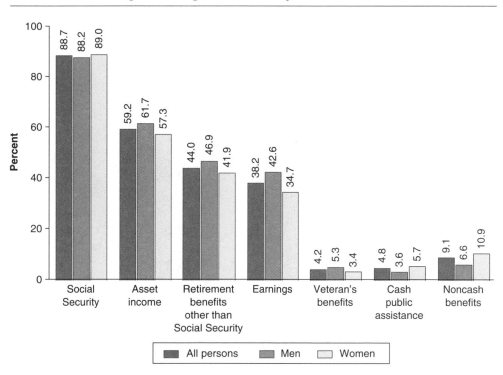

SOURCE: *Income of the Aged Chartbook, 2008,* Social Security Online (2010).

NOTE: Family incomes of persons 65 and older tends to be higher than the income of aged units 65 or older because it includes income from all family members, not just a spouse. In addition, each older adult counts individually rather than as part of a marital unit; statistics based on aged units give greater weight to nonmarried persons as compared to statistics based on persons.

over age 65, but by 1976, earnings had been surpassed by Social Security, and this trend has continued. Another trend has been the rising importance of assets and private pensions. Between 1978 and 1984, income from assets, such as savings accounts, increased dramatically, from 18% to 27% of income in later life, whereas pensions grew modestly, from 14% to 16%. Sources of income for unmarried older people followed similar trends, except that for them Social Security was even more important.

Overall, retirement income has increased significantly (see Exhibit 33). Indeed, the real per capita income and purchasing power of people over age 65 has been clearly growing over the past two decades and, in fact, has risen at a rate substantially higher than for both the working population and children (Levy, 2001). In 2009, median income for adults 65 and older was $15,282 for women and $25,866 for men; the median income for all households headed by older people was $43,072 (Administration on Aging, 2010).

Exhibit 33   Median Income, by Age and Sex, 2008

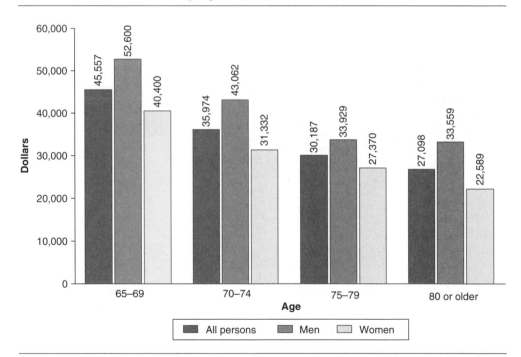

SOURCE: *Income of the Aged Chartbook, 2008,* Social Security Online (2010).

NOTE: Family income of persons 65 and older tends to be higher than the income of aged units 65 and older because it includes income from all family members, not just a spouse. In addition, each older adult counts individually rather than as part of a marital unit; statistics based on aged units give greater weight to nonmarried persons as compared to statistics based on persons.

However, older people have not shared equally in the increase in retirement income. The disadvantaged—the poorly educated, minorities, and women—have not had the jobs that allow them to collect maximum Social Security benefits or private pensions or to accumulate much wealth. For instance, Gibson (1983) found that older African American women had typically had disadvantaged work lives in which opportunity was constricted by prejudice. In previous decades, Black women too often ended up confined to the secondary labor market, where a job such as household maid would pay low wages with negligible fringe benefits and poor job security. By contrast, a White man would be much more likely to find employment in the primary labor market, in a managerial or unionized position offering job security and the opportunity to accumulate pension benefits or other assets for old age (Crystal & Shea, 1990).

Older women can be particularly vulnerable because often their financial well-being changes drastically when they lose their husbands. In fact, the most significant predictor of poverty in later life is for an older woman to become single. Most women in current cohorts of midlife and older women are likely to become widows at some point. Although widows receive Social Security benefits at age 60 based on their husband's earnings record (O'Grady-LeShane, 1990), spousal benefits under private pensions are typically less generous. Only 3% of American women receive benefits from their deceased husband's pensions. Widowhood therefore generally means a drop in income (Holden, Burkhauser, & Myers, 1986; Lopata, 1987; Smeeding, Estes, & Glasse, 1999); in 2004, 17.4% of single older women and 13.1% of single older men fell below the poverty line (Karamcheva & Munnell, 2007).

Older divorced women, even those who were long married, are worse off economically in later life as well. One consequence of no-fault divorce is that they receive little or no compensation in the form of alimony for years of investment in a marriage (Weitzman, 1985). Nor do today's older women who become "displaced homemakers," divorced after spending their prime adult years maintaining the home and raising the kids, have much paid work history to exploit in finding a job. Typically, they receive nothing from their ex-husband's pensions. They can therefore plunge easily into poverty.

Gender differences in employment also produce inequities for older women. Among people over age 65, more than twice as many men as women receive private pensions. With more women in the labor market accumulating pension rights on their own, the economic position of women could improve in the future. Women frequently leave the workforce to take care of small children or older adult relatives; however, today, around two thirds of those caring for frail older adults are women (Stone, Cafferata, & Sangl, 1987). When they leave the workforce for caregiving, they reduce their opportunity to increase income with experience or to guarantee eligibility for a pension. Finally, women reap the cumulative disadvantage of persistent patterns of pay inequity. In that regard, they are like other disadvantaged groups.

Social Security

Today, Social Security is the biggest source of income for people over age 65 in the United States. Nine out of 10 older adults rely on Social Security for some portion of income, 3 out of 10 older adults depend on it for 90% or more of their total income, and 20% get all their income from Social Security (Social Security Online, 2010). Social Security is vital for the poorest older adults.

At present, approximately 95% of the American workforce is covered by Social Security, and the trend has been toward universal coverage. Social Security payments go to 29 million retirees (ages 62 and over), along with 7 million survivors (widows and children) and 4.6 million workers with disabilities of all ages. In effect, 16% of the U.S. population comprising retired older adults, people with disabilities, and family dependents receives payments from Social Security.

To be eligible for full retirement benefits under Social Security, a person must be 66 years old (or 62 for partial benefits under an early retirement option) and must have a wage-earning history in a job covered by Social Security or be married to a spouse with such a history. In other words, both chronological age and a record of earnings are necessary to qualify.

The principle of individual **equity** would insist that people get back an amount in Social Security income proportional to whatever they contributed. Conversely, the principle of social adequacy would give lower-income retirees a larger replacement proportion to compensate for the fact that they are much less likely than wealthier people to have adequate assets or private pensions in retirement. Social Security, as it has evolved over time, incorporates principles of both equity and adequacy and tries to achieve a compromise between the two. Many debates about the fairness of Social Security have come about because of this compromise between two opposing values (see Controversy 9: What Is the Future for Social Security?).

Pensions

A pension is a contractual plan by an employer to provide regular income payments to employees after they have left employment, typically at retirement. Pensions are thus a form of deferred compensation. Pensions as a fringe benefit became widespread only in the years after World War II. Over the years, the proportion of private sector workers covered by private pension plans grew from 25% in 1950 to a peak of 50% in the mid-1980s. By contrast, 90% of state and local government employees are covered by civil service pensions. Pensions are far from universal among older Americans, however. Today, only 45% of households over age 65 have private pension income, and the proportion of workers covered by any kind of pension has not grown—a worrisome trend for the future well-being of an aging population (G. Clark & Whiteside, 2004).

Two different types of pension plans are available today. One is the **defined-benefit plan**, which promises a specific or defined amount of pension income for the remainder of life. The company then is responsible for setting aside funds to cover the benefits promised. Another type of pension is the **defined-contribution plan**. In this plan, employers, employees, or both contribute money, but the amount of pension income depends on how much is contributed to the fund over the years and also on how successfully it is invested. In the past, most workers eligible for pensions, chiefly employees of large companies, were mainly enrolled in defined-benefit plans. The current trend is to offer only a defined-contribution plan, which means that retirees in the future could face less economic security than retirees enjoy today. Still another approach is the **cash balance plan**, which combines elements of both defined-benefit and defined-contribution types (Gordon, Mitchell, & Twinney, 1997). Today, cash balance plans may erode the value of defined-benefit promises for many older workers.

Employers in the United States are not required to offer a pension plan to employees, but if they do, they must satisfy certain legal requirements. A major step in protecting workers and retirees came with passage in 1974 of the **Employee Retirement Income Security Act (ERISA)**. ERISA regulates private pension plans in the United States and provides protection against loss of benefits to retired workers. Protection is not absolute, however. A pension plan can be terminated if a company goes out of business or is merged with another company. In addition, employers with pension plans aren't required to include those who work less than 20 hours a week, a crucial omission in light of the tremendous growth of part-time employment in the United States in recent years.

Equally ominous is the financial outlook for the Pension Benefit Guaranty Corporation (PBGC), a federal agency established to protect pensions when companies cannot meet their obligations. The PBGC has a deficit of $11 billion and may face $50 billion in underfunded obligations in the future for bankrupt companies. The economic downturn that began in 2008 could make matters much worse. This hidden deficit could be a major problem for future generations.

A key point to understand about pensions is **vesting (of pension)** rights, or the period an employee must meet according to the employer's requirements to become eligible to collect the pension. If an employee leaves a job before being vested in the plan, then the worker is not entitled to receive benefits. Until recently, most pension plans required at least 10 years of employment for full vesting, but legislation following adoption of ERISA reduced this period to only 5 years, a move likely to help workers in our increasingly mobile society. Critics have urged a system of pension portability as well so workers can transfer their pension rights from one employer to another after changing jobs. Such portability, it is argued, would not only improve equity, but would also promote flexibility and better use of middle-aged or older workers, who might be more inclined to move to new opportunities if they wouldn't lose their pension rights in the process.

An increasingly important feature of private pension income is the option for early retirement benefits, that is, pension eligibility at age 55 or 60. Early retirement may no longer be a choice when the pressure builds for layoffs and downsizing the workforce. In these conditions, early retirement benefits are often enriched to give older workers an inducement to leave, thereby saving jobs for those who remain. This trade-off between early retirement and jobs for younger workers is not always fully voluntary, however, and it also poses significant hidden risks that may need to be considered carefully. For example, employees who have opted for early retirement may face a reduction in employer-sponsored retiree health benefits at a later date.

Private pension coverage is often designed to be coordinated with Social Security coverage, sometimes in ways that present retirees with an unpleasant surprise. For instance, some private pension plans unilaterally reduce the amount of the pension by whatever Social Security benefits a retiree gets, a practice that in the future may be limited to a loss of up to half the total pension promised.

Social Security benefits are indexed for inflation, meaning that income automatically increases when the cost of living does, but private pensions usually have not been fully indexed. As a result, pension income may be quite sufficient early in retirement, but may become inadequate as inflation diminishes its real value over time.

Individuals whose employers do not provide pension benefits and those who are self-employed can set aside savings to provide pensions for themselves. One vehicle for retirement savings that has been around for a long time is the **annuity,** an investment vehicle sold by life insurance companies permitting one to defer taxes on accumulated earnings. Congress has also created a number of tax incentives for people to save for their own retirement. One is the individual retirement account (IRA), in which a person can accumulate money until retirement, subject to certain limitations, but on a tax-deductible and tax-deferred basis. The Tax Reform Act of 1986 tightened eligibility for IRA contributions, which are now limited to taxpayers with incomes below $25,000 ($40,000 for couples). The so-called Keogh plan and the **401(k) pension plan** have the same basic purpose, but they are not limited to people below a certain income.

Assets and Savings

People accumulate assets over a lifetime; here again, a person's circumstances over an entire life course have consequences for financial status in old age. Some assets are tangible, such as a home, whereas others are financial, such as shares of stock. Assets are second only to Social Security as a source of income for older people; around one fourth of retirement income is based on assets. As a group, older households own many more assets and have more accumulated wealth than younger households, at least until household members are roughly 80 years old.

The net worth of an individual or a family consists of the total value of all assets—including real estate, savings, and personal property—minus the debts. Exhibit 34 shows

Exhibit 34   Median Net Worth, by Age Group: 2000

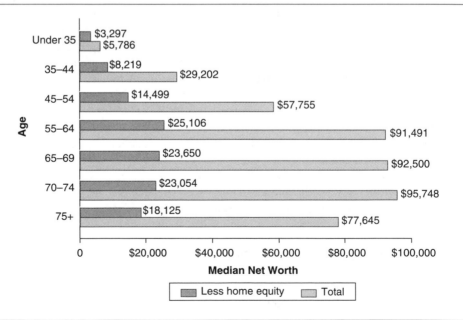

SOURCE: U.S. Census Bureau (2000).

the median net worth of households based on the age of the householder. Half of all householders under age 35 have a net worth of no more than $5,786, but among people in their late 60s, half have a net worth above $92,500. Net worth declines a bit with advancing age, but note that even among those above age 85, the figure is still $77,645. In short, older Americans have dramatically higher levels of total wealth compared with younger people.

From Exhibit 34, it is clear that the overwhelming majority of assets for those over age 35 are in the form of home equity: the market value of a home exclusive of mortgage debt. The value of home equity is even more important for older people. Eighty percent of people over 65 are homeowners, and, of that group, two-thirds have paid off their mortgage and their home free and clear. Two-fifths of all assets of older households are in the form of home equity. Typically, older homeowners bought their homes years ago, and thus they have seen an appreciation in their market value, although such gains were reduced by the drop in home values after the Great Recession beginning in 2008.

Because so much wealth is represented by home ownership, there has been repeated interest in enabling older people to convert the accumulated value of a home into a regular monthly income (Venti & Wise, 2000). One approach to home equity conversion is the **reverse mortgage**, through which a bank guarantees a monthly income for life, but usually takes ownership of the home when the homeowner dies. To date, few older people have participated in reverse mortgages, so home ownership remains an important contributor to assets, but not to income for older people.

The importance of home equity distorts the net worth of older households in average statistics. Home ownership represents about 70% of older Americans' assets. When home equity is left out of the reckoning, median net worth of households over age 65 drops to only around $23,000, compared with nearly $90,000 when the economic value of home ownership is taken into account (2000 data; see also Exhibit 35). The smaller assets are chiefly in the form of interest-bearing savings and checking accounts, with much smaller amounts in stocks and bonds or other real estate. In summary, home equity is the over-whelming part of net worth for older Americans today.

Just as general well-being differs dramatically according to social class, gender, and ethnicity, so does the distribution of assets. Note that in Exhibit 36 the median net worth for older men and women does not differ appreciably, but there is a dramatic difference between the averages for married couples and for all households—more testimony to the importance of marriage for financial well-being in later life. Racial and ethnic differences are also dramatic. The median net worth of White householders is three to four times greater than that of older Blacks. Once again, the average figures hide important differences in the aging experience.

Changing Financial Outlook

The financial picture for today's older people, at least on average, may look pretty good. They enjoy historically high Social Security and pension income and have a comparatively high net worth. Then why do so many older people feel financially insecure?

Exhibit 35   Percentage of Older Adults With Various Levels of Family Income, 2008

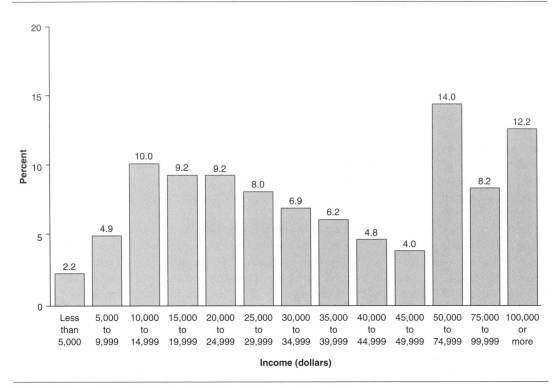

SOURCE: *Income of the Aged Chartbook, 2008,* Social Security Online (2010).

NOTES: Totals do not necessarily equal the sum of the rounded components.

Family incomes of persons 65 and older tends to be higher than the income of aged units 65 or older because it includes income from all family members, not just a spouse. In addition, each older adult counts individually rather than as part of a marital unit; statistics based on aged units give greater weight to nonmarried persons as compared to statistics based on persons.

The poorest older adults have reason to feel insecure; they are by no means living comfortably. Other questions here revolve around the economic status of the baby boom generation, who are now nearing retirement age in the first decade of the 21st century. A 1995 survey by the Merrill Lynch brokerage firm revealed that two thirds of workers ages 45 and older already had anxiety about retirement income; after the 2008 economic downturn, the anxiety only increased. The Merrill Lynch Baby Boom Retirement Index estimates that baby boomer households are saving at only a third of the rate needed to provide them with a secure retirement at age 65. But this index may underestimate the importance of home equity as a form of savings. If housing wealth is taken into account, baby boomers are actually saving at a full 84% of the rate needed to maintain current living standards.

Exhibit 36   Median Net Worth for Older Adult (65+) Households, by Type of Household: 2000

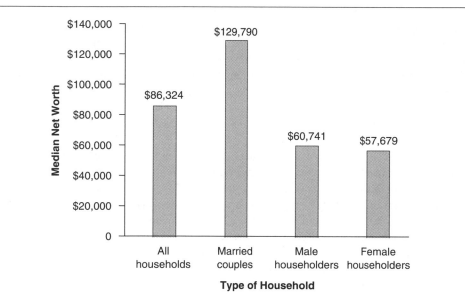

SOURCE: U.S. Census Bureau (2000).

Despite gloom about the future, baby boomers in overall terms may actually be better off than their parents (Butrica & Uccello, 2004). Their median household income and wealth are already higher than that of their parents at comparable ages. But will this be enough to ensure adequate retirement income? The **life cycle model of savings** predicts that as boomers reach later life, they will begin to put more money into savings when retirement looms on the horizon (Modigliani, 1985). Whether this happens, of course, depends on other factors besides age: Cohort and period effects also influence savings behavior.

Other unanswered questions also demand attention. Will housing prices remain high so that home equity values are not lost? What will happen to market values in the early decades of the 21st century if large numbers of retiring baby boomers try to convert mutual fund assets into cash? The problem for both home equity values and mutual funds is that a marketplace requires both buyers and sellers. Real estate, like the stock market, tends to go through cycles: What goes up may also come down. Again, there are no easy ways to avoid risk. Until the mortgage crisis and decline in real estate values beginning in 2007, many people counted on home equity as a source of savings. But that, too, proved to have risks. As individuals take on more responsibility for retirement investment decisions, each of us will have to prepare for a future that is certainly different from the past.

Perhaps the biggest source of financial uncertainty for older people is how their expenditures will change as the years go by. The turbulent U.S. economy of the last two decades has had important effects on the well-being of older Americans. Older people, who live disproportionately on fixed incomes from pensions and annuities, are affected more severely by inflation than other groups. Fortunately, government policies and economic trends have reduced the erosion of buying power through inflation. For example, 86 federal programs increase their benefits for the old based on some index for inflation (R. Clark et al., 1984). The largest of these programs is Social Security, indexed for inflation since the early 1970s. On the economic front, although the United States is currently experiencing a low inflation rate, it is accompanied by historically low interest rates, which have sharply diminished the income of the old, who rely on interest from savings more than do other groups.

Nevertheless, the overall economic position of older Americans has improved substantially in the past two decades. Improvements in pension coverage and indexing of Social Security are the main reasons for this improvement in income, whereas rising home equity is the main reason for the gain in assets. Income from earnings has declined as more older Americans have left the labor force for retirement.

Despite the good news, the stereotype that most older people are affluent is mistaken. When we look behind the average figures, we see two things: (a) enormous variation among subgroups of the aging with respect to economic circumstance, and (b) a large group of older people who have been brought above the poverty line but are still near poor. The insidious feature of old-age poverty is that it lasts longer and is more likely to be permanent than poverty among younger people. The poor older people have fewer chances for remarriage or finding better jobs. For all these reasons, income and services for poor older people have enormous importance, and so has the question of how to target those benefits to those who are least advantaged.

The change in the economic circumstances of older Americans can be both a source of pride in the success of government programs, chiefly Social Security, and a challenge for the future of those programs. These matters have become items of broad public controversy and are likely to continue to provoke vigorous debate.

We have seen that the rise in home equity is the reason for increases in wealth of most older persons, but a home is much more than a financial asset in people's lives. Until they reach a condition of needing daily care, the overwhelming majority of older people prefer to live in their own homes, in the same neighborhood where they have always lived. If the neighborhood is stable, and few homes are sold over the years to younger families, eventually the neighborhood will be dominated by long-term, older residents.

As we move into the next few decades of the 21st century, as the baby boom generation begins to enter later life, the demand for senior housing may jump dramatically. But building affordable senior housing will be a major challenge because many baby boomers will have inadequate retirement assets and because federal funding for senior housing has shrunk and is not likely to increase if the current budget-cutting climate continues. Thus, policy makers are discussing many new public and private sector strategies for senior housing.

PUBLIC POLICY ON AGING

When we look at income programs, housing subsidies, and health and long-term care, it is clear that government action has had a decisive effect on the well-being of today's generation of older people. That fact in itself marks an important historical change. When today's older Americans were born—mainly before 1935—the U.S. government gave no special attention at all to issues of old age. Yet by the 1970s, a prominent political scientist took note of the "graying of the federal budget." Trends since then have amply confirmed his point (Hudson, 1978). Today, more than 30% of the total federal budget is spent on the aged, and the percentage is rising each year.

Until the 20th century, the aged population was small, and the role of government was quite limited (Achenbaum, 1978). The big change came during the Great Depression of the 1930s, with the passage of the Social Security Act (1935). It remains the cornerstone of U.S. policy on aging. The expansion of Social Security to families (1939) and the provision of early retirement benefits (1956) were important changes in the law. After World War II, Social Security coverage slowly but steadily expanded. Revisions of Social Security in 1977 and 1983 helped put the program on a more secure financial foundation, but doubts about the future solvency of Social Security remain.

Like the 1930s, the decade of the 1960s was one of social upheaval and pressure for political change. The Great Society legislation of the 1960s included such steps as the federalization of old-age assistance and such landmark laws as **Medicare** and *Medicaid* (1965). But during the 1970s and 1980s, federal commitments to all social programs were scaled back (Estes, 1989). Cost-containment measures for Medicare were enacted in 1983, and efforts to expand Medicare had mixed results: The 1988 law was repealed, but in 2003, Congress enacted a major expansion of Medicare to cover prescription drugs. The bulk of federal funding for older people goes for Social Security and Medicare, as Exhibit 37 indicates.

Another area of concern in federal legislation has been retirement. Today, more than 40 million private sector workers in the United States participate in 800,000 different pension plans. Those plans represent retirement security for millions of workers, and for this reason, the safety of pension funds is critical. Steps to protect pensions culminated in the Employee Retirement Income Security Act of 1974, which was strengthened in 1986. Retirement benefits came under further federal protection in 1984 and 1990. In addition, the federal government is directly involved in providing pension income through civil service retirement, military retirement, and the railroad retirement system.

Exhibit 37   Social Security, Medicare, and Medicaid Outlays as a Percentage of GDP, Fiscal Years 1950–2075

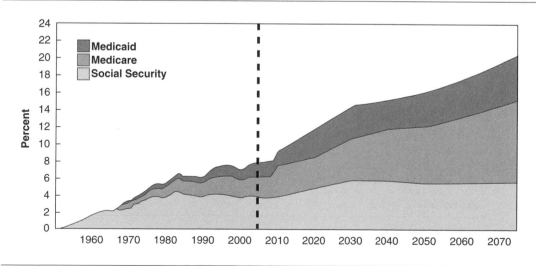

SOURCE: *Budget Crisis at the Door,* Steuerle and Carasso, The Urban Institute, 2003.

In the period since 2000 there has been a continuing decline in the number of Americans covered by defined benefit pension plans and a continuing shift toward defined contribution plans. Along with corporate pension plans, public pension plans face large funding deficits, although levels vary with different cities and states. As more and more workers shift to 401(k) and other individual plans, they have been directly exposed to risks in the stock market, as shown in the financial crisis beginning in 2008 when equity values dropped. Since that time, the stock market continues to experience volatility greater than in the past.

Federal legislation has been responsive to older workers as well as retired persons. The general trend has been to protect the rights of older workers against arbitrary dismissal on the grounds of age. A key breakthrough was the Age Discrimination in Employment Act of 1967, which gave protection to older workers; it was significantly expanded in 1978. By 1986, a long campaign to eliminate mandatory retirement was finally successful in prohibiting age discrimination in the hiring and retention of employees.

Still another area of federal government activity on behalf of older people has been the direct provision of social services. The 1978 Congregate Housing Services Act brought housing and social services for the aged together, an important step for long-term-care policy. The enactment of comprehensive social service programs was made possible by the OAA in 1965. Funding for OAA programs has been inadequate, just as it has been inadequate for comprehensive social services under Title XX of the Social Security Act. Still, the OAA has provided a strong stimulus for planning and advocacy on behalf of the aged. Exhibit 38 summarizes the high points of federal legislation on aging.

Exhibit 38   Key Federal Legislation on Aging

Social Security Act	1935
Expansion of Social Security Benefits (for spouses and children)	1939
Early Retirement Benefits under Social Security (men added in 1961)	1956
Medicare and Medicaid	1965
Older Americans Act	1965
Age Discrimination in Employment Act	1967
Older Americans Act Amendments (creating the National Aging Network)	1973
Employee Retirement Income Security Act (ERISA, Amended 1986)	1974
Congregate Housing Services Act	1978
Social Security Amendments	1983
Medicare Prospective Payment System	1983
Retirement Equity Act	1984
Abolition of Mandatory Retirement	1986
Medicare Catastrophic Coverage Act (Repealed 1989)	1988
Medicare Modernization Act (Prescription Drugs)	2003

The Aging Network

One component of government was specifically created to serve older people and advocate on their behalf. The OAA, first passed in 1965 and amended in 1973, created a national **aging network** of services for older people, such as nutrition programs, senior citizen and older adult community centers, and information and referral services (Gelfand, 2006). A key element of the aging network is local Area Agencies on Aging, generally based in city or county government and responsible for planning and organizing services to older people in that region. Each state also has a state agency on aging that plans and disburses federal funds under the OAA. At the federal level, the U.S. Administration on Aging coordinates OAA programs and provides a focal point for advocacy.

Among the key service programs under the OAA are senior centers, which have grown from the first, founded in 1943, to more than 10,000 (Krout, 1989). Despite that growth, OAA programs reach only a tiny proportion of Americans over age 65—at most, 5% to 10%. Yet the professionalized aging network provides a vehicle for planning and advocacy at all levels of government that would otherwise be lacking.

Those who criticize the professionalized aging network do so from contrasting perspectives. Analysts on the political left argue that human service programs, such as those created by the OAA, provide a meager response to social problems that older people face, doing nothing about the underlying causes of the problems (Minkler & Estes, 1984; Olson, 1982). These analysts favor an approach that attacks the discrimination, unemployment, and oppression that they believe are inherent in capitalist society. By contrast, analysts on the conservative right tend to see the current system of services for the aging as wasteful social spending by the welfare state (Rabushka & Jacobs, 1980). Still others believe older Americans are being helped at the expense of young people (Peterson, 2004) and reject arguments made by advocates for the aged.

Aging Interest Groups

Indisputably, the role of the federal government in helping older people has expanded remarkably. The federal government went from paying virtually no attention to aging in 1930 to spending more than $3 out of every $10 on older Americans. This increase in the budget has come about primarily for two reasons: (a) population aging, meaning a growing proportion of older people in the population; and (b) the enactment of programs that protect income and provide health care to older adults. But why did government respond as it did? To understand what happened, we need to appreciate the influence of aging-based interest groups (Van Tassel & Meyer, 1992).

From the beginning, the United States has thought of itself as a young country, and old age did not attract much government interest. The Great Depression, however, made Americans willing to think in new ways about the role of government. During the early years of the Great Depression, Social Security did not exist, and older people were among the poorest in America. It is not surprising that older Americans banded together to advocate for their own interests while government was expanding social welfare programs.

One of the first groups established was the Townsend Movement, founded in 1934 to eliminate old-age poverty and stimulate the economy (Holtzman, 1963). Some scholars believe that the Townsend Movement was influential in the passage of the Social Security

Act in 1935, which marked the birth of a federal government policy on aging. A similar group was the colorfully named "Ham and Eggs Movement," also based in California, which urged a pension for those over age 50 who were unemployed (J. Putnam, 1970). The passage of Social Security and the end of the Great Depression caused these age-based movements to go into eclipse. Social Security grew during the 1940s and 1950s, although not in response to political pressure or social upheaval.

Age-based movements again appeared in the United States in the 1960s. Today, the most prominent of these is AARP, formerly the American Association of Retired Persons, now the country's largest voluntary organization of older people. Founded in 1958 as an outgrowth of the National Retired Teachers Association, the AARP today is open to anyone over the age of 50. AARP provides a range of services and benefits to members, such as health and life insurance, prescription drugs, and travel service (Lynch, 2011). Its magazine is one of the most widely read periodicals in the United States. AARP is joined by other important **aging interest groups**, such as the National Council on Aging, the Gerontological Society of America, the American Society on Aging, and the National Committee to Preserve Social Security and Medicare.

This whole collection of national interest groups, along with parallel organizations at the state level, such as the "Silver-Haired Legislatures," has been called the "gray lobby" (Hess & Kerschner, 1978; Pratt, 1976, 1982). An umbrella body called the Leadership Council of Aging Organizations, based in Washington, DC, brings together many aging interest groups. Over the years, more than 1,000 aging interest groups have grown up to lobby all levels of government (Day, 1990). The professionalized aging network remains a successful instance of "interest group liberalism" (Binstock, 1972), that is, advocacy for expanded government intervention on behalf of a group—in this case, older adults.

An important question to ask is, to what extent do age-based interest groups actually serve older people (Estes, 1979)? Furthermore, should age-based advocacy groups press for laws or benefits to help older people alone as opposed to those that help people of all ages? For every person over age 65 who has disabilities, a younger person with similar disabilities could benefit from home health care. When Medicare was created in 1965, many people believed it was the first step toward creating a national health insurance program (Oberlander, 2003). But over four decades later, despite health care reforms during the Obama administration, there seems little likelihood of a new program covering all age groups.

Politics of Aging

To understand the intensely political debates about public policies on aging, we must understand something about political attitudes in old age. It is a mistake to assume that older people are bound to be more conservative just because they are older (Cutler, 1981). A body of literature in political science suggests that age in itself does not have much impact on political attitudes or behavior, and people tend to maintain their political affiliations as they age. Thus, there seems to be little likelihood of a "generational politics" developing in which people vote according to age lines (Heclo, 1988). However, interest in politics and certainly voting behavior do tend to increase with age. The most active voters are in the 65- to 75-year-old age group. Remember, however, that, although older Americans are likely to have a disproportionate voice in elections, they do not necessarily speak with one voice.

In addition, the sheer size and diversity of aging interest groups may weaken their ability to influence public policy. For instance, AARP is the largest membership group of any kind in the United States. With 40 million members, it has more people than Canada. But the power of AARP, which is viewed as speaking for mid- and later-life adults, is commonly overrated. A striking example of its limited power is the case of the Medicare Catastrophic Coverage Act of 1988, supported by AARP, which was repealed a year after it was passed. In 2003, AARP did provide crucial support for the Medicare Prescription Drug, Improvement, and Modernization Act, but the law was approved in Congress by only a slim margin and passed mainly because of strong support for it by President Bush. In 2009, AARP and other Washington-based interest groups provided crucial support to pass the Affordable Care Act. Yet that law proved controversial among older voters, partly out of fears that Medicare would be cut. Health care reform, as results in the 2010 elections, showed that older voters are a force to be reckoned with in electoral politics.

URBAN LEGENDS OF AGING

"The Gray Lobby has a stranglehold on aging policy in the United States."

This is a favorite myth of those who don't like AARP or other advocacy groups. But even Social Security is no longer the "third rail of politics," as pundits used to call it. Apparently George Bush didn't believe that in 2005, nor did the Obama White House in 2011, which proposed that Social Security be part of deficit reduction plans.

Is "Gray Power," then, a reality? Political scientists who have studied the matter tend to argue that aging interest groups are not all that effective except in preventing cuts in popular programs, such as Social Security (Binstock, 1972). Voting in high numbers does not necessarily guarantee political results (Binstock, 2006–2007). Other analysts have found that, over time, participation in politics and government programs actually reinforce each other. Perhaps the best example is Social Security, which helped transform older people from a vulnerable group in society to the most politically active age group, at least as measured by voting rates. In that respect, public policy has profoundly influenced the role of citizenship in an aging society (Campbell, 2003).

Trends in Public Policy and Aging

As we think about the future, it is important to keep in mind the diversity of the aging population, including dramatic differences in well-being among subgroups and the impact of cumulative disadvantage, whereby earlier experience over the life course influences what old age will be like. Clearly, the improved economic circumstances of older people have already prompted some reassessment of aging programs. Whatever we may think about the effectiveness of political advocacy for the aged, an increasing share of the federal government's expenditures have unquestionably been directed toward the older population. This growth in spending for older Americans has several explanations:

- Social Security and Medicare have grown rapidly because they have been available on the basis of age alone, without means testing. These programs serve all older adults without reference to need. In fact, beneficiaries of this spending are mostly the middle-class older adults, not the poor. Even more affluent older adults collect Social Security because they paid into it.

- A related point is that federal programs for older people continue to enjoy broad political support. Most recipients sincerely believe that they are getting back only what they are entitled to and that Social Security functions something like an annuity, with every dollar contributed in the working years being returned, with interest, during retirement. This view is, in fact, mistaken: Social Security does not operate like a bank account, but the majority of Americans, conservative or liberal, expect their due. At the same time, many younger people are only too happy to support programs that help their parents, even if growing numbers of young adults have doubts about whether they will collect similar benefits when they retire.

- Programs for older Americans have developed incrementally for the most part, making it more difficult to see just how much they have grown over the years. Sometimes breakthroughs have been dramatic, as when Social Security was instituted in 1935. But equally important have been the less noticed, gradual expansions of Social Security, which have had a large cumulative impact. It is true that periods of growth have been offset by cutbacks like the Social Security amendments of 1983, when Congress agreed for the first time to tax Social Security benefits. But these technical details have not been well understood by the general public.

- The growth of benefit programs for the aged reflects the growth in size of the aging population. Population aging virtually guarantees that a growing share of resources will go to older people. Age-based entitlements make such a shift natural and inevitable, not the result of special political advocacy.

The rapid growth of federal programs serving older people over the past 20 years presents a quandary. The sheer size of age-based entitlements in the budget makes planners worried (Torres-Gil, 1992). The retirement of baby boomers will dramatically increase the numbers of those eligible for Social Security and Medicare, whereas the number of working adults will be smaller. How will we find a fiscally responsible way of maintaining benefits in the face of these population pressures? One solution came in 1983, when Congress decided to raise the age of eligibility for Social Security benefits by 2 years early in the 21st century. But the evolution of Social Security is by no means over. Among the changes being advocated are measures that would make Social Security more fair to women and that would allocate Social Security funds to private investment, perhaps for individuals to make their own investment decisions. The general shape of federal aging programs is unlikely to change dramatically. Many people have a strong interest in seeing these benefit programs continue, and the political pressure to maintain them will be intense. However, precisely because aging expenditures are so large and because we have other public needs, pressure to cut costs is likely to bring age-based entitlements under continuing scrutiny.

Equity

Other serious questions remain. One of them concerns equity or fairness. The most successful federal aging programs have been the ones that serve all the aged, such as Social Security and Medicare. These programs have succeeded in improving the well-being of the average older person; indeed, they constitute a stunning example of how government can come to agreement and move decisively to solve a problem (Schwarz, 1983). But celebration of success should not obscure the fact that groups such as older women and minorities continue to have high rates of poverty.

Furthermore, the most successful aging programs have all been based on a presumption of need or dependency. Both Social Security and Medicare arose from a "permissive consensus" that depicted the aged as weak, needy, and dependent (Hudson, 1978). That image is what Kalish (1979) called the "failure model" of old age. To the extent that older Americans today are healthier, better educated, and more affluent, the permissive consensus is likely to be eroded (Binstock, 1985).

Related to this point is the issue of **generational equity**. Older adults already receive a substantial portion of the federal budget and will likely receive more as the retired population grows. At the same time, federal expenditures on children and families have decreased. Rates of homelessness, poverty, malnutrition, and poor health have increased among younger Americans. Is it fair or wise to continue distributing benefits on the basis of age alone, or do we need to take a more **needs-based benefits** approach? The great strength of social insurance programs such as Social Security and Medicare has been universality and egalitarianism. As discussed before, how should universal programs such as Social Security and OAA take account of differences in the financial well-being of men and women, Whites, and minorities? Should, for instance, women receive more in benefits because they on average have lower earnings and income in retirement?

Productivity

In the United States, as in other advanced industrialized societies, one result of higher retirement benefits has been the movement of more older workers out of the labor force through early retirement, a move that can be attractive to workers and employers alike. Today's policies in effect push many vigorous, capable "young-old" out of their productive social roles. To be sure, older people often find great personal meaning in leisure, family life, and voluntary association with churches, community groups, and the like. But can our society afford to lose their contributions? Now that boomers are celebrating their 65th birthday in growing numbers, it is significant that more and more aging boomers say they expect to continue working, by either choice or necessity, in years of traditional retirement.

Racial and ethnic differences in aging are important when we look at the future of work and retirement. Just what does retirement mean for different subgroups within the aged population (Zembek & Singer, 1990)? Older Blacks, for instance, may respond to poor employment prospects by defining themselves as retired (Gibson, 1987). People who qualify for disability benefits but have no pension might withdraw from the labor force without ever being officially labeled as retired. Finally, many minority groups exchange services among kin, but that pattern of productivity doesn't show up on official statistics about work and retirement.

Another question is how the retirement income system, both private pensions and Social Security, can take account of gains in longevity and the greater vigor and productive capability of today's older people. Age-based benefit programs such as Social Security are understood to be entitlements, just as private pensions are forms of deferred compensation. But the idea of entitlement, of benefits owed to us as a right, is different from the idea of productivity. Productivity implies that income and benefits are provided on the basis of how much a person contributes. How is it possible to maintain the integrity of retirement income while also encouraging productivity among the older population (Moody, 1990)? Questions such as these will have to be explored more seriously as society considers what leisure and "productive aging" may mean in the future (Allen & Chin-Sang, 1990; Butler & Gleason, 1985).

CONCLUSION

The overall position of older Americans has improved substantially in the past three decades. But today the picture is more mixed. Improvements in health status and life expectancy have been balanced by rising levels of chronic disease among aging boomers. Earlier gains in home equity values are threatened by the collapse of home prices after the recession beginning in 2008. The indexing of Social Security benefits to inflation is the main reason for improvement in income, but debates about the federal deficit and the future of Social Security have put the future in doubt.

Most older Americans are not affluent. In 2009, 3.4 million older Americans fell below the poverty line. There is enormous variation among subgroups of the aging and an even larger group of older people who have been brought above the official poverty line but still remain near poor and vulnerable.

The new politics of aging in America has been deeply influenced by the economic condition of older people and especially by recent improvement in their average economic status. This shift in mood marks a change from the past and provokes new controversies that will command attention for years to come. As we think about the future, it is important to keep in mind the key ideas emphasized in this discussion: (a) the diversity of the aging population, including dramatic differences in well-being among subgroups; (b) the impact of cumulative disadvantage, whereby earlier life course experience influences what old age will be like; and (c) the socially agreed-on expectations that we have about old age, including attitudes toward the meaning of retirement.

Population aging is an important and indisputable fact. But by itself, population aging does not dictate the shape of things to come, and it is certainly no cause to be gloomy about the coming of an aging society. For both individuals and society, the key point to remember is that the life one leads as a younger person will affect one's prospects for old age. The socioeconomic outlook for an aging society, then, is not simply something to be predicted, but something to be constructed. The decisions we make today depend on thoughtful consideration of the controversies that will shape the aging society of tomorrow (Steckenrider & Parrott, 1998).

Suggested Readings

Altman, Stuart H., and Shactman, David (eds.), *Policies for an Aging Society,* Baltimore: Johns Hopkins University Press, 2002.

Burkhauser, Richard, and Clark, Robert L., *The Economics of an Aging Society,* Oxford: Blackwell, 2004.

Butler, Robert N., *The Longevity Revolution: The Benefits and Challenges of Living a Long Life,* Public Affairs, 2008.

Coyle, Jean M., *Handbook on Women and Aging,* Praeger, 2001.

Schulz, James H., and Binstock, Robert H., *Aging Nation: The Economics and Politics of Growing Older in America,* Johns Hopkins University Press, 2008.

Controversy 8

SHOULD AGE OR NEED
BE THE BASIS FOR ENTITLEMENT?

Not so long ago, older adults as a group were thought of as uniformly poor and vulnerable, but image and reality have grown farther apart. In fact, the average condition of older Americans has improved markedly in recent decades when measured according to income, health status, and educational level. It is no longer surprising to see older people depicted in the media as well off or even prospering at the expense of the young, especially children. This new image leads to questions about fairness in how different generations are treated.

A TALE OF TWO GENERATIONS

The question of generational equity can be seen in the story of the Walton family. George and Martha Walton, members of the World War II generation, were proud of what they had accomplished. They were members of what Tom Brokaw has called the greatest generation. When George got out of the army, he and Martha were able to buy a house with a VA mortgage loan. George went to college on the GI Bill, and during the 1950s and 1960s, the Waltons enjoyed prosperity. When they retired, they found it possible to sell their house and use some of the capital gains, along with George's pension and Social Security benefits, to enjoy a comfortable income in retirement. The Waltons were never rich, but they are satisfied with how things turned out.

Things worked out differently for the Walton children. Carol and Robert were baby boomers born in the 1950s. They had comfortable childhoods and graduated from high school with high hopes. Both attended college and married, but they never seemed to be able to get ahead of their bills. They had trouble putting their own kids through school in the 1990s because college tuition had risen dramatically. Unfortunately, Robert lost his job because of downsizing; his oldest boy finished school, but couldn't find a job and ended up moving back home with his parents. Carol felt she needed to stay at work, but was forced to cut back her hours when her father, George, got sick. Now she helps out her mother, Martha, a lot. Now in midlife, Carol and Robert are seriously in debt and haven't managed to save much, and then the financial crisis beginning in 2008 hit them hard. The stock market went down, and home values declined. They worry now a lot about their own future.

Sometimes it seems as if life hasn't been fair to the Walton children and grandchildren. Some things that have happened to the different Walton generations are a matter of economic circumstance, such as rising home values and losing a job. Other things are a matter of government policy—for example, the impact of the GI Bill and Social Security. Regardless of whether the result is intended, circumstances and policies do affect different generations in different ways. How, then, can we sort out what is at stake in debates about justice between generations?

GENERATIONAL EQUITY

In thinking about issues of equity or fairness between age groups, it is clear that two different meanings of *generation* must be kept distinct: first, a specific age group, such as older adults (65+) or children under age 18, and, second, a historical cohort consisting of a group of people born in the same year or in a certain period (Ryder, 1965). As discussed earlier, we note that the term **cohort** describes those who experienced specific historical events (Mannheim, 1952), such as the World War II generation or the '60s generation.

Corresponding to these two meanings of generation are two different concerns about the fairness of distributing benefits on the basis of age. First, are today's older adults as a group receiving too much of society's resources in comparison to children? Second, can today's workers and their successors count on economic security—for example, Social Security benefits—in the future?

Stories like the Waltons' have become familiar. When these trends were first recognized, noted gerontologist Bernice Neugarten (1983) published a book titled *Age or Need?* This book asked a provocative question: Should government benefit programs for older people be available on the basis of need rather than chronological age? Are older people today getting too much at the expense of the young?

For many years, Americans were used to thinking of older people as the "deserving poor": financially dependent for reasons beyond their control and therefore deserving of help. This assumption was the basis for many important government programs, from Social Security to the Great Society. Not all older people fit this stereotype, of course, but a picture of the old as especially needy or vulnerable did become widespread—a form of "compassionate ageism," as some have called it (Binstock, 1985). But as government programs for the aged have become more expensive, another, less compassionate point of view was voiced. Critics began to depict the aged as a burden on society and to worry about the negative impact of population aging now and in the future. These concerns have become part of a debate about generational equity.

The generational equity debate is basically a controversy about whether older adults are receiving an excessive share of limited government resources in comparison with other age groups. There are at least four issues that fall under the label *generational equity* (Wisensale, 1999). First, there are questions about the allocation of resources between older adults and children: Are older people getting too much? Second, there is a concern about large government deficits, a concern intensified by huge deficits after the economic downturn beginning in 2008. Third, there is the issue discussed in the fourth controversy in this book about rationing health care resources. Many younger people doubt that they'll collect Social Security benefits in the future. So the question arises: Is Social Security really fair to younger people (Moody, 2009)? How can the Social Security system be made fair to all generations (Diamond & Orszag, 2005)?

To understand the many elements of the generational equity debate, we need to look at several trends, including the incidence of poverty among the old and the young, the relative burden of supporting a dependent aging population, and the impact of taxes and other public policies on different age groups. Finally, we consider the question of who among the older population are most in need of help.

Poverty Among the Old

By any measure available, there has been a dramatic reduction in the official rate of poverty among older Americans in the last three decades. As discussed in Basic Concepts III, as of 2009, there are 3.4 million older people below the poverty line, but that figure represents major progress. Note that most of the gains in income for older people took place between 1960 and 1974, long before complaints were heard that older people were getting too many benefits. For instance, in 1959, the poverty rate of older people in the United States was 35%, more than one out of three. According to U.S. census figures, the poverty rate among older adults has now dropped to a historic low of 8.9%, below the poverty level. The economic situation of children is quite different. Twenty-one percent of children under age 18 live in poverty, more than twice the rate for elders.

URBAN LEGENDS OF AGING

"Poverty among the old remains a major problem."

That was true in 1959 but not today; child poverty is substantially higher, and older people as a group are less likely to be poor than other adult groups. Near-poor older adults do face serious problems, however, and subgroups, including minorities, have high rates of old-age poverty. Social Security, above all, has been dramatically effective in reducing, though not eliminating, poverty among the old.

When Medicare is taken into account, the proportion of older people who are "poor" drops to only 6%. Here again we see the impact of successful entitlement programs. If we were to ignore all government cash transfer programs, such as Social Security, then 44% of Americans over age 65 would fall below the poverty line. This is another way of stating how important Social Security has been in preventing late-life poverty.

The official poverty line is not the whole story (Cook & Kramek, 1986). First, consider that 90% of people 65 and older receive Social Security, and that 64% get half or more of their income from Social Security (Social Security Administration, 2008). Second, if we look at the group of older adults who could be called "near-poor"—those with an income up to 1.5 times the poverty line—then more than one out of four older people are facing economic hardship. Exhibit 39 shows the degree to which older adults rely on Social Security as a major source of their later-life income.

Another group of elders has been called the 'tweeners. They are people caught "in between," the economically insecure lower-middle-class older adults. The 'tweeners are not well off economically, but they are not poor enough to be eligible for programs such as **Supplemental Security Income (SSI)**, a cash benefit program for older adults who are poor, are blind, or have disabilities that is administered by Social Security. Like food stamps (the Supplemental Nutrition

Exhibit 39   Shares of Aggregate Income, by Source, 2008

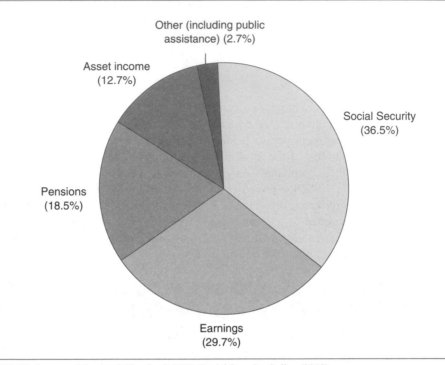

SOURCE: *Income of the Aged Chartbook,* 2008, Social Security Online (2010).

NOTE: Totals do not necessarily equal the sum of the rounded components.

Assistance Program) and welfare (Temporary Assistance for Needy Families), SSI is a **means-tested program**—that is, it is only available if your income and assets fall below a designated level.

But the 'tweeners do not qualify because they are slightly above the poverty level. When people in this in-between category encounter unexpected hardship, such as a rent increase, the death of a spouse, or an expensive illness requiring long-term care, they are extremely vulnerable.

The stereotype of the well-to-do older person is, therefore, a mistaken one. Many older people have been lifted out of poverty, but they still remain precariously perched in the "near-poor" category. In fact, the near-poverty rate among older people has hardly changed since the 1970s. Furthermore, the average statistics on poverty and aging conceal some important differences among subgroups of older people. Most discussion of generational equity revolves around averages, but these can be misleading.

Poverty Among Children and Young People

The current generation of children is clearly threatened—by everything from poor prenatal care to inadequate child care and a mediocre educational system. They have been called a "generation in crisis" (Hamburg, 1992; Moynihan, 1986). International comparison shows that U.S. poverty rates among children are higher than those of every industrialized nation in the world (Smeeding, 1990). The passage of tough welfare reform legislation in 1996 suggests that more government aid for poor children is unlikely in the near future.

Poverty among young people has been going up, whereas poverty among the old has been declining. But are government programs to help the aged responsible for that trend? It is true that government programs to help younger poor families, such as welfare and food stamps, are unpopular and have been reduced, first during the 1980s and later by federal legislation to "end welfare as we know it" passed in 1996. Public support for children at risk has declined sharply in recent years, and means-tested programs for families have been cut. By contrast, social entitlement programs benefiting retirees, such as Social Security and Medicare, have withstood most attacks on them.

Nevertheless, the basic cause of rising child poverty seems to have little to do with government entitlement programs. Defenders of old-age programs point out that the high poverty rate among younger generations is caused mainly by family structure, unemployment, and declining wages. In most countries, children living in single-parent families have poverty rates more than double those in two-parent families.

Some commentators have blamed the declining well-being of children on the voting power of older adults. Demographer Samuel Preston (1984) was one of the first to worry that the political power of older adults, along with their greater numbers, poses a threat to the well-being of American children. Many articles in the popular media have picked up this theme, but there is actually little evidence that older people as a group have voted to take resources from children (Rosenbaum & Button, 1989).

Still, it is a fact that families with children today form a smaller part of the electorate than in the past. The proportion of children in the population has declined with population aging, as the numbers in Exhibit 40 demonstrate. Even a program like Medicaid, which is intended

Exhibit 40   Percentage of Children and Older Adults in the Population, 1900, 1980, and 2030

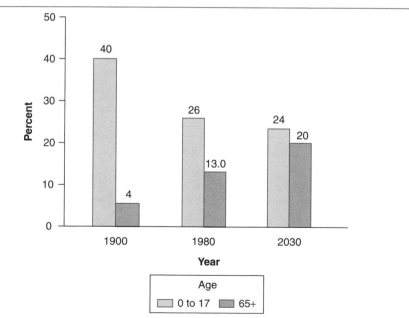

SOURCES: The 1900 figures exclude Alaska, Hawaii, and armed forces overseas. U.S. Census Bureau (1965), the 1980 and 2030 figures: U.S. Census Bureau (1989).

to provide health care for poor people of all ages, has tended to provide more and more benefits for older people as a group, whereas the benefits for children and families have remained stagnant, as the chart in Exhibit 41 makes clear. The main reason that per capita expenditure for older people under Medicaid has gone up is simply that Medicaid has become a major payer for long-term care, especially nursing homes. Thus, the reason that older people get more benefits is because more have survived into an old age with frailty.

The Dependency Ratio

The changing proportion of children and older adults in the population has prompted another question: Will we as a society be able to support such a large population of older people in the future? The support ratio, or **dependency ratio,** is a numerical measure of the economic burden imposed on members of the working population who must ultimately support people who are not in the labor force (Mirkin & Weinberger, 2000). If we look at

Exhibit 41   Average Real Medicaid Payments per Person Served, 1978–1998

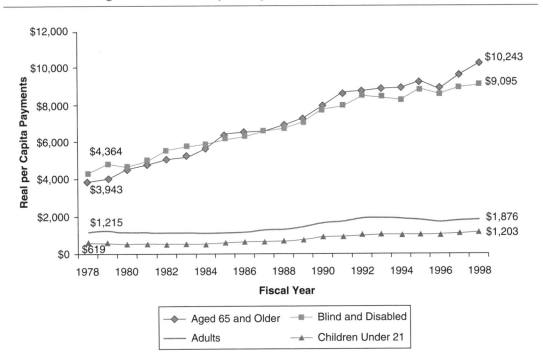

SOURCE: Centers for Medicare and Medicaid Services (2004).

NOTE: (a) Data shown above are expressed in 1998 dollars; (b) for FY 1998, "payments" describe direct Medicaid vendor payments and Medicaid program expenditures for premium payments to third parties for managed care (but exclude DSH payments, Medicare premiums, and cost sharing on behalf of beneficiaries dually enrolled in Medicaid and Medicare), whereas data from previous years only include direct vendor payments; (c) the term *adults* as used above refers to nonelderly, nondisabled adults; and (d) disabled children are included in the blind and disabled category shown above.

the aged alone, we can compute the dependency ratio by comparing the number of people over age 65 with the number in the working-age group (18–64). We can compute the dependency ratio for children by comparing the number of those under age 18 with the number in the working-age group.

Exhibits 42 and 43 show how both the youth and old-age dependency ratios have changed over time and are projected to change into the 21st century. Those who are fearful about the burden of an aging population point out that when Social Security was created in the 1930s, there were 50 workers to support each Social Security beneficiary. They point out with alarm that today, in the second decade of the 21st century, there are only 3 workers for each beneficiary. And there will be fewer and fewer workers per older beneficiary as we approach 2040 and the height of population aging. Thus, we will have a greater burden on a shrinking working-age population to support a larger group of older dependents.

To offset this scenario, it should be noted that the dependency ratio for children has declined dramatically since 1960 because of smaller family sizes. As a result, when we combine both ratios, we see an overall decline that is projected to continue declining until

Exhibit 42   Young, Older Adult, and Total Support Ratios: 1900–2050[a]

Year	65+	Under 18	Total
Estimates			
1900	7	76	84
1920	8	68	76
1940	11	52	63
1960	17	65	82
1980	19	46	65
Projections			
1990	20	41	62
2000	21	39	60
2010	22	35	57
2020	29	35	64
2030	38	36	74
2040	39	35	74
2050	40	35	75

SOURCE: U.S. Census Bureau (1988, 1989).

NOTE: a. Number of people of specified age per 100 people ages 18 to 64.

Exhibit 43   Dependency Ratios for the United States: 2010–2050

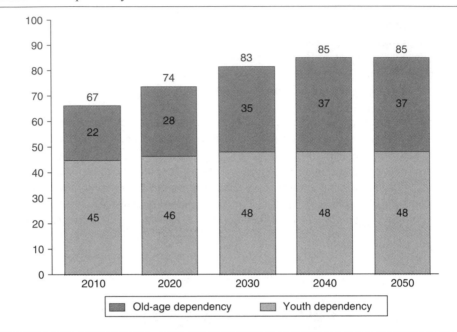

SOURCE: U.S. Census Bureau (2008).

NOTES: Total dependency = [(Population under age 20 + Population aged 65 years and over) / (Population aged 20 to 64 years)] * 100.

Old-age dependency = (Population aged 65 years and over / Population aged 20 to 64 years) * 100.

Youth dependency = (Population under age 20 / Population aged 20 to 64 years) * 100.

after 2010, when it begins to rise modestly.

The key point is that, because of the declining child-dependency ratio, the overall dependency ratio, including the burden of an aging population, is not excessive now and may not be in later decades. Although the cost of specific programs, such as Medicare, may pose problems, we should focus on the means by which such expenditures are financed and controlled, instead of making judgments based on overall dependency ratios in the abstract.

Taxation and Generational Accounting

Discussions of generational equity usually look only at the cost of government expenditures for older adults. Equally important is the role of the tax system with its far-reaching impact on different age groups and cohorts. Tax expenditures are distributed much less equally than are income or benefit entitlements such as Social Security and Medicare. For example, a number of tax breaks—sometimes called tax expenditures—go disproportionately to older people with higher incomes. For example, wealthier people with big mortgages get a much larger tax deduction than poor people who rent.

A key stimulus to the generational equity debate has been a growing awareness that not all members of the older population are poor or needy. Taking that idea seriously might mean replacing age-based entitlement programs, such as Medicare, with those that are accessible to persons of all ages. A related question is whether benefits, such as Social Security, should be taxed at progressive rates to take into account the heterogeneity of today's older population.

An illuminating method for analyzing generational equity has emerged from economics. It is a method known as **generational accounting**, as developed by economist Lawrence Kotlikoff (1992). The aim of this approach is to analyze how government tax and spending policies affect different cohorts. Generational accounting adds up all taxes paid to federal, state, and local governments over a lifetime and then subtracts benefits received, such as Social Security, Medicare, and schooling. For example, it turns out that baby boomers born in 1950 will pay an average lifetime tax rate about a third higher than their parents paid (31% as opposed to 24%). That difference means paying $200,000 more in taxes than one collects in benefits over a lifetime (Nasar, 1993). The problem has only increased with large federal budget deficits that have come since 2001 and continued into the 21st century.

Generational accounting provides a way of thinking about the long-range impact of budget decisions made today. Many controversies about Social Security can also be analyzed in terms of generational accounting. But critics of generational accounting and the generational equity framework argue that future projections about population or the economy are filled with uncertainty. Some have warned about what they call "apocalyptic demography," namely, the belief that an increasing aging population inevitably means an increasing burden on society. This scenario is closely coupled with fear of competition between age groups (Robertson, 1997). After all, the U.S. population has already aged considerably without terrible consequences. Demography alone doesn't determine the future; economic growth and government policies are also vital in shaping the world (Friedland & Summer, 1999).

Power and Competition for Scarce Resources

Does the debate about generational equity imply that the old and young are locked in conflict? We need to distinguish between the notion of conflict between generations and competition for different public programs. Conflict and competition are not the same thing. Age conflict does occur in some societies, such as when the aging patriarch of a family controls property (Foner, 1984). Similarly, it is possible to find examples of gerontocracy, or the rule of the aged, when groups cling to positions of power as they grow older. A current instance might be communist China, with its aging party leaders, yet even the Chinese case might better be described not so much as rule by the aged but rather as power through incumbency and authoritarian control (Eisele, 1979).

In the United States, there is no serious evidence of conflict or polarization between age groups. On the contrary, public support for Social Security and other benefit programs for older adults remains high. In fact, a recent survey by the National Academy of Social Insurance (2011) revealed that the majority of adults of all ages between 18 and 64 believe it is important that Social Security be preserved for future generations. But there can be competition over limited resources, and the political process may affect age groups differently, as when school bond issues are voted down or when Social Security benefits are taxed. These changes in taxes and spending are not motivated by hostility among age groups, however, and it would be a mistake to pit age groups against one another. However, it is also a mistake to overlook the fact that competition, choice, and trade-offs are always part of political life as periodic changes in Social Security have demonstrated (Light, 1985).

The fact that changes have taken place at all is proof enough that the political power of the aged has clear limits.

This older woman represented one of many generations participating in this health care demonstration.

The Least-Advantaged Older Adults

Since the 1980s, hard questions have been asked not only about how to pay for benefits to older Americans, but also about who should receive them (Binstock, 1994). Are government-funded services actually helping vulnerable older people? How should government target its resources to those most in need? How, in fact, do we agree on who the least advantaged are among older adults (Harel, Ehrlich, & Hubbard, 1990)?

Answers to this last question have included the following:

- The entire older population
- People above an advanced age
- Elders in minority ethnic groups
- Older women
- Rural or inner-city older people
- The physically or mentally frail
- Older people who are vulnerable to abuse or neglect
- Older adults living in poverty

The earliest answer to the question of who is least advantaged was that all older people should be viewed as vulnerable; in other words, age should be the criterion for need. This conviction originally inspired the Older Americans Act. Some said that older people, as a group, were subject to prejudice or bigotry and their treatment could be compared with that of minority groups. Instead of racism, one could speak of ageism, or stereotyped prejudice against older people, leading to discrimination and disadvantage (Butler, 1969). But is it helpful to think about older people as a minority group? That formulation has proved controversial (Streib, 1965).

It seems clear enough that the older population has changed in character in significant ways since the OAA was passed 40 years ago. Today, we can see many people in their 60s and 70s who are healthy and active: the so-called "well-derly" group. In contrast, because people are living longer, they often live into their 80s or beyond, and many in this group join the ranks of the "ill-derly." There is a good case to be made that disability and frailty, not chronological age, should be the basis of access to services (Fried et al., 2004).

Some have therefore suggested that the age of entitlement might perhaps be raised (Torres-Gil, 1992). For example, raising the age for Medicare eligibility from 65 to 67 would, over a period of years, save billions of dollars and prevent bankruptcy of the system. This change was made when the age of eligibility for Social Security was raised, now to age 66 and within a few years rising to age 67. But raising the age of eligibility would have a negative impact on minority groups, whose life expectancy is on average lower than that of the rest of the population.

Still another answer to the question of who is least advantaged might be the aged poor (W. Clark, Pelham, & Clark, 1988; Crystal, 1986). Indeed, most of the multiple disadvantages of the aged, regardless of gender or ethnicity, come from poverty, often a lifetime of low earning power (Nelson, 1982). Looking at disadvantage from a life course point of view underscores the importance of **socioeconomic status (SES)**, a term sociologists use to describe what is often known as *social class* (Salas, 2002). Instead of simply describing older people as rich, poor, or middle class, we can rank groups of people in terms of SES and then see how occupation, income, and educational background are interrelated over the course of a lifetime.

SES is one way to think about stratification, or the structured inequality in the distribution of power, prestige, or wealth among older people (Pampel & Williamson, 1989). Lower SES over the life course tends to produce **cumulative disadvantage**, which is perpetuated in old age. Higher SES tends to mean greater longevity and better health, as well as more income (Streib, 1984). Gender and race, along with class, create interlocking hierarchies of privilege and disadvantage, making it more difficult to identify a single characteristic that defines the least advantaged among the aged. Increasingly, we come to recognize that the condition of older people reflects the cumulative advantage and disadvantage they have experienced over an entire lifetime (Dannefer, 2003).

GLOBAL PERSPECTIVE

Vulnerable Elders in China

Traditional China was profoundly influenced by the ideas of Confucius, and the Confucian religion encouraged an attitude of filial piety—respect for elders—a practice that endured for thousands of years. Since 1949, communism has eroded the power of Confucianism. In recent decades, the impact of free markets in China has prompted far-reaching changes in the position

(Continued)

(Continued)

of older people in society. Finally, the introduction of a one-child policy means far fewer adult children are available to care for older people. In urban areas, it is less and less possible for multigenerational households to exist. In rural areas, pressing poverty means that few people are able to save for retirement. Women, in particular, face increased risk of poverty in old age.

Along with other countries around the world, China is facing its own demographic transition toward population aging. By the year 2050, it is estimated that there will be up to 100 million Chinese ages 80 or older. Yet most workers in China today do not have substantial pension coverage. The one-child policy along with falling birthrates means that in the future there will be fewer adult children to pay to support the old. In traditional China, the family was the prime protection against poverty or dependency in old age. Chinese government policy has favored rapid economic growth and international trade. But in the future, China will have to find new ways to ensure that adequate living standards also include elders in its population.

SOURCE: Richard Jackson and Neil Howe, *Graying of the Middle Kingdom: The Demographics and Economics of Retirement Policy in China,* Center for Strategic and International Studies, 2004.

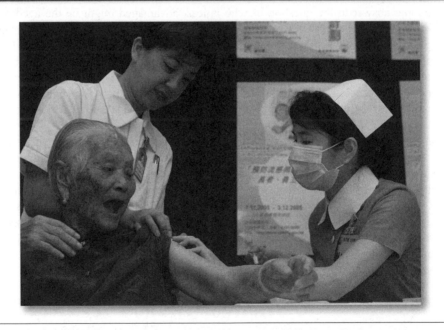

A nurse gives a free influenza vaccination to an older woman living in a nursing home.

HELP FOR THOSE MOST IN NEED

If need instead of age alone is to be the basis for distributing benefits or services, then how will we assess need as a practical matter? One popular method is to use a means test, a measure of eligibility based on whether a person's income or assets fall below a certain amount. For example, Medicaid, Supplemental Security Income, and food stamps are all programs with a means test. By contrast, Medicare and Social Security are not means tested.

A common argument against means-tested programs is that they are stigmatizing; that is, people who are forced by necessity to make use of such programs feel embarrassed and degraded. Even if eligible, many older people are reluctant to apply for Supplemental Security Income because, to them, it has the image of welfare. In the past, local governments took care of old-age poverty through a local poorhouse or almshouse, a familiar institution in American life dating back to colonial times (Achenbaum, 1978). Older people who descended into poverty or who had no family to care for them could be forced into the stigma of "outdoor" relief, a predecessor of today's welfare system.

Other methods of helping the least advantaged have been less stigmatizing, but have also taken account of different levels of need. The income tax system reflects a principle of progressive taxation or ability to pay: The higher your income, the higher the percentage of taxes owed. At the local level, where property taxes are important, many local governments have "circuit breaker" provisions or homestead exemptions that offer tax reductions or exemptions for low-income older homeowners. Since 1983, half of Social Security benefits for more affluent beneficiaries have been subject to federal taxation. Both a means test and progressive taxation recognize that ability to pay for benefits differs within the older population.

Finally, **cost sharing** is an approach that combines elements of means testing and taxation. OAA programs are not permitted to charge for their services, but Congress has never appropriated enough money to reach more than a small part of those who could be served by the programs. Some program administrators believe that, to expand the services, it is reasonable to charge recipients for services based on their ability to pay. For example, older adults with incomes at least 200% above the poverty line might pay a part of the cost of a service according to a sliding scale. Opponents of cost sharing fear that it will discourage the use of services by those least able to pay and that it will also begin to erode broad public support for universal programs (Kassner, 1992).

THE TARGETING DEBATE

There are different answers to the question: Who are the least-advantaged older people? For example, the OAA explicitly directs that the aging-service network target its services to "individuals with the greatest economic or social needs, with particular attention to low-income minority individuals." That legislative mandate has given rise to vigorous debate about targeting of benefits. A major legal case in the federal courts (*Meek v. Martinez,* 1987) disputed what criteria should be used for allocating resources to those "most in need."

The targeting debate has provoked strong differences of opinion about how universal programs, such as the OAA, can properly give preference to some needy groups (Jacobs, 1990). The debate has some similarity with disputes about financing and distributing the benefits of Social Security. On one side are those who believe that programs are most fair and effective when they are universal and open to all. On the other side are those who believe that the least advantaged need special consideration if they are to receive their fair share. The arguments here are reminiscent of debates about affirmative action programs. As the aging population becomes more diverse and as long as all public expenditures remain limited, there will be controversy about age or need as a basis for entitlement.

In health care entitlements, the age versus need debate takes on a different form when it comes to health care. As we have seen, benefit programs such as Medicare can be broadly described as universal for those meeting age eligibility requirements. Other benefits, such as Medicaid, are targeted toward the poor, including the older poor. But the division between "universal" and "targeted" programs is not so simple (Grogan & Patashnik, 2003). In 2003, Congress passed the Medicare Prescription Drug, Improvement, and Modernization Act, which includes some provision for means testing for access to Medicare benefits. The intent was to make sure that the older poor and people with high drug expenses would get more benefits under the new law. Since 1983, Social Security has included a tax on higher-income beneficiaries: again, a way of targeting resources on those in greatest need. In summary, even "universal" age-based programs such as Social Security and Medicare have come to include features that target benefits to the least advantaged.

The readings that follow present sharply different views about the fairness of social programs that help older adults. On one side is Peter Peterson, who takes aim at the politics of aging in the United States. He compares population aging to a giant iceberg that threatens to sink the ship of state if we don't change course. The political power of the old, he believes, has caused us to spend too much on older people and not enough on children. He believes that need, not chronological age, should be the basis for how government helps people. In Peterson's view, universal, age-based entitlements are a big problem.

Echoing Peterson's sentiments is economist Lester Thurow, who puts the problem of entitlements in historical perspective. Thurow believes that the United States has promised too much to older people, who are no longer a needy group. Like Peterson, he argues that older people have too much political power and may vote against the interest of other generations, as in Thurow's example of older voters who reject spending on public schools.

By contrast, Meredith Minkler believes the entire debate about generational equity is misguided. She argues that talking about "greedy geezers" is actually a new kind of victim blaming that distracts attention from the real issues of injustice arising from social class along with racial and gender inequalities. Moreover, public benefit programs directed at the aged actually help people of all ages, among other reasons because they relieve middle-aged children of the need to support aged parents.

In the last reading, John Williamson, Diane Watts-Ray, and Eric R. Kingson share Minkler's skepticism about the generational equity debate, but they offer a framework within which we can understand the conflicting values at stake in this debate. Proponents of the generational equity view are looking only at one kind of equity or fairness. Those who favor interdependence between age groups have a broader view about fairness that puts the emphasis on sharing risks through a system of social insurance.

Policy makers and service providers face difficult choices about how to distribute benefits to those in greatest social or economic need. The debate about who are the least advantaged among the older population will continue to be a challenge for the future.

FOCUS ON PRACTICE

Intergenerational Programs

Interest in intergenerational programs today is a response to changes in relationships among age groups in contemporary society. In the past, the family or the local community typically provided informal opportunities for contact across age groups, especially between the very old and the very young. But one of the consequences of modernization has been the rise of age grading and grouping. For example, the educational system introduces bureaucratic categorization and separation of children by age (Aries, 1962). Moreover, a youth culture and popular media, combined with the rapid pace of technological change, all tend to accentuate cultural separation between the young and the old.

Another important tendency has been age segregation, or the residential separation of people of different ages. Age segregation has been promoted by trends such as migration, urbanization, and income transfer programs such as Social Security, which enable aged parents to live apart from their children. Age segregation may, however, pose problems in a society where young and old are both competing for limited public resources.

The challenge of the generational equity debate has been one factor stimulating programs to bring together young and old (Wilson, 1992). Such deliberate programs seem more necessary because children today often lack the opportunity for close and frequent contact with grandparents. Intergenerational programs can meet a vital need for age integration (Haber & Short-DeGraff, 1990).

One example of a successful intergenerational program is Foster Grandparents, a federally funded effort that recruits older volunteers to work in settings such as schools, day care centers, hospitals, and homes for children with disabilities. Foster grandparents provide one-on-one nurturing for children who need affection and personal attention. Foster grandparents are trained for and supervised in their tasks. Foster grandparents are typically members of lower-income groups and therefore receive a modest but significant cash stipend, as well as a transportation allowance and a hot meal. Evaluation studies have shown that Foster Grandparents can help reintegrate older people into society and give them meaningful and appreciated roles in helping younger generations (Ziegler & King, 1982), though more recent studies have addressed the complexity of foster grandparenting—the benefits and challenges—as well as cultural dynamics that influence the experience (Hayslip & Goodman, 2007).

Intergenerational programs have been successfully demonstrated in a wide variety of settings, including nursing homes. In some instances, nursing homes have also served as sites for child care programs, sparking the idea of intergenerational day care (Kopac, 1987). Some prominent corporations, such as Stride Rite, have sponsored on-site intergenerational day care (Leibold, 1989). The National Council on Aging has sponsored an initiative known as Family Friends, a nationwide outreach program to enlist seniors to visit homeless shelters. Schools have proved to be a particularly attractive site for intergenerational programs (Aday, Rice, & Evans, 1991). The National School Volunteer Program has recruited thousands of older people to tutor children in classrooms (Tierce & Seelbach, 1987).

The real benefits of intergenerational programs seem to be of two kinds: direct service, allowing young and old to serve as resources for vulnerable members of other generations, and attitude change, overcoming some of the feelings of cultural distance between old and young. On the direct service side, older adults with time available during the day have been a source of social contact for latchkey children and at-risk youth (Ventura-Merkel & Friedman, 1988). But the young, too, have shown themselves to be a valuable source of volunteers for aging network service organizations (Firman, Gelfand, & Ventura, 1983).

With respect to attitude change, there is evidence that positive opportunities for contact between old and young can help reduce ageism and prejudice on the part of the young (Dellman-Jenkins, 1986; Dobrosky & Bishop, 1986; Peacock & Talley, 1984). For instance, in oral history

(Continued)

(Continued)

programs, older people can become vital resources for improving the education and cultural knowledge of young people. Among the most successful efforts at intergenerational communication are the Stagebridge Senior Theater Company in Oakland, California, and the Roots and Branches Theatre in New York, which enlist old and young actors in portraying intergenerational issues, relationships, and attitudes. Apart from services and attitude changes, intergenerational programs also help renew the bond between generations. That bond is needed support for important public policies, such as Social Security and public education.

Significantly, AARP and other senior advocacy groups have come to the forefront in vigorously supporting more government spending for children's programs. In addition, AARP and the National Council on Aging have been active in coalition with the Children's Defense Fund on behalf of Generations United, a national coordinating body for intergenerational programs (Thursz, Liederman, & Schorr, 1989). This interest in intergenerational programs is part of an ongoing dialogue between young and old, including the debate around generational equity. By overcoming stereotypes and misunderstanding, it is possible to strengthen the bonds between age groups in decades to come, leading to more enlightened policies for taking care of both young and old.

READING 34

——— Growing Older ———

Lester Thurow

. . . The really explosive part of the volcano pushed up by demography, however, lies in the aging of the world's population. A new class of people is being created. For the first time in human history, our societies will have a very large group of economically inactive elderly, affluent voters who require expensive social services such as health care and who rely on the government for much of their income. They are bringing down the social welfare state, destroying government finances, and threatening the investments that all societies need to make to have a successful future. . . .

Spending on the elderly is not an issue of equity or deprivation. In 1970 the percentage of the elderly in poverty was higher than for any other part of the population. Now there are fewer poor people among the elderly than any other group in the population. For many in the United States, real standards of living actually rise with retirement. . . .

The elderly obviously don't want their benefits cut. The alternative is raising taxes, but that is also a very unattractive option. Today's 15 percent social security tax rate would have to be boosted to 40 percent by 2029 to provide the benefits that have been promised. Moving out farther and assuming no changes in current laws, future payroll tax rates can rise as high as 94 percent if one is pessimistic about controlling health care spending on the elderly. What is called generational accounting leads to some very disturbing future tax rates. The tax system implodes. . . .

Technically, in the United States the elderly can argue that the pension part of their benefit package does not contribute to government debts, since the pension part of social security, but not the health care part, is running a surplus—tax revenues from earmarked taxes exceeds expenditures. But that is an illusion. To evaluate the impact of government budgets, it is necessary to look at total revenue and total expenditures as a whole. If governments are running a deficit in their overall budgets, the fact that one part of the budget has a surplus because there is an accounting convention that earmarks more taxes than this sector of the budget needs is irrelevant. Whatever the earmarking, governments are net dissavers and that is what affects the economy. What matters is what is driving the expenditure side of the budget. The driver is the elderly. . . .

While all of our economic resources are not going to be given to the elderly (there are other things such as police and fire departments that simply have to be financed), no one knows how the growth of entitlements for the elderly can be held in check in democratic societies. Even when they are only 13 percent of the population, they are so powerful that no political party wants to tangle with them. . . .

Long before they are a technical majority of the population, the elderly will be unstoppable

SOURCE: Pp. 96, 98–105 from *The Future of Capitalism* by Lester Thurow. Copyright © 1996 by Lester C. Thurow. Reprinted by permission of HarperCollins Publishers.

politically, since those under age eighteen legally cannot vote and those between eighteen and thirty tend not to vote. Democracy is going to meet its ultimate test when it has to confront the economic demands of the elderly. Can democratic governments cut benefits for a group of voters that will be close to being a majority? . . .

The political problems are not created entirely by the political power of the ever more numerous elderly. Means-tested benefits (benefits that decrease as income and wealth increase) would result in dramatic cost savings, but it is not just the elderly with above-average incomes and wealth who are opposed to it. All of us will eventually become old and all of us, especially the near elderly, would rather use our own money for the luxuries of life and let the government pay for our necessities when we are old. Less generous programs are ultimately less generous programs for *us* and not just for *them.*

Even for the young not yet worried about retirement, the shift toward making the elderly pay more of their own bills is not without a downside. The shift means that the young may have to pay, or feel the guilt associated with not paying, for some of the costs of taking care of their parents if their parents don't budget their income appropriately. More dire, for those with parents who have assets, making the elderly pay means smaller inheritances. One will not inherit the house or the stock portfolio that one expected to inherit if it was sold to pay medical bills or provide the equivalent of a monthly pension. The young would rather not lose their inheritances.

The political message is simple. Targeting benefit to low-income elderly families reduces costs and improves economic efficiency (gets the money to those who need it most), but it quickly loses political support. . . .

If one is looking for a group in need, it is not the elderly. The group with the highest proportion now in poverty is children under the age of eighteen. Yet government spends nine times as much per person on the elderly (those who do vote) as it does on the young (those who don't vote). Precisely the group that most needs investments if there is to be a successful American economy in the future is the group that is getting the least. How are they to pay taxes to support the elderly if they don't have the skills to earn their own incomes?

In the years ahead, class warfare is apt to be redefined to mean not the poor against the rich but the young against the old. . . .

The most dramatic recent example of impending social conflict occurred in Kalkaska, Michigan, a retirement haven, where elderly voters essentially robbed the school budget to pay for other things such as snow plowing and then refused to vote the funds to allow the schools to finish the school year. Schools closed months early and some of Michigan's schoolchildren missed much of a year's schooling. While the elderly are probably still interested in their own grandchildren, they no longer live in the same communities with their grandchildren. Each of those elderly Kalkaska voters could vote against educating someone else's grandchildren yet convince themselves that somehow voters elsewhere in America would treat their grandchildren differently.

The implicit post–World War II social contract has been that parents will take care of children but that society, the taxpayer, will take care of parents. Both parts of that bargain are collapsing. More and more parents are not taking care of children, and the taxpayer is going to have to retreat from his promise to take care of the elderly.

READING 35

———— Will America Grow Up Before It Grows Old? ————

Peter G. Peterson

The long gray wave of Baby Boomers retiring could lead to an all-engulfing economic crisis—unless we balance the budget, rein in senior entitlements, raise retirement ages, and boost individual and pension savings. Yet politicians of both parties say that most of the urgently necessary reforms are "off the table."

A NATION OF FLORIDAS

Been to Florida lately? You may not realize it, but you have seen the future—America's future, about two decades from now. The gray wave of senior citizens that fills the state's streets, beaches, parks, hotels, shopping malls, hospitals, Social Security offices, and senior centers is, of course, an anomaly created by our long tradition of retiring to Florida. Nearly one in five Floridians is over sixty-five. But early in the next century a figure like that won't be exceptional. By 2025 at the latest the proportion of all Americans who are elderly will be the same as the proportion in Florida today. America, in effect, will become a nation of Floridas—and then keep aging. By 2040 one in four Americans may be over sixty-five. . . .

To provide for the largest generation of seniors in history while simultaneously investing in education and opportunity for the youth of the twenty-first century, we must reject the prevailing "entitlement ethic" and return to our former "endowment ethic," which generated America's high savings, high growth, and rising living standards in the past. Endowment implies

"stewardship"—the acceptance of responsibility for the future of an institution. But given our current emphasis on individual self-fulfillment, we must, in addition to endowing the future of our nation and its institutions, endow our individual futures and those of our children, because no one else is going to do it for us. What I am talking about is self-endowment.

"Hope I die before I get old," The Who sang in their classic sixties anthem, "My Generation." That statement, like so many slogans of the Baby Boomers' youth culture, was wishful thinking. The generation that once warned "Don't trust anyone over thirty" is now passing fifty.

The real question is, Will America grow up before it grows old? Will we make the needed transformation early, intelligently, and humanely, or procrastinate until delay exacts a huge price from those least able to afford it—and confronts us with an economic and political crisis to which there is no longer a win-win solution? . . .

In 1900 only one in twenty-five Americans was over sixty-five. The vast majority of these people were completely self-supporting or supported by their families. By 2040 one out of every four or five Americans will be over sixty-five, and the vast majority will be supported to some degree by government entitlements. . . .

In 1970 children under five outnumbered Americans aged eighty-five and over by twelve to one. By 2040 the number of old old will equal the number of preschool children, according to some forecasts. . . .

Of course, the United States is not the only country facing an "age wave." Indeed, the age

SOURCE: "Will America Grow Up Before It Grows Old?" by P. G. Peterson in *The Atlantic Monthly, 277*(5), May 1996, pp. 55–86.

waves in most industrial countries are approaching faster than ours, and—to judge by official projections—could have an even worse impact on their countries' economies and public budgets. But these other countries enjoy long-term defenses that we lack. Unlike the United States, most can actually budget their public spending on health care, and so have much greater control over this potentially explosive dimension of senior dependency. Unlike the United States, most generally tax public benefits as they do any other income. . . .

Most important, unlike the United States, these other countries are unencumbered by the illusion that their people have some sort of inalienable right to live the last third of their adult lives in subsidized leisure. In other countries what government gives can be taken back if doing so is deemed to be in the public's long-term interest. . . .

The senior lobby asserts that whatever the economic consequences, future American workers are duty-bound to fulfill their side of an ill-defined "contract between generations." Yet one group's "earned right" to a benefit is another group's "unearned obligation" to pay a tax. It is to this second group that our children and grandchildren belong. Understandably, they are suspicious of a binding "contract" to which they never agreed. . . .

Our structural deficits drain our already shallow pool of private savings—and hence crowd out private investment. To the extent that we try to control these deficits by reducing "discretionary" federal spending (a category that includes most future-oriented programs), they also crowd out public investment. Out of every nondefense dollar the federal government now spends, only about five cents builds tangible things that remain after the fiscal year is over. Recently a General Accounting Office study suggested that we must invest $112 billion to bring the infrastructure of schools back to acceptable levels. But where can we find such a sum when entitlements and interest on old debts crowd out everything else? . . .

In an aging America everything will depend on the skills, education, productivity, and civic good will of younger generations—for their labor must support the elderly. Yet nothing seems less obvious than their capacity to rise to the challenge we are passing on to them. They will be relatively few in number. They will inherit a huge national debt and a high and rising payroll-tax burden. To make matters worse, many more of these future adults than today's adults are growing up in families, neighborhoods, and schools plagued by economic hardship and social dysfunction.

Since 1973 the real median income of households headed by adults aged sixty-five and over has risen by more than 25 percent, while the real median income of households under age thirty-five has fallen more than 10 percent. Counting all sources of income, poverty in America is three times as likely to afflict the very young as the very old. The United States is the global leader in the life expectancy of eighty-five-year-olds but has fallen near the bottom of the industrial world's rankings in rates of infant mortality, marital breakup, child poverty, child suicide, hours of school-assigned homework, and functional illiteracy. Meanwhile, per capita federal spending on the elderly towers eleven to one over federal spending on children. The appropriate response to the outrageous is to be outraged, yet we seem oblivious of this devastating disproportion.

How can we remain an economic superpower when nearly a third of our children are born out of wedlock and few of their fathers are willing to assume legal, financial, and moral responsibility for them? How will America prosper in a competitive technological and information-based global economy when its children grow up to exhibit school-dropout rates and rates of functional illiteracy that are among the highest in the industrial world? How do we answer Senator Daniel Patrick Moynihan's haunting question: "Will we be the first species that forgets how to raise our young?" Or, to paraphrase Churchill, "Have we ever asked so much from so few, having done so little to prepare them for their burdens?"

We're talking not about physical capital but about human and social capital: the intact families, work habits, education, and high-tech skills upon which any hope of increasing productivity ultimately rests. If we are going to rely on just 1.6 to 2.0 workers to support every retiree, as the SSA forecasts suggest, we should want today's children to become the best educated, most skilled, and most productive citizens imaginable. How does that square with our current rush to cut discretionary spending and defund social programs, from Head Start to vocational schools, that have long provided education and training? How can we generate the funding and the political support to educate our young in today's overburdened economy? How can we make the twenty-first century the century for our children?

THE POLITICAL TRANSFORMATION

Today's seniors, represented by powerful lobbies and voting in disproportionate numbers compared with the young, are already a potent political force. Will the rapid growth in the number of elderly enthrone the senior lobby as an invincible political titan? Or will the young, who must pay for tomorrow's senior benefits, find their political voice while there's still time to do something about it? Averting a destructive conflict between the generations will require a political transformation. But how can the young be encouraged to participate more aggressively in the political process? How do we merge the public interests of young and old and show how dangerous it is for them to become adversaries? . . .

Trying to achieve long-term budget balance without reforming entitlements is like trying to clean out the garage without removing the Winnebago. The following reforms, taken together, would put these programs in long-term sustainable balance well into the twenty-first century.

Subject all federal benefits to an "affluence test." The first sensible step toward long-term budget balance is to scale back entitlement subsidies flowing to people who don't need them. To this end I recommend that we enact a comprehensive "affluence test" that would progressively reduce entitlement benefits to all households with incomes over $40,000—or more than $5,000 above the U.S. median household income for 1996. Households with lower incomes would retain all government benefits. The affluence test would be applied annually—protecting the elderly in the event of an unexpected loss of income. Higher-income households would lose 10 percent of all benefits that raised their income above $40,000, and 10 percent for each additional $10,000 in income. Thus a household with $50,000 in total income and $10,000 in federal benefits would lose $1,000, or 10 percent of its benefits; a household with $100,000 in income and the same $10,000 in benefits would lose $6,000, or 60 percent; a household with more than $120,000 in income would lose $8,500, or 85 percent—the maximum benefit-withholding rate. (This 15 percent exemption would ensure that even today's most affluent beneficiaries continue to enjoy a respectable tax-free return on their personal FICA contributions.) All income brackets would be indexed for inflation.

Because the test would leave in place all benefits to lower-income households, the original "floor-of-protection" intention of nearly all federal benefits programs would continue to apply. Because such a large share of entitlements now goes to middle- and upper-income Americans (nearly 40 percent of Social Security payments go to recipients with incomes above the U.S. median), savings would be large and would compound as the population aged and the number of beneficiaries grew. Indeed, it is estimated that by 2040 annual savings would amount to more than $550 billion. Finally, because the test would also be comprehensive, covering not just Social Security and Medicare but everything from farm aid to federal pensions to veterans' benefits, this plan would not pit one special-interest constituency against another. . . .

John F. Kennedy once challenged us to ask not what our country can do for us but what we can do for our country. Today's youth see the most conspicuous interest groups in our political system busily asking what the country can do for them. But who represents the future and the general interest? The young, alas, are the new silent majority. The demographer Samuel Preston once remarked apropos of the relentless growth in senior entitlements that the political system would behave a lot differently if people were forced to live their lives backward—that is, if they had to look forward to the burdens imposed upon youth as their own future.

CONCLUSION

And what of the special role for geezers like me? Pessimists say, "Forget it"—Americans will not reform senior benefits until a severe crisis is actually upon us, but will persist in viewing them as contractual obligations that by definition are always affordable. After all, an America that acknowledges limits is an America that has lost the one illusion that makes it unique and creative. According to this view, America must always be

an unteachable force of nature that can never back away from any promise or expectation, no matter how extravagant. This, pessimists say, is why American voters repeatedly elect leaders who promise lower taxes, higher benefits, rejuvenated economic growth, and a magic bullet for every social problem—without caring how the pieces fit together.

But I have a more optimistic view. . . . German theologian Dietrich Bonhoeffer said it best for us when he observed, at the height of the Second World War, "The ultimate test of a moral society is the kind of world that it leaves to its children." . . .

We will have to balance our public budgets, trim back benefits to those who need them least, save more as households, retire somewhat later from the work force, explore innovative means of economizing on health care, take a more effective public interest in the welfare of children, and offer the rising generation some tangible evidence that we are willing to make sacrifices in their behalf. If we do so sooner, we still have time to plan for a gradual and humane transformation. If we do so later, the changes are likely to be forced upon us, suddenly and painfully, in the midst of an economic, political, and family crisis that will leave the eventual outcome much in doubt.

READING 36

—— "Generational Equity" and the New Victim Blaming ——

Meredith Minkler

This policy and reverence of age makes the world bitter to the best of our times; keeps our fortunes from us till our oldness cannot relish them. I begin to find an idle and fond bondage in the oppression of aged tyranny, who sways, not as it hath power, but as it is suffer'd.

—*King Lear,* 1, ii, 47

SOURCE: "'Generational Equity' and the New Victim Blaming" by Meredith Minkler in *Critical Perspectives on Aging* (pp. 67–79), edited by Meredith Minkler and Carroll Estes. Amityville, NY: Baywood, 1991.

Stereotypic views of the elderly in the United States as a wealthy and powerful voting block have replaced earlier stereotypes of this age group as a poor, impotent, and deserving minority (1). Capitalizing on and contributing to these changing perceptions, the mass media and a new national organization, Americans for Generational Equity (AGE), prophesize a coming "age war" in the United States, with Social Security and Medicare as the battleground (2). Like Gloucester's young son in *King Lear,* they argue that entitlement programs for today's affluent elderly "mortgage our children's future" and contribute to high poverty rates among the nation's youth.

The framing of complex public policy issues in terms of conflict between generations, however, tends to obscure other, far more potent bases of inequities in our society. Indeed, in Binstock's words (1, pp. 437–438):

> To describe the axis upon which equity is to be judged is to circumscribe the major options available in rendering justice. The contemporary preoccupation with . . . intergenerational equity blinds us to inequities within age groups and throughout society.

This [essay] will examine the assumptions underlying the concept of generational equity, with particular attention to notions of fairness and differential stake in the common good. Tendencies by policy makers, scholars, and the mass media to statistically homogenize the elderly and to utilize inadequate and flawed measures of poverty further will be examined and seen to contribute to the myth of a monolithic and financially secure elderly population. Finally, the false dichotomy created between the interests of young and old will be found to illustrate a new form of victim blaming, whose employment is inimical not only to the elderly but to the whole of society.

THE MAKING OF A SOCIAL PROBLEM

In his trenchant look at the cyclical nature of social problems, O'Conner noted that when the economy is perceived in terms of scarcity, social problems are redefined in ways that permit contracted, less costly approaches to their solution (3). Thus, while the economic prosperity of the 1960s permitted us to discover and even declare war on poverty, the recession of 1973 and its aftermath resulted in a redefinition of poverty and the subsequent generation of less costly "solutions."

The discovery of high rates of poverty and poor access to medical care among America's elderly in the economically robust 1960s and early 70s helped generate a plethora of ameliorative programs and policies including Medicare and Social Security cost of living increases (COLAs). By the mid-1970s however, amid high inflation and unemployment, Social Security and Medicare were themselves being defined as part of the problem. By implication, the elderly beneficiaries of these programs frequently were characterized by the mass media as targets of special resentment. The "compassionate ageism" which had enabled the stereotyping of the elderly as weak, poor and dependent, from the 1930s through the early 1970s, (1) gave way to new images of costly and wealthy populations whose favored programs were "busting the federal budget."

In the mid-1980s, a new dimension was added to this socially constructed problem when the rapidly increasing size of the elderly population and the costliness of programs like Social Security and Medicare were directly linked to the financial hardships suffered by younger cohorts in general and the nation's children in particular. Indeed, in a widely publicized article in *Scientific American,* Samuel Preston, then President of the Population Association of America, argued that the elderly now fare far better than children in our society (4): In the twelve years from 1970 to 1983, he pointed out, the proportion of children and elders living in poverty was reversed, with the proportion of children under fourteen living in poverty growing from 16 percent to 23 percent and the percent of elderly poor dropping from 24 percent to only 15 percent, while public outlays for the two groups had remained relatively constant through 1979. Moreover, many public programs for children were cut back in the 1980s, at the same time that programs for the politically powerful elderly were expanded. Preston's widely quoted paper went on to compare the elderly and children on a variety of parameters including suicide rates and concluded with a plea of redressing the balance of our attentions and resource allocations in favor of youth.

Following closely on the heels of Preston's analysis, the President's Council of Economic Advisors reported in February of 1985 that the elderly were "no longer a disadvantaged group," and indeed were better off financially than the population as a whole (5).

These two publications helped generate a new wave of media attention to the "costly and wealthy elderly." They further provided much of the impetus for a new, would-be national organization devoted to promoting the interests of younger and future generations in the national political process. With the backing of two prominent congressmen and an impressive array of corporate sponsors, Americans for Generational Equity (AGE) attacked head on government policies which, under pressure from a powerful gray lobby, were seen as creating a situation in which "today's affluent seniors are unfairly competing for the resources of the future elderly" (6).

The mass media, AGE, and other proponents of an intergenerational conflict framework for examining current U.S. economic problems have successfully capitalized on growing societal concern over certain facts of life which have coincided with the graying of America. These include:

- a massive federal deficit;
- alarming increases in poverty rates among children, with one in five American children now living in poverty (7); and
- a 76 million strong "baby boom" generation whose real incomes have declined 19 percent over the last fifteen years.

These statistics have been coupled with another set of facts and figures used to suggest that the growing elderly population is itself part of the problem:

- The elderly, while representing only 12 percent of the population, consume 29 percent of the national budget and fully 51 percent of all government expenditures for social services (8);
- Since 1970, Social Security benefits have increased 46 percent in real terms, while inflation-adjusted wages for the rest of the population have declined by 7 percent (9).

The picture presented is one of a host of societal economic difficulties "caused," in part, by a system of rewards that disproportionately benefits the elderly regardless of their financial status.

The logic behind the concept of generational equity is flawed on several counts, each of which will be discussed separately. In addition to these specific inaccuracies, however, a broader problem will be seen to lie in an approach to policy which lays out the issues in terms of competition among generations for scarce resources. Each of these problems will now be discussed.

THE MYTH OF THE HOMOGENEOUS ELDERLY

Basic to the concept of generational equity is the notion that elderly Americans are, as a group, financially secure. Borrowing statistics from the President's Council of Economic [Advisors], proponents of this viewpoint argue that the 1984 poverty rate for elderly Americans was only 12.4 percent (compared to 14.4 percent for younger Americans), and dropped to just 4 percent if the value of Medicare and other in-kind benefits was taken into account (5).

While the economic condition of the elderly as a whole has improved significantly in recent years, these optimistic figures obscure several important realities. First, there is tremendous income variation within the elderly cohort, and deep pockets of poverty continue to exist. Close to a third of black elders are poor, for example (32%), as are 24 percent of older Hispanics and 20 percent of all women aged eighty-five and above (10).

Minority elders and the "oldest old," aged eighty-five and above, not only have extremely high rates of poverty but also comprise the fastest growing segments of the elderly population. Thus, while only about 8 percent of blacks are aged sixty-five and over, compared to over 13 percent of whites, the black elderly population has been growing at a rate double that of the white aged group. The number of black elders further is increasing at twice the rate of the younger black population (11). In a similar fashion, the "oldest old" in America—those aged eighty-five and above—are expected to double in number by the turn of the century, from 2.5 million in 1985 to some 5 million by the year 2000 (12). The very high rates of poverty in the current generation of "old old" may reflect in part a Depression-era cohort effect. At the same time, however, the heavy concentration of women in the eighty-five and over age group, coupled with continued high divorce rates and pay and pension inequities, suggest a significant continuing poverty pocket as the population continues to age (13).

The myth of a homogenized and financially secure elderly population, in short, breaks down when the figures are disaggregated and the diversity of the elderly is taken into account.

PROBLEMS IN THE MEASUREMENT OF POVERTY

Analyses which stress the low poverty rates of the aged also are misleading on several counts. First, as Pollack has noted, comparisons which stress the favorable economic status of the aged vis-à-vis younger cohorts fail to acknowledge the use of two separate poverty lines in the United States—one for those sixty-five and above and the other for all other age groups (14). The 1987 poverty line for single persons under sixty-five thus was $5,905—fully 8 percent higher than the $5,447 poverty line used for elderly persons living alone (15). If the same poverty cutoff had been used for both groups, 15.4 percent of the elderly would have fallen below the line, giving the aged a higher poverty rate than any other age group except children.

The inadequacy of even the higher poverty index also merits attention. It is telling, for example, that Molly Orshanksy, the original developer of the poverty index, dismissed it some years ago as failing to accurately account for inflation. By her revised estimates, the number of elderly persons living in or near poverty almost doubles (16).

Discussions of the role of Social Security and other in-kind transfers in lifting the elderly out of poverty also are problematic. Blaustein thus has argued that while Social Security and other governmental transfers helped lift millions of elders out of poverty, they for the most part succeeded in "lifting" them from a few hundred dollars below the poverty line to a few hundred dollars above it (17). Indeed, some 11.3 million elders, or 42.6 percent of the elderly, live below 200 percent of the poverty line, which for a person living alone is about $10,000 per year (15).

Recent governmental attempts to reduce poverty by redefining it also bear careful scrutiny in an effort to uncover the true financial status of the elderly and other groups. The argument that poverty in the aged drops to 4 percent where Medicaid and other in-kind transfers are taken into account thus is extremely misleading. By such logic, an elderly woman earning less than $5,000 per year may be counted as being above the poverty line if she is hit by a truck and has $3,000 in hospitalization costs paid for by Medicaid. The fact that she sees none of this $3,000 and probably incurs additional out-of-pocket health care costs in the form of prescription drugs and other deductibles is ignored in such spurious calculations (14).

The continuing high health care costs of the elderly are themselves cause for concern in any attempt to accurately assess the income adequacy of the elderly. The elderly's out-of-pocket health care costs today are about $2,400 per year—three and one-half times higher than that of other age groups, and higher proportionately than the amount they spent prior to the enactment of Medicare and Medicaid more than two decades ago (18). Contrary to popular myth, Medicare pays only about 45 percent of the elderly's medical care bills, and recipients have experienced huge increases in cost sharing (e.g., a 141 percent increase in the Part A deductible) under the Reagan administration (14). Inflation in health care at a rate roughly double that of the consumer price index further suggests that the de facto income adequacy for many elderly may be significantly less than the crude figures imply.

MORAL ECONOMY AND THE CROSS-GENERATIONAL STAKE IN ELDERLY ENTITLEMENTS

Another criticism of the logic behind intergenerational equity lies in its assumption that the elderly alone have a stake in Social Security, Medicare, and other governmental programs which are framed as serving only the aged. Arguing that the nation's future "has been sold to the highest bidder among pressure groups and special interests" (6), Americans for Generational Equity and the mass media thus cast Social Security and other income transfers to the aged in a narrow and simplistic light. For even if one disregards the direct benefits of Social Security to nonelderly segments of the society (e.g., through survivors benefits to millions of persons under age sixty-five), the indirect cross-generational benefits of the program are significant. By providing for the financial needs of the elderly, Social Security thus frees adult children from the need to provide such support directly. As such, according

to Kingson et al., it may reduce inter-familial tensions while increasing the dignity of elderly family members who receive benefits (19). Research by Bengtson et al. has suggested that the family is not perceived by any of the major ethnic groups in America as having major responsibility for meeting the basic material needs of the elderly (20). Rather, families are able to provide the support they do in part because of the availability of government programs like Social Security. When these programs are cut back, the family's ability to respond may be overtaxed, to the detriment of young and old alike (21).

Programs like Social Security are not without serious flaws.... Yet despite these problems, and contrary to recent media claims, there has been little outcry from younger taxpayers to date about the high costs of Social Security and Medicare. Indeed, in an analysis of some twenty national surveys conducted by Louis Harris and Associates over a recent two-year period, Taylor found no support for the intergenerational conflict hypothesis (22). While the elderly appeared somewhat more supportive of programs targeted at them than did younger age groups, and vice versa, the balance of attitudes in all generations was solidly on the same side. The majority of both elderly people, and young people under thirty, thus opposed increasing monthly premiums for Medicare coverage, opposed increasing the deductible for Medicare coverage of doctors' bills, and opposed freezing Social Security cost of living increases. Similarly, both young and old Americans opposed cutting federal spending on education and student loans, and overwhelmingly opposed cutting federal health programs for women and children (22). In short, while young and old differ significantly on questions relating to values and lifestyle, issues of government spending and legislation affecting persons at different stages of the life course appeared to evoke intergenerational *consensus* rather than *conflict*.

The strong support of younger Americans for Social Security and Medicare is particularly enlightening in the wake of recent and widely publicized charges that these programs may be bankrupt before the current generation of young people can reap their benefits. Such charges, while poorly substantiated, have made an impact: poll data suggest that today's younger workers are pessimistic about the chances of Social Security and Medicare being there for them when they retire. In light of this pessimism, how might their continued high level of support for elderly entitlements be explained?

As noted earlier this phenomenon reflects in part the fact that younger people in the workforce prefer to have their parents indirectly supported than to shoulder this burden themselves in a more direct way.

Yet on a more fundamental level, the continued support of younger generations for old-age entitlements they believe will go bankrupt before their time reflects what Hendricks and Leedham describe . . . as a moral economy grounded in use value—one whose central goal is "to structure a society so as to maximize possibilities of a decent life for all." As these analysts go on to note, moral economies grounded in use value envision the public interest as negotiated rule structure, to which people adhere *even though it may run counter to their own immediately desired satisfactions* since it improves overall opportunities of obtaining satisfaction" (emphasis added). Further, a moral economy based on use value or individual and social utility views citizens "not as passive recipients or consumers of public policy, but as active moral agents." Within such a vision of moral economy, resource allocation would be viewed not in terms of competition between generations, but in Rawls' sense, as allocations appropriate to ourselves at various points over the life course (23). . . .

Generational equity proponent Daniel Callahan has argued that such life course perspectives are flawed in their failure to adequately address the fact that the huge demands made by unprecedented numbers of elders in recent years "threaten to unbalance any smooth flow of an equitable share of resources

from one generation to the next" (24, p. 207). Yet he and other proponents of generational equity similarly evoke a notion of moral economy in support of their arguments that the purpose of old age in society should be reformulated as involving primarily service to the young and to the future—in part through an acceptance of the need for "setting limits" on government support for lifesaving medical treatment for those over a given age.

Callahan indeed proposes a shifting of the existing moral economy and a questioning of some of its basic assumptions. In particular, a fundamental tenet of the existing moral economy—that health care not be rationed on the basis of age—should, he argues, be reconsidered in light of current and growing inequities in the distribution of scarce public resources for health care between young and old (24).

The strong public outcry against age-based rationing, coupled with the considerable evidence of the continued popularity of Social Security and Medicare among young and old alike, suggests, however, that the older moral economy notions of what is just and "due" the old have continued to hold sway. For both young and old, it appears, elderly entitlement programs like Social Security are not simply the way things are, but the way they should be. The reality, in short, stands in sharp contrast to the rhetoric which claims that Social Security is "nothing less than a massive transfer of wealth from the young, many of them struggling, to the elderly, many living comfortably" (25). Instead, the program, firmly grounded in American moral economy, is one which all generations appear to support, and from which all see themselves as receiving some direct or indirect benefit.

AGE/RACE STRATIFICATION AND INTERETHNIC EQUITY

Arguments that young and old alike have a stake in programs like Social Security and Medicare may break down, according to some analysts, when the element of ethnicity is introduced.

Indeed, Hayes-Bautista et al. suggest that there may be strong resentment of entitlement programs for the elderly thirty to fifty years from now in age/race stratified states like California (26). Within such states, the burden of support for the large, predominantly white elderly population is expected to fall heavily on the shoulders of a young workforce composed primarily of Latinos and other minorities.

Utilizing California demographic projections as a case in point, Hayes-Bautista et al. note that unless major shifts take place in fertility and immigration patterns, and/or in educational and job policy achievements, the working-age population of the future will not only be heavily minority but also comprised of individuals whose lower total wage base will require that larger proportions of their income go simply to maintaining current Social Security benefit levels for the white elderly baby boom generation. Under such conditions, they argue, the nation may be ripe for an "agerace collision."

While these investigators go on to suggest policy measures that might avert such a catastrophe, the "worst case scenarios" which they and other analysts (27, 28) describe have unfortunately received far more media attention than their recommended policy solutions. It is precisely because of the popularity of these "age wars" predictions, moreover, that a deeper look at the current reality is in order.

While it is of course impossible to project attitudinal shifts in the population in the way that demographic changes can be forecasted, current national opinion poll data on the attitudes of Latinos and other minorities toward entitlement programs for the elderly are instructive.

If the hypothesis is correct that there may be substantial resentment of Social Security and Medicare in the future on the part of a large minority working-class population left to shoulder this burden, one might expect hints of such resentment now. Contrary to expectation, however, opinion poll data show Hispanics, blacks, and other minorities to be strongly supportive of Social Security and Medicare and indeed often more opposed than whites to proposed budget cuts in these programs (22).

The hypothesis of growing minority resentment of elderly entitlement programs also may be questioned on demographic grounds. Thus, while minorities make up only about 10 percent of the aged population today, that figure is expected to increase by 75 percent by the year 2025. [The comparable increase among white elders will be only about 62 percent (29).] Elderly Hispanics, already the fastest growing subgroup within the older population, will see their numbers grow even more rapidly, quadrupling by the year 2020 (30). From 2025 to 2050, the period corresponding to the graying of the huge baby boom generation, the proportion of elderly within the nonwhite population is projected to increase another 29 percent, compared to only 10 percent for the white population (29). While the minority aged population will remain small numerically compared to the elderly cohort, the fact that greater proportions of working-aged Latinos, blacks, and other minorities will have parents and grandparents reaching old age suggests again an important phenomenon which may work against the reification of an age/race wars scenario.

A final factor which may mitigate against the likelihood of increasing minority resentment of elderly entitlement programs concerns the differential importance of programs like Social Security to the economic well-being of whites, blacks, and Hispanics. The disadvantaged economic position of elderly blacks and Hispanics relative to whites, for example, means that higher proportions rely on Social Security as their only source of income, and that far fewer have private insurance and other nongovernment resources to cover the costs of medical care. Under such conditions, more rather than less support for elderly entitlement programs might be expected among these economically disadvantaged minority groups, and that is indeed what the survey data appear to suggest.

GUNS VERSUS CANES?

A common theme throughout the inter-generational equity movement is that the elderly are not only numerous but also expensive: The high cost of Social Security, amounting to about 20 percent of the federal budget, is juxtaposed against a $2 trillion national debt and a massive federal deficit, with the implicit and often explicit message that the costly elderly are a central part of our economic crisis.

As Pollack has noted, however, such equations are misleading at best (14). Social Security, for example, is not a contributor to the deficit and in fact brings in considerably more money than it pays out. Indeed, throughout the 1990s, when the small Depression-era cohort is elderly, the system will bring in literally hundreds of billions more than it will spend (14).

Ironically, the nation's military budget, while a major contributor to the deficit, is virtually excluded from discussions by many analysts of the areas necessary for scrutiny if we are to achieve a balanced budget. AGE President Paul Hewitt indeed has spoken out against a congress which, under pressure from aging interest groups, will "weaken our national defense before it will cut cost of living allowances (COLAs) for well to do senior citizens" (6).

Tendencies by the mass media and others to overlook defense spending in discussions of the costliness of programs for the elderly continue . . . a tradition described earlier by Binstock when he noted that we are taught to think in terms of how many workers it takes to support a dependent old person, but not how many it takes to support an aircraft carrier (16). Citing an OMB fiscal analyst, Binstock further pointed out that classic political economic trade-off, "guns vs. butter," has been reframed "guns

vs. canes," in reference to the perceived costliness of the aged population.

In the relatively few years since Binstock's initial analysis, a further reframing has occurred with "canes vs. kids" constituting the new political economic trade-off.

As Kingson et al. (19) have noted, such an analysis assumes a zero sum game in which other possible options (e.g., increased taxes or decreased military spending) are implicitly assumed to be unacceptable (19). It is worthy of note that such traditionally conservative analysts as Meyer and Lewin of the American Enterprise Institute (31) have begun arguing in favor of cutbacks in defense spending as a means of balancing the federal budget without in the process decimating needed social programs. While favoring increased taxation of Social Security and some initial taxing of Medicare benefits, the AEI analysts appear to have come down hardest on the need for massive reductions in military spending if the United States is to cease "protecting sacred cows and slaughtering weak lambs" (31, p. 1).

Still other analysts have urged the closing of tax loopholes for corporations and the rich as a means of redressing huge national deficits. Noting that corporate taxes dropped from 4.2 percent of the GNP [gross national product] in 1960 to 1.6 percent in the 1980s, and that money lost to the treasury through tax loopholes grew from $40 billion to $120 billion over the period 1980–86 alone, Pollack thus has argued that tax breaks for the rich, rather than Social Security COLAs [cost of living adjustments] for the elderly, should be viewed as the real culprits in the current economic crisis (14).

Finally, national opinion poll data show overwhelming public support for cutting military spending and closing tax loopholes, rather than cutting programs for the elderly and other population groups as a means of addressing America's economic difficulties (32).

The "canes vs. kids" analytical framework, in short, does not appear to have wide credence in the larger society, and where a hypothetical "guns vs. canes" trade-off is proposed, the American public overwhelmingly supports the latter.

VICTIM BLAMING REVISITED

. . . The new victim blaming is particularly well illustrated in the application of a market theory perspective to determinations of appropriate resource allocations for the different age groups. Demographer Samuel Preston thus has argued that, "Whereas expenditure on the elderly can be thought of *mainly as consumption,* expenditure on the young is a combination of consumption and investment" (4, p. 49; emphasis added). Leaving aside the inaccuracies of such statements from a narrow, worker productivity perspective (since many of the elderly continue to be employed full- or part-time), the moral and ethical questions raised by such an approach are significant. While it is true that the elderly "consume" about a third of all health care in the United States, for example, such figures become dangerous when used to support claims of the differential "costs" of older generations, and the need for the rationing of goods and services on the basis of age (1).

The scapegoating of Social Security and Medicare as primary causes of the fiscal crisis has served to deflect attention from the more compelling and deep-seated roots of the current economic crisis. At the same time, and wittingly or unwittingly fueled by recent mass media and groups like AGE, it has been used as a political tool to stoke resentment of the elderly and to create perceptions of a forced competition of the aged and younger members of society for limited resources. As Kingson et al. have noted, (19, p. 4):

> . . . while the concept of intergenerational equity is seemingly neutral and possesses an intuitive appeal (who can be against fairness?), its application, whether by design or inadvertence, carries with it a very pessimistic view about the

implications of an aging society, which leads to particular policy goals and prescriptions.

These "policy goals and prescriptions," as we have seen, reflect a "new victim blaming" mentality in their suggestion that we must cut public resources to the elderly in order to help youth and to avert conflict between generations. For as noted elsewhere, to trade the victim blaming approaches of the 1980s for those of the 1960s and 70s is not a solution to problems which ultimately are grounded in the skewed distribution of economic and political power within the society (33). Ultimately, as Pollack has argued, the central issue is not one of intergenerational equity but of income equity (14). Cast in this light, inadequate AFDC payments and threatened cuts in Social Security COLAs which would plunge millions of elderly persons into poverty are part of the same problem. Programs and policies "for the elderly" like education and health and social services "for youth" must be redefined as being in fact not "for" these particular subgroups at all, but for society as a whole. Conversely, threatened cutbacks in programs for the elderly, under the guise of promoting intergenerational equity, must be seen as instead promoting a simplistic, spurious and victim blaming "solution" to problems whose causes are far more fundamental and rooted in the very structure of our society.

CONCLUSION

In the short time since its inception, the concept of generational equity has become a popular framework for analysis of contemporary economic problems and their proposed solutions. Yet the concept is based on misleading calculations of the relative financial well-being of the elderly vis-à-vis other groups and on questionable assumptions concerning such notions as fairness and differential stake in the common good. Predictions that young people, and particularly minority youth, may be increasingly resentful of elderly entitlement programs are unsubstantiated by current national survey data which show strong continued support for these programs across generational and ethnic lines. The reframing of political economic trade-offs in terms of "canes vs. kids" indeed appears to have little credence with the public at large, despite its popularity in the mass media, and with a growing number of scholars, policy makers and new, self-described youth advocates.

Proponents of an intergenerational equity approach to policy have performed an important service in calling attention to high rates of poverty in America's children, and to the need for substantially greater societal investment in youth and in generations as yet unborn. At the same time, however, their tendency to blame the costliness of America's aged for economic hardships experienced by her youth is both misguided and dangerous. The call for intergenerational equity represents a convenient smokescreen for more fundamental sources of inequity in American society. By deflecting attention from these more basic issues, and by creating a false dichotomy between the interests of young and old, the advocates of a generational equity framework for policy analysis do a serious disservice.

REFERENCES

1. Binstock, R. H. The oldest old: A fresh perspective or compassionate ageism revisited? *Milbank Mem. Fund Q.* 63:520–541, 1983.
2. Longman, P. Age wars: The coming battle between young and old. *The Futurist* 20:8–11, 1986.
3. O'Connor, J. *The Fiscal Crisis of the State.* St. Martin's, New York, 1973.
4. Preston, S. Children and the elderly in the U.S. *Scient. Am.* 251:44–49, 1984.
5. Annual Report of the President's Council of Economic Advisors. U.S. Government Printing Office, Washington, D.C., 1985.
6. Hewitt, P. A Broken Promise, Brochure of Americans for Generational Equity. AGE, Washington, D.C., 1986.
7. Edelman, M. W. Meeting the needs of families and children: Structural changes that require new social arrangements. (Statement before the Consumer Federation of America.) Children's Defense Fund, New York, March 15, 1990.
8. Longman, P. Justice between generations. *Atlantic Monthly,* pp. 73–81, June 1985.
9. Taylor, P. The coming conflict as we soak the young to enrich the rich.
10. Villers Foundation, On the Other Side of Easy Street: Myths and Facts about the Economics of Old Age. The Villers Foundation, Washington, D.C., 1987.
11. U.S. Bureau of the Census. Current reports: Poverty in the U.S. Series P 60, No. 163. U.S. Government Printing Office, Washington, D.C., 1989.
12. Bould, S., Sanborn, B., and Reif, L. *Eighty Five Plus: The Oldest Old.* Wadsworth Publishing Company, Belmont, California, 1989.
13. Minkler, M., and Stone, R. The feminization of poverty and older women. *Gerontologist* 25:351–357, 1985.
14. Pollack, R. F. Generational equity: The current debate. Presentation before the 32nd Annual Meeting of the American Society on Aging, San Francisco, March 24, 1986.
15. U.S. Bureau of the Census, *Statistical Abstract of the U.S.* (109th edition), U.S. Government Printing Office, Washington, D.C., 1989.
16. Binstock, R. The aged as scapegoat. *Gerontologist* 23:136–143, 1983.
17. Blaustein, A. I. (ed.). *The American Promise: Equal Justice and Economic Opportunity.* Transaction Books, New Brunswick, New Jersey, 1982.
18. Margolis, R. J. *Risking Old Age in America.* Westview Press, Boulder, Colorado, 1990.
19. Kingson, E., Cornman, J., and Hirschorn, B. *Ties That Bind: The Interdependence of Generations in an Aging Society.* Seven Locks Press, Cabin John, Maryland, 1986.
20. Bengtson, V., Burton, L., and Mangen, D. Family support systems and attributions of responsibility: Contrasts among elderly blacks, Mexican-Americans, and whites. Paper presented at the Annual Meeting of the

Gerontological Society of America, Toronto, Canada, November 1981.

21. Pilisuk, M., and Minkler, M. Social support: Economic and political considerations. *Social Policy* 15:6–11, 1985.

22. Taylor, H. Testimony before the House Committee on Aging, Washington, D.C., April 8, 1986.

23. Rawls, J. *A Theory of Justice.* Belknap Press, Cambridge, Massachusetts, 1971.

24. Callahan, D. *Setting Limits: Medical Goals in an Aging Society.* Simon and Schuster, New York, 1987.

25. Schiffres. M. The Editor's Page, "Next: Young vs. old?" *U.S. News and World Report,* p. 94, November 5, 1984.

26. Hayes-Bautista, D., Schinck, W. O., and Chapa J. *The Burden of Support: The Young Latino Population in an Aging Society.* Stanford University Press, Palo Alto, California, 1988.

27. Longman, P. The youth machine vs. the baby boomers: A scenario. *The Futurist* 20:9, 1986.

28. Lamm, R. D. *Mega-Traumas, America at the Year 2000.* Houghton Mifflin Company, Boston, Massachusetts, 1985.

29. U.S. Senate Special Commission on Aging. *Aging America: Trends and Projections, 1985–86.* U.S. Department of Health and Human Services, Washington, D.C., 1986.

30. Andrews, J. Poverty and poor health among elderly Hispanic Americans. Commonwealth Fund Commission on Elderly People Living Alone, Washington, D.C., 1989.

31. Meyer, J. A., and Lewin, M. E. Poverty and social welfare: Some new approaches. Report prepared for the Joint Economic Committee. American Enterprise Institute, Washington, D.C., 1986.

32. Ryan W. *Blaming the Victim* (1st edition). Random House, New York, 1972.

33. Minkler, M. Blaming the aged victim: The politics of retrenchment in times of fiscal conservatism. In *Readings in the Political Economy of Aging,* edited by M. Minkler and C. L. Estes, pp. 254–269. Baywood Publishing, Amityville, New York, 1984.

READING 37

The Generational Equity Debate

John B. Williamson, Diane M. Watts-Ray, and Eric R. Kingson

Two major groups, or "advocacy networks," to use a more precise term suggested by Gamson and Stuart (1992), have been competing to frame the debate over public policy toward the elderly. Since the mid-1980s this debate has been referred to as the "generational equity" debate, a designation that reflects both a symbolic victory and a rhetorical advantage for one of the two major advocacy networks and its interpretive package. The term "generational equity" has come to designate a number of assumptions, arguments, values, and beliefs associated with the more conservative of the two competing interpretive packages. Those making up the more liberal advocacy network are less unified behind a single catch phrase to designate their alternative interpretive package with its set of assumptions, arguments, values, and beliefs on this issue; but they often use terms such as "intragenerational equity" and

SOURCE: Introduction from *The Generational Equity Debate* by John B. Williamson, Diane M. Watts-Ray, and Eric R. Kingson. Copyright ©1999 Columbia University Press.

"generational interdependence" (Kingson et al. 1986). In the sections that follow, we briefly summarize the major arguments that have been made by advocates of these two interpretive packages, beginning with the generational equity framing.

THE GENERATIONAL EQUITY FRAME

Advocates for the generational equity interpretive package argue that Social Security and Medicare policy makers need to give more attention than they have been giving to the issue of fairness between generations (Longman 1987; Peterson 1996). Too much public money is being spent on the retired elderly at the expense of the rest of the population, particularly children and young adults. This is doubly unfair because those paying for the very generous Social Security and Medicare benefits enjoyed by today's elderly will not receive comparable benefits when they retire.

The retirement of the baby boom generation is going to put a very heavy burden on the Social Security and Medicare programs because there will be so few workers paying into the trust funds that support these programs relative to the number of retired workers who will be drawing benefits. One consequence of this burden is that it will not be possible to support the retired baby boomers at benefit levels enjoyed by those who are currently retired.

Current recipients of Social Security are consuming more than their fair share of societal resources. As a group, in recent decades their financial situation has improved considerably while at the same time the financial situation of children has declined dramatically. Poverty rates among the elderly have decreased while at the same time poverty rates among children have increased sharply. Over the years federal spending on the elderly has increased considerably; today such spending makes up more than 25 percent of the federal budget. Were we not spending so much on the elderly it would be possible to spend more on children and young adults, a policy that would help alleviate the unacceptably high poverty rates for children.

Advocates for the generational equity interpretive package are committed to the idea that each generation should be expected to provide for itself. It is not fair for one generation to be expected to provide for another generation at a level that the generation providing the support is itself unlikely to realize. The structure of a pay-as-you-go, social insurance-based Social Security scheme assumes that each generation will provide support to those currently retired and will in turn be supported by a younger generation when it moves into retirement. A privatized Social Security scheme, in contrast, is more consistent with the idea that each generation should be responsible for itself.

Generational equity advocates often point out that the elderly constitute a very strong voting bloc. They, and interest groups such as the AARP that represent them, have a great deal of influence on public policy. The elderly and their interest groups often use their political clout in selfish ways to support programs that benefit themselves to the exclusion of others. One example that is sometimes mentioned is voting down school bond issues in some communities. More common is their lack of support for politicians who suggest serious cuts in Social Security or Medicare as a way to balance the federal budget.

If we are going to be able to provide for baby boomers when they retire without putting a very heavy burden on those in the labor force at the time, we need to do all we can between now and then to grow the economy. The more economic growth we have between now and then, the larger the economic pie will be and the easier it will be to support a large retired population. To do this we need to increase the savings rate, which will increase the investment rate, which in turn will increase the rate of economic growth. To increase the savings rate we need cuts in current consumption, including a variety of government social programs. We also need new policy initiatives such as a shift to a partially privatized Social Security scheme.

As a society we need less emphasis on an entitlement ethic and more emphasis on a work ethic. We need to emphasize such values as thrift, self-reliance, independence, personal freedom, and limited government. In short, advocates for the generational equity interpretive package make it a point to link their arguments to the widely and deeply held values in American society that define individualism.

THE GENERATIONAL INTERDEPENDENCE FRAME

Advocates for the generational interdependence interpretive package argue that Social Security and Medicare policy makers need to take into consideration the interdependence of generations when making and changing policy. They reject the idea that each generation can or should be expected to provide for itself. In part because of demographic fluctuation and in part because of unique historical events such as the Great Depression or the Second World War, it is not possible for each generation to be assured a standard of living during retirement at least equal to that of its parents' generation. When there are demographic, economic, or other factors that make it difficult to provide for the retirement of a particular generation, the burden should be shared by both generations, those who are retired and those still in the labor force. It is unreasonable to expect either generation to bear the entire burden in such situations.

Advocates of this perspective point out that sharp cuts in benefits for the retired or the soon to be retired will also have an adverse impact on their adult children. Due to interdependence between generations at the family level, such cuts would produce pressure on their adult children to take in older family members or to supplement their Social Security and Medicare benefits. Whereas the generational equity perspective emphasizes the need for each generation to be responsible for itself, the proponents of the generational interdependence perspective argue that the generations are and should continue to be highly interdependent. This is true at both the family and the societal level. Advocates of generational interdependence accept the idea that we should be making an effort to plan for the retirement of the boomers, but they emphasize the need to balance considerations of generational equity and of generational interdependence.

A distinctive aspect of the generational interdependence perspective is its emphasis on what different generations have to offer one another as opposed to what one is or will be consuming at the expense of the other. More explicitly, it tends to emphasize the many transfers taking place among the generations, within the family and society. At the level of individual families this includes transfers of income, child care support, psychological support, and advice. In 1994, 3.7 million grandchildren were being raised in households headed by grandparents (Saluter 1996); many of the elderly also provide countless hours assisting functionally disabled family members. In addition the elderly are making major artistic, intellectual, and leadership contributions to society more generally.

There is a great deal of diversity in economic circumstances among the elderly. When we base policy decisions on what is best for average-wage workers or for the affluent, we do so at the risk of potential harm to some vulnerable groups such as low-wage workers, minority workers, and the very old, particularly very old women. The mean and median income for the elderly have increased in recent decades, but there are still millions of elderly people living in or very near to poverty. When the elderly are treated as a homogeneous aggregate for policy purposes, as is often the case in policy analysis from the generational equity perspective, the special needs of these vulnerable groups tend to be neglected (Binstock 1992).

There has been an increase in poverty among children in recent decades—the same decades that have seen a decrease in poverty rates among the elderly. Proponents of the intragenerational equity perspective do not view these two events as causally linked. Many factors have contributed to the increases in poverty rates among children, including changes in the American economy, the increase in single-parent families, and changes in public willingness to support public spending on the poor. If the current level of spending on Social Security and Medicare is a contributing factor, a claim that has not been demonstrated, it most likely is a very minor factor.

Proponents of generational interdependence and intragenerational equity argue that we need to take a close look at the values that inform our policy decisions. In contrast to the individualistic values emphasized by proponents of generational equity, proponents of generational interdependence emphasize solidaristic values such as community obligation to provide for those in need and the right of all citizens to adequate health care, food, and shelter; and goals such as reduction of poverty and inequity, and income redistribution to compensate for our economy's tendency to increase income inequality.

It is sometimes useful to assess an author's ideas about fairness with respect to the distribution of societal resources. Those working within the generational equity frame tend to emphasize fairness between generations and policies that make an effort to assure that one generation is not favored over another. Some within this same advocacy network favor fairness in the form of individual equity, for example, Social Security pension benefits that are linked as closely as possible to actual payroll tax contributions over the years with very little in the way of redistribution.

In contrast, those working within the generational interdependence frame generally support a different conception of fairness—one that sees a need for redistribution to those who reach old age with very low incomes. This group of analysts and commentators urge us to make an effort to even out some of the inequality our market economy tends to produce among those reaching old age, particularly inequality linked to race, class, and gender. They are also likely to emphasize the need under some circumstances, such as the retirement of the baby boom generation, for two generations to attempt to share the dependency burden.

REFERENCES

Binstock, Robert H. 1992. "The Oldest Old and Intergenerational Equity." Pp. 394–417 in *The Oldest Old,* edited by Richard M. Suzman, David P. Willis, and Kenneth Manton. New York: Oxford University Press.

Gamson, William A., and David Stuart. 1992. "Media Discourse as a Symbolic Contest: The Bomb in Political Cartoons." *Sociological Forum* 7:55–86.

Kingson, Eric R., Barbara A. Hirshorn, and John M. Cornman. 1986. *Ties That Bind.* Washington, DC: Seven Locks Press.

Longman, Phillip. 1987. *Born to Pay: The New Politics of Aging in America.* Boston: Houghton Mifflin.

Peterson, Peter G. 1996. *Will America Grow Up Before It Grows Old?* New York: Random House.

Saluter, Arlene F. 1996. "Marital Status and Living Arrangements: March 1994." *Current Population Reports,* Series P-20, No. 484. Washington, D.C.: U.S. Department of Commerce, Bureau of the Census.

FOCUS ON THE FUTURE

Walled Retirement Villages?

Separate living arrangements for the aged have long been a feature of nuclear families in Western societies (Laslett, 1972), but intentionally segregated housing is a more recent pattern. National surveys of retirement destinations frequently cite places like Sun City, Arizona, and Leisure World, California, as among the most attractive places for retirees to live. Such communities offer an age-segregated, recreation-oriented lifestyle that some older people find attractive.

Consider Leisure World in Laguna Hills, California, now called Laguna Woods Village, which is a private retirement community of 21,000 residents. Residents are largely conservative, middle-class, White retirees above age 75, with women outnumbering men. Security at Leisure World is tightly controlled, and the grounds and recreational facilities are attractively maintained. There are nearly 200 social or philanthropic organizations, and the local community college offers courses to Leisure World residents.

Leisure Village in Ocean County, New Jersey, is a similar community. It has nearly tripled the local county population. Changes have been made in new retirement subdivisions to keep up with housing preferences and to appeal to younger, more active retirees. Century Village in Florida has also grown despite some conflicts with neighboring towns, and it is second in size only to Sun City.

Since opening in 1960, the original Sun City has grown continuously. Successive stages of development have kept up with demands from increasingly affluent retirees. Sun City provides a comfortable home for thousands of affluent, active, and healthy residents, along with those who are ill or no longer affluent. Sun City also embraces a vast volunteer network, which enhances services provided by the government and formal organizations. Recreation centers and golf courses are always crowded. But beneath the recreation-oriented lifestyle, unanswered questions face all age-segregated communities.

One question is how a retirement village can maintain its active lifestyle. Not surprisingly, new residents tend to be among the young-old (ages 65–74). Over time, however, the population of retirement communities is "aging in," and this raises questions about long-term care and other supportive services. Retirement community residents are less likely than peers to have children, and, when they do, their adult children tend to live at too great a distance to help with chores or to assist in the event of an illness. Limited short-term assistance comes from friends and neighbors (D. Sullivan, 1986), but long-term illnesses present a problem. Some retirement complexes have become full-scale life-care communities, with levels of care ranging from housing amenities for independent elders to nursing home support for those requiring constant care and supervision.

Social critic Lewis Mumford was one of many who argued early against the whole idea of age segregation. Yet people in deliberately age-segregated communities voluntarily choose to live there. Why do they make that choice? People often hear about a retirement community through relatives, but in selecting a place to live, only a small proportion cite proximity to friends and relatives as a reason. Climate, recreation facilities, health, cleanliness, and age segregation are commonly cited as reasons for migrating to Sun City (Gober & Zonn, 1983).

Still, more than 90% of people who reach retirement age do not choose to move away, but continue to stay right where they have lived all their lives. For the most part, older people remain integrated in communities consisting of people of all ages, but some older people end up in age-segregated communities anyway. For instance, many housing developments evolve into **naturally occurring retirement communities (NORCs)**, defined as housing that was not originally planned for older residents but that evolved in that direction over time. Location, especially proximity to shopping and family or friends, is a major attraction of NORCs. In rural areas, NORCs appeal to some residents because of amenities and a natural environment remote from the big city, whereas others are attracted by convenience or proximity to relatives (Hunt, Merrill, & Gilker, 1994).

There has long been debate about the virtues of age-integrated versus age-segregated approaches to serving older adults. Arguments in favor of age-integrated services are that age integration reduces ageism and improves efficiency of service delivery. But some evidence suggests that older people prefer and benefit more from being with their peers. In age-integrated groups, the needs of older members may too easily be neglected (Lowy, 1987). Some studies have suggested that age segregation, in itself, can actually be a positive factor promoting social integration and a sense of community (Osgood, 1982).

Informal relationships play a big part in most age-segregated communities, whether planned commercial ventures like Sun City or NORCs, such as apartment buildings that have "aged in" over time. Age-segregated patterns of association do bring many benefits: a chance for self-expression, a sense of security, and a peer network that can help people cope with the problems of aging (Jerrome, 1992).

Hidden Costs of Age Segregation

Another issue is the question of the effect that high-density retirement villages have on communities around them. One study did find that apartment buildings occupied mostly by people over age 60 tended to exclude or impose restrictions on families with children. Older respondents sharing established city housing with families with young children and teenagers expressed concern for their own well-being. As a result, families with children sometimes confront a serious problem of housing availability (Margulis & Benson, 1982).

One ominous sign of intergenerational tension was the "adults-only movement" that arose in Arizona. For some older people, the appeal of age-segregated communities is that there are no schools or recreation facilities crowded with teenagers playing loud radios. But what happens when families with youngsters want to move in? In Arizona in 1973, a court case was brought by a group of neighbors against a family with a minor who had bought land in a mobile home subdivision in the town of Apache Wells, despite a restrictive covenant intended to keep out anyone under age 21. The "politics of age exclusion" caused age groups to be in opposition to one another in disturbing ways (Anderson & Anderson, 1978).

(Continued)

(Continued)

When people are forced to make difficult choices, there is a danger of generational confrontation. For instance, one study found that Florida retirees tend to vote against bond issues for public schools much more often than younger voters, but different studies in other geographic areas found no such correlation. One explanation for the difference may be that Florida has many recent migrants, whereas long-term residents in other areas are more sympathetic to schools than to new residents without children. That hypothesis is supported by studies of voting patterns for older adults (Button, 1992). Another Florida study found differences between older and younger residents around topics such as community development, zoning regulations to ban nursing homes, and lifestyle issues, such as older drivers. Negative images of older adults were rarely shared across age groups, however; in fact, different age groups often saw the older adults as a positive economic influence (Rosenbaum & Button, 1992).

There are significant generational differences around the economic impact of population aging at the local level. Immediate windfalls from attracting higher-income elders who demand few services may give way later to higher expenditures for an older population needing public services, especially health and long-term care. But targeting service programs to segregated groups of older people can present problems. In that case, polls show that younger respondents are more likely to feel that older adults in their counties or cities are a divisive influence or a selfish voting bloc (Rosenbaum, 1993). Tensions between young and old are magnified at the local level by competition for shrinking public budgets. Conversely, improved intergenerational communication could overcome some of these problems.

Another glimpse of the future may be visible in the nation's most populous state, California. Older adults in California are overwhelmingly (three-quarters) Anglo, but among children under age 17, Anglos number less than half the population. In the future, interethnic and intergenerational conflict could emerge if a well-educated, older Anglo population has to be supported in retirement by contributions from a poorer, younger Latino population. The well-being of the older population will be linked to the financial success or failure of the minority population (Hayes-Bautista, 1991), a link that suggests a need for intergenerational coalition building.

Part of the answer may be to avoid "walled retirement villages" and instead favor residential patterns to connect generations. One attractive option along these lines is **cohousing**, or planned communities that offer self-sufficient housing units tied to shared common spaces. Cohousing is already popular in Scandinavia and can be attractive to elders who want to live independently. Notable projects in the United States include Generations of Hope/Hope Meadows in rural Chicago and a more recent related project in Portland, Oregon, called Bridge Meadows. These intentional multigenerational communities target children in the foster care system and lower-income older adults. In such multigenerational environments, older people can serve as mentors, friends, or surrogate grandparents to younger people. A retirement housing complex linked to a college or university could offer similar advantages. This approach is found in mixed-use retirement communities, where services are shared and supported by the broader public and older adults do not live in isolation or at a distance from the larger community.

Questions for Writing, Reflection, and Debate

1. Some have criticized Peter Peterson and others who are worried about generational equity for being too gloomy and pessimistic about the future of U.S. society. Is this a valid criticism? How might Peterson reply to such a charge of "pessimism"?

2. If a political conservative were to agree with Peterson's picture, what would a likely conservative response to the problem be? If a political liberal were to agree with Peterson's picture, what would a likely liberal response to the problem be?

3. Lester Thurow believes that older people are quickly becoming a voting bloc whose demands are impossible to refuse. Can you offer some recent examples that support this view? What about examples that contradict Thurow's claim?

4. Advocates for the aging often argue that both young and old have a "common stake" that should overcome generational differences. Prepare a short article for a local community newspaper in which you outline ideas and examples that support this "common stake" point of view.

5. Public opinion suggests that Meredith Minkler may be right when she says the public at large does not much agree with the idea of generational equity. What does this fact suggest about public opinion or about the programs that government officials should implement for children and older adults? As an exercise, try writing opinion poll questions about generational equity and rephrase the same idea in two ways. Then test the two ways of framing the issue by asking the questions of people not enrolled in a gerontology class.

6. Consider the following difficult task. Imagine that you are in charge of a Meals on Wheels program providing home-delivered meals to older adults in your community. Your agency has just received a 50% budget cut for next year, and now you must recommend which of the current meal recipients should continue to receive food. Write a detailed memorandum explaining how you would decide who are the least advantaged older adults deserving of continued service.

7. Is there a way of targeting benefits to those in greatest social and economic need that doesn't make use of factors such as race, which tends to be divisive, and income, which tends to be stigmatizing? What factors might work in such an approach to targeting?

8. Assume that you are the new superintendent of a school district with a larger proportion of senior citizens among the voters, who are about to be asked to approve a new school bond issue that will raise taxes. Prepare a letter to be distributed to senior citizens in the community in which you set forth reasons why they should vote in favor of the school bond issue.

9. What might be the consequences of adopting what Peter Peterson calls an "affluence test" instead of a "means test"? Based on what you know about financial aid in higher education, how would you estimate the possible consequences of either an affluence test or a means test instead of an age-based entitlement? Give arguments in favor of, or against, introducing either of these tests.

10. Visit some websites with opposing views about competition between generations. A good place to begin is the Concord Coalition, a national group concerned about budget deficits for the future (http://www.concordcoalition.org/). Another group with similar concerns is the Peter G. Peterson Foundation (http://www.pgpf.org/). What issues are of greatest concern on these websites? What issues are left out? Now visit the website for Generations United (http://www.gu.org/). What issues are of greatest concern on this website? What issues are left out?

Suggested Readings

Estes, Carroll, Rogne, Leah, Hollister, Brooke, Grossman, Brian, and Solway, Erica (eds.), *Social Insurance and Social Justice,* Springer, 2009.

Freedman, Marc, *Prime Time: How Baby Boomers Will Revolutionize Retirement and Transform America,* New York: Public Affairs, 2002.

Hacker, Jacob S., *The Great Risk Shift: The New Economic Insecurity and the Decline of the American Dream,* Oxford University Press, 2008.

Hudson, Robert B. (ed.), *The New Politics of Old Age Policy,* Johns Hopkins University Press, 2010.

Kotlikoff, Laurence J., and Burns, Scott, *The Coming Generational Storm: What You Need to Know about America's Economic Future,* Cambridge: MIT Press, 2005.

Williamson, John, and Kingson, Eric (eds.), *The Generational Equity Debate,* New York: Columbia University Press, 1999.

Student Study Site

Visit the Student Study Site at **www.sagepub.com/moody7e** for these additional learning tools:

- Flashcards
- Web quizzes
- Chapter outlines

- SAGE journal articles
- Web resources
- Video and audio resources

Controversy 9

WHAT IS THE FUTURE FOR SOCIAL SECURITY?

Social Security is the public retirement pension system administered by the federal government, the largest domestic government program today. Social Security covers all U.S. workers except for some state and local government employees. The biggest controversies over Social Security revolve around this core retirement program, which is essentially a tax on wages during working years followed by a wage subsidy during retirement years.

Social Security retirement benefits averaged $14,120 a year for a single individual in 2011 (usually 50% more for a couple). Yet Social Security was never intended to be the sole income source for people in retirement. In fact, the large majority of beneficiaries have other income from a pension, savings, or continued part-time employment. For most retired people, the retirement annuity portion of Social Security is most important because it provides a foundation for retirement income. The broad outlines of Social Security are familiar to most Americans, and the program remains enormously popular and widely supported by American adults of all ages.

But many people today wonder whether Social Security will still be around later in the 21st century. A *Newsweek* survey (Longley, 2005) showed that 61% of adult Americans are "not confident" that Social Security will be there for them. At the same time, results from a recent National Academy of Social Insurance report suggest that there's widespread support for Social Security from adults 18 to 64 as well as across political parties (Reno, Lamme, & Walker, 2011). These somewhat contradictory opinions illustrate how complicated this issue is. Opinion polls aside, the federal government will be facing difficult choices as the baby boom generation moves into retirement. The number of beneficiaries will rise, but the proportion of contributors to Social Security will decline. Benefits of future Social Security beneficiaries might be lowered; alternatively, the age of eligibility for full retirement benefits could be raised; or, again, future annual cost-of-living adjustments might be diminished. Currently, the federal budget shows a surplus because of Social Security taxes. But continuing into the first part of the 21st century, deficits are likely to emerge and grow larger as tax cuts continue to take effect.

The 1983 Social Security reforms responded to earlier anxiety and succeeded in putting the program on a more secure footing (Light, 1985), but the public rhetoric of "crisis" has continued, eroding public confidence. Part of the problem is that Social Security was created over 75 years ago under different historical conditions. The Social Security Act of 1935 was a centerpiece of the New Deal, a response to the Great Depression at a time when only 5% of the U.S. population was over age 65. This legislation was notable because it acknowledged government's role in providing income support to individuals outside the labor market. Social Security from the outset was conceived as a form of social insurance. Social Security helped replace income lost and thereby provided a cushion or an incentive to leave the labor force (Myers, 1985). It also legitimized age 65 as an age when it was customary and predictable to leave the labor force.

President Franklin Delano Roosevelt signs the Social Security Act in 1935.

MAIN FEATURES OF SOCIAL SECURITY

Social Security is actually more than just a retirement income program. It includes other features important to younger age groups. For example, Social Security wage earners are covered by a disability insurance policy worth more than $200,000 on the private market, and 98% of American children are insured against the loss of a parent. For an average wage earner with two children, Social Security provides the equivalent of $500,000 in life insurance. Still, what most people think of when they hear the term *Social Security* is the retirement annuity part of the program, which is what is under consideration here.

The money to fund Social Security comes from its own separate payroll tax, which is compulsory for almost everyone who receives wages or salary, including people who are self-employed. Other kinds of earnings—such as interest and dividends, partnership income, and so on—are excluded from the tax. The tax on individual wages is 6.2%, but an additional tax is paid by the employer so that the combined payroll tax rate comes to 12.4%, with an additional tax for Medicare Part A (also known as Medicare Hospital Insurance). In 2011, the tax on individual wages was temporarily reduced to 4.2%; this "payroll tax holiday" was intended to stimulate the economy.

Critics have often pointed out that the Social Security payroll tax is a **regressive tax**. That is, a person earning $20,000 a year and another earning $106,800 a year pay at the same 6.2% rate. Even a person earning $1 million a year pays this percentage on only the first $106,800 of earnings in 2010 and 2011—and nothing on anything above that. By contrast, in a more progressive tax system, such as the federal income tax, the percentage rises as income increases so that wealthier people, at least in principle, pay a greater percentage as well as a greater absolute amount of money in taxes. In contrast, Social Security, as a flat percentage, is a heavier burden for poorer people than for those earning more money (see Exhibit 44).

However, Social Security is modestly progressive in its distribution of benefits because of the **replacement rate**, or the proportion of wages replaced by Social Security at the point of retirement. For instance, a worker earning an average wage could expect to receive Social

Exhibit 44   Shares of Aggregate Income for the Lowest and Highest Income Quintiles, by Source, 2008

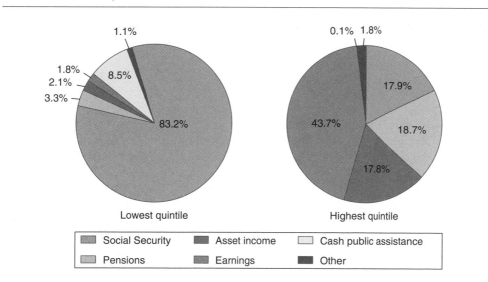

SOURCE: *Income of the Aged Chartbook*, 2008, Social Security Online (2010).

NOTES: The quintile limits for aged units for 2008 are $12,082, $19,877, $31,303, and $55,889.

Totals do not necessarily equal the sum of the rounded components.

Security payments of about 43% of salary before retirement. Someone earning only half the average would have a replacement rate of 56%; a high-wage earner would receive only 28%. Although the percentage replacement rate for higher-wage workers is lower, they do receive more back in absolute dollars.

Another attractive feature of Social Security is that, for more than two decades, benefits have been indexed to the Consumer Price Index, so benefits automatically increase with inflation. A package of the current Social Security and Medicare benefits with these inflation protection features would be difficult to buy in the private marketplace, and some features—including inflation protection—would be difficult or impossible to buy at all.

But these benefits come at a price. In 1960, less than 15% of the federal budget was spent on the older population; by the 1990s, the proportion had risen to more than 30%; by 2010, the proportion was 43% for Social Security, Medicare, and Medicaid combined. Social Security payroll taxes now bring in about $700 billion each year, an amount larger than the entire military budget.

The large reduction in the poverty rate among older Americans that has taken place since the 1960s is partly due to the increase in social insurance benefits such as Social Security. For example, 2 out of 3 people 65+ get half or more of their later-life income from Social Security. If Social Security benefits were to no longer exist, it is estimated that the poverty rate of older adults would increase to 50% (Wentworth & Pattison, 2001/2002).

The relative income of over-age-65 households has improved faster than for younger people, and government policies have begun to shift in response. Social Security was revised in 1993 to tax up to 85% of Social Security benefits for individuals with incomes above $25,000 and couples with incomes above $32,000. But, at present, Social Security benefits are subject to taxation for only about a quarter of older people, an indication that the other three quarters of older Americans, in fact, have quite modest incomes. As discussed in Basic Concepts III, older people as a group are hardly uniformly affluent: 33% of Whites, 47% of older Blacks, 44% of older Asians, and 53% of older Hispanics who receive Social Security rely on it for 90% or more of their income (Social Security Administration, 2008).

SUCCESS—AND DOUBT

Seventy-six years after its founding, Social Security remains the United States' most successful and perhaps most popular domestic government program. Social Security has long been characterized as *the system that works*, and most Americans would agree with that positive judgment.

Despite its success and popularity, Social Security has long been a subject of debate (Achenbaum, 1986). One fundamental question concerns the purpose of Social Security: Is it a welfare program designed to prevent impoverishment in old age? Or is it an annuity program that entitles everyone who pays into it to receive proportional benefits? Social Security as it actually exists has come to embody both purposes. On the one hand, Social Security provides a floor, or minimum income, for almost all older Americans. In this way, it redistributes income (Choi, 1991). Social Security helps those who are at risk of total disability and helps families with only one wage earner. On the other hand, Social Security is a universal program that benefits both poor and affluent older people. As a result, nearly everyone has a stake in the system.

Any program that tries to accomplish such fundamentally different goals is bound to have both its challenges and its critics. Thus, some people question the basic fairness and integrity of the Social Security system. For example, proponents of the generational equity idea have argued that Social Security is unfair to future generations. They argue that these future cohorts will get back less than do current beneficiaries of the system. According to their view, it might be better for people to have a private pension system instead of a compulsory one like Social Security.

Under some proposals, people could drop out of Social Security and sign up for a personal savings account, hoping to collect a better return than Social Security offers. But defenders of Social Security point out that better returns are not guaranteed by the private marketplace.

A related criticism is the idea that Social Security is not stable and may not be there for future recipients because the costs will go out of control. There has been much discussion about the dependency ratio, or the proportion of younger workers to retirees (Crown, 1985). Beginning in the mid-1980s, public opinion polls have repeatedly suggested that the public, particularly younger people, are worried about the future of the system, whether rightly or wrongly (Shaw & Mysiewicz, 2004; Upston, 1998).

PAY AS YOU GO

Social Security is funded by payroll taxes that go into a large account called the Social Security Trust Fund (which is actually several funds). Despite the name, the Trust Fund is not actually a big bank account where money is saved for the future. In fact, Social Security was originally designed to operate as a modified **pay-as-you-go system**; that is, the money collected each year mostly pays for people who receive benefits in that same year. Younger workers don't actually "save up" for their own Social Security benefits; instead, they pay for current beneficiaries. If more money is collected from payroll taxes than is paid out that year, then the Trust Fund runs a **Social Security surplus**.

URBAN LEGENDS OF AGING

"Social Security is going broke; within a few decades, there won't be any money to pay promised benefits."

People often assume that "going broke" means there will be no money in the Social Security Trust Fund to pay benefits. That's not the case. Even if we do *absolutely nothing* at all, by 2036 Social Security will still be able to pay more than 75% of benefits, which is hardly the same thing as "going broke." With modest changes Social Security could remain solvent for decades beyond that. Policy makers could address solvency, as they did in 1983, by making modest changes in benefits or revenues, keeping Social Security sustainable indefinitely.

Social Security is actually a compact or agreement between different generations. Today's workers pay for today's retirees and hope that future workers will take care of them when they retire, too. Under a pay-as-you-go system, "contributions" to Social Security are actually taxes, not investments. One result of this system, as Martin Feldstein, a conservative economist, points out, is that people feel richer than they really are. If Social Security benefits will be there, maybe people will save less for their retirement.

The fact that Social Security has been a pay-as-you-go system creates some big problems in years to come. The issue is that the baby boom generation is much bigger than the cohort before it or after it. This is a phenomenon demographers call the "pig in the python" pattern, that is, a big bulge in an otherwise skinny snake—or the age distribution—a bulge that moves along as baby boomers grow older.

The problem is that the baby boom cohort's contributions are being used to pay for generous benefits for earlier generations, but the "baby bust" group—or the generation of those born after 1964—does not have a large enough number of workers to contribute as much as will be needed under the pay-as-you-go system. Today, there are 3 workers for each retiree, but by 2050, there will only be 2 workers per retiree (see Exhibit 45). Under a pure pay-as-you-go system, this declining ratio presents a big problem, one that if not faced now could result in benefit cuts or a drastic rise in taxes. So what to do? One answer might be to increase the size of the workforce by allowing more immigration to the United States. Another answer might be to improve worker productivity so that a smaller number of workers can generate enough money to pay benefits. And yet another answer might be to continue to increase the age of retirement so that older adults are compelled to remain in the workforce longer.

Exhibit 45   Number of Workers per Social Security Beneficiary: 1970–2050

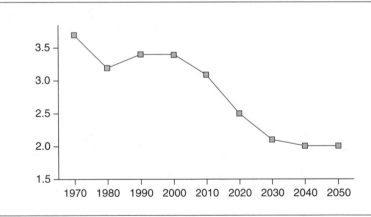

SOURCE: Steinberg (2000).

NOTE: The figures for 2010 and beyond are estimates.

The following question is often raised: Are people getting too low a return on their Social Security contributions? One reason the return is "low" is that Social Security is actually several different programs: a life insurance program, a disability insurance program, and a retirement income program. For the retirement income program, workers today and tomorrow will have to pay for the accumulated promises made in the past.

Proponents of privatizing Social Security believe that the best approach is to permit people to control more of their own savings for retirement. But private accounts alone will not take care of paying for accumulated promises—debts—from the past. Paying off those promises might involve investing part of the Social Security Trust Fund in the stock market, which is one proposal that has been made. Another approach would be to put in money from general tax revenues to make up for any shortfall.

The problem was recognized in the early 1980s. Congress could foresee that, 30 years in the future, the Social Security system could no longer run on a purely pay-as-you-go basis. In 1983, Congress raised the payroll tax rate for Social Security by two percentage points. The idea was to use the extra money to create a larger reserve fund and therefore postpone the day of reckoning when Social Security would run out of money. That tax increase, combined with the dramatic growth of the U.S. economy in the 1990s, dramatically improved the financial health of Social Security.

SOCIAL SECURITY TRUST FUND

The Social Security Administration projects that Social Security Trust Fund assets will continue to grow into the second decade of the 21st century, reaching $3.6 trillion by 2020; in other words, the payroll tax rate currently is higher than needed to pay current benefits. By law, that surplus money in the Trust Fund must go to buy U.S. Treasury notes. In effect, the federal government is borrowing the money and promising to pay it back at a certain time, with interest (Stein, 1991). The Trust Fund, in short, operates on a pay-as-you-go basis.

To understand what this means, imagine that a family is setting aside money to pay for the children's college education. As college savings accumulate, we can compare the savings with the family's debts for car payments, credit cards, and so on. We might say that the college savings fund reduces the family's outstanding debt. But putting the matter that way would be misleading because the savings fund will be rapidly depleted at some time in the not-too-distant future.

The Social Security Trust Fund works the same way. Depending on accounting assumptions, Social Security can be viewed as reducing or paying for part of the current federal budget deficit, just like a family could "borrow" from its college fund to pay off current debts. In this way, the surplus Trust Fund money reduces the need for the government to borrow the money somewhere else, but the "real" deficit is not reduced at all.

There has been considerable debate about just how the Social Security Trust Fund should be viewed: whether as part of the budget, thus reducing the combined federal deficit, or as something completely separate. Defenders of Social Security point out that the program does not increase the federal budget deficit. On the contrary, Social Security collections actually reduce the federal budget deficit because the amount of money collected in payroll taxes each year exceeds what Social Security pays out.

If the payroll tax is a burden, then we might ask, why do we need a surplus anyway? One answer can be found in the Bible's story of Joseph, in which the ancient Egyptians saved up grain during seven fat years to prepare for seven lean years. So, too, is today's Social Security surplus really a collection of "IOUs" that will be needed when the baby boom generation starts to retire in large numbers. At that point, the surplus will begin to decline, and problems will begin to arise.

Will Social Security then "go bankrupt"? The term *bankruptcy* seems extreme, but there is a serious imbalance in the system. According to the Trustees of the Social Security system, the system is expected to be able to pay full benefits until the year 2052; however, it is projected that, starting in 2023, expenditures will begin to exceed revenues, and the robust reserves will need to be drawn down to pay benefits (Social Security Administration, 2011). And, after that, the system will still be able to pay nearly 80% of benefits even if absolutely nothing is done about it (U.S. Congressional Budget Office, 2004). This situation—more than 40 years in the future—presents a problem. Various solutions have been put forward to deal with the shortfall.

For example, Peter Orszag, an economist with the liberal Brookings Institution, has compared Social Security to a car with a flat tire. He argues that it isn't necessary to throw out the car—just fix the flat tire. In a detailed book on Social Security, Diamond and Orszag (2004) offer a proposal to "share the pain" involved in fixing the flat tire. They argue that the Social Security shortfall can be corrected by reducing promised benefits in the future, modestly increasing the tax rate, and raising the maximum level of earnings subject to the payroll tax. Cuts in benefits would be progressive to protect poor people.

The baby boom cohort will probably pay higher taxes on its benefits or else receive somewhat lower benefits for two reasons that have nothing to do with the Trust Fund surplus. In the first place, the $25,000 income threshold for taxing Social Security is not indexed for inflation, so in the future a much larger proportion of Social Security recipients are likely to have benefits taxed. In the second place, Congress in 1983 decided that the age of eligibility for full Social Security benefits would rise between 2002 and 2007, from age 65, first to 66 and then to 67. A couple retiring prior to the increase in the age of eligibility could recover the money they contributed to Social Security taxes in only 4.5 years. But for those retiring after 2010, it will take 7.5 years—in short, not as good a rate of return. In other words, by waiting longer to receive full benefits, a smaller share of retirees' past earnings will be replaced.

Whether this difference in benefits from one generation to the next is seriously inequitable is another question. Increased life expectancies and different levels of wealth between historical cohorts make comparisons difficult.

None of these arguments about fairness has affected the basic popularity of Social Security with the American public, however, as demonstrated not only by polls but also by the votes of Congress in more than half a century since Social Security came into existence. Legislative support reflects the widespread public support. For example, on the 75th anniversary of Social Security, in 2010, a national survey by the Roper Center for Public Opinion Research showed that support for Social Security remains exceedingly high (see Roper Center for Public Opinion Research, 2011, for the most recent figures). Consistent with earlier surveys in 1995 and 1985, a majority of adults age 18 and older believe that Social Security is "one of the most important government programs" providing financial security to help older Americans remain independent. Despite broad support, however, in recent years, serious questions have been raised about some aspects of Social Security, and these controversies are at the center of the discussion about its future:

- Should eligibility for Social Security be restricted to people who are most in need?
- Should Social Security be changed, at least in part, into a privatized system of retirement to ensure its integrity in the future?

- Should Social Security be changed to make it more fair in its treatment of women compared with men?
- Will Social Security still be there for future generations who need it?

ELIGIBILITY

To be eligible for full benefits, an older person born in 1937 or earlier must be age 65 (62 for early retirement and reduced benefits) and have worked for 10 years in a job where Social Security taxes were deducted; the age at which one is eligible for full benefits increases for birth cohorts after 1937, to a maximum age of 67 (www.socialsecurity.gov). One way to reduce the cost of Social Security and improve its financing would be to raise the age for those eligible to receive benefits. That shift has already begun and will go into effect as the age of eligibility slowly rises from age 65 to 67 in the coming years. But some are asking whether the age of eligibility should be raised even higher, to age 69 or 70.

Some of those arguing for an increased age of eligibility cite rising average life expectancy. In 1940, a 21-year-old man had a 54% chance of living long enough to collect Social Security. But a 21-year-old man in the 1990s had a 72% chance of living to age 65, and a 21-year-old woman had an 83% chance.

What about longevity after the point of retirement? In 1940, the life expectancy for people surviving to 65 was 12.7 years for men and 14.7 years for women. Today, however, men at age 65 can expect to live 17.2 years and women 20 years (Administration on Aging, 2010). Increasing the age of eligibility by 2 or 3 years would bring the program back to the years of coverage anticipated at the time it was created. However, raising the age of eligibility could create special hardship for minority groups, whose life expectancy is lower than that of Whites. Lower-income people might have to work longer than before at difficult jobs but still be more likely to die earlier than other Americans, and, in effect, their benefits would be cut. To counter this possibility, it has been suggested that the level of the minimum Social Security benefit be increased to protect lower-paid workers from late-life poverty (National Academy of Social Insurance, 2011).

Still another approach under consideration is to reduce Social Security benefits to make them reflect the actual rate of inflation. In 1996, an official government commission examining inflation found that the Consumer Price Index has for years been overstating the rise in the cost of living by 1.1%. The commission estimated that by reducing the Consumer Price Index to a level equal to actual inflation, the federal government would save $1 trillion in Social Security benefits (which are linked to the Consumer Price Index) over a 12-year period. As a result, Social Security would be solid until the mid-21st century. However, an across-the-board reduction in benefits would create the greatest hardship for the poorest beneficiaries.

Another possible solution is to make eligibility for Social Security means tested, that is, available only to people whose income falls below some threshold, such as the Supplemental Security Income (SSI) program. However, we should note that the flat SSI benefit paid in 2008 was on average only $475 per month for an individual, which is a low amount on which to live. Another variation of this approach is the proposal that Social Security benefits be eliminated for people above some threshold—a so-called **affluence test**. This approach is in some ways an extension of the present policy of taxing benefits for higher-income beneficiaries.

Any of these approaches would either reduce expenses or increase revenues. The result might be to place Social Security on a sounder footing. All three approaches would keep Social Security a publicly financed, age-based entitlement program, but would shift eligibility in ways that might put the least advantaged at risk in unpredictable ways.

PRIVATIZATION

The term *privatization* can mean several changes from the current public system. One change might be to increase the level of advance funding, that is, to move away from a pay-as-you-go system to a **funded system** that involves a real increase in national savings. This could be accomplished by investing part or all of the Trust Fund in private savings vehicles, such as the stock market. A second meaning of privatization might be to accompany private investment with increased choice over the individual retirement savings, for example, stocks versus bonds. A third meaning of privatization might be to eliminate the way Social Security quietly redistributes benefits, either from one generation to another or from high-wage earners to low-wage earners.

In 1997, a national Social Security Advisory Council recommended several plans that involved some degree of privatization—that is, putting Social Security surplus funds into private investment in the stock market. Since 1926, stocks have earned an average of 8% after inflation—a level much higher than the 3% earned by Social Security funds now invested in Treasury bonds. The potential for greater gain from private investment is what makes privatization attractive to many people. If proponents of privatization are right, then it might be possible to reduce Social Security benefits in the future because people would be getting higher returns from investments in the stock market.

Some plans for privatization could involve keeping the traditional system intact, but investing the Trust Fund in a broad index of the stock market. Other plans involve letting individuals decide on their own investment portfolio, for example, giving them the option to invest their Social Security contributions in a small number of government-approved mutual funds. Some advocates of privatizing Social Security favor "carve-outs" or diverting a part of payroll tax revenues to personal accounts. Others favor "add-ons" or creation of government-subsidized private accounts that would not be taken out of payroll taxes going to the current Social Security program. Whatever form that privatization might take, investing in the stock market would not be cost free because the federal government would have to borrow directly from the public, thus raising taxes indirectly.

The other question about privatization is how to make the transition from today's system to a more privatized form of Social Security. To finance this, the federal government would have to borrow money to make up for retirement benefits already promised to people now retired or near retirement age. Those "transition costs" would mean higher taxes on top of those currently paid for Social Security. Under the boldest privatization plan, transition costs could require higher taxes equal to $6.5 trillion over a 70-year period. As economist Barry Bosworth (1996) has noted, what this means is that shifting to a funded system means that members of a "transition generation" would have to pay twice: for their own future retirement and for those in the current system.

One big problem with investing Social Security funds in the stock market is the size of the Trust Fund. It is estimated that by the year 2015, the federal Social Security Trust Fund would own $800 billion in shares, or 10% of the entire stock market, which would be a dramatic shift toward national influence over the capital markets. Critics of the plan note that, although investing in a broad stock index should provide some diversification and safety, putting Social Security funds at risk in the stock market is not necessarily prudent. True, stocks have risen at a high average rate since 1926, but the U.S. stock market dropped abruptly in the year 2008 and has been volatile ever since.

Conservative critics of Social Security, such as economist Martin Feldstein, have long argued that a public pension program weakens private savings and capital formation. They often point to Chile as an example of successful transition to a privatized system. Other countries, such as Finland, Brazil, and Australia, have experimented with shifting public pensions to stock market investment to improve national savings. But questions have been raised about investing Social Security funds in the stock market. What impact would forced savings plans really have on other voluntary savings and capital formation? Could federal government ownership of stocks lead to other intervention in the markets, for example, forbidding Social Security from buying tobacco stocks? What happens in a crash, such as we experienced in the Great Depression of the 1930s or the economic downturn beginning in 2008?

What about allowing workers to invest part of their Social Security taxes in private retirement accounts? On the positive side, this plan would give individuals the freedom to make their own decisions about investing for retirement, as they do with 401(k) plans now. But studies by the Employee Benefit Research Institute suggest that lower-income workers tend to be much too cautious in their investments, and, as a result, their savings do not grow as much over time. However, if people do make more risky investments, they are vulnerable to losing their retirement savings. Still another problem is how to help investors ride out the stock market's periodic ups and downs over time. What would holders of individual accounts do if the market took a long downturn?

The effect of changing Social Security into private retirement accounts would be to emphasize equity over adequacy in the operation of the system as a whole. Social Security has always embodied two different, sometimes contradictory principles: equity, which means a fair return to beneficiaries depending on how much they contribute to the system, and **adequacy**, which means maintaining a decent minimum of income for everyone. Some degree of privatization might make Americans more confident in Social Security, but by favoring equity over adequacy, privatization could be the first step toward dismantling a system that has been popular and successful for the better part of a century.

The issue of equity versus adequacy comes up when people argue that they could do better by saving and investing Social Security benefits on their own. This argument makes us realize that Social Security represents a transfer of wealth from younger to older people. Higher-wage earners get back higher Social Security benefits than lower-wage earners, but the higher-wage earners do not get back as much as they would if they invested the money and received the gains for themselves alone. In effect, Social Security involves a modest redistribution of wealth to give a more adequate level of benefits to lower-wage earners. The conflict between equity and adequacy comes up when we look at how different generations and different wage earners do under the current system.

GLOBAL PERSPECTIVE

The New Swedish National Pension System

Should pensions for retirement be given by government, or should individuals be expected to save for themselves? One answer to this question may come in the new Swedish National Pension System, introduced in 2000. The Swedish system is an individual defined-contribution plan, and participation is public and mandatory. The Swedish approach requires shared employee-employer contributions, and employees can choose from 650 different mutual funds, where these funds are invested collectively on behalf of employees. The Swedish government itself offers two stock funds, charging very low fees. Participants in all the plans also have considerable flexibility in payout choices. The new Swedish National Pension System, therefore, offers both mandatory participation and voluntary options as a way for participants to save for retirement.

The new Swedish National Pension System is a distinctive institution in a country that has a strong tradition of capitalism and free markets balanced by a strong public sector. Swedish citizens are accustomed to paying high taxes but also want to find ways to save more for their own retirement. The new public-private system in Sweden is added on to an already generous public pension provision in a country with long commitment to social welfare. The result is a blend of safety and individual choice. But the picture is not all rosy. As with any stock investment plan, there are inevitable risks. For example, Swedish investors lost money in the bear market after 2001. Perhaps as a result of that negative experience, 90% of those recently participating in the system pick the "default" option of the Swedish government's own funds. But citizens do have a choice, and the new Swedish National Pension System continues to attract interest around the world.

SOURCE: R. Kent Weaver, "Design and Implementation Issues in Swedish Individual Pension Accounts," *Social Security Bulletin* (2003–2004), 65(4): 38–56.

WOMEN AND SOCIAL SECURITY

The debate about privatization of Social Security also raises other questions: for example, how Social Security treats men and women over the life course. Some issues have come to the forefront now because of changing social conditions, such as the role of women (MacDonald, 1998). For example, a married woman is entitled to 50% of her retired spouse's earned Social Security benefits and a higher amount if the husband dies. But if a woman earned Social Security benefits on her own wage record, she would not receive any of those benefits if she receives a spousal or widow's benefit based on marriage status. As a result, Social Security as presently constructed tends to favor widows but discourages a "secondary" earner.

When Social Security first went into effect in the 1930s, it took for granted a traditional family structure. The Social Security program was originally planned for a one-earner family, but today more than 70% of women between the ages of 20 and 44 are in the labor force. Social Security seeks to treat both men and women fairly, but there are troublesome equity problems related to caregiving, divorce, and two-earner couples (B. Johnson, 1987).

The rate of poverty and near poverty among older women is shockingly high (Munnel, 2004). Women constitute almost three fourths of older people who live in poverty.

Widowhood is a distinct threat to well-being: Almost four times more widows live in poverty than do married women of comparable age (Karamcheva & Munnell, 2007).

Women's wages remain about three quarters of men's, and that gap widens in retirement. Taking time out to raise families or to care for older relatives reduces the primary insurance amount credited under Social Security (Kingson & O'Grady, 1993). Leaving the workforce to care for others, whether early or late in a career, is more costly for women with low and moderate earnings. As a result, women may fail to earn enough during their working lives to be able to save on their own for retirement. A basic problem is that the Social Security system recognizes only paid work for determining benefits. The nonmonetized work that women do in caring for family members receives no acknowledgment. These facts lead some observers to recommend earnings sharing for Social Security or even credit for years of lost wages due to caregiving.

Another equity issue arises from today's higher divorce rate. Under Social Security, women who have been married less than 10 years and who were out of the labor force during those years receive no credits toward Social Security, meaning lower benefits at retirement. As a result of past reforms, women who are divorced are eligible for a spousal benefit if they were married at least 10 years. Nevertheless, those who are divorced receive only a third of what the couple would have received had they stayed together.

Still another issue arises from the way that Social Security treats married couples. When a woman retires, she automatically receives a spousal benefit equivalent to half of that to which her husband is entitled. Unless married women have earned a great deal, their spousal benefits are higher than the level they would get from their own work history. Social Security will give benefits to a woman for work outside the home only when her accumulated benefits are greater than her husband's. Typically, then, older women's lifetime economic contribution to Social Security is totally disregarded, and two-earner couples may receive proportionately lower retirement benefits than one-earner couples under the present system. Some critics find that inequitable (U.S. Congress, 1983).

The issue here is related to the question of fairness in Social Security—namely, equity versus adequacy. Women who are eligible for Social Security receive a benefit based on their own earnings record. But if they have been married, they are also entitled to a benefit based on the earnings record of their spouse. In other words, a married woman (or widowed or divorced woman) can be dually entitled to Social Security benefits. In recent decades, more and more women have been eligible for benefits in this way. As a result, dropping out of the labor force to care for children or other dependents may be offset by a share of the spouse's earnings record.

Some suggestions have been made to reform the Social Security system to take into account a woman's life cycle (Morgan, 2000). Some critics have urged that Social Security give credit for time taken out for caregiving, and the National Academy of Social Insurance has suggested that allowing child care years to count toward Social Security benefits is an option for improving the adequacy of benefits. A related proposal is for Social Security caregivers to simply eliminate up to 5 years of low earnings if those years were devoted to caregiving responsibilities.

One problem is that a child care dropout credit might favor economically advantaged women, who can afford to take time off for child rearing more than those who are disadvantaged. But many European countries have a similar system of implicitly subsidizing child care through children's allowances. Another suggestion to help women would be the "double-decker system," under which a retiree would receive a minimum fixed amount of money along with a benefit related to work history and prior contributions (Burkhauser, 1984).

Finally, some women's advocates have urged an **earnings sharing** plan, which would divide the total earnings of the married couple between the Social Security accounts of both spouses (U.S. Congress, 1984). Earnings sharing is attractive because it treats marriage as a partnership of equals, and it also helps improve the retirement income of women who may be impoverished by divorce or widowhood. But the idea of earnings sharing brings us back to a fundamental question concerning the rationale for Social Security: Is the program to be viewed as a means for replacement of earnings or as an investment judged by equity and rate of return? Would complete privatization of Social Security be good for women (Estes, 2004; J. Williamson, 1997)? People have different ways of thinking about fairness and adequacy, and basic social values are at stake. However these debates are resolved, Social Security will remain a key issue for women.

What about the impact of privatization for women on Social Security? Here, opinions differ. Some believe that privatizing Social Security would be good for women precisely because changes in family structure have opened up more opportunities for women to control their own earnings. Privatization would encourage greater individual choice, in keeping with values of contemporary feminism (Tanner, 2004). However, one recent study concluded that privatization could have particularly severe effects for older women whose life course has not closely matched the "traditional" family model of marriage with a male breadwinner, as well as women suffering from disadvantages of ethnicity, race, and social class.

DEBATE OVER SOCIAL SECURITY

In the readings that follow, we encounter some different views about the future of the Social Security system. Privatization is rejected by Diamond and Orszag in the reading from their article "Saving Social Security: A Balanced Approach." They too acknowledge that the Social Security Trust Fund faces a long-term deficit, but they are opposed to diverting payroll tax revenues into private accounts. Instead, Diamond and Orszag argue that modest incremental changes can trim benefits and increase revenues while keeping the basic system intact. They are concerned about maintaining Social Security's role as a collective form of social insurance instead of a vehicle for individual investment.

In contrast, in "The Necessity and Desirability of Social Security Reform," Ramesh Ponnuru gives a strong case for why the current pay-as-you-go system cannot survive. He believes that personal savings accounts are an idea whose time has come and are a better way to accomplish the goals that Social Security was intended to achieve. Ponnuru's approach is a form of partial privatization, similar to what President George W. Bush proposed in the great Social Security debates of 2005.

The 2005 debate around private accounts brought to the forefront the familiar conflict between equity and adequacy as goals for Social Security. Even without further changes in the system, the 1983 reforms are now increasing the age of eligibility for full benefits from age 65 up to 66 and then 67 over a period of decades. Some voices have urged raising the age still higher. But Lawrence Thompson, in "Social Security Reform and Benefit Adequacy," points out that any rise in the retirement age is an across-the-board benefit cut that hits hardest at the least advantaged. He concludes that, to reform Social Security, we have to face difficult choices.

Citizens in favor of protecting Social Security rally on the U.S. Capitol grounds before lobbying Congress on this issue.

Finally, the role of family and gender relationships in Social Security is another topic that deserves more attention than it received in the 2005 debates. C. Eugene Steuerle and Melissa Favreault raise some critical questions about how Social Security benefits are paid out to married couples and single people. They argue that Social Security now treats beneficiaries in ways that are unequal and unfair because the system doesn't reflect the reality of family life in the 21st century. Beyond the debate over private accounts, they urge a debate over other kinds of unfairness in the Social Security system.

FOCUS ON PRACTICE

Investment Decisions for Retirement Income

"Investing for retirement? Oh, that's so far off! I've got more important things to worry about right now." Does that statement sound familiar? If you're young, it may sum up your sentiments about investing for retirement. If you're a bit older, you might have different thoughts that are no more realistic: "Investing for retirement? Oh, I'm already in a pension plan, and besides I'm covered by Social Security."

Ignoring the problem of retirement income because it's far off or assuming it's already "taken care of" is unwise. For one thing, the proportion of American workers covered by employer-paid pension plans has not been growing in recent years. As more and more employees who have pensions are under defined-contribution plans, retirement income becomes less predictable, and individuals need to plan more carefully.

(Continued)

(Continued)

In short, when it comes to pension coverage, more and more workers are on their own. In the early years of the 21st century, the shift is away from "someone else" taking care of retirement planning toward individuals making more investment decisions on their own. In fact, in 2008, only 41% of people 65+ were receiving benefits from employer-sponsored pensions (Social Security Administration, 2008).

As for Social Security, we know that a large proportion of younger Americans lack confidence in its future even while supporting its solvency in the present. Even if their pessimism is mistaken, Social Security was never intended to cover more than a portion of retirement income. Social Security benefits and private pensions, taken together, constitute an important part of the assets of older people. Decisions about retirement—such as assigning pension survivor rights and choosing whether to accept an early retirement offer—are some of the most important financial decisions a person will make over the life course.

If we take account of all these factors, the conclusion is conveyed by the title of a book by Robert Butler and Kenzo Kiikuni (1993): *Who Is Responsible for My Old Age?* The answer is, "I am responsible myself," and retirement in the future will require more careful financial planning than in the past.

A key idea for investment planning is to invest with a life course perspective in mind—that is, to think about age and tolerance for risk. Psychological studies tell us that people tend to become more risk averse as they grow older. Some risk aversiveness is quite rational. With advancing age, the bad consequences for risky behavior become more severe. If you have a skiing accident, it takes longer for bones to heal; if you're widowed or divorced, the odds of remarriage become worse; and if you make a bad investment, it's more difficult to recover.

What does risk aversiveness suggest about investing for retirement? One response might be to shift savings away from more volatile and risky assets—such as the stock market—in favor of safe, fixed-income investments, such as savings accounts, certificates of deposit, and U.S. Treasury bills.

But is that advice practical? Someone retiring at 60 years of age may have a life expectancy of 20 or 25 more years. Would a nest egg of $600,000 at age 62 be enough to last 20 or 25 years? Suppose you put the nest egg in "safe" investments (Treasury bills or corporate bonds), paying a historical yield of 5.5%—or $33,000 a year. If you decide you need a total family income of $50,000, why not just take the interest and withdraw an additional $17,000 a year from savings? Taking $17,000 over 25 years would still leave a big cushion of $175,000 in principal, wouldn't it? And you would have the advantage of safety for the principal.

Actually, this example shows one difficulty of retirement planning. To begin with, each time you withdrew money from the principal, there would be less money left to generate interest. The $33,000 yield would get smaller each year as the principal shrank. Second, the original calculation failed to take account of inflation. At a 3.14% inflation rate, the historical average for inflation, the original $600,000 would, in real dollars, be completely gone in only 14 years if you took out the $17,000 supplement each year. In other words, if you retired at age 62, you'd be completely broke by age 76.

Another problem is the rate of return on this "safe" investment. One mistake people make nearing retirement is to put too much into "safe" investments, neglecting the likelihood that they will live 20 or 25 years. A better strategy is to invest for both current income and capital gains. Fixed-income investments are part of a strategy, but they should not dominate the portfolio. Let's look at the $600,000 nest egg again, this time assuming that it is invested in a mixed stock-and-bond portfolio generating a pretax yield of 7.5%. Here's how much you could withdraw from the portfolio each year (again, assuming a 3.14% annual inflation rate):

If you withdraw	Your money will last
$30,000	40+ years
$40,000	23.7 years
$50,000	16.7 years
$60,000	12.9 years

But doesn't investing in the stock market involve too much risk for someone nearing retirement? In any given year, there may be more risk. But over a period of decades, risk goes down because volatility—the ups and downs of the stock market—tends to be absorbed by overall average gains. Historically, stocks have returned gains of around 8% since the year 1920. Of course, that historical average must take account of huge drops, such as the market downturn after 1929 and the 2008 financial crisis. But all things considered, it might make more sense to accept a higher degree of short-term risk to gain more inflation protection over the long run.

The point of this exercise is to underscore a simple point. There are several different kinds of risk: risk of losing principal, risk of inflation, and risk of not being able to convert an asset into cash when needed. There is no way to avoid all types of risk in investing. The only question is how much and what kind of risk a person is willing to take for some investment objective. To make reasonable investment decisions about risk, one needs to consider many factors, and age is certainly one of them.

In addition to overestimating how much money they will receive from a pension and Social Security, most people underestimate retirement expenses. Because people are living longer, they may end up spending as many years in retirement as they did working. They may be putting their children through college while simultaneously caring for their aging parents. The amount of money needed for retirement must take into consideration current expenses, future living expenses, life expectancy, taxes, and inflation.

Financial advisors generally recommend that people of all ages make stocks part of an investment program. They also recommend diversifying an investment portfolio and investing on a regular basis in order to soften the impact of market price swings. One favored approach is dollar cost averaging, where the same dollar amount is invested each month regardless of whether stock prices are high or low. Another way to reduce risk is asset allocation, which splits investments among stocks, bonds, and fixed-rate investments, taking account of age and other factors.

Students of financial gerontology remind us that decisions about saving for retirement are influenced by a variety of factors. Some economists have favored a life cycle model of savings, which predicts that people in their 40s will begin to put more money into savings as retirement looms on the horizon. But savings and investment also depend on cohort characteristics and the historical period or economic environment, not to mention individual variations in attitudes toward planning for the future. As people take on more responsibility for their own retirement investment decisions, the practical need for better planning will become greater in years to come.

READING 38

How to Save Social Security

A Balanced Approach

Peter Diamond and Peter Orszag

Social Security is one of America's most successful government programs. It has helped millions of Americans avoid poverty in old age, upon becoming disabled, or after the death of a family wage earner. Despite its success, however, the program faces two principal problems.

First, Social Security faces a long-term deficit, even though it is currently running short-term cash surpluses. Addressing the long-term deficit would put both the program itself and the nation's budget on a sounder footing.

Second, two decades have passed since the last significant changes in Social Security. Since then, as our economy and society have continued to evolve, some aspects of the program have become increasingly out of date. The history of Social Security is one of steady adaptation to evolving issues, and it is time to adapt the program once again.

Restoring long-term balance to Social Security is necessary, but it is not necessary to destroy the program in order to save it. To be sure, some analysts reject the view that Social Security's projected financial problems are serious enough to warrant any changes right now. Others, in contrast, exaggerate the difficulty of saving Social Security to justify proposals that would share the most valuable features of this exemplary program. Our view is that Social Security's projected financial difficulties are real and that addressing those difficulties sooner rather than later would make sensible reforms easier and more likely. The prospects are not so dire, however, as to require undercutting the basic structure of the system. In other words, our purpose is to save Social Security both from its financial problems and from some of its "reformers."

Our approach recognizes and preserves the value of Social Security in providing a basic level of benefits for workers and their families that cannot be decimated by stock market crashes or inflation, and that lasts for the life of the beneficiary. Our plan updates Social Security to reflect changes in the labor market and life expectancies. And it eliminates the long-term deficit without resorting to accounting gimmicks, thereby putting the program and the federal budget on a sounder financial footing.

Our plan to restore long-term solvency has three components, each of which addresses one of the factors that contribute to the long-term deficit in Social Security: improvements in life expectancy, increased earnings inequality, and the ongoing legacy debt that arises from the program's generosity to its early beneficiaries. Each component of our reform plan includes adjustments to both benefits and revenue to help close the long-term deficit.

The first of these components is the life expectancy component. Life expectancy at age 65 has risen by four years for men and five years for women since 1940, and it is expected to continue rising in the future. Increases in life expectancy make Social Security benefits more valuable to recipients, because the benefits are paid over more years. But for that very reason, increases in life expectancy also raise the cost of Social Security.

Many observers have recognized that it makes sense to adjust Social Security for the effects of increased life expectancy. Previous proposals to do this, however, have adopted the extreme view that all of the adjustment should occur through reductions in benefits. Instead, we propose a balanced approach in which roughly half the life expectancy adjustment occurs through changes to benefits and the rest through changes to payroll taxes.

The second component of our plan addresses earnings inequality, which has risen substantially in the past two decades. Inequality of earnings across workers in the labor force affects Social Security in several ways. For example, the payroll tax is levied on earnings only up to a certain level (in 2003 that level, the maximum taxable earnings base, was $87,000). In each year over the past two decades, about 6 percent of workers have had high enough earnings that some of their earnings were above the maximum taxable earnings base and therefore not subject to the payroll tax. These higher-income workers have enjoyed disproportionately rapid earnings growth over that period, so that the share of economy-wide earnings not taxed for Social Security has risen substantially. In 1983, when the last major reform of Social Security was undertaken, 10 percent of all earnings were above the maximum taxable earnings base. By 2002 that share had risen to about 15 percent.

In addition to having more of their earnings escape taxation by Social Security, high-income workers have enjoyed increasing life expectancies relative to other workers. This increasing difference in life expectancy tends to diminish the progressivity of Social Security (that is, its provision of relatively more generous benefits to lower-earning workers) on a lifetime basis. The life expectancy adjustments in the first component of our plan are based on average increases in life expectancy for the entire population. Since life expectancy for higher earners is increasing more rapidly than the average, an additional adjustment just for higher earners is warranted.

To address the effect of earnings inequality on Social Security, our plan again includes a balance of revenue and benefit adjustments. First, we propose gradually raising the maximum taxable earnings base until the share of earnings that is above the base—and hence escapes the payroll tax—has returned to roughly its average level over the past twenty years. This change would gradually reduce the share of earnings not subject to the payroll tax until it reaches 13 percent in 2063, roughly halfway between its current level and its level in 1983. Second, to make Social Security somewhat more progressive, and thereby offset the effects of disproportionately rapid gains in life expectancy among higher earners, we propose a benefit reduction that affects only relatively high earners. Currently, about 15 percent of workers newly eligible for Social Security benefits have sufficiently high earnings that a portion of those earnings falls in the highest tier of the Social Security benefit formula. Our benefit adjustment for income inequality consists of a gradual, modest reduction in benefits that would affect only those with earnings in this highest tier.

The third component of our plan recognizes the legacy cost stemming from Social Security's history. The first generations of beneficiaries received far more in benefits than they had contributed in payroll taxes. Beneficiaries in the earliest years of the program, for example, contributed for only a few years of their career but then received full benefits over their whole retirement. The decision to provide ample benefits to these early beneficiaries is understandable: most of them had experienced hardship during the Great Depression, many had fought in World War I or World War II, and elderly poverty rates were unacceptably high. But those benefits did not come free: the iron logic of accounting requires that since those early retirees received more in benefits than they had paid in, later generations of retirees must receive less. In other words, the system's generosity to early beneficiaries generated an implicit debt, which we refer to in this book as Social Security's legacy debt. That debt can be defined as the accumulated difference between benefits and taxes (accumulated at the market rate of interest) for past and current beneficiaries. This legacy debt imposes an ongoing cost on participants in the program, which we call the legacy cost.

We all inherit a legacy from Social Security's history. Even if we wanted to, nothing we can do now could take back what was given to Social Security's early beneficiaries. In addition, most people are unwilling to reduce benefits for those already receiving them or nearing retirement. Those two facts determine the size of Social Security's legacy debt. And once that debt is determined, its cost cannot be avoided: the only issue is how we finance that cost across different generations.

Social Security's legacy is not new. It has been with us since the origins of the program itself. But the idea of a Social Security reform based in part on explicitly recognizing the need to share the cost of that legacy is new. We propose to reform the financing of the legacy debt through three changes:

- First, we would gradually phase in universal coverage under Social Security, to ensure that all workers bear their fair share of the cost of the nation's generosity to earlier generations. Currently, about 4 million workers, almost all of them in state and local governments, are not covered by Social Security. Their nonparticipation means that those workers escape any contribution to the financing of the legacy debt.
- Second, we would impose a legacy tax on earnings above the maximum taxable earnings base, thereby ensuring that very high earners contribute to financing the legacy debt in proportion to their full earnings. The legacy tax above the base would start at 3.0 percent and gradually rise to 3.5 percent by 2080.
- Third, we would impose a universal legacy charge on future workers and beneficiaries, roughly half of which would be in the form of a benefit reduction for all beneficiaries becoming eligible in or after 2023, and the rest in the

form of a very modest increase in the payroll tax from 2023 onward. This universal legacy charge would gradually increase over time, so as to help stabilize the legacy debt as a share of taxable payroll.

This approach to financing the legacy debt reflects a reasonable balance between current and distant generations, between lower earners and higher earners, and between workers who are currently covered by the program and workers who are not.

Our three-part proposal would restore long-term balance to Social Security as that term is conventionally understood: actuarial balance over a seventy-five-year horizon. Our plan would not only eliminate the seventy-five-year deficit in Social Security, but indeed would produce a modestly growing ratio of the Social Security trust fund to annual costs at the end of the seventy-five-year period. This is important because it makes it more likely that Social Security will not again face a seventy-five-year deficit for a long time to come.

Our plan combines revenue increases and benefit reductions—the same approach taken in the last major Social Security reform, that of the early 1980s, when Alan Greenspan chaired a bipartisan commission on Social Security. That commission facilitated a reform that included adjustments to both benefits and taxes. Such a balanced approach was the basis for reaching a consensus between President Ronald Reagan and congressional Republicans on one hand and congressional Democrats led by House Speaker Thomas P. O'Neill on the other.

In addition to our three-part plan to restore long-term balance to Social Security, we propose improvements to Social Security's financial protections for certain particularly vulnerable beneficiaries. We focus on changes in four areas: benefits for workers with low lifetime earnings; benefits for widows and widowers; benefits for disabled workers and young supervisors; and further protection for all beneficiaries against unexpected inflation. These changes would significantly improve Social Security's ability to provide cost-effective social insurance while maintaining long-term financial balance.

What do these various changes imply for the benefits that individual workers will receive and for the taxes they will pay? Workers who are 55 years old or older in 2004 will experience no change in their benefits from those scheduled under current law. For younger workers with average earnings, our proposal involves a gradual and modest reduction in benefits from those scheduled under current law for successive cohorts. For example, a 45-year-old average earner would experience less than a 1 percent reduction in benefits under our plan. A 35-year-old average earner would experience less than a 5 percent reduction. And a 25-year-old with average earnings would experience less than a 9 percent reduction in benefits (Figure 1). Higher earners would experience somewhat larger reductions in benefits than the average, and lower earners would experience smaller reductions. These modest reductions in benefits are also in keeping with the tradition set in 1983. For example, the 1983 reform reduced benefits by about 10 percent for those 25 years old at the time of the reform, a slightly larger benefit reduction than under our plan for average earners age 25 in 2004.

It is important to underline that the reductions just described are relative to currently scheduled benefits; they are not absolute reductions from what retirees receive today. Although today's younger workers would experience somewhat larger percentage reductions in scheduled benefits when they retire than older workers, those benefits would still be higher, even after adjusting for inflation, than those of the older workers. An average earner who is 25 years old in 2004, for example, would receive an annual inflation-adjusted benefit at retirement that is more than 25 percent higher than the inflation-adjusted benefit of an average earner who is 55 years old in 2004. The reason is that Social Security benefits increase when career earnings rise, and today's 25-year-olds

Figure 1   Benefit Reductions Under Proposed Reform for Average Earners

Age at end of 2004	Change in benefits from scheduled benefit baseline percentage	Benefit at full benefit (2003 dollars)[a]
55	0.0	15,408
45	−0.6	17,100
35	−4.5	18,200
25	−8.6	19,400

SOURCE: Authors' calculations.

[a]For a retired worker with scaled medium preretirement earnings pattern. This scaled earnings pattern allows wages to vary with the age of the worker but ensures that lifetime earnings are approximately equal to those of a worker with the average wage in every year of his or her career.

are expected to have higher career earnings than today's 55-year-olds because of ongoing productivity gains in the economy. Even with the modest benefit reductions in our plan, the result is that inflation-adjusted benefits rise from one generation to the next.

Our plan balances its modest and gradual benefit reductions with a modest and gradual increase in the payroll tax rate. As Figure 2 shows, the employee share of the payroll tax under our plan would slowly increase from 6.2 percent in 2005 to 7.1 percent in 2055. Because employees and their employers each pay half of the payroll tax, the combined employer-employee payroll tax rate would rise from 12.4 percent today to 12.45 percent in 2015, 13.2 percent in 2035, and 14.2 percent in 2055. This gradual increase in the payroll tax rate helps ensure that Social Security continues to provide an adequate level of benefits that are protected against inflation and financial market fluctuations, and that last as long as the beneficiary lives.

Figure 2 Payroll Tax Rates Under Proposed Reform Percentage of Earnings

Year	Employee rate	Combined employer-employee rate
2005	6.20	12.40
2015	6.22	12.45
2025	6.35	12.69
2035	6.59	13.18
2045	6.84	13.68
2055	7.09	14.18

SOURCE: Authors' calculations based on memorandum from the Office of the Chief Actuary.

In summary, our plan differs from most other recent Social Security reform proposals, and in our view it represents the most auspicious way of reforming the program, for the following reasons:

- It balances benefit and revenue adjustments.
- It restores long-term balance and sustainable solvency to Social Security.

- It does not assume any transfers from general revenue.
- It does not rely on substantial reductions in disability and young survivor benefits to help restore long-term balance.
- It strengthens the program's protections for low earners and widows.
- It does not divert Social Security revenue into individual accounts.
- It preserves Social Security's core social insurance role, providing a base level of income in time of need that is protected against financial market fluctuations and unexpected inflation.

Despite our confidence in the plan's substantive merits, we are under no illusions regarding the political difficulties of enacting it. Social Security reform is controversial, as it should be. After all, Social Security plays a critical role in the lives of millions of Americans and in the federal budget. Reforms to such an important program *should* generate political interest and debate. Nonetheless, we hope that the simplicity and balance of our basic three-pronged plan demonstrate that Social Security can be mended without resorting to the most controversial and problematic elements included in some other recent reform plans.

READING 39

The Necessity and Desirability of Social Security Reform

Ramesh Ponnuru

Social Security has mostly been a "pay as you go" program. Today's workers finance, with their payroll taxes, the retirements of today's senior citizens. When those workers get old, they will in turn rely on taxes from their children's generation. But increased lifespans and decreased childbearing have made this bargain hard to sustain. In 1950, there were

16 workers supporting each retiree. There are 3.3 today. By 2040, there will be only 2.1.

Benefits are calculated using a complex formula. If you had higher wages, and thus paid more payroll taxes into the system, you'll get a higher benefit. But benefits don't rise in exact proportion to wages: There's some (somewhat haphazard) redistribution to the poor. Homemakers get a benefit, too; the system, it might be said, recognizes that they have contributed to it by nurturing the next generation of taxpayers. During your retirement, your benefits go up every year to keep up with inflation.

A TRILLION HERE, A TRILLION THERE

A worker who has earned the average American worker's wage each year will get $14,854 from Social Security this year if he retires now. Because wages grow over time, so do benefits. By 2050, the average worker's retirement benefit is expected to be $23,811 (adjusting for inflation). That's 60 percent higher than today.

That growth is what is causing the fiscal problem. In 2018, the program is expected to start sending out more checks than it collects in revenues. The system has an expected $283 billion shortfall during the following five years. By 2045, it is expected to lose $394 billion in one year. (There's an asterisk by that number, though, as we'll see below.) One way to measure the financing gap is to estimate how much additional money the government would have to devote to Social Security *now* to make the program permanently solvent. Depending on how you calculate it, that number is $10 to 11 trillion. That's larger than our entire economy.

It is true that we do not have to fix it immediately. It is also true that every year we wait, the choices get worse. We can gradually cut benefits if we get started now. If we don't, we will have to cut them (or raise taxes) very sharply. Andrew Biggs, now a commissioner at the Social Security Administration, has estimated that delaying reform for one more year will cost $600 billion—and that cost goes up every year. The *Titanic*

didn't have a crisis until it hit the iceberg, but it would have been better off gently steering a different course beforehand.

Liberals contend that the scenario I have painted above is alarmist. Social Security, they say, will not actually go bankrupt until 2042. The program can be saved with some minor adjustments. Just rolling back Bush's tax cuts would raise the necessary funds. If the economy grows better than expected, the program might have enough revenues to pay for its promises.

None of this is true. The idea that the problem does not start until 2042 depends on sleight of hand. For several decades, the program, in anticipation of the retirement of the Baby Boomers, has collected more revenues than it pays out. The surplus has been banked in a Social Security trust fund. The liberal argument is that when payouts start to outstrip revenues, in 2018, the program can just draw on the trust fund—and it can keep drawing on it until 2042, when it is scheduled to run out.

When 2018 rolls around, the government will have to find the money to pay off those IOUs. It will have to raise revenues or cut spending or borrow elsewhere to do that.

FORMS OF DENIAL

The "minor adjustments," meanwhile, aren't all that minor. The leading liberal plan on Social Security is that of economists Peter Diamond and Peter Orszag. Its backers, such as Jonathan Cohn of *The New Republic,* prefer to describe it as "a series of tweaks" that "involves both raising taxes and cutting benefits." Likewise, the *Washington Post* describes Diamond and Orszag as "'balancers,' who would use benefit cuts and tax increases in equal measure." In truth, 85 percent of the improvement in solvency their plan accomplishes would come from higher taxes. (They even propose adding some benefits.) They want to raise the payroll-tax rate. They want to raise the amount of wages to which the tax applies. Then they want to add a "surtax" to high earners.

The Congressional Budget Office (CBO) estimates that the Diamond-Orszag plan would mean that the share of the economy eaten up by Social Security taxes would grow by more than a third. (Those taxes equal 4.9 percent of the economy now but would rise to 6.6 percent by 2080 under the plan.) The cost of their plan over the next 75 years would be almost exactly the same as that of the current program. The CBO also estimates that the plan would reduce economic growth, the capital stock, and work effort.

The idea that merely repealing the Bush tax cut would pay for Social Security's promises is misleading. The people who make that claim are, again, ignoring the problem of paying off the IOUs in the trust fund. (They're assuming, that is, that the trust fund can pay for Social Security benefits painlessly.) And there's another problem. Because the income tax is progressive, its burden grows over time. As wages rise, people move into higher tax brackets. The average tax rate—the share of the economy that the government is claiming through taxes—therefore rises.

Finally, there is the illusion that we can grow our way out of the Social Security shortfall. Sure, higher growth would yield revenues. But higher growth would also increase wages—and since benefits are tied to wages, benefits would rise too. Growth is a very good thing, but it can't relieve us of the need to get the program's benefits structure in line with its revenues. Even if economic growth doubled, Social Security would still go insolvent. And to the degree that economic growth does help, it is mostly by swelling the balance of the trust fund, giving the system more IOUs to claim against future taxpayers.

CHECKING THE INDEX

Moving from "wage indexing" to "price indexing" isn't a minor change. The middle-income worker of 2050 would be getting an annual benefit worth 37.5 percent less than he would have gotten under wage indexing. Price indexing would eliminate Social Security's shortfall all by itself. Would it be a draconian cut in benefits? If wages grow over time, workers will be putting more tax money into Social Security: Shouldn't they get bigger benefits as a result?

What that question ignores is that Social Security is not capable of converting our worker's payments into those massively higher benefits. The only way he could get those benefits is if he agreed to pay more taxes over the course of his working life (and to work in the smaller economy caused by higher taxes on everyone). The system can't pay for the larger benefits without tax increases. So nothing he could actually have, on terms he would want to have, would be taken away from him. If our worker wanted more money for retirement, he would almost certainly prefer to invest additional money himself rather than have the government raise his taxes.

In the debate over Social Security reform, even knowledgeable reformers are being careless in using phrases like "promised benefits" and "current-law benefits" to describe the 60 percent increase our median worker might get under wage indexing. The Supreme Court has ruled that nobody has a legal right to a particular benefit level. And current law includes a provision that automatically slashes benefits when the "trust fund" runs out in 2042. The current projections have benefits getting cut by 27 percent—and not gradually, but in one year, with more cuts following.

THE CASE FOR PERSONAL ACCOUNTS

Solvency isn't the only problem with Social Security's current structure. It takes a lot of money from young workers without giving them much in return. The program can afford to give a middle-income 25-year-old only 91 cents for every dollar he is going to put in over the course of his working life. Price indexing recognizes that reality, but does nothing to improve it. Personal accounts are a way of softening the blow. People would be able to supplement their reduced Social Security checks with the wealth built up in their accounts.

The second way to look at it is that the accounts are a way of putting Social Security on a firm footing. We would be moving from a "pay as you go" model in which each generation funds the retirement of the next to a "pre-funding model" where each generation saves in advance for its own retirement. If the government tries to do the pre-funding for us, without letting us have the money in individual accounts, there's a greater risk that the money will be diverted to other programs. The government could also gain too much influence over the economy if it tries to invest all that money itself.

The third view is that making it possible for millions of Americans to buy a share of the economy through personal accounts would be a good thing in itself. It would give them wealth they could hand on to their heirs. It would make them worker-capitalists with a direct interest in sound economic policies. Edward Prescott, the 2004 Nobel Prize winner in economics, argues that letting people invest payroll taxes would also expand the economy by stimulating them to work more. The CBO has found that a plan including price indexing and personal accounts would lead to a larger economy. National wealth would be 10 to 12 percent larger by 2080. (That's in contrast with the wealth-destroying Diamond-Orszag plan.)

Liberal politicians and journalists persist in speaking of capital-market investment as though it were "speculation" or "roulette." But nobody is talking about having 64-year-olds take all of their Social Security money and put it all on a tech stock. We are talking about letting people invest a portion of their Social Security contributions over a long period of time. There are plenty of ways to reduce the risks—by diversifying stock portfolios, by buying index funds, by buying bonds as well as stocks. People are quite capable of taking advantage of these methods (and a law

establishing personal accounts would probably require them to do so). Between the end of 1999 and the end of 2002, broad stock-market indices fell by nearly 40 percent. But a report jointly produced by the Employee Benefit Research Institute and the Investment Company Institute found that the average 401(k) plan fell by only 10 percent. It's established fact that stocks grow less volatile over the long run: There is no 20-year period in American history in which the stock market has declined. The average annual return on stocks has been 6.7 percent above inflation.

Setting up personal accounts does, however, involve a short-term increase in the federal debt. This debt increase is the badly misunderstood "transition cost" to personal accounts. Since we have a pay-as-you-go program, most of today's payroll taxes go to today's retirees. The rest goes, notionally, to the trust fund, but really to fund other government operations (which borrow from the trust fund). If people can put some of their payroll taxes in personal accounts, the government has to find other money to fund its other commitments. The result could be $1 trillion or even $2 trillion of borrowing over the next decade.

Those numbers may look alarming. But it's important to remember that they do not represent new costs. The government is already undertaking to pay for the retirements of today's senior citizens and tomorrow's. Pre-funding part of tomorrow's retirements would not increase the government's total obligations; it would just bring some of them forward in time. If the accounts are part of a long-term deal to reduce the cost of Social Security—as in Bush's price-indexing plan—then temporary debt increases should not deter us. The markets are well aware that the future benefits from Social Security constitute a large off-budget liability for the government. An increase in on-budget debt, even a large one, will not frighten them if they have confidence in the long-term trajectory of the reform plan.

DECISIONS, DECISIONS

How much should [the government] let people invest in personal accounts? Should it vary by income, so that poorer people can invest a higher percentage of their wages? How tightly should the government regulate the investments? Should it require people to convert their accounts into annuities when they retire?

There are trade-offs attending each choice. Large accounts would give workers a reason to participate, but would also mean higher transition costs.

On the other hand, if the whole account has to be made into an annuity, then some of the appeal of personal accounts vanishes. Reformers would not be able to say that the accounts are a way of building wealth that can be handed down the generations.

There is a political theory that can help guide reformers in making some of these choices—a theory that falls under the heading of "the new investor class." The theory holds, among other things, that participation in capital markets changes people's political attitudes and behavior. Perhaps more pertinently here, it holds that

pro-investor policies tend to perfect themselves over time. So, for example, a limited tax break for retirement savings will generate a constituency that wants bigger tax breaks for saving for retirement as well as for health and educational expenses.

A relatively small personal-account option, for the same reason, will generate demands for its expansion. Reformers need not be disappointed if they don't win large accounts this year. Holders of 401(k)s have over time gotten more and more options in how they invest their money. Regulations on personal accounts will probably get looser as well.

But the theory, by itself, is not a substitute for the political judgment required to make the major choices here. The theory is capable of telling conservatives that instituting personal accounts offers the prospect of reshaping American politics in their favor. The upside from the accounts is larger than from policies such as expanded IRAs, which would not do as much to increase the number of investors. But the political risks are greater, too, and the theory cannot tell you whether a specific historical moment is ripe for reform.

READING 40

Social Security Reform and Benefit Adequacy

Lawrence H. Thompson

Over a third of all retirees, including more than half of retired women, receive monthly Social Security benefits that are less than the poverty level for a single elderly individual. Many receive benefits this low despite having worked under Social Security and contributed for more than 30

years. Any significant reduction in the benefits of these people would have serious consequences for both the adequacy of income among the elderly and for the 1935 social commitment to provide a floor of protection in retirement to those who contributed to Social Security while working.

SOURCE: "Social Security Reform and Benefit Adequacy" by Lawrence H. Thompson from *The Retirement Project,* Brief Series, No. 17, March 2004. pp. 1–7. Copyright © 2004 Urban Institute.

This brief focuses on Social Security retirement benefit payments to low-wage workers and how current changes in the retirement age and likely future cost-cutting reforms could affect these benefits. The analysis begins by comparing the average level and distribution of current benefits with several benchmarks of benefit adequacy, using data on actual benefit awards in 2001. Information from the University of Michigan's Health and Retirement Study (HRS) is then used to simulate the impact on benefit adequacy of both the retirement-age increase and the further benefit reductions that would also be required to restore long-range fiscal balance to the Social Security program.[1] The brief concludes with some observations about the implications for future program policy.

Social Security is an income-transfer program based on a social contract. In return for payroll tax contributions while employed, workers are promised income support when they retire or become disabled. Benefits are related to prior contributions, but the relationship is not proportional. Those who contribute more while working or work more years receive higher benefits when retired. But in keeping with the social aspects of the program, the benefits paid to lower earners replace a larger percentage of their previous earnings. This feature of the program's structure recognizes that lower earners are less likely to have other sources of retirement income and will be relying primarily on Social Security benefits to assure a minimally adequate living standard in retirement. In fact, among elderly households, Social Security accounts for over 80 percent of total family income in the bottom two-fifths of the income distribution.

THE ACTUAL VERSUS THE HYPOTHETICAL

The average retirement benefit is actually quite a bit lower than the hypothetical benefit illustrations used in most discussions of Social Security policy. Benefit illustrations often focus on a hypothetical worker that was employed continuously for at least 35 years, earned each year the average amount earned by all workers under Social Security that year (about $33,000 in 2002), and retired at the normal retirement age. Such a worker retiring in January 2002 would have qualified for a monthly benefit of $1,127, some 40 percent of his or her 2001 earnings (SSA 2002, Table VI.E11).[2]

Actual retirees do not fare as well. The average person retiring in 2001 was awarded a benefit of $894; the average was $1,033 for men, and $694 for women. The actual average benefit is less than 80 percent of the benefit calculated for the hypothetical average earner and is equal to 32 percent of average earnings, not the 40 percent associated with the hypothetical worker. Clearly, the hypothetical average earner commonly used in Social Security policy discussions is not representative of actual retirees.

Figure 1 shows the actual distribution of Social Security retired worker benefit awards during 2001. Workers are categorized by the amount of their January 2002 check. The first column shows the range of monthly benefits for each category in absolute terms, while the second column shows the ratio of the dollar amount at the category ceiling to the average earnings of current workers. As the figure shows, 52 percent of 2001 monthly awards were for less than $900. Over one-third were for less than $700.

STANDARDS OF BENEFIT ADEQUACY

What level of protection should Social Security offer people that have worked for many years under the program, even if much of their work was at low wages? Figure 2 shows some potential benchmarks for defining a minimally adequate level of support for such workers.

Figure 1   Cumulative Distribution of Retired Worker Awards, 2001

Monthly benefit	Percent of average earnings[a]	Men receiving this amount (%)	Women receiving this amount (%)	Total receiving this amount (%)
Below $450	16.1	8	18	12
Below $500	17.9	10	24	16
Below $700	25.1	19	58	36
Below $900	32.3	31	77	52
Below $1,000	35.8	40	84	59
Below $1,200	43.0	60	94	75

[a]Ratio at the top boundary to estimated average 2002 earnings under Social Security.

Figure 2   Potential Adequacy Benchmarks: Monthly Amount and Ratio to Average Social Security Earnings, 2002

Adequacy benchmarks	Monthly amount ($)	Percent of average earnings
Special minimum benefit (at full retirement age)	617	22
Career minimum wage retiree (at full retirement age)	729	26
Federal SSI guarantee for a single individual	575	21
Poverty line, single elderly individual	719	26
Minimum wage	893	32
Average Social Security earnings	2,790	

SOURCES: SSA (2002, 2003), U.S. Bureau of the Census.

In contrast to most other developed countries, the United States has no official policy about minimum pension guarantees other than a relatively obscure provision known as the special minimum benefit. This special minimum pays a benefit that is based on the retiree's years of creditable service, defined as years in which the worker earned at least a specified amount. Workers must have at least 20 years of such service to qualify for a benefit, but don't get credit for any more than 30 years of service in the benefit calculation. As shown in Figure 2, the maximum special minimum benefit payable in 2002 was $617 a month, or about 22 percent of average Social Security earnings.[3] Currently, about 120,000 retired workers qualify for benefits under this provision, most of them women.

The second row of Figure 2 shows the benefit awarded in 2002 to someone that retired at age 65 after being employed continuously at the federal minimum wage (SSA 2002, Table 2A26). Such a worker would have received a monthly benefit of $729, or 26 percent of average Social Security earnings.

The third row shows the minimum income guarantee provided a Social Security recipient under the means-tested Supplemental Security Income (SSI) program, which provides a safety net for the elderly and disabled with few other resources. In 2002, the combination of the SSI and Food Stamp programs guaranteed an individual also eligible for Social Security a monthly income equal to $575, or about 21 percent of average Social Security earnings.[4] Those residing in states that supplement the SSI benefit would receive more.

The next two rows in Figure 2 show two other parameters not directly related to the Social Security program but useful in establishing benefit adequacy criteria. The 2002 poverty level for a single elderly individual was equal to roughly 26 percent of average Social Security earnings, and the earnings of a full-time minimum-wage worker would have equaled roughly 32 percent of average Social Security earnings.[5]

These various parameters suggest that a reasonable benchmark for a minimally adequate retirement benefit would be in the range of 21 to 25 percent of average Social Security earnings. The full retirement benefit awarded to a career, full-time minimum-wage worker and the single-person poverty level lie just above this range. The maximum benefit under the special minimum and the income guarantee under the universal means-tested assistance program (SSI) for the elderly fall at the bottom of this range. The analysis that follows (Figure 3) focuses on the fraction of retired worker benefits that falls below 21 percent and 25 percent, respectively, of average earnings.

The retirement-age increase now being implemented will cause more benefit awards to fall below the adequacy benchmarks. Most people begin receiving benefits before reaching the normal retirement age. In fact, about one-third of the gap between the hypothetical average earner and the actual average retiree is the result of the benefit adjustment made when someone retires early.[6]

The change now being implemented increases the normal retirement age to 67 but maintains 62 as the age at which people can first receive their retirement benefits. Until recently, those retiring at age 62 had their benefits reduced permanently by 20 percent. Those who retire at age 62 in the future will face larger early retirement reductions; after 2022, the reduction will be 30 percent.[7]

Calculations using the HRS data find that the larger reductions for early retirement will increase the fraction of long-service workers receiving less than 21 percent of average earnings from 7 percent to 12 percent among men and from 22 percent to 32 percent among women, after adjusting for the expected behavior change of working an additional two months. The fraction of long-service women whose benefits fail to reach the higher adequacy benchmark will increase from 37 to 47 percent.

Predictably, the additional benefit reduction required to balance the program's finances, should that be accomplished only through benefit cuts, would further increase the incidence of low-benefit awards. That change would cause about half of all retirees to get benefits that fell short of the minimal adequacy standard used here. Even workers with at least 30 years of service fare poorly. These further benefit reductions would cause one-quarter of long-service men and three-quarters of long-service women to have benefits amounting to less than 25 percent of average earnings. Over half of long-service women would fail to reach the lower of the two adequacy benchmarks.[8]

SUMMARY AND IMPLICATIONS

A little over a third of the people now retiring receive Social Security benefits that amount to less than 25 percent of the average earnings of

Figure 3 Estimated Share of Workers With Inadequate Benefits: Monthly Benefits Below Reasonable Benchmarks, Different Benefit Reduction Scenarios (percent)

	Share of workers receiving less than 21%			Share of workers receiving less than 25%		
	Men	Women	Total	Men	Women	Total
All eligible workers						
Actual benefit in late 1990s	14	46	27	21	61	37
Effect of retirement age increase	20	58	35	24	70	42
—if work effort increases	19	57	34	22	69	40
Effect of additional 25% benefit cut	28	76	47	36	86	55
Workers with at least 20 years of service						
Actual benefit in late 1990s	12	39	22	19	54	32
Effect of retirement age increase	18	51	30	22	64	37
—if work effort increases	18	49	29	20	63	36
Effect of additional 25% benefit cut	26	71	42	34	84	52
Workers with at least 30 years of service						
Actual benefit in late 1990s	7	22	11	13	37	19
Effect of retirement age increase	12	33	20	15	48	24
—if work effort increases	12	32	19	14	47	23
Effect of additional 25% benefit cut	18	57	33	25	75	39

SOURCE: Urban Institute tabulations of the first four waves of the Health and Retirement Study.

NOTES: The calculations exclude persons born after 1934, persons who received disability insurance benefits at some point in their working lives, persons who began receiving benefits prior to age 60 for any reason, and persons who had not already retired or given an expected retirement date. The ratio is the initial benefit in the year of retirement (or the year retirement is planned) divided by the average Social Security earnings that year.

those working under the program, earnings that are less than the current poverty level for a single elderly individual. Just over a quarter receive benefits of less than 21 percent of average earnings.

The retirement-age increase now being phased in will cause more people to receive Social Security benefits at these low levels, and attempting to solve the future financing predicament entirely through benefit reductions will make the problem even worse. Taken together, the two changes would cause over half of new retirees, including 85 percent of women, to receive benefits amounting to less than 25 percent of average earnings. Many of these people would have worked under and contributed to

Social Security for many years before retiring. Among those with at least 30 years of service under the program, three-quarters of women and one-quarter of men would have benefits amounting to less than 25 percent of average earnings.

These results suggest that across-the-board benefit reductions are probably not a viable way of dealing with the currently projected financing gap. When added to the impact of the retirement-age increase that has already been legislated, reductions would lower benefits to levels that undermine the social purpose of the Social Security program. Even among long-service workers, as many as half of all retirees would have benefits that fall below reasonable benchmarks for benefit adequacy.

This analysis suggests two basic conclusions. First, any further increase in the normal retirement age needs to be accompanied by an equal (or greater) increase in the age at which people are first eligible to draw benefits. Otherwise, further increases in the reduction factors for early retirement will cause actual retirement benefits to continue to fall relative to average earnings levels. Second, absent a substantial increase in the age at which benefits can first be drawn, it is probably not possible to close the financing gap currently projected for Social Security without finding additional revenues. Attempting to close the financing gap entirely through benefit reductions would drive average benefits to such low levels that the program would fail to serve the social purpose for which it was created.

NOTES

1. The University of Michigan Health and Retirement Study surveys more than 22,000 Americans over the age of 50 every two years. Supported by the National Institute on Aging, the study paints an emerging portrait of an aging American's physical and mental health, insurance coverage, financial status, family support systems, labor market status, and retirement planning.

2. The average earnings used in these calculations and this brief is the figure used to index prior-year earnings in the Social Security benefit calculation.

3. The special minimum is calculated by multiplying the number of years of creditable earnings in excess of 10 and up to 30 by a dollar factor. In 2001, the dollar factor was $30.90 and the maximum possible full retirement benefit was therefore 20 times $30.90, or $617. The dollar factor is indexed to changes in the price level. Since 1990, a year of creditable earnings has been defined as a year with annual earnings greater than 15 percent of what the taxable maximum would have been had there been no ad hoc increases after 1973. This amounted to $9,450 in 2001, or about 31 percent of average Social Security earnings. Prior to 1990, the amount required was 25 percent of what the taxable maximum would have been (SSA 2002, Tables 2A12 and 5A8).

4. The basic SSI benefit was $545. The guaranteed income to someone receiving Social Security benefits was $565, since the first $20 of the Social Security benefit is disregarded in the SSI benefit calculation. Persons receiving SSI benefits of this level would also be eligible for $10 a month in food stamps.

5. Assuming the individual works 173 hours (4 and 1/3 weeks) each month at $5.15 an hour.

6. Another cause of the discrepancy between the theoretical benefit and the actual average benefit is the taxable maximum. The average earnings figure used to calculate the hypothetical earner's benefit is based on total earnings, whereas actual benefit calculations are based only on taxable earnings.

7. Retirement benefits for those retiring at age 62 in the late 1990s were permanently reduced by 20 percent (from the level they would have received if they waited until the "normal retirement age"). When the retirement-age increase is phased in fully, the benefit awarded at age 62 will be reduced by 30 percent.

8. These calculations assume that earnings levels and service years are the same in the future as they were for people retiring in the late 1990s. In fact, future female retirees are likely to have longer careers and higher average earnings than female retirees in the late 1990s, somewhat reducing the incidence of inadequate benefits among women.

Since the key parameters of the Social Security benefit calculation are adjusted each year to reflect changes in average Social Security earnings levels, the increase in the relative status of women will lead automatically to a decrease in the relative status of men, somewhat increasing the incidence of inadequate benefits among men.

REFERENCES

Fields, Gary S., and Olivia Mitchell. 1985. "Estimating the Effects of Changes in Social Security Benefit Formulas." *Monthly Labor Review* 108(7): 44–45.

U.S. Social Security Administration (SSA). 2002. *Social Security Bulletin, Annual Statistical Supplement, 2002.* Washington, DC: Government Printing Office.

———. 2003. *Annual Report of the Board of Trustees of the Old-Age and Survivors Insurance and Disability Insurance Trust Funds, 2003.* Washington, DC: Government Printing Office.

READING 41

——— Social Security for Yesterday's Family? ———

C. Eugene Steuerle and Melissa Favreault

Imagine a retirement program that spends an ever-larger share of its resources on people who still have 15 or more years to live and an ever-smaller share on very old, poor, or infirm people. Imagine that this same program:

- treats married couples with the same total earnings differently by granting smaller benefits to those whose earnings are more equally split between spouses;
- gives an additional benefit to spouses just for being spouses, but no such benefit to single and many divorced parents, including those who raise more children, work more, and pay more taxes;
- grants people who signed divorce papers after being married nine years and eleven months hundreds of thousands of dollars less than those who waited another month to divorce;
- makes some benefits for divorced people married more than 10 years conditional upon their former spouses dying;

- increases benefits for a high earner who brings additional earnings home but does not similarly reward many second earners who contribute to household income and pay additional taxes on their earnings; and
- increases the size and likelihood of rewards for those who have kids when older rather than when younger, marry much younger spouses, and divorce and remarry several times.

Now imagine a never-ending fight among elected officials and advocates over who can best guarantee that these imbalances will continue forever.

You've probably guessed the system in question by now. It's Social Security, and the elected officials are those who think that the Social Security debate is mainly about whether some retirement money gets put into individual accounts: specifically, whether one-seventh of Social Security taxes, and far less than one-tenth

SOURCE: "Social Security for Yesterday's Family?" by C. Eugene Steuerle and Melissa Favreault. Number 35 in series Straight Talk on Social Security and Retirement Policy. Copyright © 2004 Urban Institute.

of scheduled spending on the elderly, should be held directly by individuals or put into a government trust fund.

It's bad enough that the inequities in the remaining pie—with or without a slice taken for individual accounts—are largely being ignored. Worse, many reformers on both sides of the individual account debate are willing to lock in these disparities if that's what it takes to control the one-seventh of the pie under dispute.

How might these contradictory rewards and penalties become even more institutionalized if the debate never moves beyond the question of individual accounts versus trust fund saving?

First, some reformers, in trying to create the illusion that no one will lose or pay, have tried to guarantee benefit levels near those specified under current law. This promise, if kept, would perpetuate existing inequities. Second, most reformers are set on balancing total benefits and total taxes. But if in the rush to reconcile dollars and cents reformers ignore the system's large inequities, ignored they will stay for many years. It would be difficult after a major reform for Congress to tell the public, oops, we still got it wrong, even under the newly balanced system.

The fact is that Social Security was designed in the 1930s, a time when a married male worker and a stay-at-home wife were considered the norm. In the ensuing decades, benefits were added without much regard for the way family life was evolving. For instance, generous survivor benefits were added to help care for women, who tended to be poorer. But this benefit totally ignored many women (and men) who were single or left by former spouses after a few years of marriage.

Is it too late to bring Social Security in sync with contemporary family life? Not at all. Besides reaching financial balance and increasing saving, reform should have two main goals: providing minimum levels of income above poverty and removing obvious inequities. For example, the system should stop the obvious disparities in the way it treats single heads of household and those divorced after being married fewer than 10 years.

Most of the inequities noted here are illegal under the private pension system. But that still has not prompted the Social Security Administration to take the first reform step: providing measures that assess the success of different reform options in reducing inequities as well as poverty among the elderly.

Some commissions and congressional bills have made tentative moves in the right direction. But without an independent scorecard, there is no way to determine whether these efforts have succeeded or which approach works best. And without a good understanding of what's at stake, the temptation to choose reform options that are mainly symbolic will likely win out.

The real choices before lawmakers will test their political mettle. The temptation to guarantee that reform will create no losers defies reality, as correcting inequities always means that somebody pays more or that somebody receives less. To meet the two goals of providing minimum levels of well-being and removing clear-cut cases of unfairness, any redistribution should not be based on sex or marital status. Rather, supplemental benefits should be targeted more directly to individuals with low lifetime earnings and low income. By this standard, many married individuals will still qualify for additional benefits, but not because marriage somehow makes them more deserving than a poor single person.

Beyond need, adjustments for family circumstances, such as marriage and divorce, should be made on an actuarially fair basis to avoid stacking the cards against many divorced men and women, spouses with equal rather than unequal earnings, spouses who are approximately the same age rather than different ages, and single heads of household.

Is the current debate going to address the retirement needs of modern families? Only if it is based on how six-sevenths or more of the system meets their conditions and needs—not merely on what happens to the small piece that some want to put in, and others want to keep out of, individual accounts.

FOCUS ON THE FUTURE

Two Scenarios for the Future of Social Security

First Scenario: Trouble ahead. The year now is 2025, and the oldest of the baby boomers are in their mid-70s. But the world isn't at all what most of them expected to find in old age.

Looking back, it's easy to see signs of today's problems. First came the Medicare shortfall in 2018; then Medicare was merged with Medicaid 3 years later. Today, health care costs consume 20% of the gross national product, but still there are millions with Alzheimer's disease or kept alive in a persistent vegetative state. Of course, the worst shock was the stock market decline in 2008, followed by the big collapse in 2015 brought on when baby boomers started cashing in their mutual funds. The Dow Jones Industrial Average went down to 5,000, where it has been ever since. Most people lost their Social Security savings in the stock market crash.

After the great Baby Boomers Social Security March on Washington, Vice President Chelsea Clinton was appointed to head a special commission to restore solvency to the system. Drastic cuts were enacted, but no one's very happy with the result. Most everyone nowadays stays in the labor force until age 70 to receive full benefits, but lots of people in their mid-70s are also desperately trying to work again, part time, because cost-of-living increases have been frozen for the next 5 years. Middle-age adults have mixed feelings about the Social Security crisis. They never expected to collect benefits in the first place, but their taxes are too high, and now they're in no position to support their older parents as new laws require them to do.

The economy has been in a shambles since the Great Flooding began. For the past two decades, ocean levels have been steadily rising as a result of global warming. More recently, the collapse of the Ross Ice Shelf in Antarctica led to abandonment of Manhattan and to the resettling of the population in what used to be the Netherlands. These events have plunged industrialized countries into the worst depression in the 21st century, so people have other things to worry about besides Social Security. Older people will just have to look out for themselves.

Second Scenario: The best is yet to be. It's funny to look back now, in the year 2025, on all those gloomy predictions that were so common just after the turn of the century. People at that time were actually afraid of the coming of an aging society.

It's true, we had some close calls, but everyone learned as a result. After near collapse in 2015, Medicare was finally put on a sound footing. Through a combination of outcomes research and health promotion under managed care, costs to the system were cut, and larger numbers were served. After the turndown in 2008, there was a rebound, and before long Social Security was again bringing in surpluses because of the booming stock market and investment of part of the Trust Fund in index funds. The Dow is now at 30,000 and still climbing.

Another big factor was the swelling number of job holders, including hardworking immigrants from Asia and Latin America and growing numbers of Black college graduates. People today also routinely work past age 70, usually at new jobs for which they've been retrained, some of them working from home over the Internet. Almost as many people in their 50s and 60s are in college today as young people, so lots of people are having second or third careers.

Looking back at the gloom after the turn of the century, you can't help but be reminded of similar fears during the late 1970s about running out of oil and natural resources. Sometimes the things we fear the most don't turn out that way at all.

Questions for Writing, Reflection, and Debate

1. What are the key differences between the U.S. population now and in 1935, when Social Security was first introduced? Should these differences prompt changes in the Social Security system for the future?

2. Define at least three interpretations that can be given to what it might mean to "privatize" Social Security. Give reasons for and against each of these ideas of "privatizing."

3. If a substantial part of the Social Security Trust Fund were invested in the stock market, what are some reasons that people might be more confident in the future of Social Security? What are some reasons that this plan might make people less confident in the future?

4. Critics of Social Security argue that the program does not adequately help the least advantaged in the older population. Some have urged a means test to be sure that Social Security goes to those who are poorest. Imagine that you are a staff assistant to the U.S. commissioner on Social Security and have been asked to prepare a memorandum on this issue. Examine the evidence for or against this proposal for means testing.

5. Those who believe that it is wrong to have a large surplus in the Social Security Trust Fund say they are unhappy with the fact that money in the Trust Fund is invested in Treasury certificates. In that respect, they say, the surplus simply funds the federal budget deficit. Is this claim correct? If the funds were invested in the stock market, what might be the positive and negative results of that move?

6. Imagine that you are chair of Taxpayers United of America and that your organization is about to take a position in favor of changing the payroll tax rate for Social Security. Prepare a draft version of the position paper that you will urge your membership to adopt. Outline in detail the reasons you think it is in the interest of taxpayers to change the payroll tax.

7. One proposal to make Social Security fairer to women is to give credit for the work that mothers and housewives do at home. Is this proposal fair to women who work outside the home? Is it fair to same-sex couples or stay-at-home dads?

8. Imagine that you are president of the local chapter of the National Organization for Women. Prepare a detailed statement as the basis for a public petition campaign aiming to make Social Security fairer to women. Highlight the key arguments to be used to convince the public to sign the petition.

Suggested Readings

Achenbaum, W. Andrew, *Social Security: Visions and Revisions,* New York: Cambridge University Press, 1986.

Altman, Nancy J., *The Battle for Social Security: From FDR's Vision to Bush's Gamble,* New York: Wiley, 2005.

Clark, Robert L., Burkhauser, Richard V., Moon, Marilyn, Quinn, Joseph F., and Smeeding, Timothy M., *Economics of an Aging Society,* Malden, MA: Blackwell, 2004.

Diamond, Peer A., and Orszag, Peter R., *Saving Social Security: A Balanced Approach,* Washington, DC: Brookings Institution Press, 2005.

Kingson, Eric, and Schulz, James (eds.), *Social Security in the 21st Century,* New York: Oxford University Press, 1997.

Student Study Site

Visit the Student Study Site at **www.sagepub.com/moody7e** for these additional learning tools:

- Flashcards
- Web quizzes
- Chapter outlines

- SAGE journal articles
- Web resources
- Video and audio resources

Controversy 10

IS RETIREMENT OBSOLETE?

The 20th century could well be called the age of retirement. Never before in history had retirement in old age become such a widespread, almost universal pattern of behavior. As recently as 1950, the average retirement age was 67; by 1980, it had gone down to 63 and continued to decline still further into the 1990s. After World War II, there was a consistent decline in labor force participation with advancing age in all industrialized countries. But does this trend toward retirement represent progress, or is it a sign of problems to come (Schnore, 1985)? Will the 21st century continue the prevailing pattern, or has retirement become obsolete (Krain, 1995)?

For society, the decline in labor force participation represents a loss of productivity by older people—a group that, on average, is now living longer, is better educated, and is in better health than ever before. Individuals who withdraw from the workforce often face many years without any clearly defined purpose in society (Sheppard, 1990). As the cost of Social Security and private pensions continues to rise, it is understandable that people are asking whether retirement makes sense.

But if we ask, "Is retirement obsolete?" we are actually asking two questions. First is the question of whether retirement is a wise choice for a specific individual—for example, a person who is considering taking early retirement, which is an option to retire before some conventional age for retirement. Second is the question of whether the systematic practice of retirement in our society is good policy; that is, does it make sense for the economy or for the good of society as a whole (Blau, 1985)?

Today, the average man, age 20, can expect to spend a third of his life in retirement, but what is the positive content of these retirement years? Retirement, as Rosow (1967) observed, means entering a *roleless role.* Retirement is more than simply withdrawing from work. Whether a person reduces his or her workload gradually or stops working abruptly, retirement often leads to new options in later life: leisure pursuits, volunteering and service, or an encore or new career (American Council on Education, 2008). Typically, retirement is accompanied by reliance on Social Security and pension income instead of salary as the primary means of support. In the previous discussion of investment and preretirement planning, we stressed the importance of individual decision making.

But an emphasis on individual planning fails to take into account an important point—namely, that people have an individual choice about retiring only if they can count on enough income to support themselves without working. Someone who wins the lottery or cleans up in the stock market might have that choice at a younger age, but for most people, retirement is not an option until much later in life. Middle-aged or older workers can choose to retire only if there is a social or institutional policy supporting the choice by paying them to no longer be in the workforce. In that respect, retirement remains very much an issue for public policy debate (Munnell & Sass, 2009). As we saw in debates about the future of Social Security, critics have raised questions about whether in the 21st century we as a society can or ought to maintain retirement as it has been known in the past (W. Clark, 1988). That questioning has already resulted in some important changes. For example, in 1986, the common practice of mandatory retirement was abolished by law.

Today, people tend to take retirement for granted and assume it is a natural and appropriate pattern for later life, but, in fact, retirement as a social practice or institution is historically quite recent. We need to understand the origins of retirement as an institution and to better appreciate how work and retirement are now being transformed by changes in the U.S. economy and in the individual life course. Instead of taking retirement for granted as a natural phase of life, we need to consider current trends that will determine what work and retirement mean in the 21st century.

HISTORY OF RETIREMENT

Widespread retirement by workers became possible only after the industrial revolution of the 19th century (D. Costa, 1998). It was Prussian Chancellor Otto von Bismarck who first introduced age 65 as the basis for a pension. By the early 20th century, many European countries began to institutionalize retirement through government pension systems. The United States followed with Social Security in 1935, a development that made leaving the labor force much more attractive to people. In 1890, 68% of men over age 65 were in the labor force, but that number dropped to 54% in 1930. In 1950, after improvements in Social Security, the number of older men in the labor force dropped further to 46% and has continued to decline to less than 17% in 1989. Exhibit 46 shows the major trends over the period between 1976 and 2005. Exhibit 47 includes similar data for the period from 1990 through 2006.

Several points are clear from the data. First, beyond age 65, the overwhelming majority of both men and women are retired from work, although a not insignificant minority continues in the labor force. Second, among older men, the trend has clearly been away from work and toward retirement, and the decline in labor force participation applies equally to men in their 50s. Early retirement has become a major phenomenon in its own right, but since the 1990s, it has gradually been diminishing for a variety of reasons. For women over age 65, labor force participation has not declined, but recently it has gone up. In fact, women in the 55-to-64 age bracket have counteracted the male trend in that they have joined the workforce in larger numbers. The earlier discussion of gender and aging illuminates some reasons for these contrasting trends. Older men tend to have earned higher pension and Social Security benefits, but older women must face a longer life span with lower average income expectations. Further, there's a more recent trend in the early part of the 21st century of older men returning to the labor force full-time (Gendell, 2008).

Some insight into these trends comes from one of the first major longitudinal studies of work and retirement, the Cornell Study of Occupational Retirement, which was carried out between 1952 and 1962 (Streib & Schneider, 1971). That study helped challenge

Exhibit 46   Labor Force Participation Rates for People 55 Years and Older, by Sex, 1976–2005 (annual averages)

Year	Total 55 years and older	55 to 59 years	60 to 61 years	62 to 64 years	65 to 69 years	70 years and older
Men						
1976	47.8	83.5	74.3	56.1	29.3	14.2
1980	45.6	81.7	71.8	52.6	28.5	13.1
1985	41.0	79.6	68.9	46.1	24.4	10.5
1990	39.4	79.9	68.8	46.5	26.0	10.7
1994	37.8	76.9	64.8	45.1	26.8	11.7
1995	37.9	77.4	65.6	45.0	27.0	11.6
1996	38.3	77.9	66.3	45.7	27.5	11.5
1997	38.9	78.7	66.0	46.2	28.4	11.6
1998	39.1	78.4	67.0	47.3	28.0	11.1
1999	39.6	78.4	66.3	46.9	28.5	11.7
2000	40.1	77.0	66.0	47.0	30.3	12.0
2001	40.9	77.2	67.7	48.2	30.2	12.1
2002	42.0	78.0	67.3	50.4	32.2	11.5
2003	42.6	77.6	67.0	49.6	32.8	12.3
2004	43.2	77.6	64.9	50.8	32.6	12.8
2005	44.2	77.6	65.6	52.5	33.6	13.5
Women						
1976	23.0	48.1	39.9	28.3	14.9	4.6
1980	22.8	48.5	39.6	28.5	15.1	4.5
1985	22.0	50.3	40.3	28.7	13.5	4.3
1990	22.9	55.3	42.9	30.7	17.0	4.7
1994	24.0	59.2	45.3	33.1	17.9	5.5
1995	23.9	59.5	46.1	32.5	17.5	5.3
1996	23.9	59.8	47.2	31.8	17.2	5.2
1997	24.6	60.7	47.9	33.6	17.6	5.1
1998	25.0	61.3	47.3	33.3	17.8	5.2
1999	25.6	61.8	46.2	33.7	18.4	5.5
2000	26.1	61.4	49.0	34.1	19.5	5.8
2001	27.0	61.7	50.5	36.7	20.0	5.9
2002	28.5	63.8	52.8	37.6	20.7	6.0
2003	30.0	65.5	53.9	38.6	22.7	6.4
2004	30.5	65.0	54.0	38.7	23.3	6.7
2005	31.4	65.6	53.8	40.0	23.7	7.1

SOURCES: Monthly Labor Review, October 2006, U.S. Department of Labor, Bureau of Labor Statistics.

Exhibit 47 Civilian Labor Force Participation Rates for Persons 55 Years and Older, 1990–2006

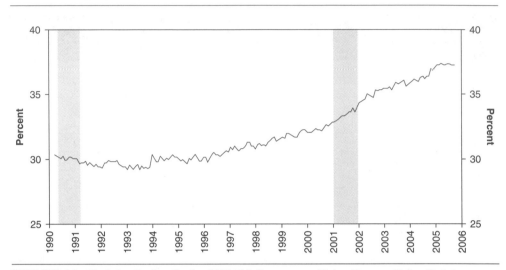

SOURCES: Monthly Labor Review, October 2006, U.S. Department of Labor, Bureau of Labor Statistics.

For women over age 65, labor force participation has recently been rising.

previously accepted but mistaken beliefs, such as the idea that retirement has a negative effect on people's health or that it causes feelings of worthlessness. The Cornell study found no evidence for the idea that retirement causes ill health, yet that idea continues to persist (Ekerdt, 1987). The study did document a drop in income upon retirement, but most retirees still reported that their income was sufficient, and a quarter even felt their standard of living was better than it had been earlier.

Retirement, in short, is not usually a negative event. Subsequent studies have confirmed the main findings of the Cornell study (Palmore et al., 1985; Parnes, 1985).

URBAN LEGENDS OF AGING

"Retirement is bad for your health."

This idea was proven false long ago by longitudinal studies of retirement, such as the work of Gordon Streib (1956) and many others since then. There are of course cases where people in ill health opt to retire and do die soon afterward. But retirement isn't the cause. Maybe there are other reasons to encourage people to delay retirement, but fear of bad health isn't one of them.

Although the transition to retirement may involve some stress, after retirement, people are generally happier with more leisure time (Pillemer et al., 2000). Because disability can give rise to retirement, a group of retirees may, on average, appear to be less healthy. But retirement in itself does not raise the risk of decline in physical or mental health. There are even suggestions that retirement for most people will be a benefit to health and happiness (Ekerdt, 2007). What is different today is that the long-range trend toward early retirement appears now to have halted and even reversed: In 2000, labor force participation for people over age 65 was higher than it had been at any time since 1979, and the trend has continued into the 21st century. As a result of the impact of the Great Recession beginning in 2008, more and more workers express a preference for working longer. Smaller and smaller numbers are able to take advantage of early retirement as a matter of personal choice.

ORIGINS OF LATE-LIFE LEISURE

Retirement has come to mean an expansion of leisure time, most of it during the retirement years. Futurist Graham Molitor, of Public Policy Forecasting, has estimated that, by 2030, more than half of an American's waking hours over a lifetime will be spent in leisure, in contrast to 41% today. But leisure can be defined in different ways. Leisure can be viewed not merely in a negative sense as time away from work, but also as free time perceived in a positive sense and containing opportunities for recreation, relaxation, personal development, and service to others (M. Kaplan, 1979).

Retirement as a time of leisure is feasible only with a certain degree of wealth. From a historical standpoint, widespread retirement first became possible when the industrial economy was productive enough to support sizable numbers of nonworking adults. At the same time, the economy no longer needed so many workers in the labor force, and companies believed that older workers were not as quick or productive as the young. Governments, corporations, labor unions, and older workers found retirement to be a desirable policy, and it soon became the normal practice (Graebner, 1980).

This older couple traveling the open road in an RV is a persistent symbol of retirement.

Since 1900, the average life expectancy for Americans has risen from 47 to nearly 78 years (in 2011). With longer lives, people have spent increased time in education, work, and retirement. But when we look at the different uses of time over the life course during the 20th century, the most significant trend has been the growth in the amount of leisure time.

Although life expectancy has increased by more than 50% since 1900, the period spent in retirement has increased even more dramatically. Work now accounts for a smaller percentage of a man's life than it did in 1900. By contrast, for women, the big change has been a substantial increase in the number of years spent in the paid workforce. But for both men and women, years in retirement constitute a larger and growing portion of life.

How did this pattern come to be? Free time, along with higher wages, was always a potential by-product of industrialization and economic progress. Trade-offs always exist: free time versus higher wages. From 1900 until the 1930s, the workweek and the workday gradually became shorter across the industrialized world. Subsequently, free time has also increased, but not through a reduction of the workweek. Instead, free time has taken different forms over the life cycle: increased vacation time, later entry into the workforce for the young, and earlier retirement for the old (Hunnicutt, 1982).

In the 1920s, there was a "shorter-hours movement," which remains a largely unexplored chapter in liberal reform movements of the 20th century. It was followed by a "share-the-work" plan during the Great Depression. Reformers favored shorter working hours and more leisure as a vehicle for self-development. This goal of self-development had much in common with ideas that would later be advocated by writers such as Fred Best, Willard Wirtz, Gosta Rehn, and Max Kaplan, who have advocated more flexible boundaries between work and leisure. When we ask, "Does retirement as it currently exists make sense?" we are asking whether a flexible and humanistic goal of self-development in adulthood is still a possibility.

In the past, economic forces quickly foreclosed any debate on that question. U.S. business leaders during the 1920s saw a combined threat to economic prosperity from overproduction and expanding leisure time. They feared that consumer markets were becoming oversaturated and that workers would demand more free time rather than continuing to work. Declining working hours along with declining production spelled lower profits and slower economic growth. Against this view were the "optimists" who argued that consumption demand could be driven higher primarily through advertising and marketing techniques to stimulate new purchasing power.

In fact, it was the consumer view that triumphed, as the emergence of the affluent society after World War II came to testify. But the consumer society of rising demand and stable working hours was also supported by government actions. The stimulus for government action was, of course, the Great Depression. With the New Deal and World War II, the U.S. government took on new responsibility for managing the economy to moderate the business cycle and ensure aggregate consumer demand.

The passage of the Social Security Act in 1935 and, later, the spread of private pensions ratified the practice of a fixed retirement age, typically age 65. Public policy promoted the idea of displacing leisure into later life and enlarging the period of education in early life, presumably as preparation for work. Purchasing power for goods and services would be maintained during retirement by transfer payments—Social Security—disbursed in such a fashion as to keep older people out of the labor market. Work would increasingly be compressed into the middle period of the life course.

With the establishment of a standard 40-hour workweek by 1945, the movement toward a shorter workweek lost popularity. Productivity gains were channeled into higher wages and fringe benefits; free-time gains became converted into longer vacations and, above all, earlier retirement. The amount of free time available in old age increased by about 5 years during the 20th century chiefly from gains in life expectancy, but also from earlier retirement. It remains to be seen what the trends will look like as the 21st century unfolds.

Because Social Security has always intended to remove people from the labor force, some suggest that retirement is forced on people and therefore not really voluntary leisure (Hunnicutt, 1982). This view reflects a debatable assumption that work is desirable and meaningful for the majority of people, but, in fact, just the opposite seems true (Terkel, 1985). Most people are eager to leave their jobs, and, contrary to popular stereotype, retirement is not dangerous to one's health. Except for displaced workers or elite groups who leave their jobs reluctantly, people generally prefer to retire rather than continue working at jobs that are unsatisfying (Boaz, 1987), if they can afford to do so. What is problematic about retirement is not that work is always better, but that the abundance of free time in later life is not adequately structured for any larger social purpose or meaning.

Overall, most people applaud the fact that the amount of time devoted to leisure has increased in the 20th century through reductions in the workweek, longer vacations, and early retirement. But a few dark clouds lurk on this pleasant horizon as we move into the 21st century.

First is a phenomenon sometimes described as "the overworked American." More and more married women have entered the labor force, and more workers are taking on second jobs to maintain their standards of living (Schor, 1991). One result is that people have less time for volunteer activities or adequate leisure during their middle years.

Second, since the 1980s, the proportion of workers covered by pension plans has remained stalled at about 50% of the workforce. If half of all workers lack pension coverage, this fact raises the prospect that future cohorts of older Americans may not be able to enjoy early retirement at levels comparable to those today. In fact, labor force data show that the long-range trend toward earlier retirement seems to have halted and perhaps begun to reverse (Rix, 2004), as can be seen illustrated in Exhibit 48.

Exhibit 48   Numeric Change in Labor Force by Age, Projected 2002–2012

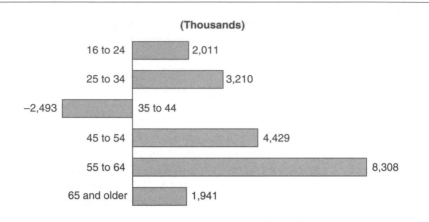

SOURCE: U.S. Department of Labor, Bureau of Labor Statistics (2003–2004).

Third, there are concerns about the base of economic productivity that supports a sizable old-age population devoted to leisure pursuits. These concerns range from future financing for Social Security to international economic competitiveness and the quality of the present and future workforce.

CHANGES IN THE AMERICAN ECONOMY

The U.S. economy is in the midst of far-reaching changes that will affect retirement in the future. A new postindustrial economy shaped by information technology and global competition has reshaped U.S. society. Large corporations routinely engage in downsizing (i.e., letting go of employees through layoffs and early retirement incentive programs). Employers have encouraged early retirement as a way of restructuring the labor forces in favor of younger, less costly, and technologically more sophisticated workers. After the recession beginning in 2008, blue-chip corporations such as IBM, General Motors, and Sears cut their workforces amid a deepening recession.

Increasingly, large companies can no longer guarantee employment or a predictable work life based on the patterns of the past. Smaller companies are even less secure employers because of the new competitive environment. This loss of security and predictability in the labor market has severe consequences for older workers. In the first place, older workers face a leveling off earlier in their careers as they compete for a diminishing number of good jobs. When companies downsize and eliminate middle-management positions, older workers easily find themselves pushed aside. Many opt for early retirement rather than face unemployment.

For the organization, corporate downsizing often means a loss of older employees' experience and skills. Ironically, companies that try to decrease operating costs by pressuring older workers to retire have already begun finding themselves with unexpected skill gaps. The U.S. Conference Board, a major industry organization, reported that few companies have age-conscious downsizing policies. However, some enlightened companies, such as

Exhibit 49   Labor Force Participation Rate of Men and Women, 1952–2002 and Projected
              2012

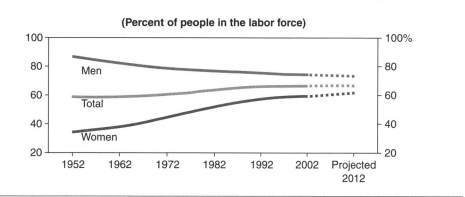

SOURCE: Bureau of Labor Statistics (2003–2004).

Travelers Insurance Company of Hartford, Connecticut, have developed ways of using older
workers by arrangements such as consulting, job sharing, flextime, and temporary work
(Shea & Haasen, 2005).

For individuals, early retirement is not necessarily good or bad. A "push-and-pull" factor
usually prompts the decision to retire (Williamson, Rinehart, & Blank, 1992). There are
many factors, including health status, perceived retirement income, financial plans of one's
spouse, peer group pressure to retire, the psychological attractiveness of the workplace, and
strength of family ties, that influence the decision to retire (LaRock, 1993). *Early retirement
incentive packages* can also be influential. The volatility of the labor market today is more
difficult for older workers to cope with than it is for younger workers. Statistically, older
workers actually have lower unemployment rates than younger workers, but when older
workers are displaced, it takes them much longer to find new jobs. Furthermore, statistics
measuring unemployment rates may be deceptive. Many older people give up trying to find
a job and become "discouraged workers"; that is, they are not even counted as part of the
official unemployment rate. Still others declare themselves "retired" so that retirement
becomes a disguised form of unemployment. Indeed, one of the prime reasons for adopting
the Social Security system in the United States during the Great Depression was to reduce
the level of official unemployment by drawing older people out of the labor market and thus
opening up jobs for young people.

Looking at the labor force as a whole, long-range changes are clearly visible (see Exhibit 49).
In 1979, total manufacturing jobs constituted 23% of the labor force, but by 2006, the proportion had shrunk to just over 10% (Marlene Lee & Mather, 2008). But the service sector
jobs on average paid less and offered fewer benefits than older jobs in manufacturing. In
addition, job growth is increasingly found not in big corporations but in small companies
that can move quickly and flexibly. Typically, the new jobs are much less likely to offer full
fringe benefits or pension coverage.

In recent years, the pattern of pension income has been changing. During the 1950s and
1960s, the most common type of retirement programs were defined-benefit programs that

guaranteed to pay a specified level of income in retirement. In the 1990s, this guaranteed approach to retirement income had changed in two ways. First, fewer employers offered pensions as part of a fringe benefits package. Second, there is a trend toward defined-contribution plans in all major industrial sectors and for all firms with pension plans. This trend may reduce the effects of an aging workforce by reducing barriers to job mobility for older workers. Displaced workers or others who wanted to move to new jobs would then not lose pension coverage as a result of moving. But the increasing shift to defined-contribution pension plans may help diminish this problem of "job lock."

Some pension changes are dramatic. For instance, personal plans, such as 401(k) pension plans, have spread rapidly. Around half of all American workers are eligible for a 401(k) or some other form of defined-contribution plan, and 86% of eligible workers are actually participating in a plan—an astonishingly high proportion. In 1984, the total amount of 401(k) plan assets was $92 billion. By 2011, 401(k) plan assets were more than $3 trillion, and the majority of workers had a defined-contribution plan of some kind.

When we look at government, we see incentives for moving older employees out of the labor force, but there is a serious question about whether such practices are financially sustainable in the future. A striking example is military personnel, who can retire after 20 years of service and receive half their salary for the remainder of their lives. These military commitments amount to nearly half of the federal government's pension debts, and the Pentagon expects to spend an increasing amount to fund these pensions. Estimates are that the cost will grow from just over $11 billion in 1992 to $296 billion by 2041.

Similar patterns of early retirement are common in state and local government, especially for police and firefighters. Faced with budget problems, state and local governments have begun using early retirement as a staff-reduction technique. Early retirement yields short-term savings, but may turn out to be an expensive proposition in the long run. There are now large costs extending into the future for inadequately funded pension plans of state and municipal governments. Federal civil service pension plans aside from the military are underfunded by more than $1 trillion. The Pension Benefit Guaranty Corporation now insures the private pensions of 40 million American workers (Schieber & Shoven, 1997).

A NEW VIEW OF RETIREMENT

Retirement in the 21st century has already started taking on new forms. In the past, employers found retirement a convenient tool for managing the labor force (Schulz, 1995). On the positive side, the availability of retirement has meant expanded leisure and opportunities for self-fulfillment in later life. On the negative side, the practice of retirement entails large hidden costs, both in funding required for pension systems and in the loss of the accumulated skills and talents of older people.

Critics of retirement as it exists today have pointed to the rigidity of retirement practices. Retirement is typically an all-or-nothing proposition. Would it not be better to have some form of flexible or phased retirement, in which employees gradually reduce their work hours or take longer vacations? Such an approach might enable older workers to adjust better to retirement and permit employers to make gradual changes instead of coping with the abrupt departure of a well-integrated employee. Retirement could be radically redefined in the future if phased retirement were to take hold (Schmahl, 1989).

Earlier criticism of **mandatory retirement** at a fixed age led to legal abolition of the practice in 1986. Mandatory retirement is still permitted for high-level executives in private industry and for jobs where age is a bona fide occupational qualification. The Age Discrimination in Employment Act of 1967 forbids older workers from being limited or treated in any way that would harm their employment possibilities. Nonetheless, most observers recognize that age discrimination in the workplace remains widespread. Yet empirical studies have not shown older workers to be less effective in their job performance (Shea & Haasen, 2005).

The average age of retirement fell throughout the 20th century, a trend that held not just in the United States, but also internationally. Between 1965 and 1995, the average age of retirement among both men and women went down in most advanced industrialized countries (Gendell, 1998). As a result, people have been spending more and more of the years of their lives in retirement.

But this trend may not continue in the future. There is evidence that the long post–World War II trend of earlier retirement may now be coming to an end. For instance, labor force participation rates in later life have held steady since the mid-1980s and even increased in the first part of the 21st century. More older people are working than earlier trends would have predicted. Some surveys suggest that new attitudes may be developing about work late in life. An AARP survey in 2011 revealed that up to 80% of aging baby boomers expect to work at least part-time after what used to be normal retirement age. Similar surveys conducted in 2008 confirmed this trend, strengthened by the economic downturn beginning in that year (American Council on Education, 2008; MetLife Foundation & Civic Ventures, 2008).

GLOBAL PERSPECTIVE

Older Workers in Japan

In the decades after World War II, employment in Japan typically involved working for the same employer with a measure of guaranteed job security, followed by retirement at reduced salary with scaled-back benefits and lower job status. Faced with the oldest population in the world, Japan has recently undertaken a variety of efforts to extend working life. For example, Japan has now raised its pension eligibility age and is providing subsidies to employers to hire or retain older workers. Unlike the United States, age discrimination in Japan is not illegal, and mandatory retirement is common. Continuing earlier practices, Japanese employers rehire selected workers who have been compelled to retire, while paying them less than they had earned before. There are a variety of reasons why the Japanese have extended their working lives: (1) desire to maintain earlier standard of living; (2) high level of self-employment; (3) positive cultural attitude toward remaining productive; (4) government initiatives, including subsidies, to promote work-life extension; and (5) good health of older Japanese people. As a result of these trends and incentives, labor force participation rates of older people in Japan are now among the highest of major industrial nations.

(Continued)

(Continued)

A notable innovation in Japan has been the creation of Silver Human Resource Centers, which are funded by both national and local governments. These Silver Human Resource agencies offer jobs at the community level for local people over age 60. They focus strongly on short-term and part-time employment favored by older people themselves. An association of Silver Human Resource groups also provides retraining, counseling, and job-matching services, with links to business groups and government employers. This combination of incentives, opportunities, and institutional leadership has put Japan in the forefront of countries making good use of older people who want to continue to work.

SOURCE: John B. Williamson and Masa Higo, *Older Workers: Lessons from Japan,* Boston College: Center for Retirement Research, 2007.

This 86-year-old Japanese woman hand spins thread from fibers of a banana plant in order to make traditional jofu cloth from the threads.

Changes in the relative attractiveness of work and retirement may encourage continued employment into later life. One problem in the past has been that policies have encouraged a trend toward earlier retirement and have discouraged older workers from continuing to work. However, there have been some positive changes in recent years. Mandatory retirement has been eliminated. Another major incentive for retirees to work took effect in 2000, when the so-called earnings penalty was repealed, allowing those over age 66 to earn as much as they can without jeopardizing their Social Security benefits. Moreover, the spread of defined-contribution plans means fewer work disincentives than the typical defined-benefit plan (J. Quinn, 1999). Under defined-contribution plans, employees can often accumulate assets even if they work on a reduced schedule, thus eliminating the "all-or-nothing" approach to retirement that defined-benefit plans have encouraged.

These trends offer promise for the future. If employers are willing to revise compensation and job structures to meet the needs of potential older workers, then they can draw on a growing pool of experienced and willing workers for years to come. Too often, companies respond to new challenges of technology or the marketplace by favoring younger employees. As a result, regardless of whether older workers stay in place or retire early, companies lose the rich life experience embodied in proven, valued workers (Buonocore, 1992).

Managing an older workforce will clearly be a challenge for the future (H. Dennis, 1988). As labor shortages arise from declining numbers of young people in the workforce, that fact could stimulate a need for older workers and reverse the trend toward early retirement. There has long been support for the idea of work-life extension, that is, adaptations of retirement rules or employment practices to enable older people to become more productive (Doering, Rhodes, & Schuster, 1983). In favor of this idea is the fact the labor force continues to grow older: The median age rose from 35 in 1980 to over 40 by 2005 and is likely to rise further in the future.

We have also seen the movement from an industrial to a postindustrial economy. For example, three quarters of employees over age 65 are in white-collar occupations in service industries, which are less physically demanding than agriculture or manufacturing jobs. In fact, the five top "encore careers" for adults 50 and older are education, health care, government, nonprofit organizations, and for-profit businesses that serve the public good (MetLife Foundation & Civic Ventures, 2008). As a result, older people can remain in productive jobs longer than in the past. However, changes brought by information systems and telecommunication technology continue to bring rapid shifts in methods of production. Older workers who are decades away from school are likely to become victims of skills obsolescence. Without retraining, older workers may not be in a position to take jobs that might open up in the future. However, trends in the early part of this century suggest that higher education is increasingly important for older adults who not only wish to engage in lifelong learning, but are compelled to learn new skills and knowledge so as to be competitive as workers.

There is an even more fundamental question about whether work-life extension is a valid goal in the first place. After all, why shouldn't the economic surplus from improved productivity be taken in the form of leisure time, either earlier in the course of life or in retirement years? The assumption that old age is unsatisfactory unless people continue to work seems an uncritical extension of the work ethic that originated in capitalist ideology, as the sociologist Max Weber argued in his classic work *The Protestant Ethic and the Spirit of Capitalism.* Although those older people who love their work may find the work ethic compelling, declining labor force participation for most older people suggests that the attraction is not universal. Late-life leisure has important meaning and deserves attention as an activity in its own right (Leitner & Leitner, 2004; Teague & MacNeil, 1992).

Productive Aging

As we have seen, the question "Does retirement make sense?" is complicated. For both individuals and society, the alternative of work versus retirement is in many ways a false choice. The real question may be how to enable older people to lead lives of greater productivity, whether in employment or in retirement. Interest in this question about productive aging has grown in recent years (Bass & Caro, 1993; Morrow-Howell, Hinterlong, & Sherraden, 2001; Morrow-Howell & Mui, 2012).

First of all, we should recognize that people over age 65 are already productive in many different ways. When we include unpaid work, such as housework, family caregiving, and volunteer roles, three quarters of older people are actually engaged in productive activity. In

1991, Louis Harris and Associates, with support from the Commonwealth Fund, carried out the most comprehensive survey done on the subject (Bass, Caro, & Chen, 1993). A national survey asked 3,000 adults over age 55 about their involvement in productive activities such as work, volunteer roles, and caregiving for a spouse, relative, neighbor, or friend who was sick or had disabilities. The survey showed that, among people over age 55, 27% were working, 26% were volunteering, 42% were helping children or grandchildren, and 29% were assisting people who were sick or had disabilities. Such findings are in stark contradiction to myths that depict older Americans as dependent, depressed, and isolated.

The Commonwealth study also found that Americans over age 55 represent an overlooked national resource. The total value of their contribution to society is equal to nearly 12 million full-time workers and, in caregiving activities alone, the equivalent of more than 7 million full-time workers. The Commonwealth study discovered that men play a bigger role as volunteers and caregivers than they are usually credited with, including helping relatives who are sick or have disabilities as well as helping children or grandchildren. Informal caregiving may be viewed as a barrier to other productive activities (O'Reilly & Caro, 1994), but it can also be considered as an important form of productivity. Some critics have charged that the ideal of productive aging may devalue the contributions of women outside the paid workforce in maintaining relationships and providing care to others, activities that enrich all of society (M. Holstein, 1992).

Another important recent finding of the Commonwealth study is the remarkable vitality of many people over age 75. Health problems do increase in this group and limit some activities (Herzog et al., 1989), but even among the old-old, more than half reported being in "excellent or good health," and nearly a quarter were still volunteering through organizations or were involved in caregiving for family or neighbors.

Finally, the Commonwealth study confirmed that, along with current productive roles, older people are eager to be even more actively involved: 31% were not yet employed but wanted to be. A similar picture has emerged for volunteers, reaffirming the importance of the older-volunteer activities documented in earlier studies (Chambre, 1987). This rise in the number of able elders has optimistic implications that deserve consideration (American Council on Education, 2008).

Organized vehicles for productive aging already exist in public service employment programs sponsored by the government. One of these is the Senior Community Service Employment Program (SCSEP), created under Title V of the Older Americans Act (1965). Volunteer activities are another channel of productive roles for older people, and here, too, government has played an important role. Senior Corps is the federal government office that engages those over 55 for volunteer roles in a variety of tasks, such as mentoring, friendly visiting, and other volunteer activities. One of the older volunteer programs run by Senior Corps is the Senior Companions Program, similar to Foster Grandparents, but directed toward impaired adults needing help to continue living in their own homes. The largest Senior Corps-sponsored initiative is the Retired Senior Volunteer Program (RSVP), offering community volunteer service opportunities in day care centers, nursing homes, libraries, and adult education programs.

A larger proportion of older people are volunteering today than was true a generation ago, and that trend seems likely to continue in the future (Chambre, 1993). Most of those who do volunteer are middle-class people who are continuing a lifetime pattern (Cnaan & Cwikel, 1992). A major national study found that personal ideals, such as altruism and the wish to be useful to others, are a key to volunteerism (Hendricks & Cutler, 1990). Several studies reveal a strong correlation between volunteerism and higher life satisfaction, but the higher health and economic status of those who volunteer may confound any conclusion about causal connections (Fengler, 1984).

Another challenge for productive aging lies in the workplace. Most older workers perform as well as younger workers—this is welcome news because one third of the U.S. workforce now consists of workers over the age of 40. This "graying" of working America poses a challenge for the economy. In the past, companies have relied on younger workers to bring in the new skills and knowledge that are key to raising productivity. If the U.S. economy is to prosper in the future, managers will have to pay more attention to making use of older workers, who will constitute an increasing proportion of the workforce during a period when technological change will be accelerating (Kieffer, 1983).

Experience in industry over many years has demonstrated that middle-aged and older workers can be trained or retrained to acquire new skills as previous skills or knowledge becomes obsolete (Cappelli & Novelli, 2010). This potential for retraining is related to the concept of **plasticity**. The positive experience in older worker retraining serves to contradict a common prejudice that "you can't teach an old dog new tricks." Laboratory tests do show a decline in test performance, but in the real world, any decline in speed or accuracy tends to be compensated for. In addition, older workers have other positive attributes, such as less absenteeism, lower job turnover, and a lower accident rate. Older people typically demonstrate some declines in motor performance and speed of comprehension as new information is acquired. Such effects sometimes discourage older workers, who may come to believe the stereotype. But modest changes in teaching techniques can serve to offset most of these problems.

DEBATE OVER RETIREMENT POLICY

The volatility of the U.S. economy and the aging of the workforce make predictions difficult, but the vitality of the aging population today already constitutes a positive sign for the future, which can be optimized if there is greater attention to the potential of productive aging. Community service employment, second and encore careers, volunteer opportunities, and higher education participation and retraining of older workers may all be part of the picture in years to come.

In the readings that follow, we hear different voices in the debate over the future of retirement. Caro, Bass, and Chen provide a historical perspective and argue that under the conditions of the later part of the 20th century, which set the stage for trends we are now experiencing, we need to think again about the meaning of retirement, and, in particular, we should design social institutions that can open up new possibilities for productive aging. Making better use of the talents of older people would represent a new social policy goal for American society.

In the excerpt from his book *Prime Time,* Marc Freedman calls our attention to a paradox about today's retirement generation. Older people represent an enormous resource to society, yet it is a resource that is hardly being used. By most standards, people now over age 65 are in better health than ever before and are among the most civic-minded groups in society. They also have more free time than any other group of adults. Yet until recently, labor force participation has gone down, and rates of volunteerism among older adults have not been increasing. Like Caro, Bass, and Chen, Freedman believes we cannot afford a society that ignores the productive contribution of its older members. He predicts that, in the future, aging baby boomers will behave in ways that make earlier ideas of retirement obsolete. In this view, civic engagement could become a new retirement role in years to come (Kaskie et al., 2008).

A key assumption in debates around work and retirement is the idea that people are somehow living better lives if they continue to be active and productive. Indeed, this moral valuation of work and productivity has deep roots in American culture in the Protestant work ethic—the idea that hard work is a virtue in itself. David Ekerdt's article highlights what he calls the "busy ethic" and shows that, for many Americans, both work and retirement are characterized by the same style of activity and productive engagement, whether in a paid job or not.

But this "busy ethic" is not the only ideal for human fulfillment. Ronald Manheimer's article provides us with a view of how retirement could be a creative period in a person's life. Just as paid work is not the only form of productivity, so recreation and relaxation are not the only forms that leisure might take. "Creative retirement" takes the view that the freedom of old age gives us an opportunity to use our imagination and become the person we have always dreamed of being, as an artist does in the act of creativity.

Through the ages, philosophers and social thinkers have dreamed about what society might look like if the burden of work were lifted from humankind. Today, that dream has come to pass for many people. In the 20th century, with mass retirement as a normal social institution, we witnessed what happens when vast numbers of people are given—and expect—leisure in their later years. The experiment goes on into the 21st century, with a new twist, increasing participation in higher education and the workforce among older adults. But the results will be unclear and debatable for some time. In fact, controversies about work and leisure demonstrate that we have not yet agreed on the key social values at stake when we ask, "Is retirement obsolete?"

FOCUS ON PRACTICE

Retirement and Life Planning

In the 20th century, retirement became a longer and more important part of the life course. In 1900, around 3% of an average man's lifetime was spent in the retirement years, but by 1980, that proportion had increased to 20%, or between 10 and 30 years. For those who retire early, the number of years in retirement can even approach the number of years in the workforce. Yet although we prepare for the world of work through schooling, few people give comparable attention to planning or preparing for retirement. Moreover, the shape and timing of retirement have changed substantially in recent years:

- Data from the Bureau of Labor Statistics show consistent year-to-year employment increases for Americans 65 and older while employment for those under 65 has been dropping.
- In 2007, 15.5% of Americans ages 65 and older were in the labor force, the highest rate since 1971 and part of an upward trend from a low of 10.4% in 1985.
- More than 56% of workers ages 65 and older were working full-time in 2007, a new high. Until 2002, part-time workers always outnumbered full-time workers among older workers.
- In the last 30 years, the number of American workers ages 65 and older grew by 101%, compared with a 59% increase in all workers. Those ages 75 and older increased by 172%.

These are trends from the recent past. What about the future? In the United States, the number of workers over age 50 is expected to increase 34% by 2012, according to the Bureau of Labor Statistics. But the number of workers under age 50 will grow by only 3% during the same period. Repeated surveys from AARP suggest that, among aging boomers, more than 70% say they expect to continue working, at least part-time, beyond the conventional retirement age of 65. What does it mean if most people see "retirement" as a time when they will continue to be working? Why has the traditional idea of retirement as a time of pure leisure begun to change?

One solution proposed has been the idea of preretirement planning. During the 1980s, for example, there was an expansion in the numbers of preretirement programs offered by business and industry. Around a third of business corporations offered some kind of formal program to their workers for preretirement planning (Morrison & Jedriewski, 1988), with higher numbers among *Fortune 500* companies. Traditional preretirement planning programs covered issues such as financial planning, housing options, use of leisure time, and adjustment to the retirement role (Giordano & Giordano, 1983). Trends in preretirement planning included the following (Dennis, 1989):

- Individualized instruction and counseling
- Use of educational technology, such as computer software, to model financial decisions
- Attention to special needs of women in retirement
- Recognition of options for positive growth and productive aging, such as second careers and volunteerism

Since the year 2000, and especially since the Great Recession beginning in 2008, the whole concept of retirement has begun to change, and preretirement planning is no longer what it used to be. Aging boomers largely express the view that they expect to continue working, full-time or part-time, during traditional retirement years. Early retirement has become much less popular, and workers are staying on the job longer. Those who do want to retire may be interested in getting new skills for part-time work, consulting, or starting their own business. There are several reasons for these trends. Foremost is the dramatic drop in confidence about having enough money for a comfortable retirement. The percentage of workers "very confident about having enough money for a comfortable retirement" went down from 27% in 2007 to 18% in 2008, the biggest 1-year drop in history. This decrease in confidence appeared across all age groups and income levels, and a big factor has been rising health care costs (Helman, VanDerhei, & Copeland, 2008).

The fading of early retirement and the rise of work-life extension may actually be a good thing for the wider economy. For instance, David DeLong, in *Lost Knowledge: Confronting the Threat of an Aging Workforce* (2004), warns of a looming problem: namely, loss of intellectual capital that will happen as growing numbers of aging boomers leave the workforce. The problem is not just hypothetical. If Americans wanted to launch another rocket to the Moon, as happened in the Moon landings beginning in 1969, it would not be an easy thing to do. The reason is that NASA threw away blueprints for the Saturn rocket, and most of the skilled engineers involved in Moon flights have now retired.

There are now many books in print insisting that we're "too young to retire," that we should "rewire, not retire" during our exciting "Power Years." One of the most inspiring of these is Marc Freedman's *Encore: Finding Work That Matters in the Second Half of Life* (2007). Freedman founded the Experience Corps, a volunteer network for Americans over age 55 now operating in 19 major

(Continued)

(Continued)

cities in the United States, including major urban centers such as Boston, San Francisco, Minneapolis, and Washington, DC. Approximately 20,000 Experience Corps members have served as tutors and mentors helping children to read. They work in urban public schools and after-school programs, where, along with basic skills, they help develop confidence for future success.

The business world has also begun to take notice of this new phase of life called "retirement-plus-work." For example, RetirementJobs.com has become the leading career website for job seekers in the 50+ age group (D. Lewis, 2006). Its website lists between 20,000 and 30,000 open positions refreshed several times each week. Most openings are in the retail industry, but many are also in financial services. Older job seekers can go directly to the site and be confident of finding employers who are eager to hire older workers. RetirementJobs.com is not the only Web-based search company focused on older workers. Workforce50.com and RetiredBrains.com both certify that employers listing jobs actually want to hire older workers—not only for entry-level positions, such as flipping burgers at McDonald's, but for higher-level positions as well. Another site, ExecSearches.com, is targeted primarily at midlevel and executive positions in the government, health, nonprofit, and education sectors, where job shortages are already evident. Even more distinctive is YourEncore.com, which aims for experienced scientists, engineers, and product developers who want time-limited assignments. Still another site calls itself DinosaurExchange.com because it favors opportunities for "dinosaurs"—that is, retirees with solid experience.

In summary, preretirement planning is now giving way to new thinking about how people will plan for longer lives. For example, the national Life Planning Network now offers "third age life planning," a completely different approach from past approaches such as executive coaching, personal financial planning, and preretirement preparation. Topics such as spirituality, lifelong learning, and family relationships are now part of the mix that life planners must consider. Retirement will probably not disappear completely from American life, but it will take different forms in the 21st century than it did in the century just past.

———— Achieving a Productive Aging Society ————

Francis G. Caro, Scott A. Bass, and Yung-Ping Chen

Older people face a prolonged period in life in which they are relatively healthy and vigorous but lack a recognized role in the economic and social life of the society. Although elders, especially older women, are encouraged to provide support to their extended families, they too frequently are left without a significant role in late adulthood. This ambiguous status of retirement and explicit devaluation may last for a period as long as twenty or thirty years, which in some cases may be as long as a working career. . . .

A so-called productive aging perspective views older people as a major and valuable resource. In the United States and certain other nations, within this large and diverse group of older individuals, many are becoming increasingly dissatisfied with a life primarily structured around leisure. But even these people too frequently experience serious barriers as they seek significant societal roles.

Most older people are relatively healthy and robust well into their sixties and seventies, and in some cases beyond. Although, as a group, they tend to experience chronic ailments, they are capable of sustaining most of the intellectual and many of the physical activities in which they participated during their fifties. . . .

DEFINING PRODUCTIVE AGING

The term *productive aging* has emerged over the past decade as a rallying cry for elder advocates, policymakers, and academicians dissatisfied

with the stereotype of older people as dependent and frail. . . .

This more positive approach to the examination of aging seeks to identify the changes associated with aging and to maximize the human potential throughout the life course. At the center of the discussion is the fact that, in many activities, chronological age, up and into the advanced ages, is not necessarily a strong predictor of performance. Compelling evidence indicates that the aging process is highly individualistic, with enormous differences in the way various individuals age and in their subsequent performance in physical and mental activities. Some individuals in their seventies and eighties may be very active and produce their most significant contributions, while others in their fifties and sixties may be unable to function fully in society or may choose to withdraw from productive activity. Age as a sole predictor of performance is simply too crude a tool to reflect the actual capability of older people.

Productive aging, not unlike other terms such as *successful aging* or *normative aging,* has reflected an intellectual direction or theme that has attempted to attract individuals from many different perspectives. . . .

More recently, A. Regula Herzog (1989) has defined productive aging as "any activity that produces goods or services, whether paid or not, including activities such as housework, child care, volunteer work, and help to family and friends."

The major difference among the definitions is the range of activities they include. The

SOURCE: "Introduction: Achieving a Productive Aging Society" by Francis G. Caro, Scott A. Bass, and Yung-Ping Chen in *Achieving a Productive Aging Society,* edited by Scott A. Bass, Francis G. Caro, and Yung-Ping Chen.

broadest includes nearly all activities of older people. The most restrictive includes only paid employment and formal volunteer work.

A definition we prefer is the following: Productive aging is any activity by an older individual that produces goods or services, or develops the capacity to produce them, whether they are to be paid for or not. This definition builds on Herzog's by including only voluntary or paid service or goods produced but excluding activities of a personal enrichment nature. Our definition expands upon Herzog's to encompass activities that provide training or skills to enhance one's capacity to perform paid or volunteer work; it does not include education for personal growth as that would not directly contribute to enhanced skills for paid or volunteer labor.

While we acknowledge the many facets of productive aging, we particularly emphasize paid employment and volunteering because they are sectors in which older people experience significant barriers. In other sectors, productive involvement of older people is usually expected. The role of older people in providing long-term care to disabled spouses is a good example. . . . [H]ealthy spouses of the disabled are *expected* to provide care to their partners. In fact, the public policy issue concerns how public intervention should complement that responsibility. In the case of employment, however, the issue is how to reverse skepticism about the capabilities of older people and even the loss of confidence of older people in their ability to be effective in the world of work. In the employment sector, the issue is also how to address a whole set of institutional forces that encourage early departure from the workforce. In the case of volunteering, the question is how volunteer roles can be made significantly more attractive so that volunteering among older people will expand.

The definition we have selected sufficiently excludes many important and constructive activities undertaken by the elderly, such as worshiping, meditation, reflection, reminiscing, reading for pleasure, carrying on correspondence, visiting with family and friends, traveling, and so forth. It is not to say that these are not valuable activities and part of healthy and fulfilling aging experience . . . ; they simply are outside the bounds of productive aging as we define it. Activities undertaken in our definition can be counted, aggregated, and assigned some economic value. Productive aging, therefore, is not for all older people, only for those who are interested, and that interest may vary at different times, ages, or even seasons. In our search for words that embody all that gives aging meaning, *productive aging* may be only one of several components. . . .

HISTORICAL ORIGINS

Nonproductive aging, as evidenced by retirement and an absence of a role in late life, is a relatively recent phenomenon in America. In other developed nations, it is a concept that dates back no earlier than the late nineteenth century. Prior to these times and dating back to antiquity, older people were engaged in some form of work until they were unable to continue with it. Throughout the centuries, the old were expected to work or to beg until they were simply too ill or enfeebled, leaving their care to the family, neighboring community, poorhouses, or no one (Axinn & Stern, 1988). . . .

The first reported social security system came into being in the 1890s in Bismarck's Germany. By 1913, Australia, Belgium, Great Britain, Denmark, France, New Zealand, and Sweden also had public pension systems for the elderly. The United States was among the last of the industrial nations to institutionalize a national pension program to provide economic security to the elderly. . . . But prior to the establishment of Social Security in 1935, older people, for the most part, were expected to work or to seek shelter and care from almshouses, or they were cared for by their families. . . .

In the late nineteenth and early twentieth centuries in the United States, the nonworking elderly were not usually accorded favorable consideration. In fact, David Hackett Fisher (1977) notes that expressions of hostility toward the poor, nonworking elderly continued to grow during the nineteenth century. The opinion at the time was that there was enough work for all who could work. Those who could not placed a burden on the family and society. Little was available other than personal family charity for those who could not work, including the frail aged.

Within the first fifty years of Social Security, a relatively short time in history, all had changed. From a situation where the elderly had no option but to continue to work or be dependent on others, we arrived at a place where retirement had become an institution rather than a luxury (U.S. Senate, 1990).

Near-universal work for the aged has been replaced by near-universal retirement. The contrast and extremes remain stark. The contemporary pattern of nonparticipation of the elderly in the work force is particularly remarkable in light of the growth of the elderly population and the number of years older people now typically live in retirement. The fact that many elderly are able to live comfortably on the basis of pensions and savings is a reflection of the strength of the economy (Schulz, Borowski, & Crown, 1991). The ability of the U.S. economy to function adequately without the presence of most older people in the work force also is consistent with the assumptions that Patten advanced nearly 100 years ago. But will these assumptions hold true as we look to the economic future of America?

APPROPRIATENESS OF THE PAST IN TODAY'S POLICY

To what extent is there a labor-supply abundance? Since the late 1960s the economy has absorbed large numbers of new women workers, and new supplies of labor for economic growth are limited as a result of the low birthrate of the 1970s and 1980s and recent restrictions on immigration (Schulz, Borowski, & Crown,

1991). If the nation faces modest economic growth, from where will labor support come? Older people are one nontraditional population to consider (McNaught, Barth, & Henderson, 1989). For the most part, they have good work habits and extensive on-the-job experience. With the older population increasingly in better health, could it not be an important economic resource to the nation (Bass & Barth, 1992)? Further, in light of the vast array of social needs and the declining public willingness to provide tax support for human service programs, could not interested older people be trained to fill certain important, but currently unmet, social needs, and if so, to what extent? We are not advocating that older volunteers replace paid workers, but we do believe that there are societal functions that may be performed by trained older volunteers or stipend workers.

Such a scenario for older people is not without its problems. There are family and societal expectations of the aged. And no equivalent of career counseling is available for those who leave their primary employment at age sixty or sixty-five and seek an alternative. Retirement is thought of as the terminal work experience; in fact, it may be a transitional one. Career planning for paid or volunteer roles after retirement is not yet common. Part-time or flexible work hours which many older workers want remain elusive and are considered unconventional. Training programs and higher educational opportunities are designed for younger people, with little thought to the needs of older workers seeking work past retirement. . . .

OBSTACLES TO PRODUCTIVE AGING

How is the limited participation of older people in paid work and volunteering to be explained? To what extent do older people prefer not to work and not to volunteer? To what extent are important opportunities for employment and volunteer work denied elders? Are elders tracked out of mainstream roles in subtle and discriminatory ways?

One hypothesis is that the limited participation of older people in paid work and volunteering is the result of "institutional ageism"—that societal institutions have structures, rewards, and sanctions that value certain cultural norms. These idealized and otherwise unspoken values and cultural traditions are inclusive of certain behaviors and groups and exclusive of others. Indeed, some individuals may overcome these barriers, but they are the exceptions rather than the rule.

The forces that exclude older people on bases other than merit are widespread. Some have been documented, such as age discrimination in employment. However, institutional ageism may be such a pervasive aspect of all major institutions in our society that we often do not recognize it. In fact, many older people themselves have internalized ageism. Too often they underestimate their own capabilities and accept the notion that older adults should leave productive roles at certain prescribed ages.

The major force at the root of institutional ageism may be the conflicting interests of the elderly and the nonelderly. Embedded in existing institutional arrangements may be management's desire to remove the elderly from attractive jobs and other positions of power and influence to facilitate greater access for younger people. Conflict theory suggests that such removal can be explained in part by economic competition between the nonelderly and elderly. It hypothesizes that pressure to exclude the elderly is affected by labor market conditions. During recessions, when jobs are scarce, pressure to remove the elderly is expected to increase. In periods of economic boom, when workers are in short supply, conflict theory predicts that older workers will be seen in a much more favorable light and the employment of older people is more likely to be actively encouraged.

Cultural lag is a second potential explanation for the limited participation of older people in attractive paid-work roles. The cultural-lag hypothesis differs from the conflict hypothesis on the basis of its assumptions about the underpinnings of institutional patterns. The cultural-lag hypothesis suggests that, as a society, we are slow to adjust our institutions in response to changing conditions. We may be slow, for example, in modifying our retirement policies moving from an era of labor oversupply to one of labor shortages. We may be slow in reorganizing our educational institutions to provide the lifelong training for work necessary in an economy characterized by sharper competition and rapid technological advances. Further, our society may take too long to recognize that people now have the potential for remaining productive later in life than in the past as a result of their improved health and of reduced physical demands in the workplace.

A third explanation, which might be called the defective-institutions hypothesis, is that employment and volunteer options are so badly flawed that people who can choose to depart from jobs as early as they can and generally avoid extensive volunteer commitments. A factor here is the quality of the work environment itself. In many fields, working people of all ages in this country complain about their work environments and compensation. Further, many of the volunteer assignments are unattractive, having unappealing tasks, insufficient challenges, or heavy demands, and training and support are inadequate.

A fourth explanation, which we will call the alternate-preferences hypothesis, is that many older people organize their lives around alternatives to the work ethic. According to this hypothesis, many older people subscribe to values other than those that lead to employment and community service. They find personal expressive activities highly attractive and, when given a choice, are not interested in paid employment or volunteering options. According to this hypothesis, some advocates for productive aging may be overestimating the number of older people currently interested in access to work and volunteer roles. . . .

Both the conflict and cultural-lag theories can provide explanations of the alternate-preferences hypothesis. Both of these theories would argue that alternate preferences are learned either through the popular culture or through direct experience with the negative aspects of work environments and volunteer opportunities. Both theories predict that older people would regard paid employment more favorably if work environments were made more attractive. Further, as a change strategy, the infusion into the society of more positive views about employment of older people might trigger increased public support for the employment of older people.

Because we suspect that both conflict theory and cultural-lag theory help explain current arrangements that discourage participation of older people in paid employment and meaningful volunteer work, we believe that both confrontation and public education are useful strategies for directed change. Political action may help break down discriminatory policies and practices and create improved employment options for the elderly. Public education also may be effective in encouraging older people to seek to remain active in both paid employment and volunteering, it may stimulate employers and voluntary organizations to be more creative in recognizing older people as resources, and it may lead educational institutions to develop attractive retraining programs.

Preferences among the elderly that cause them to focus their interests and activities on sectors other than paid employment and volunteering must be recognized. Our hypothesis is that, if work and volunteering are made more attractive as later-life options, many more older people would pursue them. How many more people actually would work or volunteer would then depend both on the attractiveness of work and volunteer options and the pull of competing alternatives.

Like any other reform movement, productive aging can have perverse effects on its intended beneficiaries. Our emphasis is on expanded opportunities for the elderly for paid employment and volunteering. . . . At this point, we would not endorse blanket proposals to expand either work obligations or reduced pensions for the elderly premised on an extended work life. A great deal has to be accomplished in extending work opportunities before there is a sound basis for debating whether work obligations should be increased.

Similarly, we advocate improved volunteer opportunities for older people to address serious human needs that currently are not being met. But we do not regard elderly volunteers as substitutes for paid workers. If efforts to recruit and retain older volunteers were spectacularly successful in some sectors, some might conclude that fewer paid workers are needed in those fields. We prefer to wait until large, effective cadres of older volunteers actually threaten to displace paid workers and debate the specific issues on their merits.

REFERENCES

Axinn, J., & Stern, M. J. (1988). *Dependency and poverty: Old problems in a new world.* Lexington, MA: Lexington Books.

Bass, S. A., & Barth, M. (1992). *The next educational opportunity: Career training for older adults.* Draft for the Commonwealth Fund. New York: The Commonwealth Fund.

Fisher, D. H. (1977). *Growing old in America.* New York: Oxford University Press.

Herzog, A. R. (1989). Age differences in productive activity. *Journal of Gerontology: Social Sciences, 44,* S129–S138.

McNaught, W., Barth, M., & Henderson, P. (1989, Winter). The human resource potential of Americans over 50. *Human Resources Management, 28*(4), 455–473.

Schulz, J., Borowski, A., and Crown, W. H. (1991). *Economics of population aging.* New York: Auburn House.

U.S. Senate, Special Committee on Aging. (1990). *Aging America.* Washington, DC: U.S. Government Printing Office.

URBAN LEGENDS OF AGING

"The U.S. introduced age 65 for retirement, following German Chancellor Otto von Bismarck, who picked that number because it was his own age."

It's true that Germany was a model considered in planning Social Security. But Germany set age 70 as the retirement age at a time when Bismarck himself was 74. It wasn't until 27 years later that the age was lowered to 65, when Bismarck had been dead for 18 years.

READING 43

Prime Time

Marc Freedman

OUR ONLY INCREASING NATURAL RESOURCE

Contrary to prevailing stereotypes, America now possesses not only the largest and fastest-growing population of older adults in our history but also the healthiest, most vigorous, and best educated. Only 5 percent of these individuals live in nursing homes, and the vast majority experience no disability whatsoever. . . .

Perhaps most important, older Americans possess what everybody else in society so desperately lacks: time. First, older Americans have time to care. As the British historian Peter Laslett observes, free time was once the exclusive province of the aristocracy; today, it is the democratic possession of millions of citizens—those in later life. Retirement frees up 25 hours a week for men and 18 hours for women, according to time-diary studies conducted by the University of Maryland's Survey Research Center. . . .

Second, older adults have more time lived. They have practical knowledge—and, often, wisdom—gained from experience. . . . They've held jobs, built social networks, raised families. They are also particularly civic-minded: Older adults vote at a higher rate than any other segment of the population. And because these individuals often carry with them a world lost to younger generations, they may well be our greatest practical repository of the "social capital" that many observers today fear is drying up.

Third, time left to live may give older adults a special reason to become involved in ways that both provide personal meaning and make a significant difference to others. The awareness in old age that death is closer than birth inspires many to reflect—and act—related to the legacy that we leave behind. According to the late psychologist Erik Erikson, the hallmark of successful late-life development is the capacity to be generative, to pass on to future generations what one has learned from life. For Erikson, this notion is encapsulated in the understanding "I am what survives of me."

For all these reasons, older Americans may well be our only *increasing* natural resource.

SOURCE: *Prime Time: How Baby-Boomers Will Revolutionize Retirement and Transform America,* by Marc Freedman, founder and CEO of Civic Ventures. Copyright © 1999 Civic Ventures. Used by permission. All rights reserved.

Nevertheless, it is not a resource in oversupply. At precisely the juncture that this country contains so much pent-up time, talent, and experience in the older population, we are in the midst of a human resource crisis of staggering proportions, a crisis that goes well beyond Silicon Valley's never-ending search for more software engineers.

According to a 1996 Gallup survey conducted for Independent Sector, the proportion of adults who volunteer declined from 54 to 49 percent between 1989 and 1995. In absolute numbers, that's a decline of 5 million volunteers. And the vast majority of participating individuals gave only a few hours a week.

These numbers are hardly surprising, given the time famine afflicting so many adults in our society. As Harvard economist Juliet Schorr documents, the average American now works 162 more hours a year than twenty years ago, the equivalent of an extra month on the job. With these individuals attempting to compress thirteen months of work into the space of twelve, something has to give. . . .

Changes in the role of women are surely one of the key factors in the overall crisis in the social sector. For most of the twentieth century, through a myriad of unpaid, undervalued, often unnoticed tasks, women have served as the glue in American communities. But today, 61.7 percent of mothers with preschoolers (and half of all mothers with infants) are working at paid jobs, up from 19 percent in 1960. And two-thirds of employed mothers work full-time. Berkeley sociologist Arlie Hochschild has shown that when child rearing, housework, and paid employment are combined, women work 15 more hours a week than men do—equivalent to an extra month of 24-hour days each year. After a workweek of 80 or even 100 hours, how many people have the time and energy for a third shift laboring on behalf of the greater good?

THE AGING OPPORTUNITY

Against this backdrop of women's changing social roles, America's burgeoning older population is poised to become the new trustees of civic life in this country. These individuals have the time to care; they have the skills and experience required; they have the personal need to contribute in new ways. Society desperately needs them, and at the same time, there is considerable reason to believe that older Americans could reap tremendous mutual benefit in the process—without many of the frustrations and sacrifices middle-aged women have faced in their historical role. This match, between the untapped resources of older Americans and the needs of American communities, constitutes *the great opportunity* presented by America's aging.

Given so many appealing factors, one might expect to see an all-out clamor to engage older Americans in new roles focused on greater involvement in communities (for example, a nationwide crusade akin to Miami's mobilization of older volunteers on behalf of schools). After all, as the Gray Panthers' late founder Maggie Kuhn observed, "We don't have a single person to waste"—much less the most experienced, stable, and available portion of the population (especially at the same time we lecture single moms on welfare that they need to contribute in some way to society). However, there is no clamor to engage older adults and community contribution falls off sharply after retirement. Although the level of involvement has improved substantially since the early 1960s, older Americans still volunteer less than any other age group—even those overwhelmed Americans in the middle generation. . . .

The lack of involvement of older adults in the civic life of communities is all the more baffling for three additional reasons. First, an accumulation of research suggests that many older people *want* to be more involved. . . .

Second, these older adults' responses to the polls likely reflect a healthy dose of enlightened self-interest. Numerous studies following people over their lives link strong social ties and community engagement to prolonged physical and mental health. The aforementioned decade-long MacArthur inquiry into successful aging produces exactly these conclusions and strongly recommends more volunteer and service opportunities for older Americans.

Third, older adults are so "civic" in other ways. They vote at a higher rate than any other segment of the population, and they are the most generous with their financial contributions. . . .

REBALANCING RESPONSIBILITIES

At present the responsibilities in our society are wildly skewed, with individuals in the middle generation overworked while older men and women are underused. One group faces a time famine; the other is adrift in a sea of discretionary time (clinging to the "busy ethic" as a way of staving off the persistent sense of uselessness). Transformation of the nature of the third age holds the potential to redress this imbalance, so that the middle agers receive a much needed respite while individuals in later life gain lives with additional purpose, meaning, structure, and significance.

Indeed, one of the principal beneficiaries of this redistribution may be the younger generation. Older adults who take on more responsibility for developing and caring for young people can provide this group with additional support while helping parents lead saner, more balanced lives.

READING 44

The Busy Ethic

Moral Continuity Between Work and Retirement

David J. Ekerdt

There is a way that people talk about retirement that emphasizes the importance of being busy. Just as there is a work ethic that holds industriousness and self-reliance as virtues so, too, there is a "busy ethic" for retirement that honors an active life. It represents people's attempts to justify retirement in terms of their long-standing beliefs and values.

The modern institution of retirement has required that our society make many provisions for it. Foremost among these are the economic arrangements and mechanisms that support Social Security, private pensions, and other devices for retirement financing. Political understandings have also been reached about the claim of younger workers on employment and the claim of older people on a measure of income security. At the same time, our cultural map of the life course has now been altered to include a separate stage of life called retirement, much as the life course once came to include the new stage of "adolescence" (Keniston, 1974).

Among other provisions, we should also expect that some moral arrangements may have emerged to validate and defend the lifestyle of retirement. After all, a society that traditionally

SOURCE: "The Busy Ethic: Moral Continuity Between Work and Retirement" by David J. Ekerdt in *The Gerontologist, 26*(3), pp. 239–244, 1986.

identifies work and productivity as a wellspring of virtue would seem to need some justification for a life of pensioned leisure. How do retirees and observers alike come to feel comfortable with a "retired" life? In this [essay] I will suggest that retirement is morally managed and legitimated on a day-to-day basis in part by an ethic that esteems leisure that is earnest, occupied, and filled with activity—a "busy ethic." The ideas in this [essay] developed out of research on the retirement process at the Normative Aging Study, a prospective study of aging in community-dwelling men (Bosse et al., 1984).

THE WORK ETHIC IN USE

Before discussing how the busy ethic functions, it is important to note a few aspects about its parent work ethic. The work ethic, like any ethic, is a set of beliefs and values that identifies what is good and affirms ideals of conduct. It provides criteria for the evaluation of behavior and action. The work ethic historically has identified work with virtue and has held up for esteem a conflation of such traits and habits as diligence, initiative, temperance, industriousness, competitiveness, self-reliance, and the capacity for deferred gratification. The work ethic, however, has never had a single consistent expression nor has it enjoyed universal assent within Western cultures.

Another important point is that the work ethic historically has torn away from its context, become more abstract and therefore more widely useful (Rodgers, 1978). When the work ethic was Calvinist and held out hope of heavenly rewards, believers toiled for the glory of God. When 19th century moralists shifted the promise toward earthly rewards, the work ethic motivated the middle class to toil because it was useful to oneself and the common weal. The coming of the modern factory system, however, with its painful labor conditions and de-emphasis on the self-sufficient worker, created a moral uncertainty about the essential nobility and instrumentality of work that made individuals want to take refuge in the old phrases and homilies all the more. As work ideals became increasingly abstract, they grew more available. Rodgers (1978) pointed out that workingmen now could invoke the work ethic as a weapon in the battle for status and self-respect, and so defend the dignity of labor and wrap themselves in a rhetoric of pride. Politicians of all persuasions could appeal to the work ethic and cast policy issues as morality plays about industry and laziness. Thus, despite the failed spiritual and instrumental validity of the work ethic, it persisted in powerful abstraction. And it is an abstract work ethic that persists today lacking, as do many other of our moral precepts, those contexts from which their original significance derived (MacIntyre, 1981). While there is constant concern about the health of the work ethic (Lewis, 1982; Yankelovich & Immerwhar, 1984), belief in the goodness of work continues as a piece of civic rhetoric that is important out of all proportion to its behavioral manifestations or utilitarian rewards.

Among persons approaching retirement, surveys show no fall-off in work commitment and subscription to values about work (Hanlon, 1983). Thus, assuming that a positive value orientation toward work is carried up to the threshold of retirement, the question becomes: What do people do with a work ethic when they no longer work?

CONTINUITY OF BELIEFS AND VALUES

The emergence of a busy ethic is no coincidence. It is, rather, a logical part of people's attempts to manage a smooth transition from work to retirement. Theorists of the life course have identified several conditions that ease an individual's transitions from one status to another. For example, transitions are easier to the extent that the new position has a well-defined role, or provides opportunities for attaining valued social goals, or when it entails a formal program of socialization (Burr, 1973; Rosow, 1974). Transitions are also easier when beliefs are continuous between two positions, that is, when action in the new position is built upon or integrated with the existing values of the person. Moral continuity is a benefit for the individual who is in transition, and for the wider social community as well.

In the abstract, retirement ought to entail the unlearning of values and attitudes—in particular, the work ethic—so that these should be no obstacle to adaptation. Upon withdrawal from work, emotional investment in, and commitment to, the work ethic should by rights be extinguished in favor of accepting leisure as a morally desirable lifestyle. Along these lines, there is a common recommendation that older workers, beginning in their 50s, should be "educated for leisure" in preparation for retirement. For example, the 1971 White House Conference on Aging recommended that "Society should adopt a policy of preparation for retirement, leisure, and education for life off the job . . . to prepare persons to understand and benefit from the changes produced by retirement" (p. 53).

But the work ethic is not unlearned in some resocialization process. Rather, it is transformed. There are two devices of this transformation that allow a moral continuity between work and retired life. One—the busy ethic—defends the daily conduct of retired life. The other—an ideology of pensions—legitimates retirees' claim to income without the obligation to work. As to the latter, a special restitutive rhetoric has evolved that characterizes pensions as entitlements for former productivity. Unlike others, such as welfare recipients, who stand outside the productive process, whose idleness incurs moral censure, and who are very grudgingly tendered financial support (Beck, 1967), the inoccupation of retirees is considered to have been *earned* by virtue of having *formerly* been productive. This veteran-ship status (Nelson, 1982) justifies the receipt of income without work, preserves the self-respect of retirees, and keeps retirement consistent with the dominant societal prestige system, which rewards members primarily to the extent that they are economically productive.

THE BUSY ETHIC: FUNCTIONS AND PARTICIPANTS

Along with an ideology that defends the receipt of income without the obligation to work, there is an ethic that defends life without work. This "busy ethic" is at once a statement of value as well as an expectation of retired people—shared by retirees and nonretirees alike—that their lives should be active and earnest. (Retirees' actual levels of activity are, as shall be explained, another matter altogether; the emphasis here is on shared values about the conduct of life.) The busy ethic is named after the common question put to people of retireable age, "What will you do (or are you doing) to keep yourself busy?" and their equally common reports that "I have a lot to keep me busy" and "I'm as busy as ever." Expressions of the busy ethic also have their pejorative opposites, for example, "I'd rot if I just sat around." In naming the busy ethic, the connotation of busyness is more one of involvement and engagement than of mere bustle and hubbub.

The busy ethic serves several purposes: It legitimates the leisure of retirement, it defends retired people against judgments of obsolescence, it gives definition to [the] retirement role, and it "domesticates" retirement by adapting retired life to prevailing societal norms. Before discussing these functions of the busy ethic, it is important to emphasize that any normative feature of social life entails endorsement and management by multiple parties. There are three parties to the busy ethic.

First, of course, are the subjects of the busy ethic—older workers and retirees—who are parties to it by virtue of their status. They participate in the busy ethic to the degree that they subscribe to the desirability of an active, engaged lifestyle. When called upon to account for their lives as retirees, subjects of the busy ethic should profess to be "doing things" in retirement or, if still working, be planning to "do things." Retirees can testify to their level of involvement in blanket terms, asserting: I've got plenty to do, I'm busier than when I was working. Or they can maintain in reserve a descriptive, mental list of activities (perhaps exaggerated or even fictitious) that can be offered to illustrate a sufficient level of engagement. These engagements run heavily to maintenance activities (e.g., tasks around the house, shopping) and involvement with children and grandchildren. Obviously, part-time jobs, volunteering, or major life projects ("I've always wanted to learn how to play the piano") can be offered as evidence of an active lifestyle. Less serious leisure pursuits (hobbies, pastimes, socializing) can also contribute to a picture of the busy life as long as such pursuits are characterized as involving and time consuming. In honoring the busy ethic, exactly what one does to keep busy is secondary to the fact that one purportedly *is* busy.

A second group of parties to the busy ethic comprises the other participants—friends, relatives, coworkers—who talk to older workers and retirees about the conduct of retired life. Their role is primarily one of keeping conversation about retirement continually focused on the topic of activity, without necessarily upholding ideals of busyness. Conversation with retirees also serves to assure these others that there is life after work. Indeed, apart from money matters, conversation about retired life per se is chiefly conversation about what one does with it, how time is filled. Inquiries about the retiree's lifestyle ("So what are you doing with yourself?") may come from sincere interest or may only be polite conversation. Inquiries, too, can be mean-spirited, condescending, or envious. Whatever the source or course of discussion, it nonetheless frequently comes to assurances that, yes, it is good to keep busy.

The third group can be called institutional conservators of the busy ethic, and their role is more clearly normative. These parties hold up implicit and explicit models of what retired life should be like, models that evince an importance placed on being active and engaged. Prominent institutional conservators of the busy ethic are the marketers of products and services to seniors, the gerontology profession, and the popular media. More shall be said about these later.

Returning to the purposes that the busy ethic serves, its primary function is to legitimate the leisure of retirement. Leisure without the eventual obligation of working is an anomalous feature of adulthood. Excepting the idle rich and those incapable of holding a job, few adults escape the obligation to work. Retirement and pension policies, however, are devised to exclude older adults from the labor force. In addition, age bias operates to foreclose opportunities for their further employment. How can our value system defend this situation—retirement—when it is elsewhere engaged in conferring honor on people who work and work hard? The answer lies in an ethic that endorses leisure that is analogous to work. As noted above, leisure pursuits can range from the serious to the self-indulgent. What legitimates these as an authentic adult lifestyle is their correspondence with the *form* of working life, which is to be occupied by activities that are regarded as serious and engaging. The busy ethic rescues retirement from the stigma of retreat and aimlessness and defines it as a succession to new or renewed foci of engagement. It reconciles for retirees and their social others the adult obligation to work with a life of leisure. This is the nature of continuity in self-respect between the job and retirement (Atchley, 1971).

In an essay that anticipates some of the present argument, Miller (1965) took a stricter view about what justifies retirement leisure. Mere activity is not meaningful enough; it must have the added rationale of being infused with aspects of work that are culturally esteemed. Activity legitimates retirement if it is, for example, economically instrumental (profitable hobbies), or contributes to the general good (community service), or is potentially productive (education or skill development). Whether people in fact recognize a hierarchy of desirable, work-correlative activities at which retirees can be busy remains to be determined. What Miller's essay and the present argument have in common, nonetheless, is the view that what validates retirement, in part, is activity that is analogous to work.

The busy ethic serves a second purpose for its subjects, which is to symbolically defend retirees against aging. Based on the belief that vigor preserves well-being, subscription to the norm of busyness can recast retirement as "middle-age like." Adherence to the busy ethic can be a defense—even to oneself—against possible judgements of obsolescence or senescence. To accentuate the contrast between the vital and senescent elder, there is an entire vocabulary of pejorative references to rocking chairs and sitting and idleness. As an illustration, a recent piece in my local newspaper about a job placement service for seniors quoted one of the program's participants, who said: "I am not working for income. I am working for therapy, to keep busy. There is nothing that will hurt an elderly person as much as just sitting alone all day long, doing nothing, thinking about nothing." It is appropriate to note here that, in scope, the busy ethic does not apply to all retirees. The busy life is more likely to be an expectation on the conduct of the "young-old" retiree, or at least the retiree who has not been made frail by chronic illness.

A third purpose of the busy ethic is that it places a boundary on the retirement role and thus permits some true leisure. Just as working adults cycle between time at work and time off, retirees too can have "time off." Because the busy ethic justifies some of one's time, the balance of one's time needs no justification. For example, if the morning was spent running errands or caring for grandchildren, one can feel comfortable with napping or a stretch of TV viewing in the afternoon. The existence of fulfillable expectations allows one to balance being active with taking it easy—one can slip out of the retirement role, one is allowed time offstage. Being busy, like working, "pays" for one's rest and relaxation.

The busy ethic serves a fourth function, and this for the wider society by "domesticating" retirement to mainstream societal values. It could be otherwise. Why not an ethic of hedonism, nonconformity, and carefree self-indulgence as a logical response to societal policies that define older workers as obsolescent and expendable? Free of adult workaday constraints, retirees could become true dropouts thumbing their noses at convention. Or why not an ethic of repose, with retirees resolutely unembarrassed about slowing down to enjoy leisure in very individual ways? Retirees do often describe retirement as a time for sheer gratification. In response to open-ended questions on Normative Aging Study surveys about the primary advantages of retirement, men overwhelmingly emphasize: freedom to do as I wish, no more schedules, now I can do what I want, just relax, enjoy life. Such sentiments, however, do not tend to serve drop-out or contemplative models of retired life because retirees will go on to indicate that their leisure is nonetheless responsibly busy. The busy ethic tames the potentially unfettered pleasures of retirement to prevailing values about engagement that apply to adulthood. For nonretirees, this renders retirement as something intelligible and consistent with other stages of life. Additionally, the busy ethic, in holding that retirees can and should be participating in the world, probably salves some concern about their having been unfairly put on the shelf.

The active domestication of retirement is the province of the institutional conservators of the busy ethic. The popular media are strenuous conservators. An article in my local newspaper last year bore the headline, "They've retired but still keep busy," which was reprised only a few months later in another headline, "He keeps busy in his retirement." Both articles assured the reader that these seniors were happily compensating for their withdrawal from work. It is common for "senior set" features to depict older people in an upbeat fashion, though in all fairness the genre of newspapers' lifestyle sections generally portrays everybody as occupied by varied and wonderful activities regardless of age. The popular media are also staunch promoters of aged exemplars of activity and achievement—Grandma Moses, Pablo Casals, George Burns, and so on through such lists (Wallechinsky et al., 1977). A current National Public Radio series on aging and creativity bears the perceptive title: "I'm Too Busy to Talk Now: Conversations with American Artists over Seventy."

Marketers, with the golf club as their chief prop, have been instrumental in fostering the busy image. A recent analysis of advertising in magazines designed specifically for older people found that the highest percentage of ads in these magazines concerned travel and more often than not portrayed older people in an active setting such as golfing, bicycling, or swimming (Kvasnicka et al., 1982). Calhoun (1978) credited the ads and brochures of the retirement home industry, in particular, with promoting an energetic image of older Americans. This industry built houses and, more importantly, built a market for those houses, which consisted of the dynamic retiree. While few retirees ever live in retirement communities, the model of such communities has been most influential in the creation of an active, if shallowly commercial, image of the elderly. One writer (Fitzgerald, 1983), visiting Sun City Center in Florida ("The town too busy to retire"), reflected:

> Possibly some people still imagine retirement communities as boarding houses with rocking chairs, but, thanks to Del Webb and a few other pioneer developers, the notion of "active" retirement has become entirely familiar; indeed, since the sixties it has been the guiding principle of retirement-home builders across the country. Almost all developers now advertise recreational facilities and print glossy brochures with photos of gray-haired people playing golf, tennis, and shuffleboard. (p. 74)

The visitor noted that residents talked a great deal about their schedules and activities. The visitor also noted how their emphasis on activities was an attempt to legitimate retirement and knit it to long-standing beliefs and values:

> Sun Citians' insistence on busyness—and the slightly defensive tone of their town boosterism—came, I began to imagine, from the fact that their philosophies, and, presumably, the [conservative, work ethic] beliefs they had grown up with, did not really support them in this enterprise of retirement. (p. 91)

The gerontological community has been an important conservator of aspects of the busy ethic. Cumming and Henry (1961) early on pointed out the nonscientific presuppositions of mainstream gerontology's "implicit theory" of aging, which include the projection of middle-aged standards of instrumentality, activity, and usefulness into later life. This implicit, so-called "activity theory" of aging entailed the unabashed value judgment that "the older person who ages optimally is the person who stays active and manages to resist the shrinkage of his social world" (Havighurst et al., 1968, p. 161). Gubrium (1973) has noted the Calvinistic aura of this perspective: "Successful aging, as the activity theorists portray it, is a life style that is visibly 'busy'" (p. 7). Continuing this orientation over the last decade, gerontology's campaign against ageism has, according to Cole (1983), promoted an alternative image of older people as healthy, sexually active, engaged, productive, and self-reliant.

Institutional conservators of the busy ethic are by no means monolithic in their efforts to uphold ideals of busyness. Rather, in pursuing their diverse objectives they find it useful to highlight particular images of retirement and later life that coalesce around the desirability of engagement.

SOURCES OF AUTHORITY

The busy ethic is useful, therefore, because it legitimates leisure, it wards off disturbing thoughts about aging, it permits retirees some rest and relaxation, and it adapts retirement to prevailing societal norms. These benefits to the participants of the busy ethic are functional only in an analytic sense. No one in daily life approves of busy retirements because such approval is "functional." It is useful at this point to ask why people ultimately assent to the notion that it is good to be busy.

The busy ethic has moral force because it participates in two great strong value complexes—ethics themselves—that axiomatize it. One, of course, is the work ethic, which holds that it is enobling to be exerting oneself in the world. The other basis for the busy ethic's authority is the profound importance placed on good health and the stimulating, wholesome manner of living that is believed to ensure its maintenance. The maintenance of health is an ideal with a deep tradition that has long carried moral as well as medical significance. Haley (1978), for example, has pointed out how Victorian thinkers promoted the tonic qualities of a robust and energetic lifestyle. The preservation of health was seen to be a duty because the well-knit body reflected a well-formed mind, and the harmony of mind and body signified spiritual health and the reach for higher human excellence. Ill, unkempt, and indolent conditions, by contrast, indicated probable moral failure. Times change, but current fashions in health maintenance still imply that a fit and strenuous life will have medical benefits and testify as well to the quality of one's will and character. Thus, admonitions to older people that they "keep busy" and "keep going" are authoritative because they advocate an accepted therapy for body and soul.

CORRESPONDENCE WITH BEHAVIOR

One crucial issue is the correspondence between the busy ethic and actual behavior. It is important to mention that not all self-reports about busy retirements are conscious presentations of conformity to a busy ethic. There are retirees who by any reckoning are very active. But in the more general case, if people believe it is important to keep busy, should they not therefore *be* busy by some standard or another?

This [essay's] argument in favor of the busy ethic has implied that belief is not necessarily behavior. On one hand, the busy ethic may—as any ethic should—motivate retirees to use their time in constructive or involving pursuits. It may get them out of the unhealthful rocking chair or away from the can-of-beer-in-front-of-the-TV. On the other hand, the busy ethic can motivate people to *interpret* their style of life as conforming to ideals about activity. An individual can take a disparate, even limited, set of activities and spin them together into a representation of a very busy life. It would be difficult to contradict such a manner of thinking on empirical grounds; "engagement" is a subjective quality of time use that simple counts of activities or classifications of their relative seriousness or instrumentality are not likely to measure. Indeed, gerontologists should be wary about the extent to which the busy ethic may shape people's responses on surveys about their leisure, frequency of activities, and experience in retirement.

In posing the question, "How busy do retirees have to be under such a set of values?" the answer is they don't objectively have to be very busy at all. Just as with the work ethic, which has been an abstract set of ideals for some time (Rodgers, 1978), it is not the actual pace of activity but the preoccupation with activity and the affirmation of its desirability that matters. After all, all of us are not always honest, but we would all agree that honesty is the best policy. The busy ethic, like the work ethic and other commonplace values, should be evaluated less for its implied link with actual behavior than for its ability to badger or comfort the conscience. The busy ethic, at bottom, is self-validating: Because it is important to be busy, people will say they are busy.

CONCLUSION

The busy ethic is an idea that people have about the appropriate quality of a retired lifestyle. It solves the problem of moral continuity: how to integrate existing beliefs and values about work into a new status that constitutes a withdrawal from work. The postulation of a busy ethic is an attempt to examine sociologically people's judgments of value and obligation regarding the conduct of daily life—their expectations of each other and of themselves.

To be sure, there are other superseding expectations on the conduct of retirees. Writing about the duties of a possible retirement role, Atchley (1976) has noted that a stability of behavior is expected, as well as self-reliance and independence in managing one's affairs. Such normative preferences are fairly vague and open-ended. Rosow (1974) surveyed the prospects for socialization to later life, in which the retirement role is nested, and found that behavioral prescriptions for older people are open and flexible, and norms are limited, weak, and ambiguous. Even admonitions to be active carry virtually no guidance about the preferred content of such activity. Perhaps this is just as well. Streib and Schneider (1971), summarizing findings from the Cornell Study of Occupational Retirement, pointed out that the vagueness of retirees' role expectations may protect retirees from demands that they might be disinclined to fulfill or from standards that diminished health and financial resources might not allow them to meet.

The busy ethic, too, comprises vague expectations on behavior. It is a modest sort of prescription—less a spur to conformity and more a way to comfortably knit a new circumstance to long-held values. Social disapproval is its only sanction. Not all retirees assent to this image of retirement, nor do they need to. Judging by the ubiquity of the idea, however, subscribers to the busy ethic are probably in the majority; one cannot talk to retirees for very long without hearing the rhetoric of busyness. The busy ethic also legitimates the daily conduct of retired life in a lower key than has been claimed by some gerontologists, who propose that work substitutes and instrumental activity are essential to indemnify retirement. While some retirees do need to work at retirement to psychologically recoup the social utility that working supplied (Hooker & Ventis, 1984), for most it is enough to participate in a rather abstract esteem for an active lifestyle and to represent their own retirement as busy in some way.

To conclude, the busy ethic, as an idealization and expectation of retired life, illustrates how retirement is socially managed, not just politically and economically but also morally—by means of everyday talk and conversation as well as by more formal institutions. Drawing its authority from the work ethic and from a traditional faith in the therapeutic value of activity, the busy ethic counsels a habit of engagement that is continuous with general cultural prescriptions for adulthood. It legitimates the leisure of retirement, it defends retired people against judgments of senescence, and it gives definition to the retirement role. In all, the busy ethic helps individuals adapt to retirement, and it in turn adapts retirement to prevailing societal norms.

REFERENCES

Atchley, R. C. (1971). Retirement and leisure participation: Continuity or crisis? *The Gerontologist, 11,* 13–17.

Atchley, R. C. (1976). *The sociology of retirement.* New York: Halsted.

Beck, B. (1967). Welfare as a moral category. *Social Problems, 14,* 258–277.

Bosse, R., Ekerdt, D. J., & Silbert, J. E. (1984). The Veterans Administration Normative Aging Study. In S. A. Mednick, M. Harway, & K. M. Finello (Eds.), *Handbook of longitudinal research. Vol. 2, Teenage and adult cohorts.* New York: Praeger.

Burr, W. R. (1973). *Theory construction and the sociology of the family.* New York: John Wiley.

Calhoun, R. B. (1978). *In search of the new old: Redefining old age in America, 1945–1970.* New York: Elsevier.

Cole, T. R. (1983). The "enlightened" view of aging: Victorian morality in a new key. *Hastings Center Report, 13,* 34–40.

Cumming, E., & Henry, W. H. (1961). *Growing old: The process of disengagement.* New York: Basic Books.

Fitzgerald, F. (1983, April 25). Interlude (Sun City Center). *New Yorker,* pp. 54–109.

Gubrium, J. F. (1973). *The myth of the golden years: A socioenvironmental theory of aging.* Springfield, IL: Charles C Thomas.

Haley, B. (1978). *The healthy body and Victorian culture.* Cambridge, MA: Harvard University Press.

Hanlon, M. D. (1983). Age and the commitment to work. Flushing, NY: Queens College, City University of New York, Department of Urban Studies (ERIC Document Reproduction Service No. ED 243 003).

Havighurst, R. J., Neugarten, B. L., & Tobin, S. S. (1968). Disengagement and patterns of aging. In B. L. Neugarten (Ed.), *Middle age and aging: A reader in social psychology.* Chicago: University of Chicago Press.

Hooker, K., & Ventis, D. G. (1984). Work ethic, daily activities, and retirement satisfaction. *Journal of Gerontology, 39,* 478–484.

Keniston, K. (1974). Youth and its ideology. In S. Arieti (Ed.), *American handbook of psychiatry. Vol. 1, The foundations of psychiatry.* 2nd ed. New York: Basic Books.

Kvasnicka, B., Beymer, B., & Perloff, R. M. (1982). Portrayals of the elderly in magazine advertisements. *Journalism Quarterly, 59,* 656–658.

Lewis, L. S. (1982). Working at leisure. *Society, 19* (July/August), 27–32.

MacIntyre, A. (1981). *After virtue: A study in moral theory.* Notre Dame, IN: University of Notre Dame Press.

Miller, S. J. (1965). The social dilemma of the aging leisure participant. In A. M. Rose & W. Peterson (Eds.), *Older people and their social worlds.* Philadelphia: F. A. Davis.

Nelson, D. W. (1982). Alternate images of old age as the bases for policy. In B. L. Neugarten (Ed.), *Age or need? Public policies for older people.* Beverly Hills, CA: Sage.

Rodgers, D. T. (1978). *The work ethic in industrial America: 1850–1920.* Chicago: University of Chicago Press.

Rosow, I. (1974). *Socialization to old age.* Berkeley: University of California Press.

Streib, G. F., & Schneider, C. J. (1971). *Retirement in American society: Impact and process.* Ithaca, NY: Cornell University Press.

Wallechinsky, D., Wallace, I., & Wallace, A. (1977). *The People's Almanac presents the book of lists.* New York: William Morrow.

White House Conference on Aging. (1971). *Toward a national policy on aging: Proceedings of the 1971 White House Conference on Aging, Vol. II.* Washington, DC: U.S. Government Printing Office.

Yankelovich, D., & Immerwahr, J. (1984). Putting the work ethic to work. *Society, 21* (January/February), 58–76.

READING 45

Moving Toward a Creative Retirement

Ronald J. Manheimer

Gale Arneson's voice communicated an urgency that convinced me I needed to make time for the out-of-town visitors that very day. "We're trying to figure out how to have a creative retirement," she announced over the phone, "and you're supposed to be the place to find out."

A handsome couple in their mid-50s, Gale and her husband, Cliff Edwards, appeared at my office later that afternoon. They had recently retired from teaching with full pensions after 30 years of dedicated service. Skeptical about the quality of life they saw among peers a few years

SOURCE: "Moving Toward a Creative Retirement" by Ronald J. Manheimer is reprinted with permission from *Where to Retire,* Fall 2000. Tel: 713-974-6903, www.WhereToRetire.com

into retirement, they had decided to blaze their own trail. . . .

I have learned all sorts of ways people are questioning and reinventing retirement. While most are not quite as ardent or explicit about the challenge as Gale and Cliff, the quest is the same. Factors such as increased longevity, income security, new work patterns and radically changed views about aging have turned retirement into a many-splendored thing, too much in flux to take for granted.

Predictable or not, Gale and Cliff were waiting for some answers. Adding complexity to their search, the Canadian couple from Edmonton, Alberta, were considering the benefits of relocating, possibly to the Sunbelt of the United States. Milder weather, however, was not their only objective. "You could be just as dissatisfied and restless on a warm, sunny day as a cloudy, chilly one," said Cliff.

Our ensuing conversation helped me crystallize my observations and synthesize basic dimensions of a creative retirement. First, a definition: Creativity is a way of looking at the world to see possibilities for making something new. Painters, writers, composers, scientists and business entrepreneurs do it all the time. Quite often this involves revisiting the theories, techniques and inventions of one's predecessors (500 years ago or yesterday) and asking, "Well, but couldn't things be otherwise?" Alexander Graham Bell, for example, admired Guglielmo Marconi's invention but thought the telephone would be an even better idea than the telegraph. Most innovators rework past accomplishments to find the new in the old.

Creative retirement is much the same. It's a way of finding new uses for our past experiences, accumulated knowledge and expertise, while taking advantage of leisure time we may not have experienced since our teens. Like a second adolescence, we may find ourselves having to rethink who we are outside the world of paid work. Unlike adolescents, we bring a tremendous amount of information and hard-won perspective to the task.

Not everyone feels a call to reinvent himself. But many, like Gale and Cliff, do. They are pathfinders because few established avenues or positive role models exist. Gale and Cliff mentioned reading former President Jimmy Carter's book, "The Virtues of Aging," learning from and admiring the Carters' enthusiastic dedication to serving others as a source of renewal and restored purpose. The Carters didn't start from scratch. They took what they knew and had done and found new outlets. Unhampered by the constraints of the presidency and public opinion polls, they found greater freedom to be themselves and do what meant the most to them.

Are the Carters able to carry out their creative retirement because they are especially privileged? To a degree, yes. But millions of Americans, neither rich nor famous, are blessed with more modest, yet similar, advantages. We can make choices, and our lives are not defined by "musts" but by "coulds." Seeing the possibilities, asking "what if," daring to re-evaluate your life while appreciating what you have been and done—these are the keys to making retirement a creative venture.

Creative retirement reminds us that our life's work—our useful place in the universe—is far from over. Creative retirement could [mean] traveling around the world, going back to school or becoming a social activist. It's not necessarily what you do but how you experience, feel and think about the ways you are engaged with the world. Filling time, avoiding boredom or keeping busy are not pathways toward creative retirement. If we do not find these activities truly satisfying, we are just avoiding facing the challenge of finding new meaning in our lives. . . .

NEW WORK

Millions of people retire, take a few months or a year off, and return to the workforce, either full or part time. Some do this because they need to stretch their dollars to cover a longer life expectancy. Others love the work they do, how it confirms their sense of personal value and identity, the social contact or the way work helps to structure their days.

A creative approach to continued paid work should involve following your sense of value. This means following the dictates of your heart in pursuing what seems right for you, and asking, "Will what I want to do make a positive difference in the world?" This is not a completely altruistic question.

Feeling that we are adding qualitatively to others' lives enriches our own. Who is to decide what is worthwhile? You have to discover the essence of what gives you satisfaction, what truly belongs to you and uniquely fits your personality. That gives you a special connection to others, whether that means neighbors, your community, co-workers, animals, nature or the Earth. . . .

Current retirement wisdom says that we should keep 75 percent to 85 percent of our preretirement income flowing to retain the lifestyle to which we are accustomed. But some folks are finding that less is better. My friends Cleve and Marion Mathews, for example, downsized to a small, tasteful home, gave many of their family heirlooms away to their adult children and sold other items they no longer needed at a yard sale.

They have found great enjoyment in inexpensive activities such as hiking, attending free college lectures and ushering at plays and concerts. "We're getting off the consumer treadmill," said Cleve. "Greater simplicity is a blessing," added Marion.

SPIRITUAL REVIVAL

As we get older, we become more aware of our limited time on the planet. We may gain a deeper appreciation for the mysteries of life and find our capacity for reverence growing. Earlier habits of religious practice may continue to satisfy some, but others find they have drifted away from the family religion or have become inactive since the children no longer connect them with religious schooling, events and celebrations. Still, their desire for a rich inner life remains.

FOCUS ON THE FUTURE

The U.S. Wisdom Corps?

Dateline: 2021. Washington, DC. United Press International.

Reporters gathered today on the White House lawn for a ceremony to hear the president proclaim the 10th anniversary of the U.S. Wisdom Corps. Former President Bill Clinton and former President George W. Bush, both now age 75, were on hand to receive an award for their part in establishing the U.S. Wisdom Corps, a group designed to enlist the talents of retired people for guiding the country during the 21st century.

In his remarks, former President Clinton noted that exactly 60 years ago, President John Kennedy had established the U.S. Peace Corps, which attracted idealistic young people. Clinton's own efforts to create the Wisdom Corps, he said, were inspired by the idea of doing something similar to what Kennedy had done, this time for seniors.

The audience was moved by Clinton's reminiscence about his own odyssey since leaving the White House:

"I was only 54 when my term as president was finished, and I have to admit that when I left the White House, it wasn't quite what I expected. I kept busy, but I was still trying to find a role for myself. I couldn't just limit myself to the golf course, like some former presidents had done. So I began to look for an organized way to make use of the talents of all former chief executives. I joined forces with former President Bush, and that's when we both started lobbying Congress to create the U.S. Wisdom Corps.

"At first we ran into a lot of criticism—you know, this was just an elephant's graveyard, a boondoggle for bored CEOs—that kind of thing. But eventually people started listening, especially when large numbers of us baby boomers began to retire. Gradually, we expanded the idea to include retired judges, ministers, all kinds of leaders. There's enormous talent out there, but we weren't attracting that talent through conventional volunteer roles. We needed to come up with roles to make use of the wisdom of a lifetime.

"The U.S. Constitution does not provide a role for former presidents, but in our history, we've seen some very interesting examples. One of my favorites is John Quincy Adams, who went on to become a congressman after leaving the presidency. I didn't feel that option would work for me, and I couldn't exactly run for my local school board—although at times I considered it! As it's turned out, the Wisdom Corps has been the most exciting thing I've done since leaving the White House."

In his speech, Clinton detailed examples of how the Wisdom Corps has used the accumulated life experience of retired leaders:

- *Conflict mediation.* By the early years of the 21st century, Americans were getting fed up with lawsuits and the breakdown of the legal system. They found an alternative approach for conflict resolution in a program originating in Boulder, Colorado, where senior citizens were being trained to serve as mediators in tenant–landlord disputes, business conflicts, and other areas where good judgment can overcome antagonism.

- *Proxy decision making.* More and more people are surviving into old age without anyone to make health care decisions on their behalf if they lose mental capacity. A program that began in New Mexico trains retired social workers and teachers to become surrogate decision makers for those with diminished mental capacity who have no family members to speak for them. Difficult end-of-life decisions require all the wisdom available.

- *Oral history.* Only a handful of people in their 90s have firsthand memories of World War II. The Wisdom Corps, in its "Alternatives to War," has been making efforts to ensure that their oral history testimonies are preserved on videotape for future generations, as an inspiration to help us find alternatives to violence in human affairs.

- *Community leadership roles.* The Wisdom Corps has a roster that furnishes board members for libraries, schools, social service agencies, and nonprofit groups of all kinds. In a particularly successful part of its program, the Wisdom Corps sends long-experienced nonprofit executives to agencies needing a temporary CEO while an executive search is taking place. These agencies are looking for a strong person who has no interest in permanent power. Retired leaders have proved to be just what was needed.

(Continued)

(Continued)

Is the U.S. Wisdom Corps just a fantasy? Not necessarily. A forerunner already exists. The Service Corps of Retired Executives (SCORE), a program by the U.S. Small Business Administration, enlists retired business executives for volunteer roles as counselors and advisers helping small business owners. SCORE now comprises 13,000 older volunteers in 750 offices around the country. SCORE volunteers rely on life experience to give advice on all aspects of business: writing a business plan, devising a marketing strategy, avoiding pitfalls in expansion, and so on. Through free counseling sessions and low-cost workshops, SCORE volunteers are now serving as mentors to more than a quarter of a million entrepreneurs each year. Building on this pattern of success, SCORE volunteers have expanded to serve new community needs: developing rural communities, assisting businesses filing papers for Chapter 13 bankruptcy, and counseling business owners in the aftermath of natural disasters.

Questions for Writing, Reflection, and Debate

1. The French poet Baudelaire once said, "Work is less boring than pleasure." Was Baudelaire right? Write a short discussion of Baudelaire's statement as you think it applies to the question of work versus leisure among older people. If possible, give examples of people you've known during their retirement years.

2. Imagine you've become the marketing director for a new $100 million residential complex called "Retirement City, USA," located in Florida. Your job is to produce a brochure describing the benefits of retirement in this unique residential environment. The idea is to attract a new group of retirees who are looking for a distinctive lifestyle. Prepare a draft of the brochure highlighting the points that would be most attractive to potential residents of Retirement City, USA.

3. The words we use to describe the same action can make a big difference in how we see it. For instance, what are the real differences among *leisure, free time,* and *recreation?* Are the distinctions just a matter of semantics?

4. Imagine a single, middle-aged individual who stays at home all day long. What is the difference between describing that person as *on a sabbatical,* *on vacation, unemployed,* or *taking early retirement?* Does the individual's age make a difference in what term might be most appropriate?

5. Assume that you've just read a long editorial in a local newspaper that calls on senior citizens to avoid the "rocking chair" approach to retirement living. You find the editorial profoundly disturbing. Write a long letter to the editor of the paper identifying the reasons that you think retirement makes sense today and in the future.

6. Based on your observation of family members or people you've seen in the workplace, what are the biggest problems that middle-aged or older workers would face if they were forced to change jobs and go into a completely new field? What steps could be taken to make such changes easier for people?

7. Some people believe that if there are not enough jobs to go around for everyone, it makes sense to encourage older people to retire and "get out of the way" to make room for the young. What are the arguments for and against this approach to older people in the labor force? What are the costs and benefits of encouraging or discouraging early retirement?

Suggested Readings

Bolles, Richard, and Nelson, John E., *What Color Is Your Parachute? for Retirement: Planning Now for the Life You Want,* Berkeley, CA: Ten Speed Press, 2007.

De Grazia, Sebastian, *Of Time, Work and Leisure,* New York: Doubleday, 1964.

DeLong, David W., *Lost Knowledge: Confronting the Threat of an Aging Workforce,* Oxford: Oxford University Press, 2004.

Morrow-Howell, Nancy, Hinterlong, James, and Sherraden, Michael, *Productive Aging: Concepts and Challenges,* Baltimore: Johns Hopkins University Press, 2001.

Weiss, Robert S., and Ekerdt, David J., *The Experience of Retirement,* Ithaca, NY: ILR Press, 2005.

Student Study Site

Visit the Student Study Site at **www.sagepub.com/moody7e** for these additional learning tools:

- Flashcards
- Web quizzes
- Chapter outlines

- SAGE journal articles
- Web resources
- Video and audio resources

Controversy 11

AGING BOOMERS

Boom or Bust?

When you hear the term *baby boomer,* what do you think of? Who are the boomers, actually? It's important to give an answer to both questions. On the one hand, we need to consider the subjective associations we have with the word *boomer*. On the other hand, we need to consider verifiable facts. The term *boomer* easily evokes stereotypes. Stereotypes are conveyed by many of the names given to boomers over the years, labels such as *the Pepsi generation, the "me" generation,* and *the '60s generation.* These phrases convey consumerism, narcissism, rebellion, and openness to change. Even the original term *baby boomer* doesn't seem quite right because people in this generation aren't babies anymore. The oldest of the boomers are already receiving Social Security benefits, and many others are thinking seriously about retirement or even entirely new careers.

Some facts are clear. There were 77 million people born in the United States between the years 1946 and 1964, and this group of people is generally referred to as the generation of the boomers. We can see this group graphically displayed as a bulge in the population pyramid featured in Exhibit 50.

As the boomer generation moves through the life course, as the boomers grow older, this demographic fact will have big implications over the coming decades. But here we should pause to consider several interrelated questions. What does the term *generation* really mean? Do all individuals who fit into this demographic group form a single generation? Are there traits they share in common? Conversely, are there differences among members of the boomer generation? This is but one set of questions we need to consider.

We will also consider what it means for a whole generation to grow older: What are the factors that influence that process, and what personal, social, and economic consequences are likely to come as a result? Finally, we need to consider how the boomers have been represented and understood by the media, by business groups, by social scientists, and by boomers. For example, some writers have been harshly critical of boomers as a group, describing them as selfish. Still other commentators see them as idealistic, which seems the exact opposite trait. Depending on how we see and represent boomers as a group, we are likely to come to different conclusions about what it will mean for this large group of Americans to move into later life. So we need to come back, again and again, to our first question: Who are the boomers, actually?

Exhibit 50 Boomer Bulge

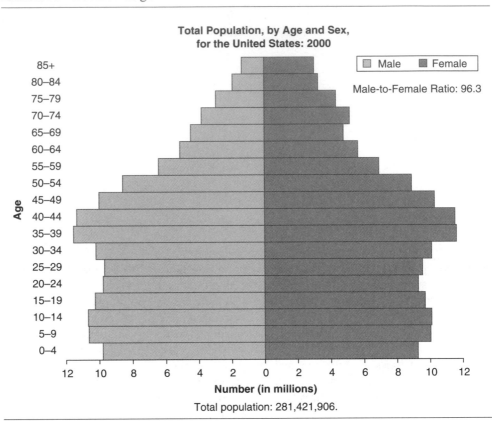

Total Population, by Age and Sex,
for the United States: 2000

SOURCE: Census 2000, 1% Public Use Micro-Sample Data.

WHAT IS A GENERATION? AGE-PERIOD-COHORT ANALYSIS

The term *generation* can mean several different things: It can mean a group of people born in a certain time period (e.g., the G.I. generation or children of the Great Depression). The term *generation* can also refer to people who are a certain age at a single point in time (e.g., the older generation or young people today). We use the technical term *cohort* to speak about people born in a certain time period (e.g., 1946–1964).

To draw objective conclusions about who the boomers are, we can make use of what is known as the age-period-cohort model in epidemiology and demography. Boomers as a group will be influenced by each of these different factors: age, period, and cohort.

Aging effects are those described earlier in this book, especially in Controversy 2: "Why Do Our Bodies Grow Old?" Chronological age effects are familiar to us all. They are effects brought about by the physiological process of aging, along with social responses by others to those effects. For example, people get gray hair, and reserve capacity tends to diminish. Declining vision or hearing may be age-associated changes, and there will be responses to those changes by other people. **Period effects** are those affecting all age groups in society at the same time: for example, the change in communications prompted by the Internet and e-mail beginning in the 1990s. In the year 1995, such a period effect probably had a different impact on, say, a 5-year-old, whose entire life experience would be to take the Internet for granted, than it did on, say, an 80-year-old, for whom learning about e-mail might be a novelty and a challenge. Although members of all age cohorts are exposed to the same period effects, the influence of these effects is modified by where in the life course one happens to be due to chronological age. Thus, period effects continue to shape our common life: For example, the political world changed in many ways after September 11, 2001.

Finally, there are **cohort effects** associated with events affecting groups of people during the same years—for example, the so-called G.I. generation whose youth was dramatically shaped by World War I. As this cohort grew older, the war had ongoing consequences (e.g., in the G.I. Bill that permitted returning veterans to go to college more easily). Another example of an enduring cohort effect is described by Glen Elder in his book *Children of the Great Depression* (1974), where he finds that early childhood experience of financial hardship shaped lifelong attitudes toward money and savings.

"Boomer Rock Band"? The Rolling Stones at a 2002 performance in Las Vegas.

All of the individuals in this picture are living during the same historical period, and thus they have shared experiences from period effects. Yet, they each have unique experiences because they represent different cohorts, as well as different ethnic backgrounds.

Are there special characteristics of boomers as a cohort? We can point to several. Distinctive characteristics of the boomer cohort are, first and foremost, sheer size. Because of higher fertility rates during the years in which they were born, the boomer cohort is simply much larger than the cohorts that came immediately before it (the silent generation and the G.I. generation) and the generation coming immediately after it (so-called Generation X). Boomers make up around a quarter of the entire U.S. population. These larger numbers, moving as a group from childhood through old age, have had consequences. First, more attention has been paid to members of the boomer generation as they crowded schools, became a target for marketers, reached eligibility to vote, and so on.

A second fact about boomers is that they have, on average, higher levels of educational attainment than earlier generations. Boomers were the first generation in U.S. history where more than half achieved some level of higher education. A third fact about boomers is that in their childhood, in the 1950s and 1960s, they experienced an extended period of postwar affluence and economic prosperity. The boomers' childhood experiences were different from the childhood experiences of their parents, many of whom were touched by the Great Depression. Boomers' experience in childhood was also different from that of Generation X, whose childhood was marked by the economic turbulence of the 1970s.

Finally, the 1960s and 1970s were a time of dramatic social upheaval, including Vietnam War protests and campaigns for civil rights, feminism, and environmental advocacy. As a result, some commentators, such as Leonard Steinhorn (2006), have praised the boomers as "the greater generation" because of this push for rights and tolerance, sometimes contrasted with the World War II generation that came to prominence during the more conservative 1950s. Yet not all the boomers were protesting against the war in Vietnam. On the contrary, many of them were fighting in that war or were opposed to the protesters. Those born between 1960 and 1964 were often too young to be influenced by that war at all. Although boomers have many shared experiences due to period effects, they are also a huge part of the total population, and there is enormous diversity within the cohort.

The heterogeneity of boomers makes it helpful to suggest further segmentation of this large group. For example, commentators often make a distinction between *leading-edge boomers* (those born between 1946 and 1954) and *late boomers* (those born between 1955 and 1964). This later group has been called Generation Jones. For leading-edge boomers, some memorable events are the assassinations of President John F. Kennedy and Martin Luther King, as well as the Vietnam War. By contrast, for late boomers, memorable events are Watergate and the oil embargoes of the 1970s. Leading-edge boomers are often characterized as optimistic and experimental, whereas Generation Jones is viewed as more distrustful.

We can apply the age-period-cohort model to thinking about boomers as they age. A good example might be attitudes toward real estate. Leading-edge boomers were children during a time of postwar prosperity, and, later in adulthood, they witnessed a long-term appreciation in the value of home ownership. But in 2008, there was a dramatic drop in home prices, brought on by subprime mortgage problems. How will boomers approaching retirement view the sale of a home?

Finally, along with age, period, and cohort effects, we need always to remember the powerful impact of social class, ethnicity, and gender. This is a topic we emphasized earlier in this book, and these ideas need to be kept in mind as we think about the different segments of the huge boomer generation. The principle of cumulative advantage and disadvantage can be used to explain variations in life course experiences among members of the baby boom cohort. Boomers will carry into old age distinctive differences due to class, ethnicity, and gender that sustain inequalities in later life. Thus, on average, both a poor Black woman and a rich White man may be boomers, but they are likely to have differences in occupation, income, health status, and so on at every point in life: youth, midlife, and old age.

In summary, it turns out to be complicated to answer the simple question, "Who are the boomers, actually?" The boomers are many different subgroups or segments, and the impact of aging on the boomers will be the result of the three factors discussed—chronological aging, cohort characteristics, and historical period effects, including the periods in years to come—as well as due to sources of diversity such as gender, ethnicity, and class.

SOCIAL CONSTRUCTION OF THE BOOMER PHENOMENON

In thinking about aging boomers, it can be helpful to apply the idea of "social construction" as that approach was developed originally by the sociologists Berger and Luckmann (1966). We can understand what is meant by *social construction* by considering some common objects of our experience. For example, the tallest mountain in the United States is Mount Whitney, and that mountain would exist regardless of whether people were living in North America. But a man-made object, say, a $20 bill or any kind of money, has the meaning it does *only* because a group of human beings has agreed on that meaning (i.e., the group members agree to accept a piece of paper as currency of a particular value). For example, a piece of Confederate money would not be accepted as currency (although it might still be valuable). Other examples of social construction show how important it is to be aware that certain facts about the world are not simply "natural." For example, the difference between *male* and *female* is a natural division in many biological species. But the difference between *husband* and *wife* is very much a social construction. Even a material object, such as Mount Rushmore, can have two aspects: The mountain would be here if human beings didn't exist, but it has a specific meaning (four American presidents carved out of the mountain), which is very much a social construction. Human alteration of a natural material object has altered its meaning to us.

Once we adopt the idea of social construction, we begin to see basic, taken-for-granted experiences in new ways. When we understand that things such as currency, traffic lights, and marriage roles are not just "facts" but are socially constructed, then we can extend that view to look at other social facts that are institutionalized in our experience of the world. Social construction does not mean that "everything is just made up." In fact, events like going to college and getting a bachelor's degree designate real events, but they maintain their meaning only through agreed-on understanding and common acceptance. Thus, if you possess a college degree, you may be treated differently in the job market, and you perform your role as a job seeker in a different way than you would otherwise. As a college graduate, you might be treated differently by an employer than someone who hasn't graduated from high school. In short, we can view all of social life as constructed in the sense that it is negotiated and subject to change and transformation as our beliefs change.

How does this idea of social construction apply to boomers? To speak about social construction does not mean that the concept of a baby boom is simply fiction or an imaginary idea. There really was a dramatic bulge in birthrates between 1946 and 1964. But observers look on that demographic fact and draw different conclusions. It's a bit like the story of the blind men and the elephant. One described the elephant's trunk, another the elephant's tusk, and still another the elephant's leg. The blind men couldn't agree because each man was examining and describing a different aspect of the elephant. So it is with different commentators who describe the **baby boom generation**. One difference concerns segmentation: For example, some are describing leading-edge boomers, others Generation Jones. Social construction becomes especially clear when value judgments are involved. For instance, if boomers are self-oriented, some commentators will look on that as striving for self-fulfillment and authenticity, whereas others will see it as purely selfish.

We can appreciate the importance of social construction about boomers when we look at the phenomenon in international terms. The baby boom is often assumed to be a U.S. phenomenon, yet a similar upsurge in birthrates was seen in Great Britain, Australia, Finland, and other countries. This variety of differences is accompanied by different political and cultural responses to the boomer generation. In that respect, the international diversity of boomers becomes a living laboratory in which we can view the influence of age, period, and cohort unfolding under different conditions. Each nation represents—socially constructs—the meaning of boomers in ways that reflect a particular history and culture.

URBAN LEGENDS OF AGING

"Boomers are the best educated, healthiest generation ever."

They are the best educated, true, but not necessarily the healthiest. Kenneth Manton documented declining rates of disability—but only for the past. Epidemiologists are now looking at rising rates of obesity and diabetes, and there is good reason to doubt that past declines will hold true in the future. In terms of arthritis and cardiovascular problems, obesity rates are an ominous sign of things to come.

It is a curious fact that, in the United States, boomers have been represented to the public in contradictory ways: on the one hand, as selfish and negative; on the other hand, as idealistic and positive. For example, David Brooks, in his book *Bobos in Paradise* (2000), portrayed boomers as aging hippies who have now "sold out" and become materialistic yuppies. A similarly harsh indictment of boomers appears in Joe Queenan's book *Balsamic Dreams* (2001), subtitled *A Short but Self-Important History of the Baby Boomer Generation.* This indictment isn't limited to books. On the Web, we can find the blog Boomer Deathwatch. A quick glance at that site confirms the general point: There are more boomers, they're getting older, and they'll cost us too much because they'll have too much political power. The celebrated idealism of the boomer generation, it seems, is just a mask for greed and narcissism. *Time* magazine offered readers an article titled "Twilight of the Boomers," denouncing, not surprisingly, greed and narcissism (Okrent et al., 2000). The denunciation is followed by typical imagery of physical and mental decline long associated with old age. Ageism, it seems, is one prejudice that just doesn't go away.

Many analysts have been unhappy with this stereotyped picture of aging boomers. Brent Green, for example, in his book *Marketing to Leading Edge Baby Boomers* (2003), warns that critics have pushed a distinctive ageism about boomers that links a self-absorbed, unpatriotic image from the 1960s with traditional negative images of aging. In response to the critics, Leonard Steinhorn wrote his book, *The Greater Generation* (2006), where he compares boomers to the G.I. generation, a cohort famously called the greatest generation. Steinhorn rejects the critics who see boomers as selfish. Instead, he offers an unapologetic defense of the boomers' legacy and contribution to society (e.g., the civil rights movement and rights for women). His political liberalism leads him to equate the boomers with positive social change, so he is optimistic about boomers approaching old age. Another optimist is Marc Freedman in his book *Prime Time* (2002), bearing the hopeful subtitle, *How Baby Boomers Will Revolutionize Retirement and Transform America.* Instead of social activism on the 1960s model, Freedman favors a role for encore careers, in which aging boomers will bring idealism into extended working lives.

One of the problems with the way aging boomers have been represented is that writers with opposing views seem to repeat stereotypes that are polar opposites (e.g., selfish vs. altruistic). On both sides, we hear loud voices defending one view or another, but little data to support opposing claims. One writer who doesn't easily line up on one side or another is Robert Putnam, author of the influential book *Bowling Alone* (2001). Like some of the negative voices, Putnam sees in the boomer generation a pattern of declining civic engagement; that is, people are spending less time on the Parent–Teacher Association, the Lions Club, and other local social groups. People may continue bowling, says Putnam, but they don't join regular bowling leagues. In that respect, Putnam is pessimistic. He contrasts boomers with the World War II generation and offers data to support his arguments. But unlike other pessimists, Putnam is not out to condemn boomers; rather, he wants to sound an alarm and guide us all in a more positive direction. For example, he's eager to encourage more volunteerism and mutual-aid groups to work on problems facing local communities. The advantage of Putnam's work is that it is based on empirical evidence and demands that students and teachers look at the facts and get beyond the name-calling that dominates the way aging boomers are represented in media accounts.

Like the blind men's description of the elephant, public treatment of aging boomers is often divided into contradictory images. The social construction of boomers has a history going back decades. As boomers approach later life, the media invoke familiar stereotypes (e.g., the '60s generation), but often fail to take account of the complexity of the phenomenon. Tools such as the age-period-cohort model can help us clarify what is at stake in contradictory representations of boomers, while class, gender, and ethnicity underscore the power of cumulative advantage and disadvantage. Generalizations always need to take account of individual differences.

GLOBAL PERSPECTIVE

Aging Boomers

Following the end of World War II, there was a rise in birthrates across countries in Western Europe and North America. This international phenomenon was recognized immediately as a baby boom, and the name has stuck. But the trend was actually quite variable in different national settings. For example, some countries, such as Finland, had a quick surge in birthrates followed by rapid decline in the early 1950s. In the United States, by contrast, high birthrates continued

(Continued)

(Continued)

until 1964. Canada and Australia followed a pattern similar to that of the United States. But in Great Britain, there were two separate peaks in birthrate, 1947 and 1964, and some analysts have doubted if this was a real baby boom at all. Germany had only a moderate increase in birthrates during this whole historical period. In short, even in purely demographic terms, the baby boom took different forms in different countries.

The meaning of the baby boom has been understood collectively in very different ways in different national settings. For example, the French have commonly spoken of the "sixty-eighters," linking boomers to those who participated in the revolutionary activities of the year 1968. But this ascription of a collective image is not the same as genuinely felt collective identity. The example of Finland, with its smaller cohort, shows how boomers display a strong sense of collective identity, reflecting different demographic and cultural influences. In Britain, by contrast, there has been relatively limited identification with the label *boomer*, but a stronger tie with the idea of the '60s generation. In Britain, as in the United States, there is an interest in new forms of positive aging associated with activity and growth.

SOURCE: Chris Phillipson, "Understanding the Baby Boom Generation: Comparative Perspectives," *International Journal of Ageing and Later Life* (2007), 2(2): 7–11.

BOOMERS IN THE YEARS AHEAD

Gerontologist Robert Butler, invoking the title of a famous film, once asked what will happen when boomers "reach Golden Pond." Is it possible to predict what will unfold as boomers age during, say, the decade 2010–2020? That particular time frame is convenient because in 2011 the oldest boomer will turn age 65 and thus become eligible for Medicare. It is predictable that during the remainder of the second decade of the 21st century, we will see a wave of boomers moving into old age. Beyond this demographic "age wave," it is less possible to make firm predictions. For example, more than four fifths of boomers repeatedly say that they plan to work in their retirement years. But that number is far higher than the proportion of people who have actually continued to work past normal retirement age in the recent past. Another question to be considered is whether the boomers will be in good health. We have seen earlier that the rate of certain chronic diseases has declined in the past two decades, resulting in lower numbers of people in nursing homes. Will that trend continue? Or will high rates of obesity that characterize many North Americans, including members of the boomer generation, mean higher rates of diabetes, cardiovascular disease, and other ailments?

Besides these characteristics of aging boomers as individuals, we need to consider the settings in which boomers will grow older. Because of high commitments to Medicare and Social Security, along with a legacy of budget deficits from the recent past, it seems likely that the period 2010–2020 will see fiscal pressure on the U.S. government. But that negative scenario could change dramatically if large numbers of aging boomers continue to work into their late 60s. We can predict a trend of financial pressure, based on demography, but we cannot predict how that trend will play out in the period ahead. To come up with a more complete scenario, we would also have to consider rising global energy prices, along with environmental trends such as climate change. These factors are likely to be part of the second decade of the 21st century. But we cannot be certain about how society and government, let alone individual boomers, will respond to this circumstance.

This uncertainty about the future for aging boomers gives plenty of room for different writers to put forward their own picture of the shape of things to come. These contrasting pictures are offered in the readings for this chapter.

In the first reading, we have excerpts from Christopher Buckley's *Boomsday*, which is a novel of political satire. Buckley portrays a world of the future where retiring baby boomers are bankrupting the country and inspiring a revolt by younger generations. In this future world, inflation is running at 30% a year, and the United States is at war with many countries around the world. Boomers are collecting Social Security, but there is no money to pay for all the benefits promised. As a result, the "U30s"—that is, people under the age of 30—are rebelling against higher taxes. Into this scene comes the heroine of the novel, a 29-year-old blogger named Cassandra Devine, who is pushing for a new law in favor of "Transitioning"—that is, committing suicide at age 75. The proposal sounds extreme, but Cassandra hopes it will spark debate. Thus, the scenario in *Boomsday* brings together issues raised in earlier chapters "What Is the Future for Social Security?" and "Should People Have the Choice to End Their Lives?"

Megan McArdle's article "No Country for Young Men" shares some of Buckley's worries about the future. She argues that as boomers grow older, the workforce will shrink. She sees slower economic growth ahead and points out that, despite their optimism, many boomers have actually saved little for their retirement. Thus, it may be that boomers will have to work, but they may not find jobs corresponding to their talents and expectations. In any case, as illness and disability come, they will have fewer adult children to rely on for care, an issue discussed in the earlier chapter "Should Families Provide for Their Own?"

Theodore Roszak, in an excerpt from *The Longevity Revolution,* paints a much rosier picture of the decades ahead, "as boomers become elders." He links the boomer cohort with the idealistic aspirations of the 1960s, described so vividly in his own earlier book, *The Making of a Counter Culture* (1968). Roszak believes there could be a flowering of what he dubs "elder culture," as aging boomers blend self-actualization with altruism and social activism. Boomers in their 60s, he hopes, will revive the ideas of the decade of the 1960s.

Jeff Goldsmith, in excerpts from *The Long Baby Boom,* shares some of this optimism, but he puts the focus less on culture and more on economics and politics. He predicts that aging boomers will work longer and thereby avoid putting a larger financial burden on society. But he believes this positive scenario won't easily come to pass unless we adopt policies that promote more savings, health promotion, and personal responsibility, but without abandoning key government programs that provide security. In Goldsmith's view, a positive future for aging boomers depends on acting now to make that future a reality.

FOCUS ON PRACTICE

Aging Boomers in the Workplace

A 2011 Harris poll revealed that aging boomers are planning to delay retirement, perhaps by several years. These results coincide with surveys by AARP suggesting that older workers have been hard hit by the Great Recession beginning in 2008. They're extending their work lives because they feel they have to.

Work-life extension could be a good thing for these older workers and for the economy. But it also means that younger people will face challenges as boomers stay longer in their jobs. If those higher up don't retire, it could mean promotions are delayed. As boomers stay longer in the workplace, it will mean that different generations will be working side-by-side.

The Multigenerational Workplace

How will twentysomethings work collaboratively with people old enough to be their parents or grandparents? Management consultants and labor experts are looking at this development. Some analysts (Johnson & Johnson, 2010) have offered practical suggestions for how younger workers can respond to aging boomers. They give some commonsense advice, like "Don't ignore older workers" and "Don't give up on them." In other words, avoid age-related stereotypes. Younger workers are also encouraged to ask older workers to make continuing contributions: for example, by becoming mentors to young people.

Other analysts acknowledge that there can be problems in the workplace when generations collide (Lancaster & Stillman, 2003). Employee productivity remains the key to success, and that means ensuring the productivity of workers both young and old. Understanding multigenerational differences in the workplace will be a major factor in organizational life as aging boomers remain longer on the job.

Surveys of employers show that organizations value attributes of older workers such as long experience, reliability, and work ethic and loyalty. But employers also have some doubts about these older workers. Specifically, they have doubts about older boomers' creativity, flexibility, and willingness to learn new things (Mermin, Johnson, & Toder, 2008). On the positive side, older workers can show strong productivity because of greater life experience. But in a fast-changing workplace, they may also find their skills outdated. Again, on the positive side, they are likely to have lower rates of absenteeism, but they also have more health problems than younger workers. A key to productivity over the life span will be whether employers can build on the positive attributes of older workers while also providing training to ensure that skills and knowledge are up to date.

Later-Life Entrepreneurship

When we think of older boomers in the workplace, we are likely to think of people as employees, perhaps of big organizations. Yet the face of the workforce could be much different from this in the future. Many workers, both young and old, yearn to be their own boss and chart their own path. Some analysts speak of a "free-agent nation," and the United States has long been a nation that prizes entrepreneurship. Can older boomers be part of this picture?

The answer seems to be yes. According to research by the Kansas City–based Ewing Marion Kauffman Foundation (2011), older baby boomers (aged 55–64) are driving a new entrepreneurship boom. Kauffman Foundation research showed that, in every year from 1996 to 2010, older boomers had a higher rate of entrepreneurial activity than adults aged 20–34. Surprisingly, the average age of the founders of high-tech companies is nearly 40, and there are twice as many over age 50 as under age 25.

The Great Recession beginning in 2008 brought about the highest unemployment levels since the Great Depression. Younger workers were disproportionately affected by the downturn, but older workers, whenever unemployed, have also found it difficult to find jobs, even as they experience a need for work-life extension. Many of these older boomers are becoming part of the "second chance revolution" (Rogoff & Carroll, 2009).

In 2010, *Forbes* magazine noted that one of the two fastest-growing tech start-ups, First Solar, was founded by a 68-year-old. It wasn't the first time an older entrepreneur hit the jackpot. Ray Kroc, founder of McDonald's, was nearly 60 years old when he took over a small hamburger restaurant and used his entrepreneurial talents to make it one of the most successful franchises of all time. The fact is that most new jobs are created by small companies, not by big corporations, so entrepreneurship at any stage of the adult life course will be key for economic growth and job creation. There is no reason why aging boomers cannot be a positive force for later-life entrepreneurship in years to come.

READING 46

Boomsday

Christopher Buckley

Prologue

"In Florida today, another attack on a gated community by youths protesting the recent hike in the Social Security payroll tax.

"Several hundred people in their twenties stormed the gates of a retirement community in the early hours this morning. Residents were assaulted as they played golf. Demonstrators seized carts and drove them into water hazards and bunkers. Others used spray paint and garden implements to write slogans on the greens.

"One such message, gouged into the eighteenth green, read: 'Boomsday Now!' The word refers to the term economists use for the date this year when the first of the nation's seventy-seven million so-called Baby Boomers began to retire with full Social Security benefits. The development has put a tremendous strain on the system that in turn has sent shock waves through the entire U.S. economy.

"A maintenance worker at the golf course said it might be, quote, weeks before residents were able to play golf.

"For the first time in history, the Bank of Tokyo declined to buy new-issue U.S. Treasury bills. Do you realize what that means?"

"They already have enough of our debt?"

"Precisely. Do you get the significance of that? The largest single purchaser of U.S. government debt just declined to finance any more of it. As in our debt. Meanwhile, and not coincidentally, the first of *your* generation have started to retire. You know what they're calling it?"

"Happy Hour?"

"Boomsday."

"Good word."

"Mountainous debt, a deflating economy, and seventy-seven million people retiring. The perfect economic storm."

A surprising 38 percent of the American public now favored having the option of being legally euthanized in return for huge tax breaks and subsidies. Posters were going up: UNCLE SAM WANTS TO KILL *YOU!*

"The chair recognizes Foggo Farquar, chairman of the President's Council of Economic Advisers."

"Mr. Farquar, good morning. You were asked by this commission to study the economic impact on the U.S. Treasury of the Transitioning proposal?"

"Yes, I was."

"And what were your findings?"

Mr. Farquar took out a large binder whose contents were projected onto a screen in the hearing room.

"The so-called Baby Boomer population cohort," he said, "numbers approximately seventy-seven million people born between 1946 and 1964. Given the rate at which they are currently retiring and withdrawing funds from Social Security, as well as the Medicare and Medicaid systems, we estimate"—the next slide showed a series of bar graphs, all the color of deep red—"that Social Security will exhaust its resources approximately two and a half months from now."

"From today?"

"As of noon. Yes."

"That's not very desirable, is it?"

"I would call it very far from desirable, Senator. But those are the numbers. They do not lie." . . .

Cass was wrangling volunteers.

She'd managed to find a few dozen sixty-something Baby Boomers who were willing to volunteer for Transitioning—though not until age seventy-five. Moreover, in return for their selfless acts of economic patriotism, they were demanding not only tax benefits well beyond the parameters of Cass's original Transitioning plan, but also subsidized burial, mausoleums, full college tuition for their children, and retroactive medical payments going back to age twenty-one. Cass estimated that the aggregate economic impact of their Transitioning to the U. S. Treasury would be a *negative* $65 billion. . . .

The Protest Against Social Security, or PASS, was held on the Mall on the Saturday before the South Carolina primary. Getting U30s to attend a political rally was like herding cats. They coalesced more readily for concerts than for political demonstrations. Still they came, and in respectable numbers. The Park Service estimated the crowd at seventy-five thousand, a good showing. Vendors did a brisk business in tuna wraps and vitamin water. Many protesters carried signs. Emergency medical crews stood ready to treat anyone stricken with self-esteem deficit. Curious Boomers who looked on from the sidelines remarked that it was just like the Vietnam protests, only completely different. "In those days," said one old-timer riding by on a Segway, "we didn't have nearly the variety of bottled waters you have today. Man, those were crazy times."

READING 47

No Country for Young Men

Megan McArdle

The retirement of the Boomers will transform the texture of our society. How will it change our economy, our culture, our politics? When it's over, will America look better, worse, or just different? . . .

As the Boomers age, they will consume fewer of the things that we produce efficiently, and more of the things that we provide relatively inefficiently. Productivity is notoriously difficult to project, but many forces will be pushing it downward as the Baby Boomers age.

Since services are labor-intensive, and the number of service consuming seniors will grow rapidly, we'll need a lot more workers (that's

bad news for those who favor restrictive immigration policies, particularly the kind that keep low-skilled workers out). And, of course, the mix of service workers that we'll need will be different from what it is today. In effect, the next 20 years will require a massive transfer of resources and people away from the care of children, who will decline in relative number, and toward the care of old people.

Economic growth derives from a very simple formula; it is basically equal to the increase in the labor.

Now come the Boomers, 80 million strong, merrily planning their retirements. Watching their generation move from childhood through adulthood and into old age on demographic charts is like watching a pig move through a python. Thanks to the Boomers' retirement, by 2020, even if immigration continues at roughly its current pace, the workforce looks likely to be only a little bigger than it is today. If immigration rates were to decline precipitously, all else being equal, the labor force would be roughly 1 million people smaller than it is now.

Though we talk about the retirement of the Boomers as an impending event, it has already started. Participation in the workforce generally peaks between the ages of 40 and 44, declines slowly throughout the next decade, and then falls off a cliff. Millions of Baby Boomers have already left the workforce, and as more of them become eligible to collect Social Security, the process will accelerate.

Slower productivity growth and (in the best case) slower workforce growth mean sluggish growth for the economy. That, in turn, will have a host of consequences, ranging from the geopolitical (slower growth could hasten the relative economic decline of the U.S. versus China, India, and other powers) to the social (as the economic historian Benjamin M. Friedman argued in his book, *The Moral Consequences of Growth,* earlier periods of economic stagnation, stretching back to the 19th century, have typically sharpened racism, intolerance, and other unsavory tendencies). But the most visible consequence will probably be on the stock market.

Many analysts have joined the debate over whether the Baby Boomers have saved enough for retirement. Optimists tend to look at the amount Boomers have saved relative to their incomes, compared with how much their parents had saved at the same age. But Boomers are living longer, so they will need more money than their parents did. And even with the new prescription drug benefit, they can expect to spend more on health care.

Perhaps most important, the Baby Boomers have fewer children. My grandmother has one daughter living nearby to help with shopping, another (my mother) overseeing her finances,

and a third who is only a plane-flight away in case of emergency; when my grandfather was dying, my mother drove up to Newark for weeks at a time. My parents' familial support network will be thinner: divorced and living in different cities, they will have to share the same two daughters. Given the vicissitudes of the modern economy, neither child may be living nearby when help is required. This is demographically typical, which means that not only will the Boomers be paying for help that family used to provide, they will also have fewer people to call on for financial assistance.

Despite all that, the Boomers themselves are remarkably optimistic about their prospects. A 2004 survey by the AARP showed that a solid majority are confident of their ability to prepare financially for retirement, which may be why 69 percent of them say that they are fairly or very optimistic about the future. More than 70 percent of them think that they will work at least part-time, but only a quarter expect to do so because they need the money; more are expecting to fulfill dreams of entrepreneurship or just to use work to break up the routine of grandkids, travel, and bingo.

But the confidence of the Boomers in their finances is hard to justify, and seems to rely on some mystical alchemy of strong stock gains, housing value increases, and government largesse.

If the Boomers really do work in large numbers, many ills will be cured. Contrary to stereotypes, seniors are valuable workers. Though research shows that some cognitive abilities, such as memory and calculation, decline with age, seniors' experience has generally endowed them with other skills, particularly "soft skills" like customer service and management. Later retirements by some workers would increase the size of the economic pie and ease the burden of providing for those who do retire. . . .

So the trillion-dollar question is, will the Boomers really work in large numbers as they hit the traditional retirement years? Whatever they say beforehand, very few people over the age of 65 do. Labor-force participation rates among seniors are far lower than they were in 1950, even though life expectancies have risen. In 1950, almost half of all men over 65 were still working. Now, that figure is less than 20 percent. . . .

One of the greatest challenges for the country will be creating good jobs for seniors, ideally ones closer to where they built their skills and knowledge than the aisles of Staples or Home Depot—ones in which they'll be the most productive. There's no national bureau that can bring about this change. It must emerge organically, from companies learning to accommodate their older workers on the one hand, and finding creative ways to mask or reduce the emotional impact of pay cuts on the other. And from changed expectations on the part of 60-somethings about career paths and hierarchy.

Until that happens—and even once it does—our politics are likely to be contentious, because to many people, spanning several generations, it may feel as if there's not enough money to go around. And indeed there's no getting around these facts: in 1945, the year before the Baby Boomers began entering the world, each retiree in America was supported by 42 workers. Now each retiree is supported by three. When the Boomers are fully retired, each of them will be supported by just two.

What happens when currently optimistic Boomers finally face the hard realities of their savings accounts? Will they ask for more from the government? At a bare minimum, seniors already struggling with their finances are not apt to look kindly on benefit cuts. Yet the cost of the benefits we've already promised them will weigh heavily on the workers expected to support a half-Boomer apiece. . . .

Aging will make the economy grow more slowly than we would like, and probably more slowly than we are used to. Social Security and Medicare will almost certainly be financed by a combination of benefit cuts, increased taxes, and higher retirement ages—which means that all of us will work longer than we want to, and pay more in taxes than we have before.

READING 48

The Longevity Revolution

As Boomers Become Elders

Theodore Roszak

We live in a world dominated by baby boomers who are growing more senior with each passing day. Yet the baby boom generation continues to be surrounded by an aura of youthfulness, much of it the invention of the mass media whose marketing practices were shaped during the 1960s when boomers were just emerging on the historical stage. . . .

Boomers often talk about themselves with a certain defensiveness, as if history has given them a reputation they must either live up to or live down. They have surely known more than their share of mean-spirited controversy. Lambasted in their youth as spoiled brats and rebels without a cause, they have grown up to be called other unflattering names: "grabby boomers," "state-of-the-art hypocrites," "the most selfish and irresponsible generation this country has ever produced." Following the 2000 census, the *Chicago Tribune* ran an editorial

SOURCE: Excerpts from *The Longevity Revolution: As Boomers Become Elders*, 2nd ed., by Theodore Roszak. Copyright © 2001 Berkeley Hills Books.

applauding the fact that boomers would now be dying off more rapidly. Good news for all those who are "irked at the boomers' collective influence and self-centeredness." Boomers, the *Tribune* reminded its readers—most of whom must be boomers—"were noisy and demanding children and adolescents. They have been noisy and demanding adults . . . and they promise to be noisy and demanding codgers." Boomers even badmouth themselves. Upon turning fifty, the humorist Dave Barry declared that "we are the self-absorbed, big, fat loudmouth of generations."

Perhaps those who make the accusations have some valid case in point—an incident, a trend, a person or two, a few slogans or graffiti they recollect with rancor. But when we set out to judge something as large and amorphous as a generation, we are rather like the blind men who tried to describe an elephant. Each had a piece of the truth, but only a piece. They were each a little right and a lot wrong.

In writing about any social group, but especially about something as large and ill-defined as a "generation," it is important to avoid casual generalizations. As they move into their senior years, boomers remain as socially and culturally diverse as they were in their youth. Not everybody on the college campuses of the 1960s joined a protest movement. Most did not. Countercultural youth of the 1960s both in the United States and elsewhere around the world were a minority within the minority of young people who went to college. For that matter, in the United States, contemporary conservatism—including its evangelical wing— traces its current following back to boomers who were offended by the dissent of that period.

Diversity will continue within the ranks of boomers as they go into retirement. Those who characterize the over-sixty-five population as a monolithic bloc of voters are indulging in political fantasies—whether to spread fear of a selfish gerontocracy, or to magnify the importance of older voters and their organizational leadership. Both motives are in play in senior politics, and both exaggerate the unity of the older population. . . .

Finally, add one more factor to this generational profile, a defining feature of the baby boom generation. Never before has an older generation been conversant with so many divergent ideas and dissenting values. These are, after all, the people who, in their teens and twenties, lived through a time of principled protest that seemed determined to subject every orthodoxy, every institution, every received idea in our society to critical inquiry. It is the generation whose cultural repertory blithely mixed and matched western political ideology with Asian religions, Native American lore with high tech, psychotherapy with psychedelic drugs. It questioned authority on the basis of sources and insights that had no precedent in the modern western world. I have sometimes suspected that the effort made by many pundits to diminish the importance of that period stems from their fear that, in age as in youth, the boomers represent an unpredictable force for sweeping change. . . .

The big boomer generation that was in its teens and twenties in the 1960s was not destined to disappear when it ceased to be a "youthful opposition." If at some point boomers passed through a "big chill," there would be plenty of time for the chill to thaw. They would live on to become fifty, sixty, seventy, still there on the stage of history, battered perhaps by the years but better prepared to change the world than they were in their high school and college years. And given their need to defend and expand the nation's entitlements programs, they might feel an even more urgent need to transform their society than they had in their youth. . . .

Not only is the older population of our day larger in numbers, but it constitutes *a different kind* of older population. Nothing we assumed about age in the past will fit these people. That is why we seem to be having some difficulty finding an appropriate name for them. . . .

The baby boom generation now entering its fifties and early sixties is most clearly characterized by not feeling old at all, and, often enough, not looking it or acting it. A more refined analysis divides the boomers into a "leading edge" (born 1946–1954) and a "trailing edge" (born 1955–1964). But both these groups are allied in not quite feeling their age. "Funny, We Don't Feel Old," proclaims a band of Americans aged fifty to seventy pictured on the front cover of the March 7, 1997 *New York Times Magazine.* Their stories are personal accounts of undiminished ambition and bustling activity, a style called "productive aging" that we will want to investigate in chapter seven.

Whether with enthusiasm or trepidation, everyone agrees that the boomers will be a senior generation of great consequence. For some, the nearly 80 million boomers are the source of fiscal nightmares. But a simple head count with a price tag attached is hardly sufficient to tell their story. These are the children of a deeply transformative episode in American culture. Just as their parents were marked by the Great Depression of the 1930s, so the boomers were shaped by Great Affluence of the postwar decades. What they learned about abundance and scarcity, wants and needs, from that remarkable episode in our history will be an indispensable part of the debate over entitlements.

Expecting more out of life than their parents, they will not easily be guilt-tripped into surrendering their share of the national wealth. If politicians think the middle old have been pushy about their rights, wait until they have to deal with the next senior generation. . . .

But there may be more in store for boomers besides social action as we have known it in the past. Especially as the discussion of our national wealth unfolds in the minds of a senior generation in search of life's higher values, we may see our politics grown more searchingly philosophical than anyone could imagine when entrepreneurial values ruled the world. It promises to be a distinguishing quality of the longevity revolution that great questions of fate and purpose will force themselves upon our awareness and into the political realm.

READING 49

The Long Baby Boom

An Optimistic Vision for a Graying Generation

Jeff Goldsmith

According to the doomsayers, the baby boom generation promises to be a gigantic albatross around society's neck. They expect the next several decades to be a time of social involution in the United States, as boomers cease working, retire to Florida, and cash in their entitlements to social support, shamelessly voting to raise taxes on their children and grandchildren to support their leisure, robbing the country of its future.

As they so often are, the pundits are going to be wrong—about the timing, the impact, and the required remedies. We call these particular pundits "catastropharians"—economists, politicians and journalists who carry on a proud,

SOURCE: Excerpts from *The Long Baby Boom: An Optimistic Vision for a Graying Generation* by Jeff Goldsmith, pp. xii–xiv, 179, 180. Copyright © 2008 The Johns Hopkins University Press. Reprinted with permission of The Johns Hopkins University Press.

masochistic tradition of viewing the American future as one of unavoidable conflict and decay. The catastropharians proffer a doomsday social policy scenario about the coming senior boom and a "take your castor oil" agenda of "painful but necessary" changes to our social programs to avert fiscal Armageddon. . . .

Catastropharians simply *assume* that baby boomers will follow their parents' and grandparents' life paths, as well as mimic their political values. They assume that longer life-spans will translate into lengthier periods of unproductive and parasitic activity on the part of older Americans. Then they superimpose this behavior on the structure of our current entitlement programs, which they assume will not adapt meaningfully to the pressures, and tote up the damage. The future will thus unfold, will past as prologue, and demography as destiny.

What's Wrong With This Picture?

The vision of the baby boomers as a gigantic societal albatross is a myth in the making. Not only are the catastropharians wrong about the next twenty years, their social prescriptions are also the wrong medicine for American society. They offer a static, zero-sum vision for what is, in fact, a dynamic, growing and creative economy and society. The crisis they envision is eminently avoidable, not by the politically untenable solutions they offer, but rather by listening to the generation itself and helping its members do what they say they intend to do.

American society does face a crisis, but not the one envisioned by catastropharians: a crisis of meaning for a new generation of older Americans for whom retirement makes neither economic nor human sense. The "golden years" vision of retirement as an extended, dry-land version of a luxury cruise is a failed social experiment (and one of relatively recent vintage). Many boomers who observed their parents' and grandparents' lengthy drift at the end of life have made a fundamentally different life plan.

Most baby boomers will probably not retire in the conventional sense. According to multiple AARP surveys, most baby boomers do not plan on ceasing to work at 65, as their parents did. Some 80 percent of baby boomers plan on working past age 65, with a plurality of boomers doing so because they enjoy working rather than because they need the money.

For some boomers, continuing to work will be a sad necessity, because millions currently lack the private pension coverage or savings to support themselves if they are not working. For millions more boomers, however, retirement simply will not be a satisfying life path. Indeed, 84 percent of workers older than age 45 plan on continuing to work "even if they are set for life." Boomers will remain taxpayers and part of the active economy far longer than most economists assume. It is likely that millions of boomers will be income-producing assets, not liabilities, on society's balance sheet well into their 70s and beyond.

Moreover, we cannot replace the boomers if they do wish to retire. The United States faces a looming and potentially crippling shortage of skilled workers that affects our vital infrastructure—schools, the health care system, government at all levels, and even manufacturing. The baby bust, which bottomed in 1975, has created a huge hole in the midcareer skilled U.S. work force. Forecasts suggest a looming and costly shortage of skilled workers in the next decade, a shortage that deepens further in the decade that follows. Knowledge-based enterprises will have a particularly difficult time replacing their older workers.

While boomers will not be able to fill all these gaps, personnel policies that push older workers out of the skilled positions in our economy will be forcefully reexamined in the next few years. In some segments of our work force, the cost of replacing experienced older workers could exceed the increased expenses related to retaining them. Rather than encouraging boomers to retire, we need to revise our tax and pension policies, as well as the Social Security and Medicare programs, to encourage boomers to remain engaged, productive (and taxpaying) citizens. . . .

There is an urgent need to change the conversation about the baby boom and its future. The expensive script written for baby boomers' next twenty years, like all the other generational life scripts written for them, is destined for the trash heap. Baby boomers, even the less-fortunate ones, are fundamentally optimistic about their *own* futures. Most boomers haven't spent fifteen minutes thinking about how Social Security or Medicare will actually affect them, for the obvious reason that they do not consider themselves "old." How they approach these topics will likely be shaped by their personal experiences over the next twenty years, not by some ironclad historical logic or the opinions or values of their elders. . . .

In a singularly unimaginative and depressing fashion, catastropharians see baby boomers following the life trajectories of their less-educated and less-energetic parents and grandparents, retiring en masse at age 65, and moving to Florida or Nevada to play golf and evening canasta. . . .

A large fraction of baby boomers, however, have other plans, and their expressed desires provide guidance on how we should begin rethinking our social policy toward older Americans. Those expressed desires are potentially highly beneficial to society as well as to them. We should begin resetting our social expectations to encourage a more active and healthier last third of life and actively support their aspirations by changing our public policies toward older people.

Along the way, there are problems that need solving. Many baby boomers lack adequate private savings and have less than optimal personal health habits. In both cases, these problems originate in a lamentable tendency to trade present pleasures against future security and health. Failure to address these concerns will impose future costs on the society and on the baby boom itself in avoidable health care bills and in income-security needs. Put more positively, the healthier and more solvent baby boomers are, the smaller the social burden they will impose in the next twenty years. . . .

Catastrophe is far from inevitable. Demography is not destiny. We have lots of room to make intelligent social and personal choices to create a happier and more prosperous society. Better health of older Americans, a changed work world, and the desire of baby boomers to continue working all point to a far better near-term outcome than most social observers believe. Encouraging longer and more productive engagement of older Americans in work as well as in their communities and postponing the onset of serious, life-changing illness are both big wins for American society.

FOCUS ON THE FUTURE

60th Anniversary of Woodstock

The year is 2029, and it's the 60th anniversary of Woodstock, the famous rock concert in 1969 when thousands converged on a farm in upstate New York to hear performers like Janis Joplin, the Grateful Dead, and other celebrated icons of rock music. This year several thousand aging boomers have come back to Woodstock to celebrate that event of their youth. There are many survivors gathered to remember the "good old days." This year, 2029, has been a banner year for a different reason. This is the year, of course, that Chelsea Clinton was inaugurated as President of the United States. Her first year in office, however, has been a time of upheaval because of the big protest marches on Washington.

(Continued)

(Continued)

While some aging boomers were gathering in Woodstock, others were marching on Washington to protest proposed cuts in Social Security under a plan tentatively agreed on by (the second) President Clinton. President Chelsea Clinton's own 82-year-old mother, former Senator Hillary Clinton, had sought the Presidency herself in 2008. Now, in 2029, she condemned these protesters from her own generation, insisting that Chelsea's cuts were really intended to save Social Security, not destroy it. The same saying was once used about villages in Vietnam during that distant war.

The big news at this 60th anniversary isn't about Social Security, but rather is about energy and the environment. At least that is what has emerged from oral history interviews conducted in 2029. Students from a nearby college at Albany, New York, converged on the old Woodstock site to take advantage of this gathering of aging boomers. They've found the gathering a great opportunity to conduct oral history interviews, just as they'd been doing with their own great grandparents, also aging boomers. They asked these aging boomers the same question: How had the world changed the most since 1969?

Many of the answers given by the aging boomers revolved around the high cost of energy, including gas for cars. "Back when I was a young person, we used to drive everywhere, sometimes just cruise down the highway for nothing. With gas now at $25 a gallon, you can't do that anymore, that's for sure." Other boomers talked about the suburbs and reminisced about the days when people lived far away from their jobs, sometimes commuting an hour or two by car each way. Earlier in the 21st century, a lot of those houses in distant suburbs were foreclosed and abandoned, eventually to be recycled as storage facilities. The elder boomers remembered how gas prices had gone up in the 1970s, but then went back down. Toward the end of the first decade of the 21st century, energy prices went up, but never came down. That changed people's lives dramatically.

Other boomers talked about climate change. Some reminisced about earlier in the 21st century when some Americans doubted whether climate change was real. Then came the big hurricanes, destroying coastal regions of the Gulf of Mexico and the southeastern United States. Floods and forest fires were also a big threat. By 2020, much of the Greenland ice cap had melted, so no one could deny the problem any longer. As a result, some aging boomers became environmental advocates, worried about the world that their grandchildren would inherit.

When students interviewed those attending the 60th anniversary gathering at Woodstock, they found many who shared these sentiments. Some jokingly called themselves "gray-haired hippies." But other boomers had become militant nationalists, pushing for the U.S. government to intervene to protect the dwindling oil supplies from the Middle East. Because "peak oil" had been recognized, it was important to protect the energy supplies we had. In 2029, these elder boomers didn't all agree about politics, but then that was also the case in 1969.

Questions for Writing, Reflection, and Debate

1. Interview an aging boomer (e.g., a relative) and ask what that individual most vividly remembers about the 1960s and 1970s. What are the biggest differences between the boomer's life experience and your own?

2. Many aging boomers seem likely to be exposed to loss of hoped-for assets in the stock market or housing values during their retirement years. Some say such people should have saved more or differently for their retirement. Is that response a

reasonable criticism? What could they do now to deal with potential problems?

3. Christopher Buckley, in *Boomsday*, puts forth the idea that in the future public opinion may be in favor of "Transitioning" (i.e., legalized euthanasia in return for tax breaks or other benefits). What are the arguments for and against treating end-of-life choices as a matter of economics?

4. In *The Long Baby Boom*, Goldsmith suggests that many problems related to aging boomers could be solved if boomers worked longer (e.g., if the average age of retirement went up from 63 to 68). What factors could promote such a change? What factors would stand in the way of such a change?

5. Many boomers report that they "don't feel old." Are there characteristics of boomers as a distinct cohort that are associated with this denial of aging? Is the feeling mostly realistic or unrealistic? If you yourself are a boomer, write from your own standpoint.

6. Think of two big unexpected world events that changed history: for example, the fall of communism in 1989, the rise of the Internet in the 1990s, or September 11, 2001. Now think about the next 10 or 15 years. What changes in the world situation—economy, environment, technology, and so forth—could dramatically change the outlook for aging boomers for the better? For the worse?

7. Theodore Roszak gives a picture of boomers as people who in their youth were rebellious and questioned authority. Can you give examples of groups of boomers for whom this was *not* at all true?

8. Imagine that you're in charge of marketing a new product or service that is to appeal to aging boomers. What factors would you consider as you plan to create a website for this campaign? Visit 20 of the sites listed under Google for "aging boomers" and evaluate their approach as you develop your own.

Suggested Readings

Freedman, Marc, *Prime Time: How Baby Boomers Will Revolutionize Retirement and Transform America,* New York: Public Affairs, 2002.

Gillon, Steven M., *Boomer Nation: The Largest and Richest Generation Ever, and How It Changed America,* New York: Free Press, 2004.

Russell, Cheryl, *Baby Boom: Americans Born 1946 to 1964,* Ithaca, NY: New Strategist Publications, 2004.

Steinhorn, Leonard, *The Greater Generation: In Defense of the Baby Boom Legacy,* New York: St. Martin's Press, 2006.

Student Study Site

Visit the Student Study Site at **www.sagepub.com/moody7e** for these additional learning tools:

- Flashcards
- Web quizzes
- Chapter outlines

- SAGE journal articles
- Web resources
- Video and audio resources

Controversy 12

THE NEW AGING MARKETPLACE

Hope or Hype?

THE NEW CUSTOMER MAJORITY

In 2011, the first boomers turned 65, attracting headlines. But that event was only the beginning of the "age wave" that is reshaping the consumer marketplace. Eight thousand boomers turn 65 each day, and consumers over age 45 already compose what business analyst David Wolfe calls "the new customer majority" (Wolfe & Snyder, 2003). As this new customer majority increases in importance, companies will ignore it at their peril, and there are signs that more and more are waking up to the potential of the new aging marketplace.

Along with their sheer numbers, the boomers possess significant purchasing power that should command our attention. According to U.S. census data, the 50-plus segment of the U.S. population controls three quarters of the total net worth and financial assets of the country, or more than $7 trillion. This segment of the market is also responsible for more than 50% of discretionary income. Their spending is far greater than that of younger consumers in the so-called silver industries of health care, travel, and financial services. Health care, of course, is one of the biggest parts of the mature marketplace, and those aged 55 and above account for a third of all health care dollars spent. But mature householders between the ages of 55 and 64 also spend more than average on housing and household furnishings, entertainment, and vehicles (Sterns & Sterns, 2006). For example, mature consumers account for nearly half of all luxury car purchases and three quarters of commercial vacation travel. More than half of all book and magazine subscriptions are sold to the 50-plus market (Purcell, 2008). *AARP The Magazine* has a circulation larger than *TV Guide* or *Reader's Digest*.

Despite the tendency to talk about the boomers and other older generations as if they are homogeneous groups, the senior market is composed of distinct segments, whether based on age or other characteristics. The question of *market segmentation* is key in thinking about older consumers. One approach to market segmentation is to divide people up according to life-stage transitions that often occur in later life, such as the empty nest, frailty of parents, and becoming a grandparent. Based on age alone, we can distinguish between preretirees (ages 55–64), the young-old (ages 65–74), and the old-old (ages 75–84). In addition, we can identify the oldest-old (ages 85 and up), smaller in numbers but the fastest-growing segment of all. It seems obvious that we cannot identify the "mature market" the same way for a 55-year-old as for an 85-year-old.

Those in the 55–64 age group are usually working, while the young-old are more likely to be purchasers of leisure services and retirement housing. The old-old are more likely to have concerns about financial security and failing health. The oldest-old are prime consumers for options like continuing-care retirement communities or assisted living. Although chronological age can be used to segment the mature market, other approaches to segmentation are likely to be more effective: for example, segmenting consumers according to health status, socioeconomic level, or psychographic characteristics. The U.S. Department of Commerce has reported that 85% of consumers 50+ purchase items by mail, and this trend has only increased with the availability of Internet shopping (Sterns & Sterns, 2006). Mail-order and Internet purchases can be especially helpful for those who are homebound.

URBAN LEGENDS OF AGING

"Ageism is the work of the advertising industry."

Actually, ageism was invented by the ancient Greeks, as a glance at their youth-oriented art will show. Youth culture was rediscovered by the Renaissance and then propagated in the 20th century for a variety of reasons. Advertising and media mainly trade on the stereotypes already widely accepted in society.

ONE MARKET SECTOR LIKELY TO GROW

As the new aging marketplace expands in the years to come, certain market sectors are likely to grow faster. These sectors, which we may term *silver industries,* include the following:

- **Financial services.** As noted, older consumers control the vast majority of financial assets. Banking, brokerage, investment, and insurance are all likely to grow in the years ahead.
- **Health care.** Rates of chronic illness go up with advancing age, and health care spending rises with age, too. Aging boomers may be especially interested in products such as vitamins and supplements, which promise to prevent disease.
- **Travel and hospitality.** The cruise industry has long understood that retirees are a key segment of its business. Others in the travel business are looking eagerly at the market now, too.
- **Retirement housing.** Traditional retirement communities such as Sun City, Arizona, are being overtaken by new growth in "active adult" communities. Only a small number of people actually move at the point of retirement, but recent retirees can be a significant market share in some parts of the country (Wolf & Longino, 2005). Those who stay in their own homes may be customers for home adaptive products and services (Lawlor & Thomas, 2008).

As the U.S. population grows older, venture capitalists are poised to profit from this profound demographic transformation. For example, North Castle Partners is a major venture-capital fund controlling a billion and a half dollars of investment funds concentrated exclusively on emerging companies in the field of healthy living and aging. Portfolio companies featured by North Castle include groups such as Grand Expeditions (luxury travel), Naked Juice, Avalon personal care products, and World Health fitness centers, among others. In sum, private investors are awakening to population aging, even if some advertisers and aging service groups haven't yet gotten the word.

What Do Older Consumers Want?

With the 25- to 44-year-old population now shrinking, companies will increasingly need to reach older age groups in order to be successful. In the silver industries, there are opportunities for new growth. Yet marketing strategies in the past have been dominated by the behavior of consumers under age 40. In other words, business has too often looked to the past, not the future. Moreover, in the advertising industry, creative copyright producers tend to be in their 20s and typically have very little understanding of what older consumers are like (Coming of Age, 2010; Nyren, 2007).

There are enduring stereotypes about older consumers, but the stereotypes aren't necessarily true. For example, some believe that older people are simply "set in their ways" and won't buy anything new. Yet those over 55 are among the fastest-growing users of Facebook, a service that didn't exist when they were younger. Evidence shows that older people can be motivated to buy and use new products or services if they see a clear benefit in what is offered.

Experience suggests that older people can take offense at advertisements that stereotype them. One reason is that older people think of themselves as younger than their actual age. Marketers who understand this fact about self-perception will produce messages that use models a bit younger than the customers they are aiming at. Even beyond such appeals, marketers need to address older consumers in ways that appreciate their lifestyle and offer products that meet genuinely perceived needs. Finally, there is the critical point about individual differences. Consumer needs can never be reduced to commonalities of age, cohort, or other segmentation strategies for a reason emphasized by David Wolfe: namely, that as we grow older, we tend to become "more ourselves," that is, more individual. Wolfe and Snyder (2003) show how marketing has often missed the target of older consumers by failing to recognize the crucial dimension of individual differences.

Another stereotype is that older consumers are loyal to familiar brands and will never consider new brands. But more recent research on brand loyalty suggests that age alone isn't the critical factor. Older consumers are willing to switch to a new brand, even a more expensive product, if they are convinced it actually meets their needs (Karani & Fraccastoro, 2010).

Population aging is certain to change the consumer marketplace in general. Aging will also increase demand for specific goods and services that are directly relevant to older consumers. Marketing analyst Laurie Orlov (2011) cites the example of a familiar item: the humble wristwatch. Computer manufacturer Hewlett-Packard has considered a new type of wireless wristwatch for the consumer market. Why would the company be looking at such a product? The answer is that between 2008 and 2010, sales of wristwatches fell by nearly 30% among young people (ages 18–24). But sales rose by 33% in the 35–44 age group and by an astounding 104% for people 65 and older. If makers of conventional wristwatches looked only at the youth market, they might give up their business. But the 65+ group is growing from 13% of the population to nearly 20% by 2030. A watch manufacturer who overlooks the older consumers could easily miss a bonanza.

GLOBAL PERSPECTIVE

The Consumer Marketplace in Great Britain

Ian Jones and colleagues have looked carefully at the consumer marketplace and old age in Great Britain. In their book, *Ageing in a Consumer Society* (2008), they point out that, in Britain, "At every stage of their adult lives, today's older people have lived through and contributed to . . . economic growth and the rise of a society based on mass consumption. They take these experiences and the material conditions accruing from these changes into later life and retirement." In the last decades of the 20th century, in the United Kingdom there was a big rise in expenditures for cars, vacations, and restaurant meals. In previous decades, spending went for consumer durables like TV sets and washing machines; more recently, it has gone for novel consumer durables like videocassette recorders and personal computers. Today's older generation grew up in this consumer society: "The once young consumers of the post-war period have grown older, all the while retaining the propensities to be active players in a society where the pursuit of lifestyle and identity" is tied to purchase of commodities. In this respect, older people have become similar to the rest of the population. In the United Kingdom, as in the United States, poverty rates among older people have declined sharply. Income inequality remains, but older people as a group are no longer defined by poverty. Instead, the "third age" in Britain is distinguished by a wide range of consumption styles, reflecting the dominance of consumerism carried into later life.

SOURCE: Jones, I. R., Hyde, M., Victor, C. R., Wiggins, R. D., Gilleard, C., & Higgs, P., *Aging in a Consumer Society: From Passive to Active Consumption in Britain*, Bristol, UK: The Policy Press, 2008.

LIMITS OF THE MARKETPLACE MODEL

"Give the customers what they want" sounds like an appealing slogan. But serious questions can be raised about this approach as well as about the limits of the marketplace model. For example, consider the rise of so-called "anti-aging medicine." We put quotation marks around this domain because, scientifically speaking, there can be no such thing as anti-aging medicine: The truth is that no interventions, aside from caloric reduction in diet, have been shown to increase maximum life span or slow the biological process of aging humans. Yet anti-aging medicine is thriving. There are more doctors who are members of the American Academy of Anti-Aging Medicine than there are members of the American Geriatrics Society. As Arlene Weintraub points out in one of the supplemental readings for this chapter, anti-aging products are often a waste of money and are sometimes actually dangerous to people's health. In any case, promotion of anti-aging products is based on age denial and tends to reinforce attitudes of ageism widespread in our society.

Is it possible to develop realistic marketing messages that do not reinforce attitudes of ageism? One approach might be to use the concept of **successful aging**, developed by Rowe and Kahn (1997). One continuing-care retirement group, Masterpiece Living, has built its entire marketing strategy around the idea of successful aging. But a careful reading of Rowe and Kahn's book actually shows two distinct versions of their idea. One definition is optimal aging, which is superior to usual aging because of a combination of good health, strong social connections, and a sense of purpose in life. Another definition is "decrement with compensation," meaning acknowledgment of losses combined with realistic interventions to adapt. Optimal aging would apply to something like Elderhostel (now Road Scholar), while decrement with compensation might apply to an assisted-living facility providing both security and opportunities for life enrichment under more limited circumstances. Each version of successful aging could be valid, as Masterpiece Living has shown in its approach to the new aging marketplace.

The positive side of the marketplace model is that it permits individual choice and can promote efficiency through competition. There lies the strength of the free enterprise system. In the aging marketplace, Moody and Sood point out that many different brands appeal to very diverse psychographic profiles of older consumers. Why shouldn't we simply permit lots of different age brands to flourish as our society grows older? One answer is that our society long ago concluded that some degree of regulation is needed to protect consumers from products that may be dangerous. Health inspectors can shut down restaurants that threaten public health. The Food and Drug Administration reviews health products before they enter the marketplace. Free choice is definitely a value, but many would argue that it must be balanced against other important values, such as personal security and protection from dangers. Many of the same issues are discussed in Controversy 6, "Should Older People Be Protected From Bad Choices?" The answer isn't always clear.

Apart from the balance between choice and protection, the new aging marketplace will have to reckon with other important dimensions of the economics of aging. The power of cumulative advantage and disadvantage means that, as we grow older, inequality is likely to increase. Inequality overall has grown significantly in the United States in recent decades, and as the population grows older, people will carry those inequalities into old age. Affluent elders will have more discretionary income, while those on the bottom will have less. Inevitably, a marketplace model will focus attention on those with money to spend, not those who are impoverished. Whatever the strength of the marketplace model, we will have to take account of inequalities as America grows older.

In addition, we have also seen that the consumer marketplace frequently fails for other reasons to meet the needs of an aging population. A classic example of this is the advertising industry, which has long ignored the buying power of older consumers partly because of mistaken ideas and stereotypes. Another reason for market failure is a tendency for mass marketing to ignore significant subgroups, such as people with disabilities, who might benefit from products redesigned to meet their needs. A positive element here is the emergence of so-called **universal design** for appliances, household devices, and other consumer products, as well as interior architecture. By making products accessible for people with disabilities, the same products can become easier to use for the whole population. With growing numbers of older consumers, the new aging marketplace could become a stimulus for business to prosper while meeting the needs of those who are growing older.

The readings excerpted below take on various aspects of the emerging issue of age-based marketing. In "Overview of the Boomer Market," Mary Furlong portrays aging boomers as a huge marketplace with great promise for the future. She reminds us that boomers are not a homogeneous group and that they are moving through distinctive stages of life. Longevity, it seems, will shape the marketplace for many years to come.

In "Age Branding," Harry Moody and Sanjay Sood analyze different motives for buying age-related products or services. Like Furlong, they urge segmenting the marketplace, and they point out that selling to an older consumer requires understanding more deeply what potential customers want to buy: For some, it may be denial of aging; for others, it may be a practical solution to problems that come in later life.

In *Selling the Fountain of Youth,* Arlene Weintraub describes the field of anti-aging medicine, now a controversial topic in health care. She shows how anti-aging doctors commonly prescribe different kinds of hormones that, it is claimed, can slow the process of aging. Anti-aging medicine has become a large and growing business enterprise, and Weintraub is sharply critical of its claims.

In "The Marketplace of Memory," Daniel George and Peter Whitehouse describe the growing demand for products that promise "brain fitness." Like Weintraub, George and Whitehouse worry that claims for brain fitness exceed what science or medicine can actually prove. Yet fears about dementia and memory loss are so great that the market for such products seems likely to grow in the future.

FOCUS ON PRACTICE

Career Opportunity in the Silver Industries

In the 1967 film *The Graduate,* there is a wonderful scene where Dustin Hoffman's character is talking to an older man who gives him advice about his future. What is the advice? In a word, the older man whispers, "Plastics." Plastics, he says, will be the growth area of the future. In similar fashion, today we can predict that the U.S. economy will be driven increasingly by population aging. To students and recent graduates today, we might whisper in their ear that "aging" is the new "plastics."

According to the U.S. Department of Labor, careers in aging are expected to be a high-growth area during the next decade. Opportunities for providing services to older people as well as those with disabilities are projected to grow by 74% between 2006 and 2016. The Bureau of Labor Statistics forecasts that new jobs in home care should grow by 55% during this period, with almost a half-million new workers needed. Similar increases are expected in community care facilities and for social and human service assistants ("Occupational Employment Projections to 2016," 2007).

One of the biggest areas for job growth will be health care, with demand for nurses, social workers, pharmacists, and medical technicians. But employment opportunities will also grow in other health-related fields, such as fitness and wellness promotion. For example, we can expect growth in the need for yoga instructors for older people, as well as professional geriatric care managers, who help families caring for older relatives.

Among the so-called silver industries we can expect to see growth in aging-related aspects of housing, financial services, and travel and hospitality. Housing is not limited to new construction of individual residences, but includes retirement communities, adaptive modification of existing housing, and related elements in real estate. For example, there is now professional certification for "Seniors Real Estate Specialists" who are qualified to address the needs of home buyers and sellers in the 50+ age group. In financial services, the Hartford insurance company has its own "corporate gerontology" group. The Association of Personal Historians provides guidance for professionals who earn a living helping older people write their life stories. Even in consumer products, new and ever-changing digital and electronic products will be needed by an aging population.

NOTE: For more on this subject see "Exploring Careers in Aging" at http://businessandaging.blogs.com/ecg/101_careers_in_aging/. See also "Resources for Developing Your Career in Aging" at http://www.aghe.org/500215. See also C. Joanne Grabinski, *101 Careers in Gerontology* (Springer, 2007).

Overview of the Boomer Market

Mary Furlong

Today, boomers around the world are reinventing their lives. They are finding new places to work, new places to travel to, new ways to spend their days, new fashions, new savings programs, new ways to spend time with their children and grandchildren, and new ways to stay vital and connected as they age. Each new choice represents a signal of enormous business opportunity. Entrepreneurs, corporate brand managers and strategists, investors, and nonprofit executives want to develop products and services that meet the boomers' changing needs. . . .

The enormous size of the boomer market—those between the ages of 42 and 60 in 2006 make up 25% of the U.S. population—translates into enormous business opportunities. But this market is misunderstood by many businesses that assume all boomers share the same tastes and passions, and that 60 "then" is the same as 60 "now." In fact, *the boomer generation represents hundreds of market segments.* The key to reaching this market lies in developing relevant services and products through understanding the demographic data, the trends, boomer psychology, and how to effectively communicate with the target audience.

Some of these areas include health and vitality; money and investing; play, passion, and travel; family, including grandparenting and elder care; entrepreneurship and technology; fashion and beauty; sexuality, romance, and intimacy; religion and spirituality; and philanthropy, along with other forms of community building.

Boomers' new values and attitudes have transformed consumer behavior at every stage of life so far, and the next stage will be no exception. Currently, 45% of the U.S. population is age 40 or older. This group is the best educated, healthiest, and wealthiest generation ever to reach midlife and beyond. It is a force to be recognized and courted over the next decades as boomers move from midlife into old age. . . .

The key to success in this marketplace is life-stage marketing because the boomers will go through more transitions in their 50s and 60s than any other phase of life. Each life-stage transition triggers business opportunities that revolve around family (empty nests, loss of parents, arrival of grandchildren); health issues (menopause, heart disease, vision and hearing loss, arthritis); housing (downsizing, rightsizing, remodeling, second homes); finances, work, retirement, and daily activities (time for passions and play); and perspective (the search for meaning).

The boomers are not a homogenous group. Many people who are in their 50s today face retirement, empty nests, grandparenting, and aging parents, but some of them have young children, and some are newlyweds. Today, millions of younger boomers (born from 1956 to 1964) are still pursuing careers and helping their kids through school. The boomers are also ethnically and economically diverse. . . .

The emergence of a large, healthy, well-educated cohort of consumers in their 60s is a social and a business revolution the world has

SOURCE: "Overview of the Boomer Market" by Mary Furlong, from *Turning Silver into Gold: How to Profit in the New Boomer Marketplace* (FT Press, 2007).

never seen before. Adults age 45 and older account for 77% of financial assets in the United States, control 70% of total wealth, and account for more than half of the nation's discretionary spending.

Longevity has already had a huge impact on the marketplace, especially in health. Already, traditional companies are repositioning themselves. Intel is no longer just a chip company; it is a technology and health care company. Philips hit a home run with its home heart defibrillator and has now moved into the home alert space with its acquisition of Lifeline. Many consumer products companies are now looking into the lucrative home spa product and service business. The anti-aging space is taking off, as are businesses catering to the activities and passions that active adults enjoy. Each of the challenges of aging—from vision and hearing loss to heart

disease, diabetes, and cancer—is also a market opportunity. The fact that many boomers have high cholesterol has led to giant sales of pharmaceuticals such as Lipitor and Crestor.

When you understand what people are coping with in their 40s, 50s, and 60s (and beyond), you can plan a wealth of new products and services targeted for this demographic. The big winners in this marketplace will be entrepreneurs who care about the fears around aging— of loneliness, of being marginalized, of not giving enough back, of not having enough money or health, of no longer being attractive or youthful. And best of all, the big winners in the aging boom will also be able to smile when they look in the mirror. Their fortunes will be based on products and services that help improve quality of life for many people.

READING 51

Age Branding

Harry R. Moody and Sanjay Sood

INTRODUCTION

Age branding is the creation of brands that are targeted to older consumers, such as AARP or Sun City. Given the demographic shifts taking place in the United States, age branding is a strategy that will continue to grow in importance. In 2000, there were 35 million Americans age 65 or above, approximately 13% of the U.S. population. That number is expected to rise to 50 million by 2010 and 70 million in 2030, or 20% of the U.S. population (Federal Interagency Forum on Aging-Related Statistics, 2000). The trend is not limited to the United States but is

evident in advanced industrialized countries of Europe and in Japan. Older people control 70% of the net worth of U.S. households at a level equal to $7 trillion. Moreover, market research suggests that the 50 to 65+ age group that commands so much designed with the older demographic in mind, what we will call *age branding,* will be a key to reaching this huge marketplace.

In this chapter, we identify four different "families" of age brands that represent distinctive cultures of aging. Age brands can be successful in different ways because each brand family meets a specific need that is relevant to

---

SOURCE: "Age Branding" by H. R. Moody and Sanjay Sood, from *The Aging Consumer: Perspectives from Psychology and Economics,* edited by Aimee Drolet, Norbert Schwarz, and Carolyn Yoon (Routledge, 2009).

the older consumer segment. This does not imply that age branding involves a new marketing strategy. Rather, traditional techniques must be designed with the older consumer as the larger market from the outset. The four families are (a) age-denial brands ("I don't have to get old"); (b) age-adaptive brands ("Age presents problems, but I can deal with them"); (c) age-irrelevant brands ("Mind over matter; if you don't mind, it doesn't matter"); and (d) age-affirmative brands ("The best is yet to be"). We also consider troubled age brands. In what follows, we describe each brand strategy, identify related academic research, and provide some best practice examples from the marketplace.

AGE-DENIAL BRANDS

An old vaudeville joke has it that "denial is not just the name of a river in Egypt." But, age-denial branding is no joke. It is a big business and likely to grow even bigger as the baby boomers move toward Golden Pond. Every age-denial brand, explicitly or not, appeals to the illusion of Peter Pan, to the fantasy of never growing up and never becoming old. Denying one's age involves seeking out brands such as Botox that help older consumers look young. Importantly, successful age-denial brands have recognized that the purchase of these age-denial brands may also be accompanied by information-processing strategies that help older consumers also *feel* young.

An interesting comparison regarding the potential effectiveness of age-denial branding can be made between the success of Botox and the failure of Geritol. Since its introduction, Botox has become a widely known brand, even a household name (Kane, 2002). Increasingly, cosmetic surgery and Botox are becoming part of mainstream American life, and age-denial branding has grown along the way. For example, today there are more physicians who are members of the American Academy of Anti-Aging Medicine than there are board certified geriatricians in the United States. One of the success factors of branding; Botox is in the use of personal and social comparisons that help make Botox the brand it is today. Typical of other medical brands, Botox emphasizes a before-and-after comparison of wrinkles to demonstrate product performance. The branding of Botox, however, also includes social comparisons by which users of the product compare themselves to other consumers in the same age range. These social comparisons help Botox users boost self-esteem by reaffirming how much younger they look with Botox as the hero.

The failure of Geritol shows what can happen if social comparisons are not managed effectively. Geritol achieved fame briefly in the late 1950s as sponsor for the fabled quiz show *Twenty One* but could not sustain its market share over time. As a brand, Geritol found itself linked to negative social comparisons that stereotyped seniors as a group.

AGE-ADAPTIVE BRANDS

A second category of brands, age-adaptive brands, are brands that are more functional in nature and help seniors to recognize age-related issues and proactively adjust to the effects of aging. It is worth noting that individual factors can lead a person to be more or less open to age-adaptive brands. One classic example would be eyeglasses or hearing aids to compensate for sensory deficits that can accompany age. Because these deficits typically come on slowly, with insidious onset, it is possible to deny for a long time that any deficit has occurred. Everyone can recognize examples of family, friends, or colleagues whose eyesight or hearing has deteriorated with age but who insist that their capacity is "as good as it's ever been." Such answers amount to a kind of therapeutic nihilism ("What's the use?") that keeps prospects from responding to age-adaptive brands.

Refusal to buy age-adaptive products has a different psychological dynamic than what we see in age-denial products. Indeed, denial is what fuels that purchase of cosmetics, plastic surgery, or anti-aging medicine. As long as denial persists, people will be unwilling to consider age-adaptive products such as hearing aids, home monitoring technology, long-term care insurance, or continuing care retirement communities. In all cases, the answer from consumers is likely to be, "I'm not ready for that yet." This delicate dance between denial and adaptation will be visible in our subsequent discussion of age-adaptive branding.

Age-adaptive brands are products that acknowledge the negative elements of aging but seek to compensate for decline through goods or services that make life better—in short, the essential definition of "successful aging" defined as "decrement with compensation" (Rowe & Kahn, 1997). Instead of denial, age-adaptive brands appeal to consumers by recognizing a problem and taking realistic steps to address the problem. Instead of self-enhancement, age-adaptive brands appeal to mature problem solving with brands that meet a functional need.

Ultimately, age-adaptive brands are problem-solving brands. Skillful marketers such as Viagra find ways to embed age-adaptive messages in ways that the frame of reference becomes positive as opposed to negative for aging consumers. Viagra is the commonly recognized name of a drug for erectile dysfunction sold by Pfizer Corporation. The drug was initially patented in 1996 and first sold in the United States in 1998. It quickly became a huge success for Pfizer.

The effectiveness of creating an appropriate frame of reference can also be contrasted in the branding of home monitoring products made by Intel and QuietCare. Home monitoring technologies are electronic devices that keep track of people at home, collecting health data or providing interventions on an emergency basis (Sixsmith et al., 2007). The potential U.S. market for home monitoring technologies is vast and growing.

The name *QuietCare* conveys the essence of the brand and does not strictly focus on the life state of the customer. QuietCare is similar to an alarm system, but instead of responding to an emergency event, like a fall, QuietCare instead silently monitors movement and uses a form of artificial intelligence to learn about individual behavior patterns. For example, the QuietCare system can indicate a possible bathroom fall by analyzing data from the activity sensors. If a client remains in the bathroom for more than 1 hour, the family caregiver or a monitoring office will receive a signal advising about the possibility of a fall.

QuietCare, like Intel, has promoted its product as a means of detecting potential risks and preventing adverse outcomes. However, it also aims to promote the idea of independence plus security, a combined message not always easy to maintain. QuietCare claims to protect privacy and dignity because it does not use cameras or microphones but rather uses artificial intelligence to identify potential health risks. These differing reference points and multiple values underscore the complexity of age-adaptive branding.

AGE-IRRELEVANT BRANDS

If the older consumer market is growing, what would be more natural than to pitch products directly to older people as such? In fact, explicit appeals linked to age can be a disaster, as many examples confirm. For instance, the venerable food giant H. J. Heinz stumbled when it tried to promote a new, easily digestible brand under the label of Senior Foods. Even marketing powerhouse Johnson & Johnson lived to regret its effort to introduce its new Affinity shampoo as just right for "older hair." Johnson & Johnson managed to recover from its poisoned Tylenol episode, but it did not recover from trying to pitch shampoo for older hair. The product was quietly abandoned. Similarly, Southwestern Bell dialed a wrong number when it published its separate Silver Pages telephone directory filled with ads aimed at older people. The list goes on, and these failures raise a natural question: How could so many marketing experts and blue-chip companies fail to understand this lucrative aging marketplace?

Holland America Line is an example of a company that has prospered by focusing on the mature market but not by doing so in an explicit or overt way. Age branding remains critical for the travel industry because of its customer base. Americans over age 50 (the threshold for AARP membership) already account for 45% of all leisure trips in the United States, and people over age 65 account for nearly a third (31%) of those trips. According to the Consumer Expenditure Survey, spending on cruises does not hit its peak until ages 65 to 74, and people over 50 account for 70% of all cruise passengers. Among household groups in this category, spending on hotels and airlines also remains above average. But, do not look for leisure travel companies like Holland America any time soon to advertise a "granny cruise." They would not get many customers if they did. The indirect approach works better. We can also speculate that senior discounts of the past are likely to be scarcer for a variety of reasons. One factor is the rising educational levels of successive cohorts of older people. Boomers, as noted, have higher levels of college attendance than generations before them, and vacation spending rises linearly with the number of years of education. But, explicit age appeals are likely to fail.

Wolfe suggested that New Balance athletic shoes also represent effective age-irrelevant branding (Wolfe & Snyder, 2003).

The branding strategy of New Balance contrasts sharply with that of competitor Nike, with its famous slogan, "Just Do It." These two brands are polar opposites. Consider their names. *Nike* conjures up a Greek god promising victory over competitors. By contrast, the name *New Balance* appeals to values better appreciated in the second half of life, such as steadiness and wisdom, rather than impulsiveness in action.

The very name New Balance might suggest a brand of yoga retreats more than success in a running competition.

AGE-AFFIRMATIVE BRANDS

Age-affirmative brands are brands that promise us a positive benefit linked to age. Age-affirmative brands do not ignore or deny age but instead focus on elements we can celebrate and affirm. Age-affirmative branding conveys a message of hope and positive aging. Some marketers are beginning to see this as an opportunity in a time of demographic change.

One way to accomplish this goal is to carefully construct the brand's personality to make certain that it reflects the personality of the consumer.

To create a long-lasting relationship with aging consumers, then, age-affirmative brands need to highlight the positive facets of aging and speak to older consumers in a tone that is credible.

One example of effective age-affirmative branding is embodied by Sun City Retirement Communities. The very first modern retirement community was Sun City, Arizona, launched on New Year's Day in 1960. It remains the most successful retirement community network of all time and offers many lessons for age-affirmative branding. The first Sun City at the time it opened included five models, a shopping center, recreation center, and golf course. At opening weekend, it drew an unexpected 100,000 people, 10 times the number expected. Traffic jams mounted for miles, and Sun City founder Del Webb found himself on the cover of *Time* magazine. Sun City quickly became a well-established age brand and clearly an age-affirmative brand. It appealed to a market niche that regarded retirement as a good thing and aging into retirement as something positive. Sun City was marketed as the leisure destination of a lifetime, and the public bought the idea. In the years since then, Sun City has been replicated many times and has proved to be a resilient and powerful brand, always with an age-affirmative orientation.

One of the most remarkable examples of age-affirmative branding was the Dove campaign calling itself "Pro-Age" and focusing on a line of personal care products. In 2007, Dove launched its "Beauty Comes of Age" Global Research Study of attitudes toward physical appearance and age. The research was conducted in nine countries (Brazil, Canada, Germany, Italy, Japan, Mexico, France, the United Kingdom, and the United States) and surveyed 1,450 women ages 50–64. As part of the campaign, a global advertising initiative was launched questioning whether ideal attributes of youthful beauty are still relevant. The ad campaign included images of women whose appearances differed from the conventional ideal. One of the most striking of these featured Irene Sinclair, age 95, with a wrinkled face, asking: "Will society ever accept old can be beautiful?" Her question might remind us of the comment by aging Italian movie actress Anna Magnani to her makeup man when he tried to smooth over her wrinkles: "Don't take a single one. I paid for them" (Hillman, 1999).

In short, Pro-Age was an entirely different kind of ad campaign, a bold and risky one, based on age-affirmative branding (Milner, 2007). Instead of the negative and fear-driven thrust of typical "anti-aging" ads, Dove stressed affirmation and hope. One of the pictures even contained the label "Too old to be in an anti-aging ad." The models used in print advertising were all women over 50 shown in natural postures. The message of the campaign was that no matter how old a woman is, she could reveal "real beauty" that transcends limits of age. The Dove campaign, in short, was much more than an effort to sell personal care products. More broadly, it was a vehicle for age resistance, a challenge to prevailing ageist attitudes in society, and an effort to replace denial and avoidance with age-affirmative branding.

Elderhostel, now known as "Road Scholar," is perhaps a paradigm case of age-affirmative branding. When one of the authors served as chairman of the national board of directors, younger people would often comment "I can't wait to be old enough to be eligible for Elderhostel!" Elderhostel is a program offering liberal education and travel for people above age 55 (Mills, 1993). Indeed, it is now the world's largest educational travel organization, attracting nearly 200,000 participants each year. Elderhostel currently offers 10,000 programs annually in all 50 U.S. states and in over 90 countries around the world. Elderhostel sometimes describes itself as "the way college was supposed to be."

Elderhostel's age-affirmative brand offers interesting lessons for marketers. We think of young people as explorers and adventurers. But, Elderhostel breaks that mold and appeals to the love of adventure at any age. Therefore, the Elderhostel brand personality closely matches the target consumer personality.

CONCLUSION: THE FUTURE OF AGE BRANDING

We have identified four types of age-branding strategies that have been used successfully by various brands. Age-denial branding involves enhancing consumer self-esteem by positioning the brands around a youth theme. Age-adaptive branding involves a focus of consumer reference points on the possibilities of what could be, providing an opportunity for solution-oriented products. Age-irrelevant branding involves that development of brand concepts that create an appropriate image that may not focus on age at all. In contrast, age-affirmative branding celebrates age and uses it as a central theme in developing a brand personality.

As these examples illustrate, one key to successful age branding lies in "positioning" of brands in the minds of consumers. There are important lessons for the future of age branding to be learned from the history of successful service innovation companies, such as companies that pioneered with service innovation, for example, overnight package delivery (Federal Express), retail stock brokerage (Charles Schwab), storefront tax services (H & R Block), drive-in fast food (McDonald's), automated teller machines (Citibank), and selling over the Internet (eBay). Our aging society is unprecedented in historical terms, and age branding will be a key area for marketing innovation in the future.

References

Federal Interagency Forum on Aging-Related Statistics. (2000). Retrieved Feb. 12, 2010 from http://www.agingstats.gov/

Hillman, J. (1999). *The force of character: And the lasting life.* New York: Random House.

Kane, M. (2002). *The Botox book.* New York: St. Martin's Press.

Mills, E. S. (1993). *The story of Elderhostel.* Hanover, NH: University of New Hampshire Press.

Milner, C. (2007). Marketing effectively to baby boomers. *Journal on Active Aging,* 6(2), 46–47.

Rowe, J. W., & Kahn, R. L. (1997). Successful aging. *Gerontologist,* 37, 433–440.

Sixsmith, A., Hine, N., Neild, I., Clarke, N., Brown, S., and Garner, P. (2007). Monitoring the well-being of older people. *Topics in Geriatric Rehabilitation,* 23, 9–23.

Wolfe, D. B., & Snyder, R. E. (2003). *Ageless marketing: Strategies for reaching the hearts and minds of the new consumer majority.* Chicago: Dearborn Trade.

READING 52

Selling the Fountain of Youth

How the Anti-Aging Industry Made a Disease Out of Getting Old—and Made Billions

Arlene Weintraub

Others have assured men that they can safely regain the sexual prowess they enjoyed in their twenties by taking a little testosterone and growth hormone every day. Still others have transformed Chinese herbs and Brazilian berries into miraculous youth-restoring elixirs—and suckered millions of patients into buying expensive subscriptions to have those concoctions delivered right to their doorsteps.

All this marketing muscle, of course, is aimed at treating a nonexistent malady. Getting old is not a disease. The U.S. Food and Drug Administration (FDA) has never approved any therapy to treat aging. And although mainstream medical organizations have issued plenty of guidelines for preventing illnesses that commonly occur late in life, such as heart disease and cancer, none has ever fully endorsed the treatment regimens that the anti-aging industry embraces.

Nevertheless, anti-aging doctors have persuaded millions of baby boomers that they can stop the inevitable march toward Metamucil mornings and Viagra nights. These capitalists have constructed a giant new industry by taking advantage of an entire generation's deep-seated aversion to getting old. Their task has been made that much easier by the American culture,

---

SOURCE: *Selling the Fountain of Youth: How the Anti-Aging Industry Made a Disease Out of Getting Old—and Made Billions,* by Arlene Weintraub (Basic Books, 2010).

which has slapped images of beautiful young people all over movie and television screens, newspapers, and magazines. For the nation's 77 million baby boomers, who are quickly racing toward the Social Security rolls, the idea of being elderly is repugnant. They are the perfect audience for the message that simple, safe substances might cure aging.

Exercise is probably the only anti-aging regimen that actually works. That was definitively demonstrated in a ten-year project called the MacArthur Foundation Study of Aging in America, which was analyzed in John Wallis Rowe and Robert L. Kahn's 1998 [*sic* 1997] book, *Successful Aging.* The study was a collection of dozens of individual research projects, all designed to "pinpoint the many factors that conspire to put one octogenarian on cross-country skis and an-other in a wheelchair." One study, for example, followed more than 1,000 high-functioning adults for eight years. Another project scrutinized hundreds of Swedish twins in an effort to determine how genetics and lifestyle played into aging. "There is a simple, basic fact about exercise and your health: fitness cuts your risk of dying. It doesn't get much more 'bottom line' than that," wrote two of the scientists who managed the project.

The study's authors debunked claims about DHEA, HGH, testosterone, and other anti-aging favorites, while laying out a litany of convincing evidence about the value of physical activity. Exercise cut the risk of developing coronary artery disease, diabetes, and colon cancer, they discovered. It slashed the chance of developing high blood pressure by half. And there was no need for seniors to engage in Mr. Universe-style body-building or to spend hundreds of dollars consulting exercise physiologists. "Moderate physical activity, such as leisurely walking or gardening, was every bit as strong as strenuous exercise," they wrote.

Other studies proved that older people could enjoy the benefits of moderate exercise regardless of how late in life they started doing it.

Many of the most famous examples of successful aging were people who shunned hormones and steroids. Fitness industry pioneer Jack LaLanne told Larry King in 2000, "I wouldn't take a steroid if you'd give me $10 trillion." He was 85 at the time and had chalked up an impressive list of achievements. In addition to founding a health-club empire (which he sold to Bally Total Fitness), he performed physical feats that most young people wouldn't dare to try.

But most people seem hardwired to hate exercise. Even working out for a scant half hour is a lot to ask: A 2009 Gallup survey found that only one in four Americans were able to get off their butts for even that short a time, five or more days per week. And while anti-aging doctors claimed to care about how fit their patients were, most didn't place nearly as much emphasis on exercise as they did on hormones.

Mainstream medical societies continued to warn patients away from the anti-aging industry and the hormones it promoted. At its 2009 annual meeting in June, the American Medical Association adopted a new policy on the use of hormones for anti-aging purposes. Based on reviews of dozens of scientific studies, the AMA's Council on Science and Public Health concluded in a report for the association's members: "Despite the widespread promotion of hormones as anti-aging agents by for-profit Web sites, anti-aging clinics, and compounding pharmacies, the scientific evidence to support these claims is lacking." It states, "Current evidence fails to support the efficacy of HGH as an anti-aging therapy and adverse events are significant." Furthermore, the AMA's statement said, "No credible scientific evidence exists on the value of so-called 'bio-identical hormones,' and there are concerns about their purity, potency and quality because they are not approved by the FDA."

READING 53

Marketplace of Memory

What the Brain Fitness Technology Industry Says About Us and How We Can Do Better

Daniel R. George and Peter J. Whitehouse

A new marketplace of technological "brain fitness" products is emerging to address the fear of brain aging and desire to enhance cognition that are prevalent in modern populations. These products are sociocultural objects deeply imbued with the values and ideologies of our age, and this article considers their material evolution while also examining limitations in their real-world contribution to cognitive health. Ultimately, a broader and more complex story of "brain health" is advanced, which goes beyond the hype and reductionism of the "brain fitness" commercial marketplace and demonstrates how local communities can play a vibrant role in supporting cognitive and psychosocial well-being across the life span.

THE BRAIN FITNESS INDUSTRY: PRODUCTS, VALUES, AND IDEOLOGIES

Broadly speaking, the "brain fitness" technology industry has been estimated to represent a $300 million marketplace and is projects (perhaps optimistically) to achieve between $2 and $8 billion in worldwide revenue by 2015 as the baby boomer generation—to which many products are explicitly marketed—move into their 60s (Fernandez, 2010). Market size estimates vary depending on what types of products are included and whether one includes professional services or just products for lay persons. Whereas pharmaceutical and "smart drug" markets offer biological products to enhance brain function, this emerging marketplace features legions of digital products including video games and computer software, mobile phone apps, and other products proclaiming to instrumentally maintain or enhance the memory, concentration, visual and spatial skills, verbal recall, and executive functions of individual users. A neologism forged by the marketplace— "neurobics"—evinces the belief that this new generation of strenuous games, puzzles, and brainteasers can encourage the growth of synapses and dendrites and enhance cognitive health just as aerobic workouts improve pulmonary health and increase cardiovascular health (Ellin, 1999; Kelly, 2006).

Indeed, in exploring the advertising language in this marketplace, one notices the word "neuro" used in protean ways. Most commonly, it serves as a prefix for terms such as "neuro-enhancing" or "neuro-boosting" that focus consumer attention on how products tangibly benefit the function of a single organ—the brain. Concepts such as "neuro-plasticity" (the brain's capacity to rewire itself throughout life by creating neural connections in response to mental activity) and "cognitive-" or "neural-reserve"

SOURCE: "The Marketplace of Memory" by Daniel George and Peter Whitehouse (*The Gerontologist,* May 2011).

(the brain's built-up resilience to age-related pathological changes) give the impression of scientific certainty that products are capable of physically impacting the brain at the molecular level. Frequently, products are said to be "clinically proven" to improve cognitive performance in users of all ages, whereas other marketing campaigns boast that their products are "designed by neuroscientists" or endorsed by medical professionals (such as the Japanese physician Dr. Ryuta Kawashima, who gives his imprimatur to the best-selling Nintendo's Brain Age games). Occasionally, more lofty claims will surface, such as the promise that brain fitness technology can affect the brain to the point of preventing, slowing, or reversing dementia.

Mass marketing slogans such as "Give your brain the workout it needs!" (Nintendo Brain Age) and "Flex your brain the fun way!" (Big Brain Academy) allude to the rather demanding relationship these technological products forge between consumers and their brains. Posit Science, a leading software company, even suggests to potential consumers that, in return for playing brain fitness games, "Your brain will thank you"—a slogan that bifurcates "self" and "brain" while imputing equal agency to both.

Modern brain fitness technology products have been physically shaped by the neoliberal ideologies of the marketplace. Although some products may feature and encourage multi-person functionality, many are sold in single units and marketed for individual consumption on personal computers, individual video game consoles, PDA-like devices, or mobile phones. Although some long-term care facilities and other organizations such as hospitals and schools have purchased multiple copies of programs and are beginning to foster interaction through peer collaboration and some emerging web-based products are designed to encourage social networking, brain fitness products themselves have dictated that brain-healthy activities generally occur during private sedentary moments in the seclusion of one's home rather than in the context of group interaction. As discussed above,

the brain is most often treated as a symbiotic source of selfhood for the atomized consumer: an organ that must be constantly maintained and improved through personal labor if one is to reap the benefit of continued soundness of mind. The prevailing meaning conveyed by marketing departments is that when one uses products in a disciplined manner, consumers can enhance their neural pathways, thereby perfecting themselves from the molecular level outwards and slowing or preventing the encroachment of neurodegeneration.

PROBLEMS IN THE MARKETPLACE

At present, the brain fitness technology industry is being met with increasing scientific, if not cultural, scrutiny (Fernandez, 2010). Empirical support for the efficacy of brain fitness training programs in meaningfully improving cognition is generally insufficient. Relatively few products have been rigorously evaluated using scientific methods or reported in peer-reviewed journals, and most existing studies generally evaluate low-intensity interventions in which community-dwelling adults may engage in the intervention for 3 or fewer hours per week. When they have been evaluated, task performance is relatively easy to demonstrate, some cognitive generalizability is sometimes reported, but measuring improvement in daily life has rarely even been attempted. Moreover, it is reasonable to ask whether we should expect a profound transfer to activities of daily living when most interventions occur across such short time intervals at such low "doses."

In fact, it is not surprising that a human being who practices performance in any task will improve over time. The crucial question, which investigators have explored for several decades, is whether the increments generalize to other domains of thinking and whether they improve activities of daily living (Ball, Wadley, & Edwards, 2002; Detterman & Sternberg, 1982). Future research on brain fitness products most certainly merits continued investment—particularly studies that might address the dosing issues by embedding brain fitness activities in regular life activities. However, as long as evidence remains lacking, the industry's claims represent the hype of marketing departments rather than a genuine hope earned through thorough scientific inquiry.

How We Can Do Better—Brain Health in the Context of Communities

Advancing a broader understanding of brain health requires us to go beyond the dominant reductionism of the current brain fitness commercial marketplace and ask how our most proximate relationships and local communities can play a role in cognitive and psychosocial wellness.

Of course, measuring the effects of these complex activities is much more difficult than the classic method of swallowing a pill and comparing it to a similar-looking inactive product (i.e., placebo). Testing a brain fitness video game or computer program is also much easier than testing a complex social intervention, and intellectual property can be more easily assigned per pill or software program than a project undertaken in a shared community space.

Despite the challenges associated with evaluating these community-based interventions, when one considers the extant data (Buettner, 2009; Poulain et al., 2004) on longevity and well-being, those societies that produce the most centenarians feature some combination of the following attributes: strong families and community affiliations; an overriding sense of purpose, engagement, and contribution in the population; structures to manage and relieve stress; accessible and walkable living areas; low incidence of smoking; opportunities for daily ambulation and natural movement (e.g., walking, gardening, play); humane treatment of the elderly; low meat, plant-based diet with legumes,

etc. It is not clear where brain fitness technology might fit on this list, if assessed.

CONCLUSIONS

Ultimately, brain fitness technology has a role to play in the complex project of cognitive well-being and may make even more profound contributions to acute recovery from such conditions as stroke and traumatic brain injury. However, their current limitations—not to mention the excesses of the marketing departments currently promoting the products worldwide, especially to baby boomers—must be deeply scrutinized and matched by a counter-narrative that reinforces broader multifactorial notions of brain health and the interdependency of human populations and pushes us to think more imaginatively about cognitive health in the context of communities. We have no doubt that learning technologies have a significant potential role to play in improving our thinking and valuing of each other and of nature that likely far exceeds the capacity of biological products. We do hope, however, that the promise of these technologies is not weakened by the excess expectations created for the rather limited contemporary approaches to brain fitness.

REFERENCES

Ball, K., Wadley, V. G., & Edwards, J. D. (2002). Advances in technology used to assess and retrain older drivers. *Gerontechnology, 1,* 251–261.

Buettner, D. (2009). *The blue zones: Lessons for living longer from the people who've lived the longest* (first paperback ed.). Washington, DC: National Geographic.

Detterman, D. K., & Sternberg, R. J. (Eds.). (1982). *How and how much can intelligence be increased?* Norwood, NJ: Albex.

Ellin, A. (1999, October 3). Can 'neurobics' do for the brain what aerobics do for lungs? *The New York Times.* Retrieved from http://query.nytimes.com/gst/fullpage.html?res=9A01E3DC103EF9

30A35753C1A96F958260&sec=&spon=&page
wanted=all

Fernandez, A. (2010, July). *Transforming brain health with digital tools to assess, enhance and treat cognition across the lifespan: The state of the brain health market.* Retrieved from http://www.sharpbrains.com/executive-summary

Kelly, C. (2006). Neurobics: A way to exercise your brain. Retrieved from http://media.www.conycampus.com/media/storage/paper832/news/2006/03/06/Healthhealth/Neurobics.A.Way.To.Exercise.Your.Brain-1638823.shtml

Poulain, M., Pes, G. M., Grasland, C., Carru, C., Ferucci, L., Baggio, G., et al. (2004). Identification of a geographic area characterized by extreme longevity in the Sardinia Island: The AKEA study. *Experimental Gerontology, 39,* 1423–1429.

FOCUS ON THE FUTURE

Everything Money Can Buy

In this "futuristic" forecast of life in the year 2030, caregiving has been completely monetized, Medicare has been privatized, and everything is for sale. A few people reminisce about the "old days." But in this scenario of the future, the marketplace has taken over much of everyday life, raising provocative questions about where we should draw the line for what can, or should, be for sale.

* * * * *

The year is 2030, and Grandpa Harry is having a talk with his 20-year-old granddaughter, Carolyn, who is majoring in gerontology and business at State University.

"How are you today, Grandpa?" asks Carolyn.

"I'm pretty good," says Harry. "Got my Social Security check yesterday. This year was my 85th birthday, and I keep getting the checks. Feels good to know that I get a predictable amount each month. Then again, I'm still covered under Old Social Security. It will be different for you, I know."

"Yeah, Grandpa. I don't think much about my own Social Security at my age. Anyway, I'm under New Social Security, and I really like that. I'm already paying into it from my job at the university. Under New Social Security, you can't be sure what you'll ever get. But at least you know you'll get something because it's all your own money. I like that."

Grandpa Harry replies: "Amazing. When I was growing up, nobody ever imagined Social Security would become private. Same thing with Medicare. Of course, I'm still covered by Old Medicare, but people just 10 years younger than me are under the New Medicare. They get vouchers to buy private insurance. I guess I'm lucky that I got in under the wire with both Old Social Security and Old Medicare. Of course, the problem is that my benefits are being gradually cut back each year."

"Well," responds Carolyn, "the great thing about New Social Security is that we've got so many choices. And the financial services industry is booming these days. I've got an interview with a hedge fund for an internship in one of those companies that's focused on the retirement market. I'm also looking at some health insurance companies, and they like people with a background in gerontology and business."

Grandpa Harry asks, "You said something about your job at the university. What are you doing there?"

"Oh, I do a lot of different jobs, anything to avoid getting too many student loans, with high interest rates. I also sell my blood every two months, and that helps. The University Blood Bank pays good money for my special blood type. Some of my classmates, the girls, are sex workers during summer vacation or inter-session. That's been a big money-maker ever since the Libertarian Party got in and they made prostitution legal in Texas and now in our state, too."

"Did you say prostitution is legal right here? I guess I hadn't heard that story."

"Oh, just since last year," replies Carolyn. "But it's a good business for college girls, even if it's not for me. I do know some of the boys who've sold a kidney—that's where you can really make money. But of course there's foreign competition, too."

"Seems like everything's for sale these days," says Grandpa Harry. "It's all different from when I was growing up."

"Well, in the aging business these days, there are lots of opportunities because everything's for sale," says Carolyn. "You know, before I was born, I heard that senior centers were run by the government. Not anymore. I'm really interested in the national Senior Franchise Group and also in the private home health care businesses. The aging boomers have made these a really hot area for growth. Last week there was an initial public offering for the Senior Franchise Group, and their stock has nearly doubled. Our student investment club should have bought that stock for sure."

Grandpa Harry sighs: "I don't know. I've got mixed feelings about how these days everything is for sale. It's true that there are a lot more private senior centers than there used to be, but a lot of them aren't even called 'senior centers' at all. You know, I have friends who would never go to something called a 'senior' center. But they do go on that Road Scholar Program for traveling. Same program used to be called 'Elderhostel' in my day, but they changed the name."

Carolyn adds, "Yeah, Grandpa, you've got the right idea on that. There's another great business opportunity I'm looking into. It's called the Elder Companions. They sell social networking to old people who are isolated and lonely, sometimes over the telephone, sometimes in person. People will pay a lot for companionship, just like they pay for sex. It's a great opportunity."

Grandpa Harry comments, "Sounds like you're majoring in the right subject."

Carolyn agrees: "Absolutely. Gerontology is an awesome field now that they've tied it to business administration. The double major is a real advantage for me. There are so many aging boomers, people just like you. I've got a good professional future ahead of me."

NOTE: For further reading, see D. Satz, *Why Some Things Should Not Be for Sale: The Moral Limits of Markets,* New York: Oxford University Press, 2010.

Questions for Writing, Reflection, and Debate

1. When do older people commonly use the phrase *senior moment,* and how could this kind of event be linked to a propensity to buy certain products or services?

2. As they grow older, adults report that "old age" always seems to be farther and farther into the future. For example, older boomers report that old age only begins at age 80. How might marketers take account of this attitude as they approach consumers?

3. Interview an older consumer (someone 60-plus) and ask what things were most on his or her mind

the last time he or she bought a car. Then interview several people in their 20s, asking the same question. Compare the responses. What conclusions can be drawn?

4. In "The Marketplace of Memory," George and Whitehouse seem to suggest that most interventions to prevent cognitive decline don't work or aren't proven. What guidance could you give consumers who are looking for helpful products to prevent cognitive decline?

5. In her article on the boomer market, Mary Furlong makes the point that aging boomers are not a homogeneous group. What are some ways to segment or separate aging boomers into distinct groups for marketing purposes?

6. Moody and Sood argue that branding is important in marketing to an older population. Interview an older person, such as a family member, and ask him or her about a product he or she has bought that might be an "age brand." What characteristics of that product are appealing to this person?

7. Imagine that you're an official of the Food and Drug Administration charged with regulating anti-aging products. What arguments would you come up with for protecting consumers who want to buy such products? What arguments could be offered against such government regulation?

8. Suppose an anti-aging pill suddenly came on to the marketplace. Would you be interested in taking the pill? What would you hope to gain? What worries would you have about taking the pill?

Suggested Readings

Bradley, D. E., and Longino, C. F., Jr., "How Older People Think about Images of Aging in Advertising and the Media," *Generations* (2001), 25, 17–21.

Moody, H. R., "Silver Industries and the New Aging Enterprise," *Generations* (Winter, 2004–2005), 28(4): 75–78.

Smith, J. W., and Clurman, A. S., *Rocking the Ages: The Yankelovich Report on Generational Marketing,* New York: HarperBusiness, 1997.

Stroud, D., *The 50-Plus Market: Why the Future Is Age Neutral for Reaching the Hearts and Minds of the New Consumer Majority*, London: Kogan Page, 2005.

Thornhill, M., and Martin, J., *Boomer Consumer: Ten New Rules for Marketing to America's Largest, Wealthiest and Most Influential Group,* Great Falls, VA: Linx, 2007.

Wolfe, D., & Snyder, R., *Ageless Marketing: Strategies for Reaching the Hearts and Minds of the New Customer Majority*, Chicago: Dearborn Trade, 2003.

Student Study Site

Visit the Student Study Site at **www.sagepub.com/moody7e** for these additional learning tools:

- Flashcards
- Web quizzes
- Chapter outlines
- SAGE journal articles
- Web resources
- Video and audio resources

APPENDIX

How to Research a Term Paper in Gerontology

Research and writing can be intimidating to many students, especially in a field such as gerontology, which is a new subject to most. But research and writing needn't be frightening. Skillful research is the key to good writing, and curiosity and careful thinking are the foundation for both.

Doing the background research for a term paper in gerontology is more than half the task of actually writing the paper itself. If you are successful in the research, you end up having other people do your work for you! Of course, that does *not* mean plagiarism or simply copying what other people have written without giving proper credit. When you find text that fits what you need, rewrite it in your own language. The trick in research and writing is to save yourself the trouble of "reinventing the wheel." You want to avoid floundering around trying to rediscover a fact or an idea that someone else has already worked out before you. Wasting time that way is not necessary at all. In fact, it detracts from the real job of research and writing, namely, *thinking* about what others have written and deciding what to take and then put into your own work.

The key is not to work harder, but to work smarter. By building on other people's work and giving credit to them where credit is due, you save yourself time and devote your best efforts to expressing what you really have to say. The process is the same as what takes place in science. All science and all scholarship stand on the work of others. This point holds true for the beginning student no less than for more experienced scholars. Indeed, the great physicist Isaac Newton himself once said, "If I have seen further than others, it is because I have stood on the shoulders of giants."

How does this approach apply to your writing a term paper? Conducting library research for a term paper is a bit like looking for buried treasure. If you don't know *exactly* where the treasure is buried, you end up spending a lot of time digging in places where you imagine the treasure *might* be. The key is to rely on guesswork when you're not sure exactly what you're looking for or where to find it. Once you have a hunch about where the treasure lies, then the actual digging takes practically no time at all. It is just the same with research. Searching on the Web can make this process much easier. But it also presents a danger: You can get too many leads and actually get overwhelmed. To avoid that trap, you need to clearly define and refine your search process. Once you have developed your search strategy—your map for where treasure might be found—then the information sources at your fingertips will guide you quickly to where the treasure lies. The rest of the work—including writing up your findings—will perhaps take less time than you feared, because you have developed the foundation upon which to build your own ideas.

DEFINING YOUR TOPIC

At every stage in the research process, you need to ask yourself: What is the question I am asking? What information am I trying to find? You ask these questions not only once but over and over again as you learn more. For example, suppose you are trying to find out what percentage of people between the ages of 60 and 70 are retired. At first, that question may seem simple: Can't you just look up the number? But as you dig deeper, you find that there is uncertainty about how to count people as *retired* instead of *unemployed* or *having disabilities.* As you look into the statistics, you discover that behind the solid numbers, differing assumptions are involved. In effect, you ask your basic question over and over again as you look through bibliographic sources. And you refine your question over and over again, as well. You might come up with several different numbers of people who are *retired* depending on how the term is defined.

When you are planning the topic for your term paper, you might find it helpful first to free-associate, or let your mind wander. You need to think about points related to your topic but also think about other subject terms and ideas related to your topic. This process of divergent thinking is at the heart of research and creativity. For example, suppose you're interested in writing a paper on retirement. Retirement is a big subject, maybe too big for one paper. Social scientists have written whole books on the subject; some have devoted their entire careers to it. But stay with that big subject for a while. Then, think about all the other subjects—the keywords—that are related to retirement: *work, leisure, pensions, Social Security,* and so on. Each of these could also be a term paper or indeed a whole book.

As you look over all your keywords, look for connections that interest you; for instance, maybe you see a connection between *pensions* and *retirement.* (Or, perhaps you are interested in how a particular group of people, Hispanic women, for example, experience retirement.) You might begin to put together a hypothesis or a theme—for example, what is the relationship between pensions and retirement behavior? When you ask yourself research questions, it is helpful if you write down some tentative answers. That's the first step toward making an outline, or a plan, for your work. Carrying out research is a bit like building a house. In constructing the house, it pays to put time into planning and thinking. You don't wait to draw up blueprints until you are halfway finished constructing the house. To write a term paper, you also need a plan. Brainstorm and write down your ideas first without worrying too much about whether your plan is adequate or complete; you're likely to change it later anyway. Then, start consulting other sources.

STARTING YOUR SEARCH

In constructing a fruitful search strategy, you have a problem. You can't really narrow your research question until you know the subject matter better. But you can't define the subject matter without carving it down to size with an adequately narrow research question.

Imagine how discouraging your task would be if you didn't realize that pension levels and retirement behavior were related. In gerontology, as in all fields, the amount of knowledge is simply too vast for you to master all of it. To make matters worse, gerontology is a multidisciplinary field, involving specialized subjects such as economics, biology, psychology, and so on. Without a clear plan for research, you can simply get lost.

The secret of research is to keep widening your search process while narrowing it at the same time. For example, the topic you've picked has two key ideas: *pensions* and *retirement.* Some of the references you find may lead you in directions that don't interest you—for instance, *pension fund investments* or *mandatory retirement.* But other references will be right on target and will lead you to refine your topic even further. There lies the real process of thinking: testing your ideas against a "map" of knowledge that sums up facts about the world. The mistake that people often make is to construct, at the beginning, a search that is either too narrow or too broad.

As you review what you find, you'll begin to see connections among concepts. But the connected concepts may or may not be exactly the ones listed in computer printouts or abstract summaries. You have to develop a sixth sense, constantly looking for clues. The result of this process is a more complete cross-referencing of your subject matter; in effect, you're creating a dense network of concepts that fully captures your topic and prepares you to write your paper.

A number of resources are available to help you build your network of concepts. One is the library's own classification system; another is the librarians themselves. The Library of Congress subject headings present a uniform method of classifying documents, and that can be a useful place to begin. But the real clues will come as you examine the books and journals themselves. *Don't* simply go to the library card catalog or start browsing through the latest issue of a periodical related to your subject. Doing that will just waste your time, unless you have done some preliminary planning (though when you are new to a subject and have some time on your hands, browsing the latest periodicals can be a fun and informative experience!). By all means, enlist librarians to help you, but don't rely exclusively on librarians. They can't be specialists in all subject matters, and they can help you the most if you've already done some thinking about the question you want to pursue. If you've thought about your question, then a librarian can help guide you to the information sources you need.

Another kind of resource that might be helpful is the computerized online database. But because searching and researching are not mechanical processes, a computer search won't solve all your problems, and it may even give the illusion of comprehensiveness. Computer searches also present the student with certain dangers. There are two general kinds of dangers in online searching. The first is summarized in a slogan familiar to computer specialists: Garbage in, garbage out. That is, you can only get an answer to the question you ask; if your question or hypothesis is badly framed—for example, if it's too vague—then you won't get useful information. The second danger is that you may get too much information, including lots of references that are irrelevant or useless. For both dangers, the cure is the same: good strategies for searching and for eliminating what is extraneous to your search. The main message here is that you can't do bibliographic research just by looking for simple terminology, by looking up words in an index, a card catalog, or a database. One reason is that there are so many related but distinct terms in gerontology: *aged, older persons, older adults, elderly, senior citizens,* and so on. But if you can formulate a research question and remain alert to the meanings of the terms you encounter, you can find the sources that will help you answer your question. Once you find the spot you've been looking for, the buried treasure will be lying at your feet.

NINE STEPS FOR CARRYING OUT LIBRARY RESEARCH

Step 1. Consult *The Encyclopedia of Aging* or a comparable source for the lead article on your subject. Be sure to make note of the relevant bibliography citations.

Step 2. Consult one of the handbooks on aging (from biology, the social sciences, the humanities, etc.) or a current textbook to see if there is a chapter or a section of a chapter devoted to your subject. The handbook's index can be useful here. (Be sure also to check the more detailed resource list provided at the end of this appendix.)

Step 3. Review the bibliography references you have found and organize them by date, starting with the most recent. Look for titles that focus directly on your topic but approach the subject in a broad way. A literature review article is often an excellent way to get started. Many published articles begin with a literature review or "state-of-the-art" summary of what is known about a topic.

Step 4. Consult the AgeLine Database with the keywords related to your topic, if you can get access to AgeLine without charge. Read the abstracts for publications to find the most up-to-date literature on your subject. Looking at abstracts is a quick and handy way to see a summary of what's in a possible reference without spending time reading the entire article. You get more than just a title, and you can find out quickly if the publication could have value for you.

Step 5. It's time to dive into the World Wide Web using a search engine such as Google. But remember that the search is likely to turn up many more hits than will be useful. Work carefully to make your search terms as relevant as possible. Test results with different search terms. Assess results in terms of what you judge to be the reliability and objectivity of the source (website, periodical, etc.). Try using Google Scholar for more academic sources.

Step 6 (optional). To be truly comprehensive and up-to-date, ask a friendly librarian to conduct a computerized search on your subject through appropriate bibliographic databases. From your previous bibliographic work, you should have a good collection of keywords or authors to help the librarian focus on your topic as precisely as possible. Don't neglect to ask librarians for help.

Step 7. You may find much of what you need on the Web, and you will want to save the citations, abstracts, and other useful bits of knowledge to use for your term paper. You may also want to go to library stacks to find relevant books and articles on your tentative topic. But note: As in your Web search, do not "judge a book by its cover" or accept a reference solely by its title. Remember to browse through any citation you find: Look carefully at the table of contents, the index, the introduction, a summary chapter, and so on. If Amazon gives access to an entire book, this can be easy. Don't make the mistake of reading straight through the entire text of what looks like the "perfect" book or article on your subject. It could be a waste of time. Instead, quickly zero in on the essential information and let the rest go. You can always come back later if you need to. It's good to get other points of view on your topic.

Step 8. When you are browsing through books or articles, be sure to check their bibliographies or reference lists for interesting titles. Using ideas from these books or articles, you will then be able to fine-tune your topic while taking notes and picking up additional ideas that you can incorporate into your paper.

Step 9. In many cases, you will find the full sources you need in your local college or university library. The library will often provide ways of getting the full text of articles you want. Don't hesitate to request books or articles on interlibrary loan, for example, from a wider university system or from other libraries. But don't fall into the trap of the perpetual scholar, who keeps searching forever and never quite finds the "perfect" reference source. In most cases, you will find what you need close to home. When writing a term paper, you have a deadline to meet.

ENDING THE SEARCH

The search must come to an end. At some point in this process, you are likely to find yourself coming up again and again with the same books, articles, and author names as you look through new information sources. Don't be discouraged by this. It isn't a sign of failure or that you are "going around in circles." On the contrary, it could be a sign of success. If you have gone far enough in your search, it may mean that you've struck pay dirt. When you have gone really deeply into any subject area, you are bound to start seeing the same authors' names coming up again and again.

At that point, it is time to look through the references on hand and decide which ones are high quality and which ones are relevant for your now refined topic area. Decide which ones are really the most useful to you and gather the key ideas, always giving credit but putting the ideas into your own words and properly citing sources for directly quoted materials. When you have found the treasure you are looking for, go home and start writing.

RESOURCES FOR RESEARCH PAPERS IN GERONTOLOGY

Encyclopedias and Handbooks

There are two definitive encyclopedias that should always be consulted: *The Encyclopedia of Aging* (Two-Volume Set), edited by Richard Schulz, Linda Noelker, Kenneth Rockwood, and Richard Sprott, New York: Springer, 2006. It contains hundreds of entries written by leading authorities in each field. This volume is accessible to students as well as more advanced scholars, and its vast list of references makes it extremely useful. Equally valuable is the *Encyclopedia of Gerontology* (Two-Volume Set, Second Edition), edited by James E. Birren, Academic Press, 2006. Of special value for the Schulz et al. *Encyclopedia of Aging* is that for this book's Amazon listing, one can search for specific index terms (*retirement, euthanasia,* etc.) and read the pages on which the search term occurs.

Also very useful are the following: *Encyclopedia of Ageism,* edited by Erdman B. Palmore, Laurence Branch, and Diane Harris, Routledge, 2005 (entries here go beyond prejudices of ageism); and *Encyclopedia of Aging and Public Health,* edited by Sana Loue and Martha Sajatovic, Springer, 2008.

Among the most useful single-volume reference works are the many handbooks that focus on aging and the biological sciences, the social sciences, psychology, and human services. These include the following:

Robert H. Binstock and Linda K. George (eds.), *Handbook of Aging and the Social Sciences* (7th ed.), San Diego, CA: Academic Press, 2010. Contains updated versions of articles that appeared in earlier editions as well as material in new areas. Covers the life course and social context, stratification and generational relations, work and economy, politics and policy analysis, and applied topics of aging and social intervention.

James E. Birren and K. Warner Schaie (eds.), *Handbook of the Psychology of Aging* (7th ed.), San Diego, CA: Academic Press, 2010. Covers theory and measurement, influences of behavior and aging, perceptual and cognitive processes, and applications to the individual and society.

Thomas R. Cole, Robert Kastenbaum, and Ruth Ray (eds.), *Handbook of the Humanities and Aging,* New York: Springer, 2000. Covers aging through history, comparative religion, arts and literature, and contemporary topics in humanistic gerontology.

Karen L. Fingerman, Cynthia Berg, Jacqui Smith, and Toni C. Antonucci (eds.), *Handbook of Life-Span Development,* New York: Springer, 2010. A comprehensive treatment integrating psychology into a broader ecological context, covering infancy through old age.

Edward J. Masoro and Steven N. Austad (eds.), *Handbook of the Biology of Aging* (7th ed.), San Diego, CA: Academic Press, 2010. Covers all aspects of biogerontology from molecular biology, cell biology, and genetics through the physiology of major organic systems of the human body.

Abstracts and Databases

Abstracts in Social Gerontology: Current Literature on Aging. Published by EBSCO, Quarterly. Each issue contains hundreds of abstracts (short summaries) of important articles, books, government reports, legislative research studies, and other recent literature, cross-indexed and organized by topic, along with other (nonannotated) bibliographic citations.

http://www.ebscohost.com/academic/abstracts-in-social-gerontology

AgeLine is the world's only bibliographic database exclusively devoted to aging and covering all aspects of the social sciences, health care, and human services. The database includes books, articles, government documents, and dissertations as well as reports on government-sponsored research in gerontology. Published by EBSCO, the database is not free unless accessed through your university library system.

http://www.ebscohost.com/public/ageline

For references to biomedical subjects and other health-related topics, another good source is PubMed, produced by the National Library of Medicine. A college or other research-oriented library will be able to provide computer search services to this and other databases.

Statistics

The conventional sources for U.S. statistics are the publications of the U.S. Census Bureau, for example, *Current Population Reports,* which updates information from the 2010 census. Census documents are available in most college libraries. The **Centers for Medicare and Medicaid Services** and the special committees on aging of the U.S. Senate and House also publish periodic reports, which can often be obtained by writing to these agencies or by visiting a large library. Documents from specialized sources may be difficult to obtain in local libraries, and they are not always easy to understand.

http://www.census.gov/main/www/cprs.html

For the beginning student, the best quick sources for statistics may be the comprehensive data published by the U.S. Administration on Aging, *A Profile of Older Americans: 2010,* available at:

http://www.aoa.gov/aoaroot/aging_statistics/Profile/index.aspx

Other Useful Reference Works

Barbara Berkman (ed.), *Handbook of Social Work in Health and Aging,* Oxford University Press, 2006.

Joseph Kandel and Christine Adamec, *The Encyclopedia of Senior Health and Well-Being,* Facts on File, 2003.

Donald H. Kausler, *The Graying of America: An Encyclopedia of Aging, Health, Mind, and Behavior* (2nd ed.), University of Illinois Press, 2001.

Jacqueline L. Longe (ed.), *Gale Encyclopedia of Senior Health,* Gale, 2008.

Barbara Wexler, *Growing Old in America,* Information Plus, 2010.

Finally, the federal government is an important source of reliable information. See, for example, the U.S. Senate Special Committee on Aging, Publications List (Washington, DC: Government Printing Office).

Textbooks

Current textbooks on aging and gerontology are valuable sources of information and further reference for students. The following is a partial list of textbooks that may prove useful:

Robert C. Atchley and Amanda Barusch, *Social Forces and Aging,* Belmont, CA: Wadsworth, 2003.

John C. Cavanaugh and Fredda Blanchard-Fields, *Adult Development and Aging,* Wadsworth, 2010.

Harold G. Cox, *Later Life: The Realities of Aging* (6th ed., Prentice-Hall, 2005.

Paul W. Foos and M. Cherie Clark, *Human Aging* (2nd ed.), Prentice-Hall, 2008.

Nancy Hooyman and H. Asuman Kiyak, *Social Gerontology: A Multidisciplinary Perspective* (9th ed.), Prentice-Hall, 2010.

William Hoyer and Paul Roodin, *Adult Development and Aging* (6th ed.), McGraw-Hill, 2008.

Cary S. Kart, *The Realities of Aging: An Introduction to Gerontology* (6th ed.), Boston: Allyn & Bacon, 2001.

Jill Quadagno, *Aging and the Life Course: An Introduction to Social Gerontology,* New York: McGraw-Hill, 2010.

Robbyn R. Wacker and Karen A. Roberto, *Community Resources for Older Adults: Programs and Services in an Era of Change,* Thousand Oaks, CA: Sage, 2007.

Important Journals and Periodicals

For an overview of important periodicals in the field of gerontology, see "Journals on Aging" compiled by Prof. Monika Deppen Wood, available at:

http://crab.rutgers.edu/~deppen/journals.htm

Among the many notable periodicals available, the following can be recommended:

Ageing and Society. New York: Cambridge University Press, eight times a year. Edited in Great Britain with an international and interdisciplinary perspective; strong on humanities and social science.

http://journals.cambridge.org/action/displayJournal?jid=ASO

Educational Gerontology. Published by Taylor & Francis, 12 times a year. Covers both gerontology instruction and education for older adults.

http://www.tandf.co.uk/journals/UEDG

Generations. Journal of the American Society on Aging (ASA). Quarterly. Each issue is devoted to a specialized topic, such as the politics of aging, AIDS and aging, the environment and aging, and so on.

http://www.asaging.org/blog/content-source/15

The Gerontologist. Washington, DC: Gerontological Society of America, bimonthly. Interdisciplinary and focused on social gerontology.

http://gerontologist.oxfordjournals.org/

International Journal of Aging and Human Development. Farmingdale, NY: Baywood, eight times annually. Interdisciplinary in the social sciences, with attention to the psychological and human dimensions of aging.

http://www.baywood.com/journals/previewjournals.asp?id=0091-4150

Journal of Aging and Social Policy. Published by Taylor & Francis, quarterly. Focused on policy issues and readable by beginning students.

http://www.tandfonline.com/loi/wasp20

Journal of the American Geriatrics Society. Published monthly by Wiley-Blackwell. A technical medical journal for geriatricians. Many articles are above the level of the beginning student, but some are accessible.

http://www.wiley.com/bw/journal.asp?ref=0002-8614

Journal of Gerontological Nursing. Published by Slack Journals, the Wyanoke Group, monthly. Leading periodical that covers clinical health care issues of interest to many health care providers.

http://www.slackjournals.com/JGN

Journal of Gerontological Social Work. Published by Taylor & Francis, eight issues per year. The leading periodical covering all aspects of social welfare policy and clinical practice in the field of aging.

http://www.tandfonline.com/loi/wger20

Journal of Women and Aging. Published by Taylor & Francis, quarterly. The only periodical covering all aspects of gender and aging.

http://www.tandfonline.com/loi/wjwa20

The Journals of Gerontology. Gerontological Society of America, bimonthly. Actually four separate journals. The most relevant of these is likely to be Series B, covering psychological sciences and social sciences. Very technical and specialized; only for very advanced inquiry.

http://psychsocgerontology.oxfordjournals.org/content/early/recent

Psychology and Aging. Washington, DC: American Psychological Association, quarterly. Covers all aspects of adult life span development and aging, including behavioral, clinical, and experimental psychology.

http://www.apa.org/pubs/journals/pag/index.aspx

Research on Aging. Thousand Oaks, CA: Sage, bimonthly. Covers a broad range of inquiry for gerontology in the social sciences. Accessible for the educated reader but contains mainly specialized articles.

http://roa.sagepub.com/

BIBLIOGRAPHY

Aaron, Henry J., and Schwartz, William B., *The Painful Prescription: Rationing Hospital Care,* Washington, DC: Brookings Institution, 1984.

Aaron, Henry J., and Schwartz, William B. (Eds.), *Coping With Methuselah: The Impact of Molecular Biology on Medicine and Society,* Washington, DC: Brookings Institution, 2004.

AARP, "Baby Boomers Envision What's Next?" (June 2011). Retrieved September 30, 2011 (http://www.aarp.org/work/retirement-planning/info-06-2011/boomers-envision-retirement-2011.html).

Achenbaum, W. Andrew, *Old Age in the New Land: The American Experience Since 1790,* Baltimore: Johns Hopkins University Press, 1978.

Achenbaum, W. Andrew, *Social Security: Visions and Revisions,* Cambridge, UK: Cambridge University Press, 1986.

Adams, Rebecca, and Blieszner, Rosemary, "Aging Well With Friends and Family," *American Behavioral Scientist* (1995), 39(2): 209–224.

Adams-Price, Carolyn, *Creativity and Successful Aging: Theoretical and Empirical Approaches,* New York: Springer, 1998.

Aday, R., Rice, C., and Evans, E., "Intergenerational Partners Project: A Model Linking Elementary Students with Senior Center Volunteers," *The Gerontologist* (1991), 31(2): 263–266.

Administration on Aging, *A Profile of Older Americans: 2010,* Washington, DC: Author, 2010.

Aleman, Sara, *Hispanic Elders and Human Services,* New York: Garland Publishing, 1997.

Alexander, C. N., Chandler, H. M., Langer, E. J., Newman, R. I., and Davies, J. L., "Transcendental Meditation, Mindfulness, and Longevity: An Experimental Study With the Elderly," *Journal of Personality and Social Psychology* (December 1989), 57(6): 950–964.

Allen, Katherine R., and Chin-Sang, Victoria, "A Lifetime of Work: The Context and Meaning of Leisure for Aging Black Women," *The Gerontologist* (1990), 30: 734–740.

Alliance for Aging Research, *Seven Deadly Myths: Uncovering the Facts About the High Cost of the Last Year of Life,* Washington, DC: Author, 1997.

American Council on Education, *Mapping New Directions: Higher Education for Older Adults,* Washington, DC: Author, 2008.

American Medical Association, "AMA Statement on End-of-Life Care," 2011. Retrieved September 16, 2011 (http://www.ama-assn.org/ama/pub/physician-resources/medical-ethics/about-ethics-group/ethics-resource-center/end-of-life-care/ama-statement-end-of-life-care.page).

Anderson, W. French, "Human Gene Therapy," *Science* (1992), 256: 808–813.

Anderson, William A., and Anderson, Norma, D., "The Politics of Age Exclusion: The Adults Only Movement in Arizona," *The Gerontologist* (1978), 18: 6–12.

Anetzberger, G. J., *The Etiology of Elder Abuse by Adult Offspring,* Springfield, IL: Charles C Thomas, 1987.

Anstey, Kaarin J., Lord, Stephen R., and Smith, Glen A., "Measuring Human Functional Age: A Review of Empirical Findings," *Experimental Aging Research* (September 1996), 22(3): 245–266.

Aries, Philippe, *Centuries of Childhood,* New York: Random House, 1962.

Arking, Robert, "Modifying the Aging Process," in Rosalie Young and Elizabeth Olson (Eds.), *Health, Illness and Disability in Later Life: Practice Issues and Interventions,* Newbury Park, CA: Sage, 1991.

Arking, Robert, *Biology of Aging: Observations and Principles* (2nd ed.), Sunderland, MA: Sinauer Associates, 1998.

Arling, Greg, Buhaug, Harold, Hagan, Shelley, and Zimmerman, David, "Medicaid Spenddown in Nursing Home Residents in Wisconsin," *The Gerontologist* (1991), 31(2): 174–182.

Arling, Greg, and McAuley, William J., "The Feasibility of Public Payments for Family Care-Giving," *The Gerontologist* (1983), 23: 300–306.

Armstrong, D., Sohal, R. S., Cutler, R. G., and Slater, T. F. (Eds.), *Free Radicals in Molecular Biology, Aging, and Disease,* New York: Raven, 1984.

Atchley, Robert, *Social Forces and Aging: An Introduction to Social Gerontology* (4th ed.), Belmont, CA: Wadsworth, 1985.

Atchley, Robert, *Spirituality and Aging: Expanding the View,* Johns Hopkins University Press, 2009.

Austad, Steven N., *Why We Age: What Science Is Discovering About the Body's Journey Through Life,* New York: John Wiley, 1997.

Avorn, Jerome, "Benefit and Cost Analysis in Geriatric Care: Turning Age Discrimination Into Health Policy," *New England Journal of Medicine* (1984), 310: 1294–1301.

Avorn, Jerome, "The Life and Death of Oliver Shay," in Alan Pifer and Lydia Bronte (Eds.), *Our Aging Society,* New York: Norton, 1986.

Baker, F. M., "Suicide Among Ethnic Minority Elderly: A Statistical and Psychosocial Perspective," *Journal of Geriatric Psychiatry* (1994), 27(2): 241–264.

Baltes, Margaret M., and Baltes, Paul B. (Eds.), *The Psychology of Control and Aging,* Hillsdale, NJ: Lawrence Erlbaum, 1986.

Baltes, Paul B., "Wise, and Otherwise," *Natural History* (February 1992).

Baltes, Paul B., "Aging Mind: Potential and Limits," *The Gerontologist* (October 1993), 33(5): 580–594.

Baltes, Paul B., and Baltes, Margaret M., "Psychological Perspectives on Successful Aging: The Model of Selective Optimization and Compensation," in Paul B. Baltes and Margaret M. Baltes (Eds.), *Successful Aging: Perspectives From the Behavioral Sciences,* New York: Cambridge University Press, 1990, pp. 1–34.

Baltes, P. B., and Schaie, K. W., "Aging and IQ: The Myth of the Twilight Years," *Psychology Today* (1974), 10: 35–40.

Baltes, Paul B., and Staudinger, Ursula M. (Eds.), *Interactive Minds: Life-Span Perspectives on the Social Foundation of Cognition,* Cambridge, UK: Cambridge University Press, 1996.

Banner, Lois W., *In Full Flower: Aging Women, Power, and Sexuality: A History,* New York: Vintage Books, 1993.

Barer, Barbara M., and Johnson, Colleen L., "Problems and Problem Solving Among Aging White and Black Americans," *Journal of Aging Studies* (August 2003), 17(3): 323–340.

Barinaga, Marcia, "Mortality: Overturning Received Wisdom," *Science* (October 16, 1992), 258: 398–399.

Bass, Scott, and Caro, Francis J. (Eds.), *Toward a Productive Aging Society,* Westport, CT: Auburn House, 1993.

Bass, Scott A., Caro, Francis G., and Chen, Yung-Ping (Eds.), *Achieving a Productive Aging Society,* Auburn House, 1993.

Bateson, M. C., *Composing a Further Life: The Age of Active Wisdom,* New York: Knopf, 2010.

Battin, Margaret P., "Choosing the Time to Die: The Ethics and Economics of Suicide in Old Age," in Stuart F. Spicker and Stanley Ingman (Eds.), *Ethical Dimensions of Geriatric Care,* Dordrecht, The Netherlands: Reidel, 1987.

Bauer, J. A., "Resveratrol Improves Health and Survival of Mice on a High-Calorie Diet," *Nature* (November 16, 2006), 444: 337–342.

Bayor, Ronald, *Columbia Documentary History of Race and Ethnicity in America,* New York: Columbia University Press, 2004.

Becker, Ernest, *The Denial of Death,* New York: Free Press, 1973.

Belbin, E., and Belbin, R. M., *Problems in Adult Retraining,* London: Heinemann, 1972.

Belbin, R. M., *The Discovery Method: An International Experiment in Retraining,* Paris: OECD, 1969.

Bell, I. P., "The Double Standard Age," in J. Freeman (Ed.), *Women: A Feminist Perspective* (4th ed.), Mountain View, CA: Mayfield, 1989, pp. 236–244.

Bengtson, Vern L., and Lowenstein, Ariela (Eds.), *Global Aging and Challenges to Families,* New York: Gruyter, 2003.

Berger, Peter L., and Luckmann, Thomas, *The Social Construction of Reality,* Doubleday, 1966.

Berman, Harry J., "To Flame With Wild Life: Florida Scott-Maxwell's Experience of Old Age," *The Gerontologist* (1986), 26: 321–324.

Bertman, Stephen (Ed.), *The Conflict of Generations in Ancient Greece,* Atlantic Highlands, NJ: Humanities Press, 1976.

Best, Fred, "Work Sharing: Issues, Policy Options, and Prospects," Kalamazoo, MI: Upjohn Institute for Employment Research, 1981.

Bianchi, Eugene C., *Aging as a Spiritual Journey,* New York: Crossroads, 1982.

Binstock, Robert, "Interest-Group Liberalism and the Politics of Aging," *The Gerontologist* (1972), 12: 265–280.

Binstock, Robert, "The Oldest-Old: A Fresh Perspective on Compassionate Ageism Revisited," *Milbank Memorial Fund Quarterly* (1985), 63: 420–451.

Binstock, Robert, "Old-Age-Based Rationing: From Rhetoric to Risk?" *Generations* (Winter 1994), 18(4): 37–41.

Binstock, Robert H., "Older People and Political Engagement: From Avid Voters to 'Cooled-out Marks,'" *Generations* (Winter 2006–2007), 30(4): 24–36.

Binstock, Robert, Cluff, Leighton, and Mering, Otto, *The Future of Long-Term Care: Social and Policy Issues,* Baltimore: Johns Hopkins University Press, 1996.

Birkhill, W. R., and Schaie, K. W., "The Effect of Differential Reinforcement of Cautiousness in Intellectual Performance Among the Elderly," *Journal of Gerontology* (1975), 30: 578–583.

Birren, James E., and Deutschman, Donna E., *Guiding Autobiography Groups for Older Adults: Exploring the Fabric of Life,* Baltimore: Johns Hopkins University Press, 1991.

Blackhall, L. J., Murphy, S. T., Frank, G., Michel, V., and Azen, S., *Journal of the American Medical Association* (September 13, 1995), 274(10): 820–825.

Blau, Zena S., *Aging in a Changing Society,* New York: Franklin Watts, 1981.

Blau, Zena S. (Ed.), *Work, Retirement and Social Policy,* Greenwich, CT: JAI, 1985.

Blazer, Dan G., "Spirituality and Aging Well," *Generations* (Winter 1991), 15(1): 61–65.

Blazer, Dan G., *Depression in Late Life* (2nd ed.), St. Louis, MO: C. V. Mosby, 1993.

Bloom, J., Ansell, P., and Bloom, M., "Detecting Elder Abuse: A Guide for Physicians," *Geriatrics* (1989), 44(6): 40–44.

Blumenthal, H. T. (Ed.), *Handbook of Diseases of Aging,* New York: Van Nostrand, 1983.

Blumenthal, Herman T., "Aging-Disease Dichotomy: True or False?" *Journals of Gerontology: Series A: Biological Sciences and Medical Sciences* (February 2003), 58A(2): 138–145.

Boaz, Rachel F., "Early Withdrawal From the Labor Force: A Response Only to Pension Pull or Also to Labor Market Push?" *Research on Aging* (1987), 9(4): 530–547.

Bodnar, A. G., Ouellete, M., Frolkis, M., Holt, S. E., Chiu, C.-P., Morin, G. B., Harley, C. B., Shay, Jerry W. Lichtsteiner, S., and Wright, W. E., "Extension of Life-Span by Introduction of Telomerase Into Normal Human Cells," *Science* (1998), 279: 349–352.

Bolles, Richard N., *The Three Boxes of Life and How to Get Out of Them,* Berkeley, CA: Ten Speed, 1981.

Bonifazi, Wendy, "Who Pays for Long Term Care?" *Contemporary Long Term Care* (October 1998), 21(10): 76–78.

Bonnie, Richard J., and Wallace, Robert B. (Eds.), *Elder Mistreatment: Abuse, Neglect, and Exploitation in an Aging America,* Washington, DC: National Academies Press, 2003.

Boron, Julie Blaskewicz, Turiano, Nicholas A., Willis, Sherry L., and Schaie, K. Warner, "Effects of Cognitive Training on Change in Accuracy in Inductive Reasoning Ability," *Journals of Gerontology: Series B: Psychological Sciences and Social Sciences* (May 2007), 62B(3): 179–186.

Bosworth, Barry P., "Fund Accumulation: How Much? How Managed?" in Peter A. Diamond, David C. Lindeman, and Howard Young (Eds.), *Social Security: What Role for the Future?* Washington, DC: National Academy of Social Insurance, 1996, pp. 89–113.

Bosworth, Barry, and Burtless, Gary (Eds.), *Aging Societies: The Global Dimension,* Washington, DC: Brookings Institution, 1998.

Boyle, Joan, and Morriss, James, *The Mirror of Time: Images of Aging and Dying,* Westport, CT: Greenwood, 1987.

Braun, Kathryn, Pietsch, James H., and Blanchette, Patricia L. (Eds.), *Cultural Issues in End-of-Life Decision Making,* Thousand Oaks, CA: Sage, 2000.

Braun, P., and Sweet, M., "Passages: Fact or Fiction?" *International Journal of Aging and Human Development* (1984), 18: 161–176.

Brennan, Penny L., and Steinberg, L. D., "Is Reminiscence Adaptive? Relations Among Social Activity Level, Reminiscence, and

Morale," *International Journal of Aging and Human Development* (1983–1984), 18: 99–110.

Breyer, Friedrich, Kliemt, Hartmut, and Thiele, F. (Eds.), *Rationing in Medicine: Ethical, Legal and Practical Aspects,* Berlin: Springer, 2002.

Brickner, Philip W., Lipsman, R., Lechich, Anthony J., & Scharer, Linda K., *Long-Term Health Care: Providing a Spectrum of Services to the Aged,* New York: Basic Books, 1987.

Brim, Orville Gilbert, Ryff, Carol D., and Kessler, Ronald C. (Eds.), *How Healthy Are We? A National Study of Well-Being at Midlife,* Chicago: University of Chicago Press, Chicago, 2004.

Brody, Elaine, "Parent Care as a Normative Family Stress," *The Gerontologist* (1985), 25: 19–29.

Brody, Elaine, *Women in the Middle: Their Parent-Care Years* (2nd ed.), New York: Springer, 2004.

Brody, Jane, "Secrets of Keeping Aging's Effects at Bay," *New York Times* (May 29, 2001), p. F-8.

Brogden, Michael, *Geronticide: Killing the Elderly,* London: Jessica Kingsley, 2001.

Bronson, R. T., and Lipman, R. D., "Reduction in Rate of Occurrence of Age-Related Lesions in Dietary Restricted Laboratory Mice," *Growth, Development and Aging* (1991), 55: 169–184.

Brookings/ICF, "Long-Term Care Financing Model" (unpublished data), Washington, DC: Author, 1990.

Brooks, David, *Bobos in Paradise: The New Upper Class and How They Got There,* New York: Touchstone, 2000.

Brooks, Jeffrey D., "Living Longer and Improving Health: An Obtainable Goal in Promoting Aging Well?" *American Behavioral Scientist* (January 1996), 39(3): 272–287.

Brown, Helen Gurley, *The Late Show,* New York: William Morrow, 1993.

Brown, Randall S., Bergeron, Jeanette W., Clement, Dolores Gurnick, Hill, Jerrold W., and Retchin, Sheldon, *Does Managed Care Work for Medicare? An Evaluation of the Medicare Risk Program for HMOs,* Princeton, NJ: Mathematica, 1993.

Brubaker, Timothy H., *Later Life Families,* Beverly Hills, CA: Sage, 1985.

Brubaker, Timothy H., *Family Relationships in Later Life,* Newbury Park, CA: Sage, 1990.

Buchanan, R. J., "Medicaid: Family Responsibility and Long Term Care," *Journal of Long Term Care Administration* (1984), 12(3): 19–25.

Budish, Armond D., *Avoiding the Medicaid Trap: How to Beat the Catastrophic Cost of Nursing Home Care,* New York: Henry Holt, 1989.

Buonocore, A. J., "Older and Wiser: Mature Employees and Career Guidance," *Management Review* (September 1992), 81(9): 54–57.

Burbank, P. M., "Exploratory Study: Assessing the Meaning in Life Among Older Adult Clients," *Journal of Gerontological Nursing* (September 1992), 18(9): 19–28.

Burke, Gerald, "Changing Health Needs of the Elderly Demand New Policies," *Journal of American Health Policy* (1993), 3(5): 22–26.

Burkhauser, Richard V., "Alternative Social Security Responses to the Changing Roles of Women and Men," in Colin D. Campbell (Ed.), *Controlling the Cost of Social Security,* Lexington, MA: D.C. Heath and Company, 1984, pp. 141–162.

Burrows, James, *The Ages of Man,* New York: Oxford University Press, 1986.

Burton, Linda (Ed.), *Families and Aging,* Amityville, NY: Baywood, 1993.

Butler, Robert N., "The Life Review: An Interpretation of Reminiscence in the Aged," *Psychiatry* (1963), 26: 65–76.

Butler, Robert N., "Age-ism: Another Form of Bigotry," *The Gerontologist* (1969), 9: 243–246.

Butler, Robert N., "Successful Aging and the Role of the Life Review," *Journal of the American Geriatrics Society* (1974), 22: 529–535.

Butler, Robert N., "Strategies to Delay Dysfunction in Later Life," in J. L. C. Dall et al. (Eds.), *Adaptations in Aging,* San Diego, CA: Academic Press, 1995, pp. 289–297.

Butler, Robert N., "Dangers of Physician-Assisted Suicide," *Geriatrics* (July 1996), 51(7): 14–15.

Butler, Robert N., Fossel, Michael, Harman, S. Mitchell, Heward, Christopher B., Olshansky, S. Jay, Perls, Thomas T., Rothman, David J., Rothman, Sheila M., Warner, Huber R., West, Michael D., and Wright, Woodring E., "Is There an Antiaging Medicine?" *The Journals of Gerontology Series A: Biological Sciences and Medical Sciences* (2002), 57: B333–B338.

Butler, Robert N., and Gleason, Herbert P. (Eds.), *Productive Aging: Enhancing Vitality in Later Life,* New York: Springer, 1985.

Butler, Robert N., and Kiikuni, Kenzo, *Who Is Responsible for My Old Age?* Springer, 1993.

Butler, R. N., and Lewis, Myrna I., *Love and Sex After 60* (rev. ed.), New York: Ballantine, 1993.

Butrica, Barbara A., and Uccello, Cori, *How Will Boomers Fare at Retirement?* Washington, DC: AARP Public Policy Institute, 2004.

Button, James W., "Sign of Generational Conflict: The Impact of Florida's Aging Voters on Local School and Tax Referenda," *Social Science Quarterly* (December 1992), 73(4): 786–797.

Byers, Bryan, and Hendricks, James, *Adult Protective Services: Research and Practice,* Springfield, IL: Charles C Thomas, 1993.

Callahan, Daniel, "What Do Children Owe Elderly Parents?" Hastings Center Report (April 1985), 15(2): 32–33.

Callahan, Daniel, *Setting Limits,* New York: Simon & Schuster, 1987.

Callahan, Daniel, "Setting Limits: A Response," *The Gerontologist* (June 1994), 34(3): 393–398.

Campbell, Andrea Louise, *How Policies Make Citizens: Senior Political Activism and the American Welfare State,* Princeton, NJ: Princeton University Press, 2003.

Cantor, Marjorie H., "The Informal Support System: Its Relevance in the Lives of the Elderly," in Edgar F. Borgatta and Neil McCluskey (Eds.), *Aging and Society: Current Research,* Beverly Hills, CA: Sage, 1980, pp. 131–144.

Cantor, Marjorie H., "Families and Caregiving in an Aging Society," *Generations* (Summer 1992), 67–70.

Capitman, J., "Case Management in Long-Term and Acute Medical Care," *Health Care Financing Review* (1988), Annual Supplement, pp. 75–81.

Cappelli, Peter, and Novelli, Bill, *Managing the Older Worker: How to Prepare for the New Organizational Order,* Harvard Business Press, 2010.

Cassel, Christine, Rudberg, M., and Olshansky, J., "The Price of Success: Health Care in an Aging Society," *Health Affairs* (1992), 11(2): 87–99.

Centers for Disease Control and Prevention, "Achievements in Public Health, 1900–1999: Control of Infectious Diseases," *MMWR* (1999), 48(29): 621–629.

Centers for Disease Control and Prevention, *Healthy Aging: Preventing Disease and Improving Quality of Life Among Older Americans,* Atlanta, GA: Author, 2003.

Centers for Disease Control and Prevention, *The State of Aging and Health in America 2007,* Washington, DC: Author, 2007.

Centers for Disease Control and Prevention, *Heart Disease and Stroke Prevention Addressing the Nation's Leading Killers: At A Glance 2011,* Washington, DC: Author, 2011.

Centers for Disease Control and Prevention, National Center for Health Statistics, *Vital and Health Statistics,* Series 10, No. 189, Washington, DC: Government Printing Office, February 1994.

Centers for Disease Control and Prevention, National Center for Health Statistics, *Health: United States 1994,* Hyattsville, MD: Author, 1995a.

Centers for Disease Control and Prevention, National Center for Health Statistics, *Supplement on Aging Study and Second Supplement on Aging Study,* Atlanta, GA: Author, 1995b.

Centers for Disease Control and Prevention, National Center for Health Statistics, *Vital and Health Statistics,* Series 13, Washington, DC: Government Printing Office, February 2005.

Centers for Medicare and Medicaid Services, "Chapter Three: Public Programs: Medicare, Medicaid, SCHIP," *An Overview of the U.S. Healthcare System: Two Decades of Change, 1980–2000,* 2004.

Centers for Medicare and Medicaid Services, "National Health Expenditure Fact Sheet," 2009. Retrieved September 20, 2011 (http://www.cms.gov/NationalHealthExpendData/25_NHE_Fact_Sheet.asp).

Chambre, Susan M., *Good Deeds in Old Age: Volunteering by the New Leisure Class,* New York: Free Press, 1987.

Chambre, Susan M., "Volunteerism by Elders: Past Trends and Future Prospects," *Gerontologist* (April 1993), 33(2): 221–228.

Chellis, Robert D., Seagle, James F., and Seagle, Barbara M. (Eds.), *Congregate Housing for Older People,* Lexington, MA: Lexington, 1982.

Cheung, Monit, "Elderly Chinese Living in the United States: Assimilation or Adjustment?" *Social Work* (September 1989), 34(5): 457–461.

Chinen, Allan B., *In the Ever After: Fairy Tales and the Second Half of Life,* Wilmette, IL: Chiron, 1989.

Choi, N. G., "Does Social Security Redistribute Income? A Tax-Transfer Analysis," *Journal of Sociology and Social Welfare* (1991), 19(3): 21–38.

Chu, Cyrus, "Age-Distribution Dynamics and Aging Indexes," *Demography* (November 1997), 34(4): 551–563.

Chudakoff, Howard P., *How Old Are You? Age Consciousness in American Culture,* Princeton, NJ: Princeton University Press, 1989.

Clark, Brian, *Whose Life Is It Anyway? A Play,* London: Samuel French, 1978.

Clark, Gordon L., and Whiteside, Noel, *Pension Security in the 21st Century: Redrawing the Public-Private Debate,* Oxford, UK: Oxford University Press, 2004.

Clark, Robert L., Burkhauser, Richard V., Moon, Marilyn, Quinn, Joseph F., and Smeeding, T. M., *Economics of an Aging Society,* Malden, MA: Blackwell Publishing, 2004.

Clark, Robert L., Maddox, G., Schrimper, R., and Sumner, D., *Inflation and the Economic Well-Being of the Elderly,* Baltimore: Johns Hopkins University Press, 1984.

Clark, Roger, "Modernization and Status Change Among Aged Men and Women," *International Journal of Aging and Human Development* (1992–1993), 36(3): 171–186.

Clark, William F., Pelham, Anabel O., and Clark, Marleen L., *Old and Poor: A Critical Assessment of the Low Income Elderly,* Lexington, MA: Lexington, 1988.

Cnaan, R., and Cwikel, J. G., "Elderly Volunteers: Assessing Their Potential as an Untapped Resource," *Journal of Aging and Social Policy* (1992), 4: 125–147.

Cockerham, William C., *This Aging Society* (2nd ed.), Upper Saddle River, NJ: Prentice Hall, 1997.

Cohen, Marc A., Tell, Eileen, Greenberg, Jan N., and Wallack, Stanley S., "The Financial Capacity of the Elderly to Insure for Long-Term Care," *The Gerontologist* (1987), 27(5): 494–502.

Cole, Ardra L, and Knowles, J. Gary, *Lives in Context: The Art of Life History Research,* Walnut Creek, CA: Rowman and Littlefield, 2001.

Cole, S., "Age and Scientific Performance," *American Journal of Sociology* (1979), 84: 958–977.

Cole, Thomas, *The Journey of Life: A Cultural History of Aging in America,* Cambridge, UK: Cambridge University Press, 1992.

Cole, Thomas, and Gadow, Sally (Eds.), *What Does It Mean to Grow Old? Views From the Humanities,* Durham, NC: Duke University Press, 1986.

Coleman, P. G., "Measuring Reminiscence Characteristics From Conversation as Adaptive Features of Old Age," *International Journal of Aging and Human Development* (1974), 5: 281–294.

Coles, Robert, *The Old Ones of New Mexico,* Albuquerque: University of New Mexico Press, 1974.

Coming of Age, "How Seniors Think," 2010. Retrieved September 22, 2011 (http://www.comingofage.com/senior-marketing/).

Cook, Fay L., and Kramek, Lorraine M., "Measuring Economic Hardship Among Older Americans," *The Gerontologist* (1986), 26: 38–47.

Corbin, J. M., and Strauss, A., *Unending Care and Work: Managing Chronic Illness at Home,* San Francisco: Jossey-Bass, 1988.

Cornelius, Steven W., "Aging and Everyday Cognitive Abilities," in Thomas Hess (Ed.), *Aging and Cognition: Knowledge Organization and Utilization* (*Advances in Psychology,* No. 71), Amsterdam: North Holland, 1990, pp. 411–459.

Cornelius, Steven W., and Caspi, A., "Everyday Problem Solving in Adulthood and Old Age," *Psychology and Aging* (1987), 2: 14–153.

Costa, Dora, *Evolution of Retirement: An American Economic History, 1880–1990,* Chicago: University of Chicago Press, 1998.

Costa, Paul T., Jr., and McCrae, Robert R., "Still Stable After All These Years: Personality as a Key to Some Issues in Aging," in P. B. Baltes and O. G. Brim (Eds.), *Life-Span Development and Behavior* (Vol. 3), New York: Academic Press, 1980, pp. 65–102.

Costa, Paul T., Jr., Metter, M., and McCrae, Robert R., "Personality Stability and Its Contribution to Successful Aging," *Journal of Geriatric Psychiatry* (1994), 27(1): 41–59.

Couzin, J., "Low-Calorie Diets May Slow Monkey's Aging," *Science* (1998), 282: 1018.

Covey, H. C., "Perceptions and Attitudes Toward Sexuality of the Elderly During the Middle Ages," *The Gerontologist* (February 1989), 29(1): 93–100.

Cowgill, Donald, *Aging Around the World,* Belmont, CA: Wadsworth, 1986.

Cowley, Malcolm, *The View From 80,* New York: Viking Press, 1980.

Coyle, Jean M. (Ed.), *Handbook on Women and Aging,* Westport, CT: Greenwood, 1997.

Crimmins, E. M., & Beltran-Sanchez, H., "Mortality and Morbidity Trends: Is There Compression of Morbidity?" *Journal of Gerontology: Social Sciences* (2010), 66B(I): 75–86.

Crimmins, Eileen M., and Ingegneri, Dominique G., "Trends in Health Among the American Population," in Anna M. Rappaport and

Sylvester J. Schieber (Eds.), *Demography and Retirement: The Twenty-First Century,* Westport, CT: Praeger, 1993, pp. 225–253.

Crown, William H., "Some Thoughts on Reformulating the Dependency Ratio," *The Gerontologist* (1985), 25: 166–171.

Cruzan v. Director of Missouri Department of Health, 497 U.S. 261, 110 S. Ct. 2841 (1990).

Crystal, Stephen, *America's Old Age Crisis: Public Policy and the Two Worlds of Aging,* New York: Basic Books, 1982.

Crystal, Stephen, "Measuring Income and Inequality Among the Elderly," *The Gerontologist* (1986), 26: 56–59.

Crystal, Stephen, and Shea, Dennis, "Cumulative Advantage, Cumulative Disadvantage, and Inequality Among Elderly People," *The Gerontologist* (1990), 30: 437–443.

Crystal, Stephen, Shea, Dennis, and Schaie, K. Warner (Eds.), *Focus on Economic Outcomes in Later Life: Public Policy, Health, and Cumulative Advantage,* New York: Springer Publishing, 2003.

Cumming, Elaine, and Henry, William E., *Growing Old: The Process of Disengagement,* New York: Basic Books, 1961.

Cunningham, Walter R., and Torner, Adrian, "Intellectual Abilities and Age: Concepts, Theories and Analyses," in Eugene A. Lovelace (Ed.), *Aging and Cognition: Mental Processes, Self-Awareness, and Interventions* (*Advances in Psychology,* No. 72), Amsterdam: North-Holland, 1990, pp. 379–406.

Cutler, Neal, "Political Characteristics of Elderly Cohorts in the Twenty First Century," in S. B. Kiesler (Ed.), *Aging and Social Change,* New York: Academic Press, 1981.

Cutler, Neal, Gregg, Davis W., and Lawton, M. Powell, *Aging, Money, and Life Satisfaction: Aspects of Financial Gerontology,* New York: Springer, 1992.

Cutler, Richard, "Species Probes, Longevity and Aging," in *Intervention in the Aging Process* (Part B), New York: Alan Liss, 1983.

Czaja, Sara J., and Barr, Robin, "Technology and the Everyday Life of Older Adults," *American Academy of Political and Social Science: Annals* (May 1989), 503: 127–137.

Czaja, Sara J., Guerrier, J. H., Nair, S. N., and Landauer, T. K., "Computer Communication as an Aid to Independence for Older Adults," *Behaviour and Information Technology* (July–August 1993), 12(4): 197–207.

Daly, Jeanette M., Jogerst, Gerald J., Brinig, Margaret F., and Dawson, Jeffrey D, "Mandatory Reporting: Relationship of APS Statute Language on State Reported Elder Abuse," *Journal of Elder Abuse and Neglect* (2003), 15(2): 1–21.

Dannefer, Dale, "Adult Development and Social Theory: A Paradigmatic Reappraisal," *American Sociological Review* (1984), 49: 100–116.

Dannefer, Dale, "Cumulative Advantage/Disadvantage and the Life Course: Cross-Fertilizing Age and Social Science Theory," *Journals of Gerontology: Series B: Psychological Sciences and Social Sciences* (November 2003), 58B(6): S327–S337.

Davies, D. R., and Sparrow, P. R., "Age and Work Behaviour," in N. Charness (Ed.), *Aging and Human Performance,* Chichester, UK: Wiley, 1985.

Davis, Richard, and Davis, Jim, *TV's Image of the Elderly,* Lexington, MA: Lexington, 1985.

Davis, T. J., "Seniors Reach Out to Troubled Youth," *Secure Retirement* (November–December, 1992), 1(5): 30–31.

Davis, T. J., "Investment Scam Line," *Secure Retirement* (September 1993), 2(6): 34–36.

Day, Christine, *What Older Americans Think: Interest Groups and Aging Policy,* Princeton, NJ: Princeton University Press, 1990.

DeBerry, S., Davis, S., and Reinhard, K. E., "Comparison of Meditation-Relaxation and Cognitive/Behavioral Techniques for Reducing Anxiety and Depression in a Geriatric Population," *Journal of Geriatric Psychiatry* (1989), 22(2): 231–247.

de Grey, Aubrey, and Rae, Michael, *Ending Aging: The Rejuvenation Breakthroughs That Could Reverse Human Aging in Our Lifetime*, St. Martin's, 2008.

Deikman, Arthur J., "De-Automization and the Mystic Experience," *Psychiatry* (1966), 29: 324–338. Reprinted in Charles Tart (Ed.), *Altered States of Consciousness* (3rd ed.), San Francisco, CA: Harper, 1990.

Delgado, Melvin (Ed.), *Latino Elders and the Twenty-First Century,* New York: Haworth Press, 1998.

Dellman-Jenkins, M., "Old and Young Together: Effect of Educational Programs on Pre-schoolers," *Childhood Education* (1986), 62: 206–208.

DeLong, David, *Lost Knowledge: Confronting the Threat of an Aging Workforce,* Oxford: Oxford University Press, 2004.

Demetrius, Lloyd, "Caloric Restriction, Metabolic Rate, and Entropy," *The Journals of Gerontology Series A: Biological Sciences and Medical Sciences* (2004), 59: B902–B915.

Demos, V., and Jache, A., "When You Care Enough: An Analysis of Attitudes Toward Aging in Humorous Birthday Cards," *Gerontologist* (April 1981), 21(2): 209–215.

Dennis, Helen, *Fourteen Steps in Managing an Aging Work Force,* Lexington, MA: Lexington, 1988.

Dennis, Helen, "The Current State of Preretirement Planning," *Generations* (1989), 13(2): 38–41.

Dennis, W., "Creative Productivity Between the Ages of 20 and 80 Years," *Journal of Gerontology* (1966), 21: 1–8.

Diamond, Arthur M., Jr., "The Life-Cycle Research Productivity of Mathematicians and Scientists," *Journal of Gerontology* (1986), 41: 520–525.

Diamond, Peter A., and Orszag, Peter, *Saving Social Security: A Balanced Approach,* Washington, DC: Brookings Institution, 2004.

Diamond, Peter A., and Orszag, Peter R., *Saving Social Security: A Balanced Approach,* Washington, DC: Brookings Institution, 2005.

Dill, Ann E. P., et al., "Coercive Placement of Elders: Protection or Choices," *Generations* (1987), 11(4): 48–66.

Dilworth-Anderson, Peggye, "Extended Kin Networks in Black Families," *Generations* (Summer 1992), 16: 29–32.

Dixon, Roger A., Backman, Lars, and Nilsson, Lars-Goran (Eds.), *New Frontiers in Cognitive Aging,* Oxford, UK: Oxford University Press, 2004.

Dobrosky, B., and Bishop, J., "Children's Perceptions of Old People," *Educational Gerontology* (1986), 12: 429–439.

Doering, Mildred, Rhodes, Susan R., and Schuster, Michael, *The Aging Worker: Research and Recommendations,* Beverly Hills, CA: Sage, 1983.

Doty, Pamela, *Cost-Effectiveness of Home and Community-Based Long-Term Care Services,* Washington, DC: U.S. Department of Health and Human Services, 2000.

Dowd, James J., "Aging as Exchange: A Preface to Theory," *Journal of Gerontology* (1975), 30: 584–594.

Dowd, James J., and Bengtson, Vern L., "Aging in a Minority Population: An Examination of the Double Jeopardy Hypothesis," *Journal of Gerontology* (1978), 33: 427–436.

Dranove, David, *What's Your Life Worth? Health Care Rationing—Who Lives? Who Dies? Who Decides?* New York: Prentice Hall, 2003.

Dressel, Paula L., "Gender, Race, and Class: Beyond the Feminization of Poverty in Later Life," *The Gerontologist* (1988), 28(2): 177–180.

Duke University, Center for the Study of Aging and Human Development, *Multidimensional Functional Assessment: The OARS Methodology* (2nd ed.), Durham, NC: Author, 1978.

Duncan, G., "The Volatility of Family Income Over the Life Course," in P. B. Baltes, D. L. Featherman, and R. M. Lerner (Eds.), *Life-Span Development and Behavior* (Vol. 9), Hillsdale, NJ: Lawrence Erlbaum, 1988, pp. 37–358.

Durkheim, Émile, *Suicide* (J. A. Spaulding and G. Simpson, Trans.), Glencoe, IL: Free Press, 1897/1951.

Ehmann, C., "The Age Factor in Religious Attitudes and Behavior," *Gallup News Service* (July 14, 1999).

Eisele, F. R., "Origins of Gerontocracy," *The Gerontologist* (1979), 19: 4.

Ekerdt, David J., "Busy Ethic: Moral Continuity Between Work and Retirement," *The Gerontologist* (June 1986), 26(3): 239–244.

Ekerdt, David J., "Why the Notion Persists That Retirement Harms Health," *The Gerontologist* (1987), 27(4): 454–457.

Ekerdt, D. J., "Work, Health, and Retirement," in K. S. Markides (Ed.), *Encyclopedia of Health and Aging*, Thousand Oaks, CA: Sage, 2007.

Elder, Glen H., *Children of the Great Depression,* Chicago: University of Chicago Press, 1974.

Ellison, James M., and Verma, Sumer (Eds.), *Depression in Later Life: A Multidisciplinary Psychiatric Approach,* New York: Marcel Dekker, 2003.

Erikson, Erik, *Childhood and Society,* New York: Macmillan, 1963.

Estabrook, Madeleine A., "False Economies: Downsizing as Cure Costs Valuable Older Workers," *Pension World* (May 1993), 29(5): 10–12.

Estes, Carroll, *The Aging Enterprise,* San Francisco: Jossey-Bass, 1979.

Estes, Carroll, "Aging, Health, and Social Policy: Crisis and Crossroads," *Journal of Aging and Social Policy* (1989), 1(1–2): 17–32.

Estes, Carroll L., "Social Security Privatization and Older Women: A Feminist Political Economy Perspective," *Journal of Aging Studies* (February 2004), 18(1): 9–26.

Ewing Marion Kauffman Foundation, *The Coming Entrepreneurship Boom,* 2011. Retrieved September 22, 2011 (http://www.kauffman.org/research-and-policy/the-coming-entrepreneurial-boom.aspx).

Fagerlin, Angela, and Schneider, Carl E., "Enough: The Failure of the Living Will," *Hastings Center Report* (2004), 34(2): 30–42.

Falkner, Thomas, and de Luce, Judith, "A View From Antiquity," in Thomas Cole, David Van Tassel, and Robert Kastenbaum (Eds.), *Handbook of the Humanities and Aging,* New York: Springer, 1992.

Fama, T., and Kennell, D. L., "Should We Worry About Induced Demand for Long-Term Care Services?" *Generations* (Spring 1990): 37–41.

Farber, Jeff, et al., "Myths of the High Medical Cost of Old Age and Dying," *International Journal of Health Services* (2008), 38:2.

Featherstone, Mike, and Hepworth, Mike, "Images of Ageing," in John Bond, Peter Coleman, and Sheila Peace (Eds.), *Ageing in Society: An Introduction to Social Gerontology,* Thousand Oaks, CA: Sage, 1993, pp. 304–332.

Featherstone, Mike, and Wernick, Andrew (Eds.), *Images of Aging: Cultural Representations of Later Life,* New York: Routledge, 1995.

Feder, Judith, Komisar, Harriet L., and Niefeld, Marlene, "Long-Term Care in the United States: An Overview," *Health Affairs* (May–June 2000), 19(3): 40–56.

Federal Interagency Forum on Aging-Related Statistics, *Older Americans 2004: Key Indicators of Well-Being,* Washington, DC: U.S. Government Printing Office, 2004.

Feinberg, Lynn, Reinhard, Susan C., Houser, Ari, and Choula, Rita, "Valuing the Invaluable: 2011 Update," Washington, DC: AARP Public Policy Institute, 2011. Retrieved September 16, 2011 (http://www.aarp.org/relationships/caregiving/info-07-2011/valuing-the-invaluable.html).

Fengler, A. P., "Life Satisfaction of Sub-Populations of Elderly: The Comparative Effects of Volunteerism, Employment, and Meal Site Preparation," *Research on Aging* (1984), 6: 208.

Ferraro, K. F., and LaGrange, R. L., "Are Older People Most Afraid of Crime? Reconsidering Age Differences in Fear of Victimization," *Journals of Gerontology* (1992), 47(5): S233–S244.

Filene, Peter G., *In the Arms of Others: A Cultural History of the Right-to-Die in America,* Chicago: Ivan R. Dee, 1998.

Finch, C. E., *Longevity, Senescence and the Genome,* Chicago: University of Chicago Press, 1990.

Finch, C. E., and Kirkwood, Thomas B. L., *Chance, Development, and Aging,* New York: Oxford University Press, 2000.

Findsen, Brian, *Learning Later,* Malabar, FL: Krieger Publishing, 2005.

Firman, J., Gelfand, D., and Ventura, C., "Students as Resources to the Aging Network," *The Gerontologist* (1983), 23(2): 185–191.

Fischer, David Hackett, *Growing Old in America,* New York: Oxford University Press, 1977.

Fleck, C., "Multigenerations Under One Roof," in *AARP Bulletin,* Washington, DC: AARP, 2010.

Folstein, M. F., Folstein, S. E., and McHugh, P. R., "'Mini-Mental State': A Practical Method for Grading the Cognitive State of Patients for the Clinician," *Journal of Psychiatric Research* (1975), 12: 189–198.

Folts, W. Edward, and Streib, Gordon F., "Leisure-Oriented Retirement Communities," in W. Edward Folts and Dale E. Yeatts (Eds.), *Housing and the Aging Population: Options for the New Century,* New York: Garland, 1994, pp. 121–144.

Foner, Nancy, *Ages in Conflict: A Cross-Cultural Perspective on Inequality Between Old and Young,* New York: Columbia University Press, 1984.

Fowler, James W., *Stages of Faith: The Psychology of Human Development and the Quest for Meaning,* New York: Harper & Row, 1981.

Freedman, Marc, *Prime Time: How Baby Boomers Will Revolutionize Retirement and Transform America* (paperback ed.), Public Affairs, 2002.

Freedman, Marc, *Encore: Finding Work That Matters in the Second Half of Life,* Public Affairs, 2007.

Freedman, Robert M., Lomasky, L. E., and May, M. I., "Why Won't Medicaid Let Me Keep My Nest Egg?" (case study), *Hastings Center Report* (April 1983), 13(2): 23–25.

Freeman, Scott M., Whartenby, Katharine, and Abraham, George N., "Gene Therapy: Applications to Diseases Associated With Aging," *Generations* (1992), 16: 45–48.

Fried, Linda P., Ferrucci, Luigi, Darer, Jonathan, Williamson, Jeff D., and Anderson, Gerard, "Untangling the Concepts of Disability, Frailty, and Comorbidity: Implications for Improved Targeting and Care," *Journals of Gerontology: Series A: Biological Sciences and Medical Sciences* (March 2004), 59A(3): 255–263.

Friedland, Robert, and Summer, Laura, *Demography Is Not Destiny,* Washington, DC: National Academy on an Aging Society, 1999.

Friedman, Howard S., and Martin, Leslie R., *The Longevity Project: Surprising Discoveries for Health and Long Life from the Landmark Eight-Decade Study*, Hudson Street Press, 2011.

Fries, James F., "Aging, Illness, and Health Policy: Implications of the Compression of Morbidity," *Perspectives in Biology and Medicine* (Spring 1988), 31: 3.

Fuchs, Victor R., "Health Care for the Elderly: How Much? Who Will Pay for It?" *Health Affairs* (January–February 1999), 18(1): 11–21.

Furlong, Mary, *Turning Silver Into Gold: How to Profit in the New Boomer Marketplace,* Upper Saddle River, NJ: FT Press, 2007.

Furlong, Mary S., and Lipson, Stefan B., *Young @ Heart: Computing for Seniors,* Berkeley, CA: Osborne McGraw-Hill, 1996.

Gale, W. G., and Scholz, J. K., "Intergenerational Transfers and the Accumulation of Wealth," *Journal of Economic Perspectives* (Fall 1994), 8(4): 145–160.

Gambria, L. M., "Daydreaming About the Past: The Time Setting of Spontaneous Thought Intrusions," *The Gerontologist* (1977), 17: 35–38.

Garber, Alan M., MaCurdy, Thomas, and McClellan, Mark, "Diagnosis and Medicare Expenditures at the End of Life," in David A. Wise (Ed.), *Frontiers in the Economics of Aging,* Chicago: University of Chicago Press, 1998, pp. 247–274.

Gardner, Howard, *Frames of Mind: The Theory of Multiple Intelligence,* New York: Basic Books, 1985.

Garrett, W. W., "Filial Responsibility Laws," *Journal of Family Law* (1980), 18: 793–818.

Gee, E. M., "Historical Change in the Family Life Course of Canadian Men and Women," in V. Marshall (Ed.), *Aging in Canada* (2nd ed.), Markham, ON: Fitzhenry & Whiteside, 1987, pp. 265–287.

Gelfand, Donald, *The Aging Network: Programs and Services* (5th ed.), New York: Springer, 1999.

Gelfand, Donald, *Aging and Ethnicity: Knowledge and Services,* New York: Springer, 2003.

Gelfand, Donald, *The Aging Network: Programs and Services* (6th ed.), Springer, 2006.

Gendell, Murray, "Trends in Retirement Age in Four Countries, 1965–95," *Monthly Labor Review* (August 1998), 121(8): 20–30.

Gendell, M., "Older Workers: Increasing their Labor Force Participation and Hours of Work," *Monthly Labor Review,* 2008.

Generations Policy Initiative, *Age Explosion: Baby Boomers and Beyond,* Cambridge, MA: Harvard Generations Policy Initiative, 2004.

George, Linda K., "Depressive Disorders and Symptoms in Later Life," *Generations* (Winter–Spring 1993), 17(1): 35–38.

Gergeron, L. Rene, and Gray, Betsey, "Ethical Dilemmas of Reporting Suspected Elder Abuse," *Social Work* (January 2003), 48(1): 96–105.

Gibson, Rose C., "Work Patterns of Older Black and White and Male and Female Heads of Household," *Journal of Minority Aging* (1983), 8(1–2): 1–16.

Gibson, Rose C., "Reconceptualizing Retirement for Black Americans," *The Gerontologist* (1987), 27: 691–698.

Gilleard, Christopher J., and Higgs, Paul, *Cultures of Ageing: Self, Citizen and the Body,* Harlow, England: Prentice Hall, 2000.

Giordano, J. A., and Giordano, N. H., "A Classification of Preretirement Programs: In Search of a New Model," *Educational Gerontology* (1983), 9: 123–137.

Gist, Yvonne, and Velkoff, Victoria, *Gender and Aging: Demographic Dimensions,* Washington, DC: U.S. Department of Commerce, Bureau of the Census, 1997.

Glaser, B. G., and Strauss, A. L., *Awareness of Dying,* Chicago: Aldine, 1965.

Glick, Henry, *The Right to Die,* New York: Columbia University Press, 1992.

Glick, Paul C., "Updating the Life Cycle of the Family," *Journal of Marriage and the Family* (1977), 39: 5–13.

Gober, P., and Zonn, L., "Kin and Elderly Amenity Migration." *Gerontologist* (June 1983), 23(3): 288–294.

Goffman, Erving, *Asylums,* Garden City, NY: Anchor, 1961.

Gokhale, Jagadeesh, and Kotlikoff, Laurence J., "Baby Boomers' Megainheritance—Myth or Reality?" *Federal Reserve Bank of Cleveland—Economic Commentary* (October 1, 2000).

Gold, Marsha, "Can Managed Care and Competition Control Medicare Costs?" *Health Affairs* (May–June, 2003), 22(3): W3176–W3188.

Goleman, Daniel, *The Meditative Mind: The Varieties of Meditative Experience,* Los Angeles: Jeremy Tarcher, 1988.

Gomez, Carlos, F., *Regulating Death: Euthanasia and the Case of the Netherlands,* New York: Free Press, 1991.

Gordon, Michael, Mitchell, Olivia, and Twinney, Marc (Eds.), *Positioning Pensions for the Twenty-First Century,* Philadelphia: University of Pennsylvania, 1997.

Gorelick, P. B., "Stroke Prevention: An Opportunity for Efficient Utilization of Health Care Resources During the Coming Decade," *Stroke* (1994), 25: 220–224.

Grace Plaza of Great Neck, Inc., v. Elbaum, 82 N.Y.2d 10, 623 N.E.2d 513, 603 N.Y.S.2d 386 (1993).

Graebner, William, *A History of Retirement: The Meaning and Function of an American Institution,* New Haven, CT: Yale University Press, 1980.

Gratton, Brian, "Familism Among the Black and Mexican-American Elderly: Myth or Reality?" *Journal of Aging Studies* (1987), 1(1): 19–32.

Gray, C. H. (Ed.), *The Cyborg Handbook,* New York: Routledge, 1995.

Green, Brent, *Marketing to Leading-Edge Baby Boomers,* New York: Paramount Market, 2003.

Gresham, G. E., and Labi, M. L. C., "Functional Assessment Instruments Currently Available for Documenting Outcomes in Rehabilitation Medicine," in C. V. Granger and G. E. Greer (Eds.), *Functional Assessment in Rehabilitation Medicine,* Baltimore: Williams and Wilkins, 1984.

Grijalva v. Shalala, 946 F.Supp. 747 D.Ariz. (1996).

Grogan, Colleen M., and Patashnik, Eric M., "Universalism Within Targeting: Nursing Home Care, the Middle Class, and the Politics of the Medicaid Program," *Social Service Review* (March 2003), 77(1): 51–71.

Gruman, Gerald, "Modernization of the Life Cycle," in S. Spicker, K. Woodward, and D. Van Tassel (Eds.), *Aging and the Elderly: Humanistic Perspectives on Gerontology,* Atlantic Highlands, NJ: Humanities Press, 1978, pp. 359–387.

Gubrium, Jay F., *Living and Dying at Murray Manor,* New York: St. Martin's, 1975.

Gullette, Margaret, *Aged by Culture,* Chicago: University of Chicago Press, 2004.

Guttchen, David, and Pettigrew, Mary, "LTC Insurance: The Missing Link in Retirement Planning," *Employee Benefit Plan Review* (November 1998), 53(5): 38–40.

Haber, Carole, *Beyond Sixty-Five,* Cambridge, UK: Cambridge University Press, 1983.

Haber, David, *Health Promotion and Aging* (2nd ed.), New York: Springer, 1999.

Haber, E., and Short-DeGraff, M., "Intergenerational Programming for an Increasingly Age Segregated Society," *Activities, Adaptation, and Aging* (1990), 14(3): 35–49.

Hagen, Stuart A., *Financing Long-Term Care for the Elderly,* Washington, DC: Congressional Budget Office, 2004.

Haight, Barbara K., "Reminiscing: The State of the Art as a Basis for Practice," *International Journal of Aging and Human Development* (1991), 33(1): 1–32.

Hall, Stephen S., *Merchants of Immortality: Chasing the Dream of Human Life Extension,* Boston, MA: Houghton Mifflin, 2003.

Halper, Thomas, *The Misfortunes of Others: End-Stage Renal Disease in the United Kingdom,* New York: Cambridge University Press, 1989.

Hamburg, David, *Today's Children: Creating a Future for a Generation in Crisis,* New York: Random House, 1992.

Harel, Z., Ehrlich, P., and Hubbard, R., *The Vulnerable Elderly: People, Services, and Policies,* New York: Springer, 1990.

Hareven, Tamara, and Adams, Kathleen (Eds.), *Aging and Life Course Transitions,* New York: Guilford Press, 1982.

Harrison, Harry, *Make Room! Make Room!* New York: Doubleday, 1966.

Harrison, Stephen, and Hunter, David J., *Rationing Health Care,* London, UK: Institute for Public Policy Research, 1994.

Hastings Center, *Guidelines on the Termination of Life-Sustaining Treatment and the Care of the Dying,* Bloomington: Indiana University Press, 1988.

Hatch, Laurie R., *Beyond Gender Differences: Adaptation to Aging in Life-Course Perspective,* Amityville, NY: Baywood, 2000.

Havighurst, Robert J., Neugarten, Bernice L., and Tobin, Sheldon S., "Disengagement and Patterns of Aging," in Bernice L. Neugarten (Ed.), *Middle Age and Aging,* Chicago: University of Chicago Press, 1968, pp. 161–172.

Hayes-Bautista, David E., "Young Latinos, Older Anglos, and Public Policy: Lessons From California," *Generations* (Fall–Winter 1991), 15(4): 37–40.

Hayflick, Leonard, "The Limited in Vitro Lifetime of Human Diploid Cell Strains," *Experimental Cell Research* (1965), 37(3): 614–636.

Hayflick, Leonard, *How and Why We Age,* New York: Ballantine, 1996.

Hays, Judith A., "Aging and Family Resources: Availability and Proximity of Kin," *The Gerontologist* (1984), 24: 149–153.

Hayslip, Bert, Jr., and Goodman, Catherine, "Grandparents Raising Grandchildren: Benefits and Drawbacks?" *Journal of Intergenerational Relationships* (2007), 5(4): 117–119.

Hebert, L. E., Scherr, P. A., Bienias, J. L., Bennett, D. A., and Evans, D. A., "Alzheimer Disease in the US population: Prevalence Estimates Using the 2000 Census," *Archives of Neurology* (August 2003), 60(8): 1119–1122.

Heclo, H., "Generational Politics," in J. L. Palmer, T. Smeeding, and B. B. Torrey (Eds.), *The Vulnerable,* Washington, DC: Urban Institute, 1988, pp. 381–442.

Held, T., "Institutionalization and De-Institutionalization of the Life Course," *Human Development* (1986), 29: 157–162.

Heller, Jonathan, "Examination of Elder Abuse Reports and Data Collection Systems Across the United States," *Victimization of the Elderly and Disabled* (November–December 2000), 3: 4.

Helman, Ruth, VanDerhei, Jack, and Copeland, Craig, "2008 Retirement Confidence Survey: Americans Much More Worried About Retirement, Health Costs a Big Concern," *EBRI Issue Brief* (April 2008), 316: 1–22.

Helson, R., Mitchell, V., and Moane, G., "Personality and Patterns of Adherence and Non-Adherence to the Social Clock," *Journal of Personality and Social Psychology* (1984), 46: 1079–1096.

Hemingway, Harry, Nicholson, Amanda, Stafford, Mai, Roberts, Ron, and Marmot, Michael, "Impact of Socioeconomic Status on Health Functioning as Assessed by the SF-36 Questionnaire: The Whitehall II Study," *American Journal of Public Health* (September 1997), 87: 1484–1490.

Hendricks, J., and Cutler, S. J., "Leisure and the Structure of Our Life Worlds," *Ageing and Society* (March 1990), 10: 85–94.

Hendricks, J., and Peters, C. B., "The Times of Our Lives: An Integrative Framework," *American Behavioral Scientist* (1986), 29(5): 662–676.

Henretta, J. C., and Campbell, R. T., "Status Attainment and Status Maintenance: A Study of Stratification in Old Age," *American Sociological Review* (1976), 41: 981–992.

Herzog, Regula A., Kahn, Robert L., Moergan, James N., Jackson, James S., and Antonucci, Toni C., "Age Differences in Productive Activities," *Journal of Gerontology: Social Sciences* (1989), 44: S129–S138.

Hess, Clinton, and Kerschner, P., *Silver Lobby,* Los Angeles: University of Southern California Press, 1978.

High, Dallas, "Why Are Elderly People Not Using Advance Directives?" *Journal of Aging and Health* (November 1993), 5(4): 497–515.

Hillman, Jennifer L., *Clinical Perspectives on Elderly Sexuality,* New York: Plenum, 2000.

Hilton, James, *Lost Horizon,* London: Macmillan, 1933.

Himes, C. L., "Elderly Americans," *Population Bulletin* (2001), 56(4): 4.

Hochschild, Arlie R., "Disengagement Theory: A Critique and Proposal," *American Sociological Review* (1975), 40: 553–569.

Hockey, Jennifer Lorna, and James, Allison, *Social Identities Across the Life Course,* New York: Palgrave Macmillan, 2003.

Hogan, C., et al, "Medicare Beneficiaries' Costs of Care in the Last Year of Life," *Health Affairs* (July 2001), 20(4): 188–195.

Hogan, D. P., and Astone, N. M., "The Transition to Adulthood," *Annual Review of Sociology* (1986), 12: 109–130.

Hogeboom, David L., and Bell-Ellison, Bethany A., "Internet Use and Social Networking Among Middle Aged and Older Adults," *Educational Gerontology* (2010), 36(2): 93–111.

Holden, K. C., Burkhauser, R. V., and Myers, Daniel A., "Income Transitions at Older Stages of Life: The Dynamics of Poverty," *The Gerontologist* (1986), 26: 292–297.

Holiday, Robin, "The Multiple and Irreversible Causes of Aging," *The Journals of Gerontology Series A: Biological Sciences and Medical Sciences* (2004), 59: B568–B572.

Holmes, Ellen, and Holmes, Lowell, *Other Cultures, Elder Years* (2nd ed.), Thousand Oaks, CA: Sage, 1995.

Holmes, Oliver Wendell, "The Deacon's Masterpiece or the Wonderful One-Hoss Shay: A Logical Story," Boston: Houghton, Mifflin, 1858/1891. Retrieved September 10, 2011 (http://www.ibiblio.org/eldritch/owh/shay.html).

Holmes, T. H., and Rahe, R. H., "The Social Readjustment Rating Scale," *Journal of Psychosomatic Research* (1967), 11: 213–218.

Holstein, James A., "Discourse of Age in Involuntary Commitment Proceedings," *Journal of Aging Studies* (1990), 4(2): 111–130.

Holstein, Martha, "Productive Aging: A Feminist Critique," *Journal of Aging and Social Policy* (1992), 4(3–4): 17–34.

Holtzman, Abraham, *The Townsend Movement,* New York: Bookman Associates, 1963.

Horn, J. L., "The Theory of Fluid and Crystallized Intelligence in Relation to Concepts of Cognitive Psychology and Aging in Adulthood," in F. I. M. Craik and S. Trehub (Eds.), *Aging and Cognitive Processes,* New York: Plenum, 1982, pp. 237–278.

Horn, J. L., and Donaldson, G., "Faith Is Not Enough: A Response to the Baltes-Schaie Claim That Intelligence Does Not Wane," *American Psychologist* (1977), 32: 369–373.

Hudson, Robert, "The 'Graying' of the Federal Budget and Its Consequences for Old-Age Policy," *The Gerontologist* (1978), 28: 428–440.

Hudson, Robert, "Tomorrow's Able Elders: Implications for the State," *The Gerontologist* (1987), 27(4): 405–409.

Humphry, Derek, *Final Exit: The Practicalities of Self-Deliverance and Assisted Suicide for the Dying* (3rd ed.), New York: Delta, 2002.

Hunnicutt, Benjamin K., "Aging and Leisure Politics," in Michael L. Teague, Richard D. MacNeil, and Gerald L. Hitzhusen (Eds.), *Perspectives on Leisure and Aging in a Changing Society,* Columbia: University of Missouri Press, 1982, pp. 74–108.

Hunt, Michael E. (Ed.), *Retirement Communities: An American Original,* New York: Haworth, 1983.

Hunt, Michael E., Merrill, J. L., and Gilker, C. M., "Naturally Occurring Retirement Communities in Urban and Rural Settings," in W. Edward Folts and Dale E. Yeatts (Eds.), *Housing and the Aging Population: Options for the New Century,* New York: Garland, 1994, pp. 107–120.

In re Conroy, 98 N.J. 321 A 2d (1985).

In re Earle Spring, 380 Mass. 629 (1980).

In re Guardianship of Estelle M. Browning, State of Florida v. Doris F. Herbert, 568 S0.2d 4 (Fla. 1990).

In re Quinlan, 70 N.J. 10, 355 A.2d 647 (1976).

In re Westchester County Medical Center, 72 N.Y.2d 517, 534 N.Y.S.2d 886, 531 N.E.2d 607 (1988).

Isenberg, Sheldon, "Aging in Judaism: 'Crown of Glory' and 'Days of Sorrow,'" in Thomas Cole, Robert Kastenbaum, and Ruth Ray (Eds.), *Handbook of the Humanities and Aging,* New York: Springer, 2000, pp. 114–141.

Iso, A. S. E., Jackson, E. L., and Dunn, E., "Starting, Ceasing, and Replacing Leisure Activities Over the Life-span," *Journal of Leisure Research* (1994), 26(3): 227–249.

Jackson, J. S. (Ed.), *The Black American Elderly,* New York: Springer, 1988.

Jacobs, Bruce, *Targeting Benefits for the Elderly: The Public Debate,* New York: Ford Foundation, 1990.

Jacobs, Lawrence R., and Skocpol, Theda, *Health Care Reform and American Politics: What Everyone Needs to Know,* Oxford University Press, 2010.

Jacobson, Solomon G., "Equity in the Use of Public Benefits by Minority Elderly," in Ron C. Manuel (Ed.), *Minority Aging: Sociological and Social Psychological Issues,* Westport, CT: Greenwood, 1982, pp. 161–170.

Jacques, Elliot, "Death and Midlife Crisis," *International Journal of Psychoanalysis* (1965), 46: 502–514.

Jerrome, Dorothy, "'That's What It's All About': Old People's Organizations as a Context for Aging," in J. Gubrium and K. Charmaz (Eds.), *Aging, Self and Community,* Greenwich, CT: JAI, 1992, pp. 225–235.

Jogerst, Gerald J., Daly, Jeanette M., Brinig, Margaret F., Dawson, Jeffrey D., Schmuch, Gretchen A., and Ingram, Jerry G., "Domestic Elder Abuse and the Law," *American Journal of Public Health* (December 2003), 93(12): 2131–2136.

John, Randy A., *Social Integration of an Elderly Native American Population,* New York: Garland, 1995.

Johnson, Barbara B., "The Changing Role of Women and Social Security Reform," *Social Work* (1987), 32(4): 341–345.

Johnson, Colleen L., and Grant, Leslie A., *The Nursing Home in American Society,* Baltimore: Johns Hopkins University Press, 1986.

Johnson, Meagan, and Johnson, Larry, *Generations, Inc.: From Boomers to Linksters—-Managing the Friction Between Generations at Work,* AMACOM, 2010.

Johnson, Sandy, *Book of Elders: The Life Stories of Great American Indians,* San Francisco: Harper, 1994.

Johnson, T. E., "Increased Life-Span of Age-1 Mutants in *Caenorhabditis Elegans* and Lower Gompertz Rate of Aging," *Science* (1990), 249: 908.

Jones, Ian R., and Higgs, Paul, "Health Economists and Health Care Provision for the Elderly: Implicit Assumptions and Unstated Conclusions," in Kevin Morgan (Ed.), *Gerontology: Responding to an Ageing Society,* London: J. Kingsley, 1992, pp. 118–135.

Jones, Ian R., Hyde, Martin, Victor, Christina R., Wiggins, Richard D., Gilleard, Chris, and Higgs, Paul, *Ageing in a Consumer Society: From Passive to Active Consumption in Britain,* Bristol, UK: The Policy Press, 2008.

Jordan, Lamar, "Law Enforcement and the Elderly: A Concern for the 21st Century," *FBI Law Enforcement Bulletin* (May 2002), 71(5): 20–23.

Kaiser, Fran, Morley, John, and Coe, Rodney, *Cardiovascular Disease in Older People,* New York: Springer, 1997.

Kalish, Richard A., "The New Ageism and the Failure Models: A Polemic," *The Gerontologist* (1979), 19: 398–402.

Kane, Robert L., and Kane, Rosalie A. (Eds.), *Assessing Older Persons: Measures, Meaning, and Practical Applications,* Oxford, UK: Oxford University Press, 2000.

Kane, Rosalie A. (Ed.), "Legacy" (special issue), *Generations* (1996), 20: 3.

Kane, Rosalie A., and Kane, Robert L., *Long-Term Care: Principles, Programs and Policies,* New York: Springer, 1987.

Kao, Rudolf Sing Kee, and Lam, Mary Leong, "Asian American Elderly," in Evelyn Lee (Ed.), *Working With Asian Americans: A Guide for Clinicians,* New York: Guilford Press, 1997, pp. 208–223.

Kaplan, George, "Epidemiologic Observations on the Compression of Morbidity: Evidence From the Alameda County Study," *Journal of Aging and Health* (May 1991), 3(2): 155–171.

Kaplan, Max, *Leisure: Lifestyle and Lifespan,* Philadelphia: W. B. Saunders, 1979.

Kapp, Marshall B., "Health Care Rationing Affecting Older Persons: Rejected in Principle but Implemented in Fact," *Journal of Aging and Social Policy* (2002), 14(2): 27–42.

Karamcheva, N., and Munnell, A. H., *Why Are Widows So Poor?* Boston: Trustees of Boston College, Center for Retirement Research, 2007.

Karani, K. G., and Fraccastoro, K. A., "Resistance to Brand Switching: The Elderly Consumer," *Journal of Business and Economics Research* (December 2010), 8(12): pp. 77–84.

Kaskie, Brian, Imhof, Sara, Cavanaugh, Joseph, and Culp, Kennith, "Civic Engagement as a Retirement Role for Aging Americans," *Gerontologist* (June 2008), 48(3): 368–377.

Kassner, Enid, "The Older Americans Act: Should Participants Share in the Cost of Services?" *Journal of Aging and Social Policy* (1992), 4(1–2): 51–71.

Kastenbaum, Robert, "Time Course and Time Perspective in Later Life," in C. Eisdorfer (Ed.), *Annual Review of Gerontology and Geriatrics* (Vol. 3), New York: Springer, 1983, pp. 80–102.

Kastenbaum, Robert, "When Aging Begins: A Lifespan Developmental Approach," *Research on Aging* (March 1984), 6(1): 105–117.

Kastenbaum, Robert, and Candy, S. E., "The Four-Percent Fallacy: A Methodological and Empirical Critique of Extended Care Facility Population Statistics," *International Journal of Aging and Human Development* (1973), 4: 15–21.

Katz, S., "Studies of Illness in the Aged: The Index of ADL: A Standardized Measure of Biological and Psychosocial Function," *Journal of the American Medical Association* (1963), 185: 914–919.

Katz, S., *Disciplining Old Age: The Formation of Gerontological Knowledge,* Charlottesville: University Press of Virginia, 1996.

Katz, S., and Akpom, C. A., "A Measure of Primary Socio-biological Functions," *International Journal of Aging and Human Development* (1976), 6: 493–506.

Katzman, Robert, and Bick, Katherine L., *Alzheimer Disease: The Changing View,* San Diego, CA: Academic Press, 2000.

Kaufman, Alan S., "WAIS-III IQs, Horn's Theory, and Generational Changes From Young Adulthood to Old Age," *Intelligence* (2001), 29(2): 131–167.

Kaufman, Sharon R., *The Ageless Self: Sources of Meaning in Later Life,* Madison: University of Wisconsin Press, 1986.

Kay, M. M. B., and Makinodan, T. (Eds.), *Handbook of Immunology in Aging,* Boca Raton, FL: CRC Press, 1981.

Kaye, Richard, "Sexuality in the Later Years," *Aging and Society* (September 1993), 13(Pt. 3): 415–426.

Kearl, Michael C., "Dying Well: The Unspoken Dimension on Aging Well," *American Behavioral Scientist* (January 1996), 39(3): 336–360.

Kieffer, Jarold A., *Gaining the Dividends of Longer Life: New Roles for Older Workers,* Boulder, CO: Westview, 1983.

Kim, K. C., Kim, S., and Hurh, W. M., "Filial Piety and Intergenerational Relationship in Korean Immigrant Families," *International Journal of Aging and Human Development* (1991), 33: 233–245.

King, Nancy M. P., *Making Sense of Advance Directives,* Washington, DC: Georgetown University Press, 1996.

Kingson, E. R., and O'Grady, LeShane R., "Effects of Caregiving on Women's Social Security Benefits," *The Gerontologist* (April 1993), 33(2): 230–239.

Klaus, Patsy A., *Crimes Against Persons Age 65 or Older, 1992–97,* Washington, DC: U.S. Department of Justice, Office of Justice Programs, January 2000.

Klein, S. M., *In-Home Respite Care for Older Adults: A Guide for Program Planners, Administrators and Clinicians,* Springfield, IL: Charles C Thomas, 1986.

Koenig, Harold G., *Aging and God: Spiritual Pathways to Mental Health in Midlife and Later Years,* Binghamton, NY: Haworth, 1994.

Koff, Theodore, *Long-Term Care: An Approach to Serving the Frail Elderly,* Boston: Little, Brown, 1982.

Koh, James, and Bell, William, "Korean Elders in the United States: Intergenerational Relations and Living Arrangements," *The Gerontologist* (1987), 27(1): 66–71.

Kohli, Martin, "Retirement and the Moral Economy: An Historical Interpretation of the German Case," *Journal of Aging Studies* (1987), 1: 125–144.

Kongstvedt, Peter R., *Managed Care: What It Is and How It Works,* Boston: Jones and Bartlett, 2004.

Kopac, C., "Bring Together the Young and Old With Intergenerational Day Care," *Pediatric Nursing* (1987), 13(4): 227–229.

Kotlikoff, Lawrence, *Generational Accounting,* New York: Basic Books, 1992.

Kotre, John, *Outliving the Self: Generativity and the Interpretation of Lives,* Baltimore: Johns Hopkins University Press, 1984.

Kowald, Axel, "Lifespan Does Not Measure Ageing," *Biogerontology* (2002), 3(3): 187–190.

Krain, Mark A., "Policy Implications for a Society Aging Well: Employment, Retirement, Education, and Leisure Policies for the 21st Century," *American Behavioral Scientist* (November–December 1995), 39(2): 131–151.

Krout, John, *Senior Centers in America,* New York: Greenwood, 1989.

Kübler-Ross, Elisabeth, *On Death and Dying,* New York: Macmillan, 1969.

Kuder, Linda, and Roeder, Phillip W., "Attitudes Toward Age-Based Health Care Rationing," *Journal of Aging and Health* (May 1995), 7(2): 301–327.

Kunitz, S. J., and Levy, J. E., *Navajo Aging,* Tucson: University of Arizona Press, 1991.

Lachs, Mark, Williams, Christiana, O'Brien, Shelley, Hurst, Leslie, and Horwitz, Ralph, "Risk Factors for Reported Elder Abuse and Neglect: A Nine-Year Observational Cohort Study," *The Gerontologist* (August 1997), 37(4): 469–474.

Lamberts, S. W. J., van den Beld, A. W., and van der Lely, A.-J., "The Endocrinology of Aging," *Science* (1997), 278: 419–424.

Lamm, Richard D., "Intergenerational Equity in an Age of Limits: Confessions of a Prodigal Parent," in Gerald Winslow and James Walters (Eds.), *Facing Limits: Ethics & Health Care for the Elderly,* Boulder, CO: Westview, 1993, pp. 15–28.

Lammers, W., and Klingman, D. "Family Responsibility Laws and State Politics: Empirical Patterns and Policy Implications," *Journal of Applied Gerontology* (July 1986), 5: 5–25.

Lancaster, Lynne, and Stillman, David, *When Generations Collide: Who They Are. Why They Clash. How to Solve the Generational Puzzle at Work,* Harper, 2003.

LaPuma, John, Orentlicher, David, and Moss, Robert J., "Advance Directives on Admission: Clinical Implications and Analysis of the Patient Self-Determination Act of 1990," *Journal of the American Medical Association* (July 17, 1991), 266: 404.

LaRock, S., "Neither Federal 'Carrot' nor 'Stick' Alters Early Retirement Patterns, 1979–1992," *Employee Benefit Plan Review* (July 1993), 48(1): 10–12.

Laslett, Peter, *The World We Have Lost,* New York: Scribner, 1965/1971.

Laslett, Peter, *Household and Family in Past Time,* Cambridge, UK: Cambridge University Press, 1972.

Laslett, Peter, *A Fresh Map of Life: The Emergence of the Third Age,* Cambridge, MA: Harvard University Press, 1991.

Lawlor, Drue, and Thomas, Michael A., *Residential Design for Aging In Place,* Wiley, 2008.

Lee, Marlene, and Mather, Mark, "U.S. Labor Force Trends," *Population Bulletin* (June 2008), 63(2), Population Reference Bureau.

Lee, Melinda A., and Ganzini, Linda, "Depression in the Elderly: Effect on Patient Attitudes Toward Life-Sustaining Therapy," *Journal of the American Geriatrics Society* (October 1992), 40(10): 983–988.

Lee, Shelley A., "Poor on Purpose," *Journal of Financial Planning* (June 2003), 16(6): 1–7.

Leenaars, Antoon, A., Maris, Ronald, McIntosh, John L., and Richman, Joseph (Eds.), *Suicide and the Older Adult,* New York: Guilford Press, 1992.

Lehman, Harvey, "More About Age and Achievement," *The Gerontologist* (1962), 2: 141–148.

Leibold, K., "Employer-Sponsored On-Site Intergenerational Daycare," *Generations* (1989), 13(3): 33–34.

Leitner, Michael, and Leitner, Sara, *Leisure in Later Life* (3rd ed.), New York: Haworth Press, 2004.

Levin, Jeffrey S., "Age Differences in Mystical Experience," *Gerontologist* (August 1993), 33(4): 507–513.

Levinson, Daniel J., *The Seasons of a Man's Life,* New York: Knopf, 1978.

Levy, Jay, "Contrary Opinion: The Importance of the Elderly," *Challenge* (May–June 2001), 44(3): 97–114.

Lewis, Diane E., "A Job-Search Site for Those Who Eschew Retirement: Website Caters to a Graying Population," *Boston Globe* (September 11, 2006).

Lewis, P. A., and Charny, M., "Which of Two Individuals Do You Treat When Only Their Ages Are Different and You Can't Treat Both?" *Journal of Medical Ethics* (March 1989), 15(1): 28–32.

Light, Paul, *Artful Work: The Politics of Social Security Reform,* New York: Random House, 1985.

Lindauer, Martin S., *Aging, Creativity, and Art: A Positive Perspective on Late-Life Development,* New York: Kluwer Academic/Plenum Publishers, 2003.

Litwak, Eugene, *Helping the Elderly: The Complementary Roles of Informal Networks and Formal Systems,* New York: Guilford Press, 1985.

Liu, Korbin, and Manton, Kenneth, "Nursing Home Length of Stay and Spenddown in Connecticut, 1977–1986," *The Gerontologist* (April 1991), 31(2): 165–173.

Liu, William T., and Kendig, Hal (Eds.), *Who Should Care for the Elderly? An East-West Value Divide,* Singapore: Singapore University Press, 2000.

Locke, S. "Neurological Disorders of the Elderly," in William Reichtel (Ed.), *Clinical Aspects of Aging,* Baltimore: Williams and Wilkins, 1983.

London, William, and Morgan, John, "Living Long Enough to Die of Cancer," *Priorities* (1995), 7(4): 6–9.

Longino, Charles F., and Kart, C. S., "Explicating Activity Theory: A Formal Replication," *Journal of Gerontology* (1982), 37: 713–722.

Longley, Robert, "Newsweek Poll: Americans Split on Bush Social Security Plan" (February 2005). Retrieved September 30, 2011 (http://usgovinfo.about.com/od/socialsecurity/a/ssnewsweekpoll.htm).

Lopata, Helena Z. (Ed.), *Widows,* Durham, NC: Duke University Press, 1987.

Lowenthal, Marjorie F., and Chiriboga, David, "Transitions to the Empty Nest: Crisis, Change or Relief?" *Archives of General Psychiatry* (1972), 26: 8–14.

Lowy, Louis, "Major Issues of Age-Integrated Versus Age-Segregated Approaches to Serving the Elderly," *Journal of Gerontological Social Work* (1987), 10(3–4): 37–46.

Lubitz, J., and Prihoda, R., "The Use and Costs of Medicare Services in the Last Two Years of Life," *Health Care Financing Review* (1984), 5(3): 117–131.

Luce, Gay, *Your Second Life: Vitality and Growth in Maturity and Later Years From the Experiences of the SAGE Program,* New York: Dell, 1979.

Ludwig, Frederic, *Lifespan Extension: Consequences and Open Questions,* New York: Springer, 1991.

Lynch, Frederick, *One Nation Under AARP: The Fight Over Medicare, Social Security, and America's Future,* University of California Press, 2011.

MacDonald, Martha, "Gender and Social Security Policy: Pitfalls and Possibilities," *Feminist Economics* (1998), 4(1): 1–25.

Mace, Nancy, and Rabins, Peter, *The 36-Hour Day,* Baltimore: Johns Hopkins University Press, 1981.

MacKinlay, Elizabeth, and McFadden, Susan H., "Ways of Studying Religion, Spirituality, and Aging: The Social Scientific Approach," *Journal of Religious Gerontology* (2004), 16(3–4): 75–90.

Mankoff, Robert, "See, the Problem With Doing Things to Prolong Your Life ..." [Cartoon], *The New Yorker* Cartoon Bank (August 8, 1994). Retrieved September 10, 2011 (http://www.cartoonbank.com/1994/see-the-problem-with-doing-things-to-prolong-your-life-is-that-all-the-extra-years-come-at-the-/invt/107940/).

Mannheim, Karl, "The Problem of Generations," in K. Mannheim (Ed.), *Essays in the Sociology of Knowledge,* New York: Oxford University Press, 1952.

Manton, Kenneth G., Gu, XiLiang, and Lamb, Vicki L., "Change in Chronic Disability From 1982 to 2004/2005 as Measured by Long-Term Changes in Function and Health in the U.S. Elderly Population," *Proceedings of the National Academy of Sciences of the United States of America* (November 28, 2006), 103(48): 18374–18379.

Manton, Kenneth G., and Soldo, Beth J., "Dynamics of Health Changes in the Oldest Old: New Perspectives and Evidence," *Milbank Memorial Fund Quarterly* (1985), 63(2).

Margolis, Robin, "Will Funneling Seniors Into Managed Care Reduce Costs?" *Health Span* (March 1995), 12(3): 16–17.

Margulis, H. L., and Benson, V. M., "Age-Segregation and Discrimination Against Families With Children in Rental Housing," *Gerontologist* (December 1982), 22(6): 505–512.

Markides, K. S., and Black, S. A., "Race, Ethnicity, and Aging: The Impact of Inequality," in R. H. Binstock and L. K. George (Eds.), *The Handbook of Aging and the Social Sciences* (4th ed.), San Diego: Academic Press, 1996.

Markson, Elizabeth Warren, and Hollis-Sawyer, Lisa A. (Eds.), *Intersections of Aging: Readings in Social Gerontology,* Los Angeles: Roxbury, 2000.

Marmor, Theodore, "Fact, Fiction, and Faction: The Politics of Medical Care Reform in Canada as It Appears South of the Border," *Canadian Journal on Aging* (Summer 1995), 14(2): 426–436.

Martin, Linda G., and Soldo, Beth J. (Eds.), *Racial and Ethnic Differences in the Health of Older Americans,* Washington, DC: National Academy Press, 1997.

Martz, Sandra Haldeman (Ed.), *When I Am an Old Woman, I Shall Wear Purple,* Watsonville, CA: Papier-Mache, 1987.

Masters, W., and Johnson, V., *Human Sexual Response,* New York and Tokyo: Ishi Press, 2010.

McCabe, Kimberly A., and Gregory, Sharon S., "Elderly Victimization: An Examination Beyond the FBI's Index Crimes," *Research on Aging* (May 1998), 20(3): 363–372.

McConnell, Stephen R., and Usher, Carolyn E., *Intergenerational House Sharing: A Research Report and Resource Manual,* Lexington, MA: D.C. Heath, 1980.

McCrae, Robert R., Arenberg, David, and Costa, Paul T., "Declines in Divergent Thinking With Age: Cross-Sectional, Longitudinal, and Cross-Sequential Analyses," *Psychology and Aging* (June 1987), 2(2): 130–137.

McCrae, Robert R., and Costa, Paul T., *Personality in Adulthood,* New York: Guilford Press, 1990.

McFalls, Joseph A., Jr., "Population: A Lively Introduction, 4th Edition," *Population Bulletin* (December 2003), 58(4). Published by the Population Reference Bureau.

McGoon, Dwight C., *Parkinson's Handbook,* New York: Norton, 1990.

McGuire, Francis A., Boyd, Rosangela, and Tedrick, Raymond E., *Leisure and Aging: Ulyssean Living in Later Life* (3rd ed.), Champaign, IL: Sagamore Publications, 2004.

McIntosh, John, Santos, John, Hubbard, Richard, and Overholser, James, *Elder Suicide: Research, Theory and Treatment,* Washington, DC: American Psychological Association, 1994.

McKusick, David, "Demographic Issues in Medicare Reform," *Health Affairs* (January–February 1999), 18(1): 194–207.

McMahan, Shari, and Lutz, Rafer, "Alternative Therapy Use Among the Young-Old (ages 65 to 74): An Evaluation of the MIDUS Database," *Journal of Applied Gerontology* (June 2004), 23(2): 91–103.

McMahon, A. W., and Rhudick, P. J., "Reminiscing in the Aged: An Adaptational Response," in S. Levin and R. J. Kahana (Eds.), *Psychodynamic Studies on Aging: Creativity, Reminiscing, and Dying,* New York: International Universities Press, 1967.

Meacham, J. A., "The Loss of Wisdom," in R. J. Sternberg (Ed.), *Wisdom: Its Nature, Origin, and Development,* Cambridge, UK: Cambridge University Press, 1990, pp. 160–177.

Mechanic, David, "Cost Containment and the Quality of Medical Care: Rationing Strategies in an Era of Constrained Resources," *Millbank Memorial Fund Quarterly* (1985), 63: 453–475.

Medawar, Peter B., *Aging: An Unsolved Problem of Biology,* London: H. K. Lewis, 1952.

Medicare Payment Advisory Commission, *Healthcare Spending and the Medicare Program: A Data Book,* Washington, DC: Author, 2004.

Medvedev, Z. A., "Repetition of Molecular-Genetic Information as a Possible Factor in Evolutionary Change of Life-Span," *Experimental Gerontology* (1972), 7: 227–234.

Meek v. Martinez, 724 F. Supp. 888, 903–04 SD Fla. (1987).

Menzel, Paul T., *Strong Medicine: The Ethical Rationing of Health Care,* New York: Oxford University Press, 1990.

Mermin, Gordon, Johnson, Richard W., and Toder, Eric, *Will Employers Want Aging Boomers?* Washington, DC: Urban Institute, 2008.

MetLife Foundation and Civic Ventures, *Americans Seek Meaningful Work in the Second Half of Life,* San Francisco: Author, 2008. Retrieved September 16, 2011 (http://www.encore.org/files/Encore_Survey.pdf).

Meyrowitz, Joshua, *No Sense of Place: The Impact of Electronic Media on Social Behavior,* New York: Oxford University Press, 1985.

Mieskiel, S., "Inheritance," in L. Vitt and J. Siegenthaler (Eds.), *Encyclopedia of Financial Gerontology,* Westport, CT: Greenwood, 1996.

Miller, Melvin E., and Cook-Greuter, Susanne R. (Eds.), *Transcendence and Mature Thought in Adulthood: The Further Reaches of Adult Development,* Lanham, MD: Rowman & Littlefield, 1994.

Minkler, Meredith, and Estes, Carroll (Eds.), *Readings in the Political Economy of Aging,* Farmingdale, NY: Baywood, 1984.

Minkler, Meredith, and Estes, Carroll (Eds.), *Critical Gerontology: Perspectives From Political and Moral Economy,* Amityville, NY: Baywood, 1998.

Minkler, Meredith, and Stone, R., "The Feminization of Poverty and Older Women," *The Gerontologist* (1985), 25: 351–357.

Minois, Georges, *History of Old Age: From Antiquity to the Renaissance* (Sarah Hanbury Tenison, Trans.), Chicago: University of Chicago Press, 1989.

Mirkin, Barry, and Weinberger, Mary Beth, *Demography of Population Ageing,* New York: United Nations, 2000.

Modigliani, Franco, "Life Cycle, Individual Thrift and the Wealth of Nations," Nobel Prize lecture, 1985.

Mollica, Robert L., *State Assisted Living Policy, 2000,* Portland, ME: National Academy for State Health Policy, 2000.

Montgomery, Rhonda, "Respite Services for Family Caregivers," in M. D. Petersen and D. L. White (Eds.), *Health Care of the Elderly: An Information Sourcebook,* Newbury Park, CA: Sage, 1989.

Moody, Harry R., *Abundance of Life: Human Development Policies for an Aging Society,* New York: Columbia University Press, 1988.

Moody, Harry R., "The Politics of Entitlement and the Politics of Productivity," in Scott A. Bass, Elizabeth A. Kutza, and Fernando Torres-Gil (Eds.), *Diversity in Aging,* Glenview, IL: Scott, Foresman, 1990, pp. 129–149.

Moody, Harry R., "Restoring Confidence in Social Security: Our Obligation to Future Generations," in Carroll Estes, Leah Rogne, Brooke Hollister, Brian Grossman, and Erica Solway (Eds.), *Social Insurance and Social Justice,* Springer, 2009.

Moody, Harry R., and Carroll, David, *The Five Stages of the Soul: Charting the Spiritual Passages That Shape Our Lives,* New York: Doubleday Anchor Books, 1997.

Moody, Harry R., and Hayflick, Leonard, *Has Anyone Ever Died of Old Age?* Occasional Paper, New York: International Longevity Center, 2003.

Moore, Crystal Dea, Sparr, Jennifer, Sherman, Susan, and Avery, Lisa, "Surrogate Decision-Making: Judgment Standard Preferences of Older Adults," *Social Work in Health Care* (2003), 37(2): 1–16.

Moreno, Jonathan, *Euthanasia: The Controversy over Mercy Killing, Assisted Suicide, and the "Right to Die,"* New York: Simon & Schuster, 1995.

Morgan, L. A., Eckert, J. K., and Lyon, S. M., *Small Board-and-Care Homes,* Baltimore: Johns Hopkins University Press, 1995.

Morgan, Leslie A., "Continuing Gender Gap in Later Life Economic Security," *Journal of Aging and Social Policy* (2000), 11(2–3): 157–165.

Morrison, M., and Jedriewski, M. K., "Retirement Planning: Everybody Benefits," *Personnel Administrator* (January 1988), 74–80.

Morrison, R. Sean, and Diane Meier, *Geriatric Palliative Care,* New York: Oxford University Press, 2003.

Morrow-Howell, Nancy, Hinterlong, James, and Sherraden, Michael, *Productive Aging: Concepts and Challenges,* Baltimore: Johns Hopkins University Press, 2001.

Morrow-Howell, Nancy, and Mui, Ada (Eds.), *Productive Engagement in Later Life: A Global Perspective,* Routledge, 2012.

Moses, Hamilton, III, Dorsey, E. Ray, Matheson, David H. M., and Their, Samuel O., "Financial Anatomy of Biomedical Research," *JAMA* (2005), 294(11): 1333–1342. doi: 10.1001/jama.294.11.1333

Moskowitz, Roland, and Haug, Marie, *Arthritis and the Elderly,* New York: Springer, 1985.

Moynihan, Daniel P., *Family and Nation,* San Diego: Harcourt Brace Jovanovich, 1986.

Munnell, Alicia H. (Ed.), *Retirement and Public Policy,* Dubuque, IA: Kendall/Hunt, 1991.

Munnell, Alicia H., *Why Are So Many Older Women Poor?* Boston: Boston College, Center for Retirement Research, 2004.

Munnell, Alicia H., and Sass, Steven A., *Working Longer: The Solution to the Retirement Income Challenge,* Washington, DC: Brookings Institution, 2009.

Myers, Robert J., *Social Security* (3rd ed.), Homewood, IL: Irwin, 1985.

Nasar, Sylvia, "The Spend-Now, Tax-Later Orgy," *The New York Times* (January 14, 1993), p. D-2.

Natanson v. Kline, 186 Kan. 393, 350 P.2d 1093, reh. Den. 187 Kan. 186, 354 P.2d 670 (1960).

National Academy of Social Insurance, *Social Security Benefits, Finances, and Policy Options: A Primer,* 2011. Available from www.nasi.org

National Center on Elder Abuse, *The National Elder Abuse Incidence Study,* Washington, DC: Author, 1998.

National Institute of Mental Health, "Older Adults: Depression and Suicide Facts," 2010. Retrieved September 16, 2011 (http://www.nimh.nih.gov/health/publications/older-adults-depression-and-suicide-facts-fact-sheet/index.shtml).

Naylor, C. D., "A Different View of Queues in Ontario," *Health Affairs* (1991), 10(3): 111–128.

Neal, M. B., and Hammer, L. B., *Working Couples Caring for Children and Aging Parents: Effects on Work and Well-Being,* Mahwah, NJ: Lawrence Erlbaum, 2007.

Neikrug, Shimshon M., "New Grandparenting: Dialogue and Covenant Through Mentoring, *Journal of Gerontological Social Work* (2000), 33(3): 103–117.

Nelson, G., "Social Class and Public Policy for the Elderly," *Social Service Review* (1982), 56: 85–107.

Nelson, Gary M., "Tax Expenditures for the Elderly," *The Gerontologist* (1983), 23: 471–478.

Nelson, Todd (Ed.), *Ageism: Stereotyping and Prejudice Against Older Persons,* Cambridge, MA: MIT Press, Bradford Books, 2002.

Neugarten, B. L. (Ed.), *Personality in Middle and Late Life,* New York: Atherton, 1964.

Neugarten, Bernice L., *Middle Age and Aging,* Chicago: University of Chicago Press, 1968.

Neugarten, Bernice L. (Ed.), *Age or Need? Public Policies for Older People,* Beverly Hills, CA: Sage, 1983.

Newcomer, Robert J., Lawton, M. Powell, and Byerts, Thomas O. (Eds.), *Housing an Aging Society,* New York: Van Nostrand Reinhold, 1986.

Nord, Erik, *Cost-Value Analysis in Health Care: Making Sense Out of QALYs,* New York: Cambridge University Press, 1999.

Nusselder, Wilma, and Mackenbach, Johan, "Rectangularization of the Survival Curve in the Netherlands, 1950–1992," *The Gerontologist* (December 1996), 36(6): 773–782.

Nyren, Chuck, *Advertising to Baby Boomers Revised,* Paramount Market Publishing, 2007.

Oberlander, Jonathan, *Political Life of Medicare,* Chicago: University of Chicago, 2003.

Oberlander, Jonathan, Marmor, Theodore, and Jacobs, Lawrence, "Rationing Medical Care: Rhetoric and Reality in the Oregon Health Plan," *Canadian Medical Association Journal* (May 29, 2001), 164(11): 1583–1587.

O'Brien, Ellen, *Medicaid's Coverage of Nursing Home Costs: Asset Shelter for the Wealthy or*

Essential Safety Net? Georgetown University Long-Term Care Financing Project, May 2005. Retrieved September 16, 2011 (http://www .mountainagents.com/UserFiles/4096/nursing homecosts.pdf).

O'Brien, Ellen, and Elias, Risa, *Medicaid and Long-Term Care,* Washington, DC: Kaiser Commission on Medicaid and the Uninsured, 2004.

"Occupational Employment Projections to 2016," *Monthly Labor Review,* November 2007.

O'Conner, Colleen, "Empirical Research on How the Elderly Handle Their Estates," *Generations* (1996), 20(3): 13–20.

O'Grady-LeShane, Regina, "Older Women and Poverty," *Social Work* (September 1990), 35: 422–424.

Okrent, Daniel, Booth, Cathy, Buechner, Maryanne Murray, Park, Alice, and Philadelphia, Desa, "Twilight of the Boomers, *Time* (June 12, 2000). Retrieved September 19, 2011 (http://www.time .com/time/magazine/article/0,9171,997133,00.html).

Okun, M., "The Relation Between Motives for Organizational Volunteering and Frequency of Volunteering by Elders," *Journal of Applied Gerontology* (1994), 13: 115–126.

Older Women's League, "Path to Poverty: An Analysis of Women's Retirement Income," in Carroll L. Estes and Meredith Minkler (Eds.), *Critical Gerontology: Perspectives From Political and Moral Economy,* Amityville, NY: Baywood, 1998, pp. 299–313.

Olshansky, S. J., and Carnes, Bruce A., *The Quest for Immortality: Science at the Frontiers of Aging,* New York: Norton, 2002.

Olshansky, S., Carnes, B., and Cassel, C., "In Search of Methuselah: Estimating the Upper Limits to Longevity," *Science* (1990), 250: 634–640.

Olshansky, S. J., Rudberg, M. A., Carnes, B. A., Cassel, C. K., and Brody, J. A., "Trading Off Longer Life for Worsening Health: The Expansion of Morbidity Hypothesis," *Journal of Aging and Health* (May 1991), 3(2): 194–216.

Olson, Laura K., *The Political Economy of Aging,* New York: Columbia University Press, 1982.

O'Reilly, Patrick, and Caro, Francis G., "Productive Aging: An Overview of the Literature," *Journal of Aging & Social Policy* (1994), 6(3): 39–71.

Orlov, Laurie, "The Real Elderly Are Hidden Behind Demographic Murkiness," *Real Experts* (April 26, 2011).

Orwoll, L., and Perlmutter, M., "The Study of Wise Persons: Integrating a Personality Perspective," in R. J. Sternberg (Ed.), *Wisdom: Its Nature, Origin, and Development,* Cambridge, UK: Cambridge University Press, 1990, pp. 181–211.

Osgood, Nancy J., *Senior Settlers: Social Integration in Retirement Communities,* New York: Praeger, 1982.

Osgood, Nancy J., *Suicide in Later Life: Recognizing the Warning Signs,* New York: Lexington Books, 1992.

Palmore, Erdman B., *The Honorable Elders,* Durham, NC: Duke University Press, 1975.

Palmore, Erdman B., *Social Patterns in Normal Aging,* Durham, NC: Duke University Press, 1981.

Palmore, Erdman B., *Ageism: Negative and Positive* (2nd ed.), New York: Springer, 1999.

Palmore, Erdman B., Burchett, B. M., Filenbaum, G. G., George, L. K., and Wallman, L. M., *Retirement: Causes and Consequences,* New York: Springer, 1985.

Pampel, Fred C., and Williamson, John B., *Age, Class, Politics, and the Welfare State,* Cambridge, UK: Cambridge University Press, 1989.

Panser, L. A., Rhodes, T., Girman, C. J., Guess, H. A., and Chute, C. G., "Sexual Function of Men Ages 40 to 79 Years: The Olmsted County Study of Urinary Symptoms and Health Status Among Men," *Journal of the American Geriatrics Society* (October 1995), 43(10): 1107–1111.

Parnes, Herbert S., *Retirement Among American Men,* Lexington, MA: Lexington, 1985.

Peacock, W., and Talley, W., "Intergenerational Contact: A Way to Counteract Ageism," *Educational Gerontology* (1984), 10(1–2): 13–24.

Peck, Robert C., "Psychological Development in the Second Half of Life," in B. L. Neugarten (Ed.), *Middle Age and Aging,* Chicago: University of Chicago Press, 1968.

Perlmutter v. Florida Medical Center, 11 Jul 1978, Fla Suppl. 47: 190–19 (1978).

Peters, R., Schmidt, W., and Miller K., "Guardianship of the Elderly in Tallahassee, Florida," *The Gerontologist* (1985), 25(5): 532–538.

Peterson, Pete, *Running on Empty: How the Democratic and Republican Parties Are Bankrupting Our Future and What Americans Can Do About It,* New York: Farrar, Straus and Giroux, 2004.

Peterson, Robin T., and Sautter, Elise Truly, "Review of the Depiction of Senior Citizen Instrumental and Congenial Behavior in Television Commercials," *Journal of Hospitality and Leisure Marketing* (2003), 10(1–2).

Pew Research Center, *The Return of the Multi-generational Family Household,* Washington, DC: Author, 2010.

Pfeifer, Susan K., and Sussman, Marvin B. (Eds.), *Families: Intergenerational and Generational Connections,* Binghamton, NY: Haworth, 1991.

Pierson, Charon A., "Public and Private Problem of Euthanasia: A Comparison of the Netherlands and the United States," *Geriatric Nursing* (November–December 1998), 19(6): 309–314.

Pillemer, Karl A., and Finkelhor, D., "The Prevalence of Elder Abuse: A Random Sample Survey," *The Gerontologist* (1988), 28(1): 51–57.

Pillemer, Karl, and Luscher, Kurt (Eds.), *Inter-generational Ambivalences: New Perspectives on Parent-Child Relations in Later Life,* New York: Elsevier, 2004.

Pillemer, Karl A., Moen, Phyllis, Wethington, Elaine, and Glasgow, Nina (Eds.), *Social Integration in the Second Half of Life,* Baltimore: Johns Hopkins University Press, 2000.

Portnow, Jay, *Home Care for the Elderly: A Complete Guide,* New York: McGraw-Hill, 1987.

Post, Stephen, "Filial Morality in an Aging Society," *Journal of Religion & Aging* (1989), 5: 15–30.

Post, Stephen G., and Binstock, Robert H. (Eds.), *Fountain of Youth: Cultural, Scientific, and Ethical Perspectives on a Biomedical Goal,* New York: Oxford University Press, 2004.

Postman, Neil, *The Disappearance of Childhood,* New York: Delacorte, 1982.

Pratt, Henry J., *The Gray Lobby,* Chicago: University of Chicago Press, 1976.

Pratt, Henry J., "The 'Gray Lobby' Revisited," *National Forum* (1982), 62: 31–33.

Preston, Samuel H., "Children and the Elderly in the U.S.," *Scientific American* (1984), 251(6): 44–49.

Purcell, Patrick, *Consumer spending by older Americans,* New York: Nova Science Publishers, 2008.

Putnam, Jackson, *Old Age Politics in California: From Richardson to Reagan,* Stanford, CA: Stanford University Press, 1970.

Putnam, Robert, *Bowling Alone: The Collapse and Revival of American Community,* New York: Simon and Schuster, 2001.

Putnam, S. M., "Nature of the Medical Encounter," *Research on Aging* (March 1996), 18(1): 70–83.

Quadagno, Jill, *Aging in Early Industrial Society,* New York: Academic Press, 1982.

Queenan, Joe, *Balsamic Dreams: A Short but Self-Important History of the Baby Boomer Generation,* New York: Picador, 2001.

Quinn, Joseph F., "Retirement Income Rights as a Component of Wealth in the United States," *Review of Income and Wealth* (1985), 31: 223–236.

Quinn, Joseph F., "The Economic Status of the Elderly: Beware the Mean," *Review of Income and Wealth* (March 1987): 63–82.

Quinn, Joseph F., "Retirement Patterns and Bridge Jobs in the 1990s," *EBRI Issue Brief* (February 1999), 206: 1–22.

Quinn, Joseph F., Segal, J., Raisz, H., and Johnson C. (Eds.), *Coordinating Community Services for the Elderly: The Triage Experience,* New York: Springer, 1982.

Quinn, Mary Joy, "Undue Influence and Elder Abuse: Recognition and Intervention Strategies," *Geriatric Nursing* (January–February 2002), 23(1): 11–16.

Quinn, Mary Joy, and Heisler, Candace J., "Legal Response to Elder Abuse and Neglect," *Journal of Elder Abuse and Neglect* (2002), 14(1): 61–77.

Quinn, M. J., and Tomita, S. K., *Elder Abuse and Neglect: Causes, Diagnosis and Intervention Strategies* (2nd ed.), New York: Springer, 1987.

Rabushka, A., and Jacobs, B., *Old Folks at Home,* New York: Free Press, 1980.

Rachels, James, *The End of Life: Euthanasia and Morality,* New York: Oxford University Press, 1986.

Radner, Daniel B., "Money Incomes of the Aged and Nonaged Family Units," *Social Security Bulletin* (1987), 50(8): 5–21.

Radner, Daniel B., "Economic Status of the Aged," *Social Security Bulletin* (Fall 1992), 55(3): 3–23.

Raschick, Michael, and Ingersoll-Dayton, Berit, "Costs and Rewards of Caregiving Among Aging Spouses and Adult Children," *Family Relations* (April 2004), 53(3): 317–325.

Regan, John, "Protecting the Elderly: The New Paternalism," *Hastings Law Journal* (1981), 32(5): 1111–1132.

Rein, Jan Ellen, "Preserving Dignity and Self-Determination of the Elderly in the Face of Competing Interests and Grim Alternatives: A Proposal for Statutory Refocus and Reform," *George Washington Law Review* (1992), 60(6): 1818–1887.

Reisberg, Barry, *Alzheimer's Disease,* New York: Free Press (Macmillan), 1983.

Relman, Arnold S., "The Trouble With Rationing," *New England Journal of Medicine* (September 27, 1990), 323(13): 911–913.

Reno, Virginia P., Lamme, Elizabeth, and Walker, Elisa A., "Social Security Finances: Findings of the 2011 Trustees Report," *Social Security Brief No. 36*, National Academy of Social Insurance, May 2011. Retrieved September 22, 2011 (http://www.nasi.org/research/2011/social-security-finances-findings-2011-trustees-report).

Riggs, Karen E., *Mature Audiences: Television in the Lives of Elders,* New Brunswick, NJ: Rutgers University Press, 1998.

Riley, Matilda White, and Riley, John W., Jr., "Age Integration and the Lives of Older People," *The Gerontologist* (February 1994), 34(1): 110–115.

Rindfuss, R. R., Swicegood, C. G., and Rosenfeld, R. A., "Disorder in the Life Course: How Common and Does It Matter?" *American Sociological Review* (1987), 52: 785–801.

Ritter, K. P., "Preparation for Guardianship Cases Poses a Constant Challenge," *Journal of Long Term Care Administration* (1995), 23(3): 14–17.

Rivlin, Alice, and Wiener, Joshua, *Caring for the Disabled Elderly: Who Will Pay?* Washington, DC: Brookings Institution, 1988.

Rix, Sara E., *Aging and Work: A View from the United States,* Washington, DC: AARP Public Policy Institute, 2004.

Robertson, Ann, "Beyond Apocalyptic Demography: Towards a Moral Economy of Interdependence," *Ageing and Society* (July 1997), 17(4): 425–446.

Robine, Jean-Marie, *Jeanne Calment: From Van Gogh's Time to Ours: 122 Extraordinary Years,* New York: W. H. Freeman, 1998.

Robinson John P., "Freeing Up the Golden Years," *American Demographics* (October 1997), 19(10): 20–24.

Robinson, John P., and Godbey, Geoffrey, *Time for Life: The Surprising Ways Americans Use Their Time,* State College: Pennsylvania State University Press, 1997.

Rodin, J., and Langer, E., "Aging Labels: The Decline of Control and the Fall of Self-Esteem," *Journal of Social Issues* (1980), 36: 12–29.

Rodin, J., Timko, C., and Harris, S., "The Construct of Control: Biological and Psychosocial Correlates," in M. P. Lawton and G. Maddox (Eds.), *Annual Review of Gerontology and Geriatrics* (Vol. 5), New York: Springer, 1985, pp. 3–55.

Rogers, Andrei, and Raymer, James, "Regional Demographics of the Elderly Foreign-Born and Native-Born Populations in the United States Since 1950," *Research on Aging* (January 1999), 21(1): 3–35.

Rogoff, Edward, and Carroll, David, *The Second Chance Revolution: Becoming Your Own Boss After 50,* Rowhouse, 2009.

Roose, Steven P., and Sackeim, Harold A. (Eds.), *Late-Life Depression,* New York: Oxford University Press, 2004.

Roper Center for Public Opinion Research, *Topics at a Glance: Social Security,* University of Connecticut, 2011. Retrieved September 22, 2011 (http://www.ropercenter.uconn.edu/data_access/tag/social_security.html#.TnYsIuxvAuw).

Rose, Michael, *Evolutionary Biology of Aging,* Oxford, UK: Oxford University Press, 1994.

Rosenbaum, Walter A., "Unquiet Future of Intergenerational Politics," *The Gerontologist* (August 1993), 33(4): 481–490.

Rosenbaum, Walter A., and Button, James W., "Is There a Gray Peril? Retirement Politics in Florida," *The Gerontologist* (1989), 29: 300–306.

Rosenbaum, Walter A., and Button, James W., "Perceptions of Intergenerational Conflict: The Politics of Young vs. Old in Florida," *Journal of Aging Studies* (Winter 1992), 6(4): 385–396.

Rosenfeld, Barry, *Assisted Suicide and the Right to Die: The Interface of Social Science, Public Policy, and Medical Ethics,* Washington, DC: American Psychological Association, 2004.

Rosenmayr, L., *Die Spaete Freiheit* [The Late Freedom], Vienna: Severin, 1984.

Rosenthal, C. J., "Kinkeeping in the Familial Division of Labor," *Journal of Marriage and the Family* (1985), 47: 965–974.

Rosenthal, Gary E., and Fortinsky, Richard H., "Differences in the Treatment of Patients With Acute Myocardial Infarction According to Patient Age," *Journal of the American Geriatrics Society* (August 1994), 42(8): 826–832.

Rosow, Irving, *Social Integration of the Aged,* New York: Free Press, 1967.

Rosow, Irving, *Socialization to Old Age,* Berkeley: University of California Press, 1974.

Roszak, Theodore, *The Making of a Counter Culture,* University of California Press, 1968.

Roszak, Theodore, *America the Wise: The Longevity Revolution and the True Wealth of Nations,* Boston: Houghton Mifflin, 1998.

Roush, Wade, "Live Long and Prosper?" *Science* (July 5, 1996), 273(5271): 42–46.

Rowe, John W., and Kahn, R. L., *Successful Aging,* New York: Pantheon, 1997.

Rudinger, Georg, and Thomae, Hans, "Bonn Longitudinal Study of Aging: Coping, Life Adjustment, and Life Satisfaction," in P. B. Baltes and M. M. Baltes (Eds.), *Successful Aging: Perspectives From the Behavioral Sciences,* New York: Cambridge University Press, 1990, pp. 265–295.

Ruiz, Dorothy Smith, "Demographic and Epidemiologic Profile of the Ethnic Elderly," in Deborah K. Padgett (Ed.), *Handbook on Ethnicity, Aging, and Mental Health,* Westport, CT: Greenwood Press, 1995, pp. 3–21.

Ryder, Norman, "The Cohort as a Concept in the Study of Social Change," *American Sociological Review* (1965), 30: 843–861.

Ryff, Carol D., "Beyond Ponce de Leon and Life Satisfaction: New Directions in Quest of Successful Ageing," *International Journal of Behavioral Development* (1989), 12(1): 35–55.

Sacher, George A., "Longevity, Aging and Death: An Evolutionary Perspective," *The Gerontologist* (1978), 18: 112–119.

Salas, Christian, "On the Empirical Association Between Poor Health and Low Socioeconomic Status at Old Age," *Health Economics* (April 2002), 11(3): 207–220.

Salthouse, T. A., "Effects of Age and Skill in Typing," *Journal of Experimental Psychology: General* (1984), 113: 345–371.

Salthouse, T. A., "Speed of Behavior and the Implications for Cognition," in J. E. Birren and K. W. Schaie (Eds.), *Handbook of the Psychology of Aging* (2nd ed.), New York: Van Nostrand Reinhold, 1985a, pp. 400–426.

Salthouse, T. A., *A Theory of Cognitive Aging,* Amsterdam: North Holland, 1985b.

Sattler, J. M., "Age Effects on Wechsler Adult Intelligence Scale-Revised Tests," *Journal of Consulting and Clinical Psychology* (1982), 50: 785–786.

Sauvy, Alfred, *Zero Growth,* New York: Praeger, 1976.

Schachter-Shalomi, Zalman, and Miller, Ron, *From Ageing to Sageing,* New York: Time Warner, 1995.

Schaie, K. Warner, "Midlife Influences Upon Intellectual Functioning in Old Age," *International Journal of Behavioral Development* (1984), 7: 463–478.

Schaie, K. Warner, *Intellectual Development in Adulthood: The Seattle Longitudinal Study,* Cambridge, UK: Cambridge University Press, 1996.

Schieber, Sylvester, and Shoven, John (Eds.), *Public Policy Toward Pensions,* Cambridge, MA: MIT Press, 1997.

Schmahl, W. (Ed.), *Redefining the Process of Retirement,* New York: Springer-Verlag, 1989.

Schneider, Edward L., and Guralnik, Jack, "The Aging of America: Impact on Health Care Costs," *Journal of the American Medical Association* (May 2, 1990), 263(17): 2335–2340.

Schnore, M., *Retirement: Bane or Blessing,* Atlantic Highlands, NJ: Humanities Press, 1985.

Schooler, C., and Schaie, K. W. (Eds.), *Cognitive Functioning and Social Structure Over the Life Course,* Norwood, NJ: Ablex, 1987.

Schor, Juliet, *The Overworked American,* New York: Basic Books, 1991.

Schulz, James H., *The Economics of Aging* (6th ed.), Dover, MA: Auburn House, 1995.

Schwartz, William B., "Inevitable Failure of Current Cost-Containment Strategies: Why They Can Provide Only Temporary Relief," *Journal of the American Medical Association* (January 9, 1987), 257(2): 220–224.

Schwarz, John, *America's Hidden Success,* New York: Norton, 1983.

Scitovsky, Anne A., "Medical Care in the Last Twelve Months of Life: The Relation Between Age, Functional Status, and Medical Care Expenditures," *Milbank Quarterly* (1988), 66(4): 640–660.

Seligman, Martin E. P., *Helplessness: On Depression, Development and Death,* San Francisco: Freeman, 1975.

Seltzer, Mildred M. (Ed.), *Impact of Increased Life Expectancy: Beyond the Gray Horizon,* New York: Springer, 1995.

Seltzer, M. M., and Troll, L. E., "Conflicting Public Attitudes Toward Filial Responsibility," *Generations* (1982), 7(2): 26–27, 40.

Settersten, Richard A., Jr. (Ed.), *Invitation to the Life Course: Toward New Understandings of Later Life,* Amityville, NY: Baywood Publishing Company, 2003.

Shanas, Ethel, "The Family as a Social Support System in Old Age," *The Gerontologist* (1979), 19: 169–174.

Shanas, Ethel, "Older People and Their Families: The New Pioneers," *Journal of Marriage and the Family* (1980), 42(9): 9–15.

Shapiro, Marla, "Menopause: Current Controversies in Hormone Replacement Treatment," *Geriatrics and Aging* (January 2003), 6(1): 30–33.

Sharpe, Charles C., *Frauds Against the Elderly,* Jefferson, NC: McFarland, 2004.

Shaw, Greg, and Mysiewicz, Sarah E., "The Polls-Trends: Social Security and Medicare," *Public Opinion Quarterly* (Fall 2004), 68(3): 394–423.

Shea, Gordon F., and Haasen, Adolf, *The Older Worker Advantage: Making the Most of Our Aging Workforce,* Praeger, 2005.

Shelanski, Vivien, "Assisted Suicide and the Courts: Spotlight on Palliative Care," *Journal of Long Term Home Health Care* (Winter 1998), 17(1): 17–28.

Sheppard, Harold, "The 'New' Early Retirement: Europe and the United States," in Irving Bluestone, Rhonda Montgomery, and John Owen (Eds.), *The Aging of the American Work Force,* Detroit, MI: Wayne State University Press, 1990, pp. 158–178.

Sherwood, S., Morris, S. A., Ruchlin, H. S., and Sherwood, C. C., *Continuing-Care Retirement Communities,* Baltimore: Johns Hopkins University Press, 1997.

Shock, Nathan, "The Physiology of Aging," *Scientific American* (1962), 206: 100–110.

Shock, Nathan, Greulich, R. C., Cosa, P. T., Jr., Andres, R., Lakata, E. G., Arenberg, D., and Tobin, J. D., *Normal Human Aging: The Baltimore Longitudinal Study of Aging,* Washington, DC: Government Printing Office, 1984.

Silva, Timothy, W., "Reporting Elder Abuse: Should It Be Mandatory or Voluntary?" *HealthSpan* (April 1992), 9(4): 13–15.

Silverman, Phyllis, *Widow to Widow,* New York: Springer, 1986.

Simon-Rusinowitz, Lori, Mahoney, Kevin J., and Benjamin, A. E., "Payments to Families Who Provide Care: An Option That Should Be Available," *Generations* (Fall 1998), 22(3): 69–75.

Simonton, Dean K., "Career Paths and Creative Lives: A Theoretical Perspective on Late Life Potential," in Carolyn E. Adams-Price (Ed.), *Creativity and Successful Aging: Theoretical and Empirical Approaches,* New York: Springer, 1998, pp. 3–18.

Singleton, J. F., Forbes, W. F., and Agwani, N., "Stability of Activity Across the Lifespan," *Activities, Adaptation and Aging* (1993), 18(1): 19–27.

Skloot, R., *The Immortal Life of Henrietta Lacks,* New York: Crown, 2010.

Smeeding, T. M., "Children and Poverty: How U.S. Stands," *Forum for Applied Research and Public Policy* (Summer 1990), 5(2): 65–70.

Smeeding, T. M., Estes, C. L., and Glasse, L., *Social Security Reform and Older Women: Improving the System,* Income Security Policy Series Paper No. 22, Center for Policy Research, Maxwell School, Syracuse, NY: Syracuse University, 1999.

Smith, Andrew H., "Age-Based Rationing: A Wrong Turn on the Road to Reform," *Ageing International* (September 1993), 20(3): 7–11.

Smith, Denise, *The Older Population in the United States: March 2002,* Washington, DC: U.S. Census Bureau Current Population Reports, 2003, pp. 20–546.

Smolensky, E., Danziger, S., and Gottschalk, P., "The Declining Significance of Age in the United States: Trends in the Well-Being of Children and the Elderly Since 1939," in J. L. Palmer, T. Smeeding, and B. Torrey (Eds.), *The Vulnerable,* Washington, DC: Urban Institute Press, 1988, pp. 29–54.

Snowden, David, *Aging with Grace: What the Nun Study Teaches Us About Leading Longer, Healthier, and More Meaningful Lives,* Bantam, 2002.

Social Security Administration, *Income of the Population 55 and Older,* Washington, DC: Author, 2008.

Social Security Administration, *Annual Trustees Report,* Washington, DC: Author, 2011.

Social Security Online, *Income of the Aged Chartbook, 2008,* Washington, DC: U.S. Social Security Administration, 2010. (http://www.ssa.gov/policy/docs/chartbooks/income_aged/).

Solomon, M., O'Donnell, L., Jennings, B., Guilfoy, V., Wolf, S. M., Nolan, K., Jackson, R., Koch-Weser, D., and Donnelley, S., "Decisions Near the End of Life: Professional Views on Life-Sustaining Treatments," *American Journal of Public Health* (January 1993): 14–23.

Somers, Anne R., and Spears, Nancy L., *Continuing Care Retirement Community: A Significant Option for Long-Term Care?* New York: Springer, 1992.

Springer, D., and Brubaker, T. H., *Family Caregivers and Dependent Elderly: Managing Stress and Maximizing Independence,* Beverly Hills, CA: Sage, 1984.

Sprott, Richard L., and Roth, George S., "Biomarkers of Aging: Can We Predict Individual Life Span?" *Generations* (1992), 16(4): 11–14.

Stagner, R., "Aging in Industry," in James Birren and K. Warner Schaie (Eds.), *Handbook of the Psychology of Aging* (2nd ed.), New York: Van Nostrand Reinhold, 1985, pp. 789–817.

Stanley, Jean F., Pye, David, and MacGregor, Andrew, "Comparison of Doubling Numbers Attained by Cultured Animal Cells With Life Span of Species," *Nature* (May 8, 1975).

Stearns, Peter N. (Ed.), *Old Age in Pre-industrial Societies,* New York: Holmes and Meier, 1982.

Sterns, Ronni L., and Sterns, Harvey S., "Consumer Issues," in Richard Schulz, Linda Nolker, Kenneth Rockwood, and Richard Sprott (Eds.), *Encyclopedia of Aging,* New York: Springer, 2006, pp. 263–265.

Steckenrider, Janie, and Parrott, Tonya (Eds.), *New Directions in Old-Age Policies,* Albany: State University of New York Press, 1998.

Stein, Bruno, "Pay-as-You-Go, Partial Prefunding, and Full Funding of American Social Security," *History of Political Economy* (Spring 1991), 23: 79–83.

Steinberg, Ed., *Understanding the Federal Budget,* Federal Reserve Bank of New York, Public Information Department, 2000. Retrieved July 8, 2005 (http://www.newyorkfed.org/education/addpub/budg4.pdf).

Steinberg, Maurice, and Youngner, Stuart J. (Eds.), *End-of-Life Decisions: A Psychological Perspective,* Washington, DC: American Psychiatric Press, 1998.

Steinhorn, Leonard, *The Greater Generation: In Defense of the Baby Boom Legacy,* New York: St. Martin's Press, 2006.

Stephan, Paula E., and Levin, Sharon, G., *Striking the Mother Lode in Science: The Importance of Age, Place and Time,* New York: Oxford University Press, 1992.

Stephens, S., and Christianson, J., *Informal Care of the Elderly,* Lexington, MA: D. C. Heath, 1986.

Sterling, Bruce, *Holy Fire,* New York: Bantam, 1996.

Sternberg, Robert J., "Older But Not Wiser? The Relationship Between Age and Wisdom," *Ageing International* (Winter 2005), 30(1): 5–26.

Stoller, Eleanor P., and Gibson, Rose C., *Worlds of Difference: Inequality in the Aging Experience,* Thousand Oaks, CA: Pine Forge, 2000.

Stone, R., Cafferata, G., and Sangl, J., "Caregivers of the Frail Elderly: A National Profile," *The Gerontologist* (1987), 27: 616–626.

Stone, R. I., and Kemper, P., "Spouses and Children of Disabled Elders: How Large a Constituency for Long-Term Reform?" *Millbank Quarterly* (1989), 67: 485–506.

Streib, Gordon F., "Morale of the Retired," *Social Problems* (April 1956): 270–276.

Streib, Gordon, "Are the Aged a Minority Group?" in Alvin Gouldner and S. Miller (Eds.), *Applied Sociology,* New York: Free Press, 1965.

Streib, Gordon, "Socioeconomic Strata," in Erdman Palmore (Ed.), *Handbook on the Aged in the United States,* Westport, CT: Greenwood, 1984, pp. 77–92.

Streib, Gordon, "Social Stratification and Aging," in Robert H. Binstock and Ethel Shanas (Eds.), *Handbook of Aging and the Social Sciences,* New York: Van Nostrand Reinhold, 1985, pp. 339–368.

Streib, Gordon, and Bourg, C. F., "Age Stratification Theory, Inequality, and Social Change," in R. F. Thomason (Ed.), *Comparative Social Research,* Greenwich, CT: JAI, 1984.

Streib, Gordon, Folts, W. Edward, and Hilker, Mary, *Old Homes—New Families: Shared Living for the Elderly,* New York: Columbia University Press, 1984.

Streib, Gordon, and Schneider, C., *Retirement in American Society,* Ithaca, NY: Cornell University Press, 1971.

Sullivan, Amy D., Hedberg, Katrina, and Fleming, David W., "Legalized Physician-Assisted Suicide in Oregon—The Second Year," *New England Journal of Medicine* (February 24, 2000), 342(8): 598–604.

Sullivan, D. A., "Informal Support Systems in a Planned Retirement Community: Availability, Proximity, and Willingness to Utilize," *Research on Aging* (June 1986), 8(2): 249–267.

Swift, J., *Gulliver's Travels,* London: Benjamin Motte, 1726.

Szilard, Leo, "On the Nature of the Aging Process," *Proceedings of the National Academy of Sciences USA* (1959), 45: 30.

Tanner, Michael (Ed.), *Social Security and Its Discontents: Perspectives on Choice,* Washington, DC: Cato Institute, 2004.

Teague, Michael L., and MacNeil, Richard D., *Aging and Leisure: Vitality in Later Life* (2nd ed.), Dubuque, IA: Brown and Benchmark, 1992.

Terkel, Studs, *Working,* New York: Ballantine, 1985.

Thompson, Paul, "'I Don't Feel Old': The Significance of the Search for Meaning in Later Life," *International Journal of Geriatric Psychiatry* (August 1993), 8(8): 685–692.

Thone, R. R., *Women and Aging: Celebrating Ourselves,* New York: Haworth, 1992.

Thursz, D., Liederman, D., and Schorr, L., "Generations Uniting," *Perspectives on Aging* (1989), 8(1): 3–23.

Tierce, J., and Seelbach, W., "Elders as School Volunteers," *Educational Gerontology* (1987), 13: 33–41.

Tilly, J., and Brunner, D., *Medicaid Eligibility and Its Effects on the Elderly,* Washington, DC: American Association of Retired Persons, 1987.

Tornstam, Lars, "Quo Vadis of Gerontology: On the Scientific Paradigm of Gerontology," *The Gerontologist* (1992), 32(3): 318–326.

Tornstam, Lars, "Gerotranscendence: The Contemplative Dimension of Aging," *Journal of Aging Studies* (Summer 1997), 11(2): 143–154.

Torres-Gil, Fernando, *The New Aging: Politics and Change in America,* New York: Auburn House, 1992.

Torres-Gil, Fernando, "Latinos and the Future of Social Security: A Time to Act," UCLA Center for Policy Research on Aging, March 2006. Retrieved September 16, 2011 (http://www.chicano.ucla.edu/press/briefs/documents/LSSPB1March06.pdf).

Treas, Judith, "Older Americans in the 1990s and Beyond," *Population Bulletin* (1995), 50(2).

Tsuchiya, Aki, "QALYs and Ageism: Philosophical Theories and Age Weighting," *Health Economics* (January 2000), 9(1): 57–68.

Uchitelle, Louis, "Stanching the Loss of Good Jobs," *The New York Times* (January 31, 1993): 1–3.

Uhlenberg, Peter, Cooney, Teresa, and Boyd, Robert, "Divorce for Women After Midlife," *Journal of Gerontology* (1990), 45(1): S3–SII.

Upston, Amy, "Recent Public Opinion Polls on Social Security," *EBRI Notes* (October 1998), 19(10): 5–9.

U.S. Census Bureau, "Projections of the Population of the United States, by Age, Sex, and Race: 1988 to 2080," *Current Population Reports,* Series P-25, No. 1018, Washington, DC: Government Printing Office, January 1989.

U.S. Census Bureau, "Money Income and Poverty Status in the United States: 1989," *Current Population Reports,* Series P-60, No. 168, Washington, DC: Government Printing Office, September 1990.

U.S. Census Bureau, "65+ in the United States," *Current Population Reports,* Series P-23, No. 190, Washington, DC: Government Printing Office, April 1996.

U.S. Congress, House Task Force on Social Security and Women, *Inequities Toward Women in the Social Security System,* Washington, DC: Government Printing Office, 1983.

U.S. Congress, House Task Force on Social Security and Women, *Earnings Sharing Implementation Plan,* Washington, DC: Government Printing Office, 1984.

U.S. Congressional Budget Office, *Older Americans Reports,* Washington, DC: Author, April 29, 1988.

U.S. Congressional Budget Office, *Baby Boomers' Retirement Prospects: An Overview,* Washington, DC: Author, 2003.

U.S. Congressional Budget Office, *Outlook for Social Security,* Washington, DC: Author, June 2004.

U.S. General Accounting Office, *Long-Term Care Insurance: Risks to Insurance Should Be Reduced,* Washington, DC: Author, December 1991.

U.S. Office of Technology Assessment, *Life Sustaining Technologies and the Elderly,* Washington, DC: Government Printing Office, 1987.

Vaco, Attorney General of New York, v. Quill, 117 S. Ct. 2293, 138 L.ed.2d (1997).

Vaillant, George E., *The Wisdom of the Ego: Sources of Resilience in Adult Life,* Cambridge, MA: Belknap Press, 1995.

Van Gennep, A., *Rites of Passage,* Chicago: University of Chicago Press, 1960.

Van Tassel, David D., and Meyer, J. E. W., *U.S. Aging Policy Interest Groups,* Westport, CT: Greenwood, 1992.

Venti, Steven F., and Wise, David A., *Aging and Housing Equity,* Cambridge: National Bureau of Economic Research, 2000.

Ventura-Merkel, C., and Doucette, D., "Community Colleges in an Aging Society," *Educational Gerontology* (March–April 1993), 19(2): 161–171.

Ventura-Merkel, C., and Friedman, M., "Helping At-Risk Youth Through Intergenerational Programming," *Children Today* (1988), 17(1): 10–13.

Verbrugge, L., Lepkowski, J., and Imanaka, Y., "Co-Morbidity and Its Impact on Disability," *Millbank Quarterly* (1989), 67(3–4): 450–484.

Vladeck, Bruce, *Unloving Care: The Nursing Home Tragedy,* New York: Basic Books, 1980.

Wagner, Lynn, "Alzheimer's Disease Tests Challenged," *Provider* (February 1996), 22(2): 65–66.

Walford, Roy, *Maximum Life Span,* New York: Norton, 1983.

Walford, Roy, *The 120 Year Diet: How to Double Your Vital Years,* New York: Pocket Books, 1986.

Walsh, R., and Vaughan, Frances (Eds.), *Paths Beyond Ego: The Transpersonal Vision,* Los Angeles: Tarcher, 1993.

Warner, Huber R., "Current Status of Efforts to Measure and Modulate the Biological Rate of Aging," *Journals of Gerontology: Series A: Biological Sciences and Medical Sciences* (July 2004), 59A(7): 692–696.

Washington v. Glucksberg, 117 S. Ct. 2258, 138 L.ed.2d (1997).

Watkins, S. C., Menken, J. A., and Bongaarts, J., "Demographic Foundations of Family Change," *American Sociological Review* (1987), 52: 346–358.

Watson, J. D., *Recombinant DNA* (2nd ed.), New York: Freeman, 1992.

Weibel-Orlando, Joan, "Grandparenting Styles: Native American Perspective," in J. Sokolovsky (Ed.), *The Cultural Context of Aging: Worldwide Perspectives,* New York: Bergin and Garvey, 1990.

Weismann, A., *Collected Essays on Heredity and Kindred Biological Problems,* Oxford, UK: Claredon Press, 1889.

Weitzman, Lenore, *The Divorce Revolution: The Unexpected Social and Economic Consequences for Women and Children in America,* New York: Free Press, 1985.

Welch, H. Gilbert, "Comparing Apples and Oranges: Does Cost-Effectiveness Analysis Deal Fairly With the Old and Young?" *Gerontologist* (June 1991), 31(3): 332–336.

Wentworth, Seyda G., and Pattison, David, "Income Growth and Future Poverty Rates of the Aged," *Social Security Bulletin* (2001/2002), 64(3).

Whitbourne, Susan K., "Test Anxiety in Elderly and Young Adults," *International Journal of Aging and Human Development* (1976), 7: 201–210.

Whitbourne, S. K., "Sexuality in the Aging Male," *Generations* (Summer 1990), 14(3): 28–30.

White, L., et al., "Geriatric Epidemiology," in Carl Eisdorfer (Ed.), *Annual Review of Gerontology and Geriatrics* (Vol. 6), New York: Springer, 1986.

Whitehouse, Peter, and George, Daniel, *The Myth of Alzheimer's: What You Aren't Being Told About Today's Most Dreaded Diagnosis,* New York: St. Martin's, 2008.

Wiener, Joshua, "Long-Term Care Financing," *Nursing Homes Long Term Care Management* (February 1998), 47(2): 19–21.

Williams, G. C., "Pleiotropy, Natural Selection, and the Evolution of Senescence," *Evolution* (1957), 11: 398–411.

Williams, Richard H., and Wirths, Claudine G., *Lives Through the Years,* New York: Atherton, 1965.

Williamson, John, "Should Women Support the Privatization of Social Security?" *Challenge* (July–August, 1997), 40(4): 97–108.

Williamson, R. C., Rinehart, A. D., and Blank, T. O., *Early Retirement: Promises and Pitfalls,* New York: Plenum Press, 1992.

Wilson, Janet, *Intergenerational Readings: 1980–1992,* Pittsburgh: Generations Together, 1992.

Windle, Gill, and Woods, Robert T., "Variations in Subjective Well-Being: The Mediating Role of a Psychological Resource," *Ageing and Society* (July 2004), 24(4): 583–602.

Wisensale, Steven K., "Grappling With the Generational Equity Debate: An Ongoing Challenge for the Public Administrator," *Public Integrity* (Winter 1999), 1(1): 1–19.

Wisensale, Steven, "Aging Societies and Intergenerational Equity Issues," *Journal of Feminist Family Therapy* (2005), 17(3–4): 79–103.

Wolf, D., and Longino, C., "Our 'Increasingly Mobile Society'? The Curious Persistence of a False Belief," *The Gerontologist* (2005), 45(1): 5–11.

Wolf, R. S., and Pillemer, K., *Helping Elderly Victims: The Reality of Elder Abuse,* New York: Columbia University Press, 1989.

Wolfe, D. B., and Snyder, R. E., *Ageless Marketing: Strategies for Reaching the Hearts and Minds of the New Consumer Majority,* Chicago: Dearborn Trade, 2003.

Wolfe, John R., *Coming Health Crisis: Who Will Pay for Care for the Aged in the Twenty-First Century?* Chicago: University of Chicago Press, 1993.

Woodworth, R. S., *Psychology: A Study of Mental Life,* New York: Henry Holt, 1921.

Wu Ke Bin, *Income and Poverty in the United States in 1995: A Chart Book,* Washington, DC: American Association of Retired Persons, Public Policy Institute, 1998.

Wykle, May L., and Ford, Amasa B. (Eds.), *Serving Minority Elders in the 21st Century,* New York: Springer, 1999.

Yankelovich, Daniel, and Vance, Cyrus R., "Final Request," *American Demographics* (April 2001), 23(4): 22.

Yee, Barbara W. K., "Elders in Southeast Asian Refugee Families," *Generations* (Summer 1992), 24–27.

Zarit, S., Orr, N. K., and Zarit, J. M., *The Hidden Victims of Alzheimer's Disease,* New York: New York University Press, 1985.

Zembek, B. A., and Singer, A., "The Problem of Defining Retirement Among Minorities: The Mexican Americans," *The Gerontologist* (1990), 30(6): 749–757.

Ziegler, S., and King, J., "Evaluating the Observable Effects of Foster Grandparents on Hospitalized Children," *Public Health Reports* (1982), 97(6): 550–557.

Zimmerman, Sheryl, Sloane, Philip, Eckert, Kevin, and Lawton, M. Powell, *Assisted Living: Needs, Practices, and Policies in Residential Care for the Elderly,* Baltimore: Johns Hopkins University Press, 2001.

Zimny, George H., and Grossberg, George T. (Eds.), *Guardianship of the Elderly: Psychiatric and Judicial Aspects,* New York: Springer, 1998.

Zweibel, N. R., Cassel, C. K., and Karrison, T., "Public Attitudes About the Use of Chronological Age as a Criterion for Allocating Health Care Resources," *Gerontologist* (February 1993), 33(1): 74–80.

PHOTO CREDITS

GLOSSARY/INDEX

Aging and Creativity (Levy and Langer), 121–125

Aging-clock theory of aging: the idea that aging results from a preprogrammed sequence, as in a clock, built into the operation of the nervous or endocrine system of the body, 58–59

Aging effects: effects brought about by the physiological process of aging, along with social responses by others to those effects, 442

Aging interest groups: associations or other organizations that seek to influence government policies on behalf of older adults, 319–320

Aging network: the national array of service programs helping older Americans, ranging from senior citizen centers and Area Agencies on Aging to programs offered by the federal government, 319

Aim Not Just for Life, but Expanded "Health Span" (Perry and Butler), 191–193

Alexander the Great, 87

Altersstil, 125

Altruistic suicide, 158–159

Alzheimer's disease, 138–139
   Down's syndrome and, 139
   genetic screening for, 228–230

Ameche, Don, 16

Annuity: an investment vehicle permitting a lump sum of money to be paid out annually to projected life expectancy at a given age, 312

Anomic suicide, 159

Antagonistic pleiotropy: the idea that some genetically determined trait can be beneficial early in life but harmful in later life, 18

Anti-aging medicine, 62

Antioxidant: a substance that destroys free radicals, thereby preventing damage to cell structures, 60